physica status solidi (a)

AKADEMIE-VERLAG · BERLIN

ISSN 0031-8965 * VOL. 54 · NO. 2 · AUGUST 1979

THEORIA CUM PRAXI

Classification Scheme

1. Structure of Crystalline Solids
   - 1.1 Perfectly Periodic Structure
   - 1.2 Solid-State Phase Transformations
   - 1.3 Alloys. Metallurgy
   - 1.4 Microstructure (Magnetic Domains See 18; Ferroelectric Domains See 14.4.1)
   - 1.5 Films
   - 1.6 Surfaces
2. Non-Crystalline State
3. Crystal Growth
4. Bonding Properties
5. Mössbauer Spectroscopy
6. Lattice Dynamics. Phonons
7. Acoustic Properties
8. Thermal Properties
9. Diffusion
10. Defect Properties (Irradiation Defects See 11)
    - 10.1 Metals
    - 10.2 Non-Metals
11. Irradiation Effects (X-Ray Diffraction Investigations See 1 and 10)
12. Mechanical Properties (Plastic Deformations See 10)
    - 12.1 Metals
    - 12.2 Non-Metals
13. Electron States
    - 13.1 Band Structure
    - 13.2 Fermi Surfaces
    - 13.3 Surface and Interface States
    - 13.4 Impurity and Defect States
    - 13.5 Elementary Excitations (Phonons See 6)
      - 13.5.1 Excitons
      - 13.5.2 Plasmons
      - 13.5.3 Polarons
      - 13.5.4 Magnons
14. Electrical Properties. Transport Phenomena
    - 14.1 Metals. Semi-Metals
    - 14.2 Superconductivity. Superconducting Materials and Devices
    - 14.3 Semiconductors
      - 14.3.1 Films
      - 14.3.2 Surfaces and Interfaces
      - 14.3.3 Devices. Junctions (Contact Problems See 14.3.4)
      - 14.3.4 High-Field Phenomena, Space-Charge Effects, Inhomogeneities, Injected Carriers (Electroluminescence See 20.3; Junctions See 14.3.3)
    - 14.4 Dielectrics
      - 14.4.1 Ferroelectrics
15. Thermoelectric and Thermomagnetic Properties
16. Photoconductivity. Photovoltaic Effects
17. Emission of Electrons and Ions
    - 17.1 Field Emission Microscope Investigations
18. Magnetic Properties
    - 18.1 Paramagnetic Properties
    - 18.2 Ferromagnetic Properties
      - 18.2.1 Ferromagnetic Films
    - 18.3 Ferrimagnetic Properties
    - 18.4 Antiferromagnetic Properties

(Continued on cover three)

physica status solidi (a)

applied research

Volume 54 · Number 2 · Pages 437 to 782, K85 to K176, and A9 to A16

August 16, 1979

PSSA 54(2) 437—782, K85—K176, A9—A16 (1979)

ISSN 0031-8965

AKADEMIE-VERLAG · BERLIN

Subscriptions and orders for single copies should be directed

in the DDR:

to the Postzeitungsvertrieb,
to a book-shop, or
to the Akademie-Verlag, DDR-108 Berlin, Leipziger Straße 3 – 4;

in the other socialist countries:

to a book-shop for foreign language literature or
to the competent news-distributing agency;

in the BRD and BERLIN (WEST):

to a book-shop or
to the wholesale distributing agency Kunst und Wissen, Erich Bieber, D-7000 Stuttgart 1, Wilhelmstr. 4 – 6;

in ÖSTERREICH:

to the Globus-Buchvertrieb, A-1201 Wien, Höchstädtplatz 3;

in the other Western European countries:

to Kunst und Wissen Erich Bieber GmbH, CH-8008 Zürich, Dufourstr. 51;

in USA and CANADA:

to Verlag Chemie International, Inc., 175 Fifth Avenue, New York, N.Y. 10010, USA;

in other countries:

to the international book- and journal-selling trade,
to Buchexport, Volkseigener Außenhandelsbetrieb der Deutschen Demokratischen Republik, DDR-701 Leipzig, Postfach 160, or
to the Akademie-Verlag, DDR-108 Berlin, Leipziger Straße 3 – 4.

Editorial Note:

"physica status solidi (a)" undertakes that an original paper *accepted for publication* before the 23rd of any month will be published within 50 days of this date unless the author requests a postponement. In special cases there may be some delay between *receipt* and *acceptance* of a paper due to the review and, if necessary, revision of the paper.

Schriftleiter und verantwortlich für den Inhalt:
Professor Dr. Dr. h. c. P. Görlich, DDR-102 Berlin, Neue Schönhauser Straße 20 bzw. DDR-69 Jena, Schillbachstraße 24.

Verlag:
Akademie-Verlag, DDR-108 Berlin, Leipziger Straße 3–4; Fernruf 2 23 62 21 und 2 23 62 29; Telex-Nr.: 114 420; Bank: Staatsbank der DDR, Berlin, Kto.-Nr.: 6836-26-20 712.

Chefredakteur:
Dr. H.-J. Hänsch.

Redaktionskollegium:
Prof. Dr. E. Gutsche, Dr. H.-J. Hänsch, Dr. H. Lange, Dr. S. Oberländer.

Anschrift der Redaktion:
DDR-102 Berlin, Neue Schönhauser Straße 20; Fernruf: 2 82 33 80.

Veröffentlicht unter der Lizenznummer 1620 des Presseamtes beim Vorsitzenden des Ministerrates der Deutschen Demokratischen Republik.

Gesamtherstellung:
VEB Druckerei „Thomas Müntzer", DDR-582 Bad Langensalza.

Erscheinungsweise:
Die Zeitschrift "physica status solidi (a)" erscheint jeweils am 16. eines jeden Monats. Jährlich erscheinen 6 Bände zu je 2 Heften.

Bezugspreis:
je Band 165,— M zuzüglich Versandspesen (Preis für die DDR: 130,— M).

Bestellnummer dieses Bandes: 1085/54

Printed in the German Democratic Republic.

AN (EDV) 20 735

Contents

Original Papers

Short Notes

Pre-Printed Titles

Systematic List

The Author Index of Volume 54 Begins on Page 783

(It will be delivered together with Volume 55, Number 1.)

phys. stat. sol. (a) 54, 447 (1979)

Subject classification: 1.5 and 10

Institute of Semiconductors, Academy of Sciences of the Ukrainian SSR, Kiev

Account of Dislocation Structure at Stress Measurements in Heteroepitaxial Systems

I. Theory

By

L. S. Khazan, L. A. Matveeva, G. N. Semenova, and Yu. A. Tkhorik

Measurement of the curvature of epitaxial systems with the following use of Stoney's formula is the most popular technique for its stress determination. This formula is valid if plastic deformation in the system is absent, but numerous experimental data reflect the occurrence of such a deformation even at rather low temperatures. Thus, the contribution of dislocations due to curvature should be taken into account. The influence of such dislocations on curvature is analysed and the inadequacy of Stoney's formula for plastically deformed systems is shown.

Наиболее распространенным методом определения внутренних напряжений в эпитаксиальных системах является измерение кривизны системы с последующим использованием формулы Стоуни, предполагающей отсутствие пластической деформации системы. В то же время многочисленные экспериментальные факты свидетельствуют, что такая деформация происходит при относительно низких температурах, в силу чего необходимо учитывать вклад в кривизну эпитаксиальных систем возникающих при этом дислокаций. Проанализировано влияние дислокаций на кривизну и показано, что при наличии в системе пластической деформации (дислокаций) формула Стоуни становится неадекватной.

1. Introduction

Internal stresses are always present in epitaxial systems and influence their physical properties and parameters of semiconductor devices. For instance, a variation of the band structure induce the corresponding variation in optical and electrical properties of films. High stresses can lead to the destruction of a heteroepitaxial system (HES) (cracking, delamination) or to degradation of devices [1, 2] while relaxing with defect generation. Therefore, the determination of the stress magnitude is an actual problem. There exists a large number of corresponding works [1 to 4].

The most popular method is based on the relationship between the stress σ (or elastic strain ε) and the curvature $\varkappa$ of the two-layer epitaxial system expressed by the so-called Stoney formula (SF) [4] or somewhat more general formulae [2, 5]. There are also similar formulae for multilayer systems [2, 6]. Papers devoted to improve the precision of both, SF [2, 5, 6] and $\varkappa$-measurement methods [3], appear up to now, but there are serious objections against the adequacy of all such formulae. The point is that the SF (and all other similar formulae) was derived in the framework of the classical theory of elasticity [7]. In epitaxial systems, however, in most cases there occurs relaxation of stresses by means of plastic deformation connected with dislocation (particularly, misfit dislocations (MD)) generation and movement [2, 8, 9].

The above described formulae do not take into account these dislocations and HES bending connected with them; at the same time such an account may drastically change the results. In some works [5, 7] it was proposed to use the difference between HES curvature and that of the substrate after film removal as value of $\varkappa$ in SF but, as will be shown below, this procedure does not reflect all aspects of the dislocation effect on bending and stresses in epitaxial layers. Moreover, up to now there has been given an inconsistent and even incorrect interpretation of the role of dislocations in the relation between stress and curvature [6, 10]. This paper is devoted to the study of the above-mentioned questions.

2. General Equations

Internal stresses in epitaxial systems are caused by the incompatibility of elastic strains that arises due to a number of factors: the difference of the impurity concentrations in individual layers, different thermal expansion coefficients, etc. [1, 2, 4]; in particular, in heteroepitaxial systems (which we shall consider below) it is caused mainly by the misfit

$$f^0_{i,i+1(\gamma)} = \frac{a_{i+1,\gamma} - a_{i,\gamma}}{a_{i,\gamma}} \qquad (\gamma = X, Y)$$

of the lattice parameters of the layers i and $i + 1$ (see Fig. 1) (for a two-layer system we shall write f rather than $f_{1,2}$). The starting point of our analysis is the well-known compatibility equation for elastic strains $\boldsymbol{\epsilon}$ [11, 12],

$$\nabla \times \boldsymbol{\epsilon} = \boldsymbol{\varkappa} - \boldsymbol{\alpha}\,, \tag{1a}$$

or equivalently

$$\nabla \times \boldsymbol{\epsilon} \times \nabla = \boldsymbol{\eta}\,, \tag{1b}$$

where $\boldsymbol{\alpha}$ is the dislocation density tensor defined by[1])

$$\int_S \mathrm{d}\mathbf{S} \cdot \boldsymbol{\alpha} = \boldsymbol{b}\,,$$

where $\boldsymbol{b}$ is the Burgers vector of dislocations passing through the surface S, $\boldsymbol{\eta}$ is the incompatibility tensor connected with self-strain (plastic one) $\boldsymbol{\epsilon}^0$ of the system,

$$\boldsymbol{\eta} = -\nabla \times \boldsymbol{\epsilon}^0 \times \nabla\,,$$

or with the dislocation density

$$\boldsymbol{\eta} = \tfrac{1}{2}\,(\boldsymbol{\alpha} \times \nabla + \nabla \times \tilde{\boldsymbol{\alpha}}) \tag{2}$$

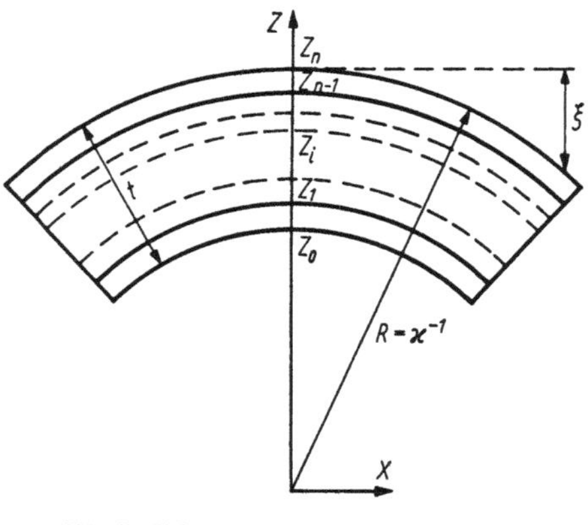

Fig. 1. Geometry of the n-layer HES

[1]) The dislocation density tensor used here is connected with the usually measured length N of dislocation lines per unit volume by the equation $N = \boldsymbol{\tau} \cdot \boldsymbol{\alpha} \cdot \boldsymbol{b}/b^2$, where $\boldsymbol{\tau}$ is the unit tangent vector of a dislocation line. Besides, everywhere in this paper $\boldsymbol{\varkappa}$ and $\boldsymbol{\alpha}$ mean the nondiagonal part of the corresponding tensors because just these define bending.

($\tilde{\alpha}$ means the conjugate tensor). These equations together with the equations of equilibrium

$$\nabla \cdot \boldsymbol{\sigma} = 0 , \tag{3}$$

Hooke's law

$$\boldsymbol{\sigma} = \hat{\mathbf{c}} : \boldsymbol{\epsilon} ,$$

and boundary conditions

$$\sigma_{\gamma z}\Big|_{\substack{Z=Z_0 \\ Z=Z_n}} = 0 \quad (\gamma = X, Y, Z) \tag{4}$$

permit us to solve the problem (bending is assumed to be small, i.e. $\xi \ll t$, where $t = |Z_n - Z_0|$ is the total thickness of the system).[2])

If bending is purely elastic, i.e. dislocations are absent in the system, it is natural to assume stresses to be dependent only on Z. Then (3) can be reduced to the form

$$\bar{\sigma} = \int\limits_{Z_0}^{Z_n} \frac{\sigma \, dZ}{t} = 0 , \qquad \overline{Z\sigma} = \int\limits_{Z_0}^{Z_n} \frac{\sigma Z \, dZ}{t} = 0 , \tag{5}$$

and it follows from (1a) that $\varepsilon^i_{XX} = \varepsilon^i_{YY} = \varkappa Z + g_i$ — is linear in Z (g_i = const within the layer). In this approximation we come to the SF; all other papers are restricted by the same result also.

When dislocations are present in the system, $\alpha(X, Y, Z)$ together with $\varkappa$ determines the distribution of strain. So, for instance, one can see from (1a) that for $\varepsilon = 0$

$$\varkappa = \alpha , \tag{6}$$

i.e. the unstrained system may be also bent (plastic bending). In principle, stresses can differ from those obtained by means of the curvature alone even in their signs. But the role of dislocations is not limited by the fact that (1a) contains the difference $\varkappa - \alpha$; the presence of dislocations makes the problem, generally speaking, not one-dimensional. In virtue of this, the linear dependence of ε on Z can be violated even in the simplest cases, and one has to operate with the general equilibrium equation (3) instead of (5). For instance, in the case of a two-layer HES, one-dimensional misfit $f_X \neq 0$, $f_Y = 0$, and a system of straight equidistant MD in the interface ($Z = Z_1$),

$$\alpha_{Xy} = b_X \, \delta(Z - Z_1) \, \delta(X - mX_0) \tag{7}$$

($m = 0, \pm 1, \pm 2, \ldots$; X_0 is the distance between dislocations; $\delta(Z)$ is the δ-function), stresses depend exponentially on Z and periodically on X [8] (in [8] bending was not considered). Due to the usual condition $X_0 \gg a_{i,X}$ stresses are macroscopically inhomogeneous in X. They are also inhomogeneous in real systems when dislocations are spaced unequidistantly (in slip bands) or in more complicated cases of three-dimensional dislocation morphology (arising, for instance, under stress relaxation by generation of dislocation half-loops).

As it follows from all mentioned above, one has to know the dislocation distribution in the system as well as its curvature to obtain correct stress values in epitaxial systems. Averaging the dislocation density over coordinates X, Y normal to R, that reduces the problem to a one-dimensional one, is not always correct, but if dislocations are spaced sufficiently closely and uniformly in the mean at the interface, this proce-

[2]) For the sake of simplicity, the case of isotropic solids will be considered later on, and the following notations will be used: $\alpha \equiv \alpha_{XY} = \alpha_{YX}$, $\varkappa \equiv \varkappa_{XY} = \varkappa_{YX}$, $\varepsilon \equiv \varepsilon_{XX} = \varepsilon_{YY}$, $\varepsilon_{XY} = 0$; and in the plane-stress approximation $C_i = E_i/(1 - \nu_i)$, where E_i is the Young modulus and ν_i the Poisson ratio of the i-th layer.

dure may be justified to some extent (for example, if dislocations are spaced in some layer, then at a distance from the layer large compared with the distance between neighbouring dislocations such an averaging is justified by Saint Venaint's principle). However, as it will be demonstrated further, taking dislocations into account drastically changes the stress distribution in comparison with the simplest cases [2, 4 to 6] even in the one-dimensional approximation.

3. The One-Dimensional Approximation

In this case[3])

$$\alpha = \alpha(Z) = \frac{\mathrm{d}}{\mathrm{d}Z}\,\varepsilon^0(Z)\,, \tag{8}$$

and (1b) can be written in the form [12]

$$\frac{\mathrm{d}^2}{\mathrm{d}Z^2}\,(\varepsilon + \varepsilon^0) = 0\,,$$

which leads to

$$\varepsilon(Z) = -\varepsilon^0(Z) + g + \mathrm{d}Z\,; \quad \mathrm{d} = \varkappa\,. \tag{9}$$

We draw the attention to the fact that in the one-dimensional case the total deformation $\varepsilon + \varepsilon^0$, but not the elastic one (ε), is linear in Z.

From (4), (5), and (9) we obtain

$$-\overline{C\varepsilon^0} + g\overline{C} + \mathrm{d}\overline{CZ} = 0\,, \qquad -\overline{CZ\varepsilon^0} + g\overline{CZ} + \mathrm{d}\overline{CZ^2} = 0\,. \tag{10}$$

For n-layer HES

$$C = C(Z) = C_1 + \sum_{i=1}^{n-1} (C_{i+1} - C_i)\,\theta(Z - Z_i)\,,$$

where

$$\theta(Z) = \begin{cases} 1, & Z > 0\,, \\ 0, & Z < 0\,, \end{cases}$$

$Z_i = Z_0 + \sum_{k=1}^{i} t_k$, t_k is the thickness of the k-th layer. From these equations one can obtain $\varepsilon(Z)$.

Note that for most of HES the difference $C_{i+1} - C_i$ is not large. Hence, we can take the elastic constants of all layers to be equal ($C_i = \text{const}$); it provides the final formulae to be more compact.

In this case it follows from (9) and (10) that

$$\left.\begin{aligned} \varepsilon(Z) &= -\varepsilon^0(Z) + \overline{\varepsilon^0} + \varkappa(Z - \overline{Z})\,; \qquad \overline{Z} = \tfrac{1}{2}\,(Z_n + Z_0)\,, \\ \varkappa &= \frac{\overline{Z\varepsilon^0} - \overline{Z}\,\overline{\varepsilon^0}}{\overline{Z^2} - \overline{Z}^2} = \frac{12}{t^3}\left(\int_{Z_0}^{Z_n} \varepsilon^0 Z\,\mathrm{d}Z - \frac{Z_n + Z_0}{2}\int_{Z_0}^{Z_n} \varepsilon^0\,\mathrm{d}Z\right). \end{aligned}\right\} \tag{11}$$

Choosing different forms for $\varepsilon^0(Z)$ (or $\alpha(Z)$) one can obtain corresponding expressions for $\varkappa$ and $\varepsilon(Z)$.

Let us consider some special cases.

---

[3]) From (8) one can see that $\varepsilon^0(Z)$ is the $B_X(B_Y)$ component of the total Burgers vector of dislocations passing through the surface bounded by the coordinates Z_0, Z and by the unit interval of the $X(Y)$ axis.

3.1 *Multilayer system without dislocations*

In this case

$$\varepsilon^0 = \sum_{i=1}^{n-1} f_{i,i+1}\theta(Z - Z_i) \tag{12}$$

where $f_{i,i+1}$ is the misfit between the i-th and $(i + 1)$-th layers.[4]) From (8) and (12) one can see that stresses in HES in the case considered can be described by a system of fictitious interfacial dislocations with density

$$\alpha(Z) = \sum_{i=1}^{n-1} f_{i,i+1}\,\delta(Z - Z_i)\,. \tag{13}$$

This analogy between f and α follows from [12],

$$\boldsymbol{n} \times [\boldsymbol{\epsilon}(Z_i + 0) - \boldsymbol{\epsilon}(Z_i - 0)] \times \boldsymbol{n} = \tfrac{1}{2}\,[\boldsymbol{n} \times \tilde{\boldsymbol{\alpha}}(Z_i) - \boldsymbol{\alpha}(Z_i) \times \boldsymbol{n}]\,,$$

as far as $f_{i,i+1}$ is just the jump of strain ε_{XX} on the interface $Z = Z_i$ (n is the normal to the interface), and it becomes evident from the comparison of the Burger's contours passed in the bicrystal and in the delaminated system (Fig. 2).

From (11) and (12),

$$\varkappa = \frac{6}{t^3}\sum_{i=1}^{n-1} f_{i,i+1}(Z_n - Z_i)\,(Z_i - Z_0)\,, \tag{14a}$$

$$\varepsilon(Z) = \varkappa Z + \sum_{i=1}^{n-1} f_{i,i+1}\left\{\frac{Z_n - Z_i}{t}\left[1 - 3\,\frac{Z_n + Z_0}{t^2}\,(Z_i - Z_0)\right] - \theta(Z - Z_i)\right\}. \tag{14b}$$

These expressions can be easily transformed to the form given in [6]. Particularly, for the two-layer system[5])

$$\varkappa = \frac{6ft_1t_2}{t^3}, \tag{15a}$$

$$\varepsilon(Z) = -f\left[\theta(Z) - \frac{t_2}{t}\right] + \varkappa\left[Z + \frac{t_1 - t_2}{2}\right]. \tag{15b}$$

For $t_2 \ll t_1$ we obtain the Stoney formula

$$\bar{\varepsilon}_2 \approx -f \approx -\frac{t_1^2\varkappa}{6t_2}, \tag{15c}$$

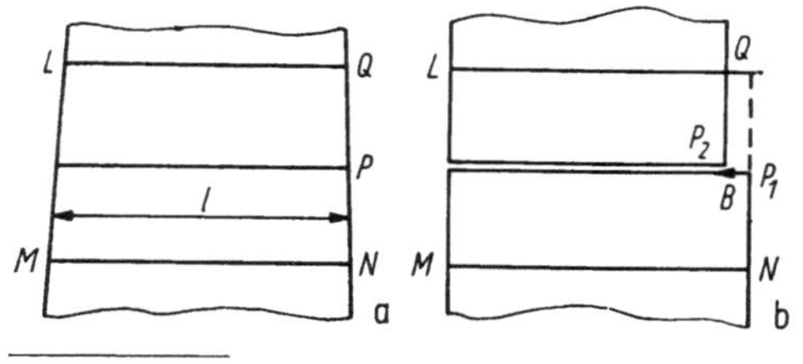

Fig. 2. Burgers contours LMNPQ and $LMNP_1P_2Q$ in a) initial and b) separated into layers system. $\boldsymbol{B}_X = f_X\boldsymbol{l}$ is the total Burgers vector ($f_Y = 0$)

[4]) Here f can be the misfit (jump of ε) after relaxation and can differ from f^0.

[5]) Here and later the formulae for $\varepsilon(Z)$ in a two-layer HES are given for the origin of coordinates at the interface (i.e. $Z_1 = 0$). If one accounts for the inequality of elastic constants, one obtains instead of (15) the formulae presented in [2, 5].

where

$$\bar{\varepsilon}_2 = \frac{1}{t_2} \int_{Z_1}^{Z_2} \varepsilon(Z)\,\mathrm{d}Z$$

is the mean strain in the film.

3.2 Multilayer system with initially bent layers

Let us note the initial curvatures of free unstrained layers as $\varkappa_i$ ($i = 1, 2, \ldots, n$). From (6) it follows that it is equivalent to the uniform distribution of dislocations with density $\alpha_i = \varkappa_i$ within each layer. When the layers are joined, strain jumps on the interfaces appear which are equivalent to fictitious interfacial dislocations with density $f_{i,i+1}\,\delta(Z - Z_i)$. So, in the considered HES

$$\alpha(Z) = \alpha_1 + \sum_{i=1}^{n-1} [(\alpha_{i+1} - \alpha_i)\,\theta\,(Z - Z_i) + f_{i,i+1}\,\delta(Z - Z_i)] \tag{16}$$

from which

$$\varepsilon^0(Z) = \alpha_1 Z + \sum_{i=1}^{n-1} [(\alpha_{i+1} - \alpha_i)\,(Z - Z_i) + f_{i,i+1}]\,\theta(Z - Z_i)\,. \tag{17}$$

Then

$$\varkappa = \alpha_1 + \frac{1}{t^3} \sum_{i=1}^{n-1} (Z_n - Z_i)\,\{(\alpha_{i+1} - \alpha_i)\,(Z_n - Z_i)\,[Z_n - Z_0 + 2(Z_i - Z_0)] + 6f_{i,i+1}(Z_i - Z_0)\}\,; \tag{18a}$$

$$\varepsilon(Z) = -\varepsilon^0 + \varkappa Z + \frac{1}{t} \sum_{i=1}^{n-1} \left\{ f_{i,i+1}(Z_n - Z_i) \left[1 - 3\,\frac{(Z_n + Z_0)\,(Z_i - Z_0)}{t^2}\right] + \frac{\alpha_{i+1} - \alpha_i}{2}\,(Z_n - Z_i)^2 \left[1 - \frac{(Z_n + Z_0)\,\big(Z_n - Z_0 + 2(Z_i - Z_0)\big)}{t^2}\right]\right\}. \tag{18b}$$

For a two-layer system[6])

$$\varkappa = \frac{1}{t^3}\,[6ft_1t_2 + \varkappa_1 t_1^2(t_1 + 3t_2) + \varkappa_2 t_2^2(t_2 + 3t_1)]\,, \tag{19}$$

$$\varepsilon(Z) = Z(\varkappa - \varkappa_1) - [(\varkappa_2 - \varkappa_1)\,Z + f]\,\theta(Z) + \frac{\varkappa}{2}\,(t_1 - t_2) + \frac{1}{t}\,[ft^2 - \tfrac{1}{2}\,(\varkappa_1 t_1^2 - \varkappa_2 t_2^2)/2]\,. \tag{20}$$

---

[6]) Taking into account the inequality of elastic constants we obtain

$$\varkappa = \frac{6fC_1C_2t_1t_2(t_1 + t_2) + C_2\varkappa_2t_2^2\,(C_2t_2^2 + 4C_1t_1t_2 + 3C_1t_1^2) + C_1\varkappa_1t_1^2\,(C_1t_1^2 + 4C_2t_1t_2 + 3C_2t_2^2)}{C_1^2t_1^4 + 4C_1C_2t_1^3t_2 + 6C_1C_2t_1^2t_2^2 + 4C_1C_2t_1t_2^3 + C_2^2t_2^4}\,,$$

$$\bar{\sigma}_2 = -\frac{C_1(\varkappa - \varkappa_1)\,t_1^3 + C_2(\varkappa - \varkappa_2)\,t_2^3}{6t_2t_1}\,.$$

If $t_2 \ll t_1$, then

$$\bar{\sigma}_2 \approx -\frac{C_1(\varkappa - \varkappa_1)\,t_1^2}{6t_2}$$

coincides with the Stoney formula (15), which contains the difference between the system curvature and that of the unstressed substrate.

3.3 *System with homogeneously distributed dislocations in layers neighbouring the interface*

The dislocation density in the i-th layer is equal to (Fig. 3a)

$$\alpha_i^-\{\theta(Z - Z_{i-1}) - \theta[Z - (Z_{i-1} + h_i^-)]\} + \alpha_i^+\{\theta[Z - (Z_i - h_i^+)] - \theta(Z - Z_i)\} .$$

Taking into account the strain jumps at IF the total dislocation density in HES is given by

$$\alpha(Z) = \sum_{i=1}^{n-1} [\alpha_i^+\{\theta[Z - (Z_i - h_i^+)] - \theta(Z - Z_i)\} + f_{i,i+1}\,\delta(Z - Z_i) + \\ + \alpha_{i+1}^-\{\theta(Z - Z_i) - \theta[Z - (Z_i + h_{i+1}^-)]\}] . \tag{21}$$

Accordingly,

$$\varepsilon^0(Z) = \sum_{i=1}^{n-1} \{\alpha_i^+[Z - (Z_i - h_i^+)]\,\theta[Z - (Z_i - h_i^+)] - \alpha_{i+1}^-[Z - (Z_i + h_{i+1}^-)] \times \\ \times \theta[Z - (Z_i + h_{i+1}^-)] + \\ + [(\alpha_{i+1}^- - \alpha_i^+)(Z - Z_i) + f_{i,i+1}]\,\theta(Z - Z_i)\} . \tag{22}$$

Let us consider for the sake of simplicity the two-layer HES ($n = 2$). With the notations

$$\alpha_1^+ \equiv \alpha_1 , \qquad \alpha_2^- \equiv \alpha_2 , \qquad h_1^+ \equiv h_1 , \qquad h_2^- \equiv h_2$$

we obtain

$$\varkappa = \frac{1}{t^3}\{6ft_1t_2 + \alpha_1h_1[6t_1t_2 - 2h_1^2 + 3h_1(t_1 - t_2)] + \\ + \alpha_2h_2[6t_1t_2 - 2h_2^2 - 3h_2(t_1 - t_2)]\} \tag{23a}$$

and, if $t_i < h_i$, one must set $h_i = t_i$ (with this we come to the previous case). The expression for the strain is

$$\varepsilon(Z) = -\alpha_1(Z + h_1)\,\theta(Z + h_1) + \alpha_2(Z - h_2)\,\theta(Z - h_2) - \\ - [(\alpha_2 - \alpha_1)Z + f]\,\theta(Z) + \varkappa\left(Z + \frac{t_1 - t_2}{2}\right) + \\ + \frac{1}{t}\left[\alpha_1h_1\left(t_2 + \frac{h_1}{2}\right) + \alpha_2h_2\left(t_2 - \frac{h_2}{2}\right) + ft_2\right]. \tag{23b}$$

One can express α_1 and α_2 by means of the substrate and film curvatures, $\varkappa_1$ and $\varkappa_2$:

$$\alpha_i = \frac{t_i^3\varkappa_i}{h_i^2(3t_i - 2h_i)} \qquad (i = 1, 2) .$$

This yields

$$\varkappa = \frac{1}{t^3}\left\{6ft_1t_2 + \varkappa_1t_1^3\left[1 + \frac{3t_2(2t_1 - h_1)}{h_1(3t_1 - 2h_1)}\right] + \varkappa_2t_2^3\left[1 + \frac{3t_1(2t_2 - h_2)}{h_2(3t_2 - 2h_2)}\right]\right\}. \tag{23c}$$

If one layer of HES is removed and the remaining layer (number i) is etched from its interface side, then its curvature is equal to

$$\tilde{\varkappa}_i = \alpha_i\,\frac{3t_i(h_i - \tilde{h}_i)^2 + 3h_i^2\tilde{h}_i - \tilde{h}_i^3 - 2h_i^3}{(t_i - \tilde{h}_i)^3} \quad \text{if} \quad \tilde{h}_i \leqq h_i \tag{24}$$

and

$$\tilde{\varkappa}_i = 0 \quad \text{if} \quad \tilde{h}_i \geqq h_i \qquad (i = 1, 2) .$$

Here $\tilde{h}_i$ is the depth of the removed layer.

3.4 Model of misfit dislocations

The dislocation density in van der Merwe's theory [8] is given by (7). In reality, however, total stress relaxation does not occur. This yields a non-equidistant spacing of dislocations in IF and also a distribution of dislocations not in the IF itself, but in the layer of finite thickness neighbouring it with dislocation density decreasing with distance from the interface.

As

$$\delta(Z) = \lim_{h\to\infty} \frac{1}{\pi} \frac{h}{h^2 + Z^2},$$

we adopt for the dislocation density averaged over X, Y the relation

$$\alpha(Z) = \\ = \sum_{i=1}^{n-1} \left\{ \alpha_i^+ \frac{h_i^+}{h_i^{+2} + (Z - Z_i)^2} [\theta(Z - Z_{i-1}) - \theta(Z - Z_i)] + f_{i,i+1}\,\delta(Z - Z_i) + \right. \\ \left. + \alpha_{i+1}^- \frac{h_{i+1}^-}{h_{i+1}^{-2} + (Z - Z_i)^2} [\theta(Z - Z_i) - \theta(Z - Z_{i+1})] \right\}, \tag{25}$$

where α_i^-, h_i^- and α_i^+, h_i^+ are the parameters defining magnitude and width of the dislocation distribution in i-th layer at the bottom (near Z_{i-1}) and at the top (near Z_i) interface, respectively, see Fig. 3b; $f_{i,i+1}$ means, as it was already mentioned, the jump of ε (misfit) at the interface between the i-th and $(i + 1)$-th layers. Then

$$\varepsilon^0 = \sum_{i=1}^{n-1} \left\{ \alpha_i^+ \arctan \frac{Z - Z_i}{h_i^+} [\theta(Z - Z_{i-1}) - \theta(Z - Z_i)] + \right. \\ \left. + f_{i,i+1}\theta(Z - Z_i) + \alpha_{i+1}^- \arctan \frac{Z - Z_i}{h_{i+1}^-} [\theta(Z - Z_i) - \theta(Z - Z_{i+1})] \right\}. \tag{26}$$

It is easy to obtain expressions for $\varkappa$ and $\varepsilon(Z)$ from (11). However, as such formulae are rather complex, we shall restrict ourselves to the two-layer system ($n = 2$). With

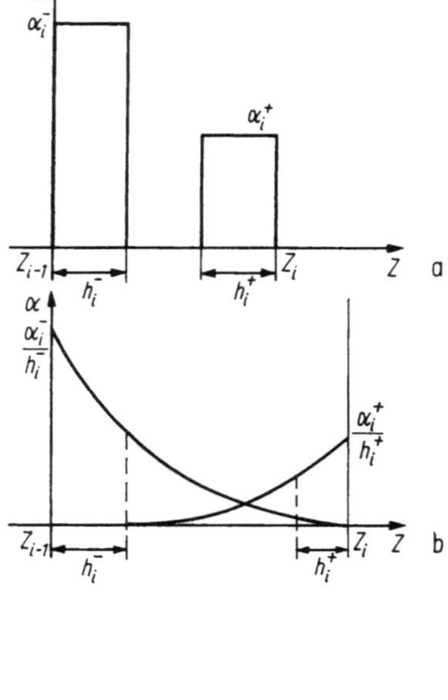

Fig. 3. Dislocation density in the i-th layer of an epitaxial system versus Z-coordinate. $h_i^{-(+)}$ in Fig. 3a is the width (in Fig. 3b halfwidth) of the dislocation layer adjoining to HES layer with next lower (higher) number; $\alpha_i^{-(+)}$ defines the magnitude of the corresponding distribution

the notations $\alpha_1^+ \equiv \alpha_1$, $\alpha_2^- = \alpha_2$, $f_{1,2} = f$ we obtain

$$\varepsilon^0 = \alpha_1 \arctan \frac{Z - Z_1}{h_1} + \left(\alpha_2 \arctan \frac{Z - Z_1}{h_2} - \alpha_1 \arctan \frac{Z - Z_1}{h_1} + f\right) \theta(Z - Z_1) \,. \tag{27}$$

In this case

$$\varkappa = \frac{6}{t^3} \left\{ f t_1 t_2 + \alpha_1 \left[(h_1^2 + t_1 t_2) \arctan \frac{t_1}{h_1} + \frac{h_1}{2} (t_1 - t_2) \ln \left(1 + \frac{t_1^2}{h_1^2}\right) - h_1 t_1 \right] + \alpha_2 \left[(h_2^2 + t_1 t_2) \arctan \frac{t_2}{h_2} + \frac{h_2}{2} (t_2 - t_1) \ln \left(1 + \frac{t_2^2}{h_2^2}\right) - h_2 t_2 \right] \right\}, \tag{28a}$$

$$\varepsilon(Z) = -\varepsilon^0 + \varkappa \left(Z - \frac{Z_2 + Z_0}{2}\right) + \frac{1}{t} \left\{ f t_2 - \alpha_1 \left[t_1 \arctan \frac{t_1}{h_1} - \frac{h_1}{2} \ln \left(1 + \frac{t_1^2}{h_1^2}\right)\right] + \alpha_2 \left[t_2 \arctan \frac{t_2}{h_2} - \frac{h_2}{2} \ln \left(1 + \frac{t_2^2}{h_2^2}\right)\right]\right\}. \tag{28b}$$

In principle, one can obtain the parameters α_i and h_i from the minimum of elastic energy $\overline{W} = t^{-1} \int_{Z_0}^{Z_2} W \, dZ$, where $W = \frac{1}{2} \sigma\varepsilon = \frac{1}{2} C\varepsilon^2(Z)$. Such an approach correlates with that of van der Merwe [8] and, in addition, allows to account for the HES bending. However, as the total relaxation (energy minimum) is not reached in practice, it is more reliable to obtain these parameters from a comparison with experiment. In particular, setting in (28a) $\varkappa = \varkappa_2$ at $t_1 = 0$ (curvature of the second layer after removal of the first one), and $\varkappa = \varkappa_1$ at $t_2 = 0$, $t_1 \neq 0$ (curvature of the first layer after removal of the second one) one can express (28a) as

$$\varkappa = \frac{1}{t^3} \left[6 f t_1 t_2 + \varkappa_1 t_1^3 \left(1 + \frac{t_2}{h_1} T_1\right) + \varkappa_2 t_2^3 \left(1 + \frac{t_1}{h_2} T_2\right)\right], \tag{28c}$$

where

$$T_i = \frac{\tau_i \arctan \tau_i - \frac{1}{2} \ln (1 + \tau_i^2)}{\arctan \tau_i + \frac{1}{2} \tau_i \ln (1 + \tau_i^2) - \tau_i}; \qquad \tau_i \equiv \frac{t_i}{h_i}. \tag{28d}$$

It should be noted here that if dislocations are absent in the system, i.e. $\alpha_1 = \alpha_2 = 0$, then $\varkappa_1 = \varkappa_2 = 0$, too, and (28c) reduces to (15a). If dislocations are homogeneously distributed in the layers, i.e. $h_1, h_2 \to \infty$, $\alpha_1/h_1 = \varkappa_1 = \text{const}$, $\alpha_2/h_2 = \varkappa_2 = \text{const}$ in (25) then, (28c) reduces to (19). The substrate curvature $\tilde{\varkappa}_1$, after film removing and at further etching from the IF side (cf. with (24)) is

$$\tilde{\varkappa}_i = \frac{\varkappa_i t_i^3}{(t_i - \tilde{h}_i)^3} \, \frac{(1 - \tau_i \tilde{\tau}_i) \arctan (\tau_i - \tilde{\tau}_i) - \tau_i + \tilde{\tau}_i + \dfrac{\tau_i + \tilde{\tau}_i}{2} \ln \dfrac{1 + \tau_i^2}{1 + \tilde{\tau}_i^2}}{\arctan \tau_i - \tau_i + \dfrac{\tau_i}{2} \ln (1 + \tau_i^2)}, \tag{29}$$

where $\tilde{\tau}_i \equiv \tilde{h}_i/h_i$, $\tilde{h}_i$ is the depth of the removed layer from the IF side, $i = 1, 2$, $\tilde{\varkappa}_2$ is an analogous value for the film.

4. Conclusion

We have considered the effect of dislocations on the system curvature and have shown that the Stoney formula is invalid when dislocations generated under the stress relaxation are present in the system.

It should be noted that the present consideration is not restricted to the case of heteroepitaxial systems and stresses caused by lattice misfit. It is possible to think of strain incompatibility caused by non-coincidence of thermal expansion coefficients or nonhomogeneous doping and to use all the obtained results in the case of thermal stresses or homoepitaxial systems including graded ones. For instance, in [2] a linear dependence of ε^0 on Z in a homoepitaxial film (a special case of (17) with $n = 2$ and $\alpha_1 = 0$) was considered and the corresponding stress distribution was obtained (let us note here that, as it follows from our results, the stress distribution after plastic deformation in an epitaxial system presented in Fig. 2 of [2] is not fully correct).

In the second part of this paper we want to present an experimental evidence of the statements given.

References

[1] M. G. MILVIDSKII and V. B. OSVENSKII, Kristallografiya **22**, 431 (1977).
[2] V. M. USTINOV and B. G. ZAKHAROV, Obzory po elektronnoi tekhnike, Ser. Materialy **4**, (1977).
[3] D. S. CAMPBELL, in: Handbook of Thin Film Technology, McGraw-Hill Publ. Co., 1970.
[4] R. W. HOFFMAN, in: Physics of Thin Films, Vol. 3, Academic Press, New York/London 1966.
[5] K. RÖLL, J. appl. Phys. **47**, 3224 (1976).
[6] G. H. OLSEN and M. ETTENBERG, J. appl. Phys. **48**, 2543 (1977).
[7] S. TIMOSHENKO, J. Opt. Soc. Amer. **25**, 233 (1925).
[8] J. H. VAN DER MERWE, J. appl. Phys. **34**, 123 (1963).
[9] J. W. MATTHEWS, J. Vacuum Sci. Technol. **12**, 126 (1975).
[10] M. S. ABRAHAMS, R. L. WEISBERG, and J. J. TIETJEN, J. appl. Phys. **40**, 3754 (1969).
[11] V. L. INDENBOM and A. N. ORLOV, Uspekhi fiz. Nauk **76**, 557 (1962).
[12] V. L. INDENBOM, in: Fizika kristallov s defektami, Vol. 1, Tbilisi 1966.

(Received May 3, 1979)

phys. stat. sol. (a) **54**, 457 (1979)

Subject classification: 11; 10.2; 19; 22.1.2

Institute of Semiconductor Physics, Academy of Sciences of the USSR, Siberian Branch, Novosibirsk (a) and Krasnoyarsk State University (b)

Efficiency of Defect Build-up in Ion-Implanted Polycrystalline Silicon Films

By

A. V. Dvurechenskii (a), L. P. Potapova (b), and V. V. Kalinin (a)

Implantation of polycrystalline silicon films (poly-Si) with H^+, He^+, N^+, and Ar^+ ions at doses between 10^{13} and 2×10^{17} ions cm^{-2} are studied. The number of paramagnetic vacancy-type defects (VV-centres) ΔN_{VV} are measured by EPR technique at 77 K. Also the structure of poly-Si and effective grain size are controlled by electron diffraction. Implantation of poly-Si with H^+, H_2^+, and He^+ ions is found to lead to the appearance of VV-centres in contrast to the data on Si single crystals. There are two stages in the dose dependence of ΔN_{VV}. When ΔN_{VV} is saturated at high doses the poly-Si layers are not completely amorphized by H^+ and H_2^+ irradiation as distinguished from the samples, implanted with N^+ and Ar^+ ions, where the amorphization dose D_a is less than that of Si single crystals by a factor of 100. D_a is found to increase for the implanted samples of poly-Si annealed before. The discussion of the results obtained is based on the assumption of the important role of intrinsic stresses in the accumulation of defects in poly-Si up to amorphization.

Исследовались поликристаллические пленки кремния (поли-Si), облученные ионами H^+, He^+, N^+ и Ar^+ в интервале доз 10^{13} до 2×10^{17} cm^{-2}. Методом ЭПР измерялось число парамагнитных дефектов (VV-центров) ΔN_{VV} при 77 K. Структура исследуемых образцов и средний размер зерна монокристаллитов контролировались с помощью электронной дифракции. Обнаружено, что в отличие от монокристаллического кремния облучение поли-Si ионами H^+, H_2^+ и He^+ приводит к появлению VV-центров. Дозовая зависимость $\Delta N_{VV}(D)$ состоит из двух стадий. При больших дозах $\Delta N_{VV}(D)$ выходит на плато, однако полной аморфизации облучаемого слоя по данным электронной дифракции не происходит. При бомбардировке поли-Si ионами N^+ и Ar^+ наблюдается аморфизация всего облучаемого слоя. При этом доза аморфизации D_a примерно на два порядка меньше, чем в моно-Si. Установлено также, что D_a растет с увеличением температуры отжига ($T_{отж}$), проводимого перед облучением образцов ионами. При $T_a \geqq 1000$ °C D_a совпадает со значением, получающимся для моно-Si. Анализ полученных результатов проводится на основе предположения об определяющей роли упругих напряжений в образовании аморфной фазы в поли-Si.

1. Introduction

The defect accumulation in ion-implanted semiconductors until the bombarded layer becomes amorphous can proceed by the following mechanisms [1, 2]:

1) amorphous phase production directly by a single disordered region,
2) interaction (overlap) of disordered regions which are not amorphous,
3) trapping of vacancies (or interstitial atoms) and their accumulation in the vicinity of the strained crystal regions (interfaces and as-grown imperfections) to the critical concentration.

At room-temperature bombardment of silicon the first mechanism occurs for heavy-ion implantation and the second one takes place for light- and medium-ion implantation. The kinetics of the amorphous state growth for these two mechanisms has been well studied. The third one has been used in [3 to 6] to explain the appearance of a surface damage layer in the structures of light-ion-implanted Si and Si–SiO_2. The amorphization kinetics for this mechanism was not investigated.

Polycrystalline silicon (poly-Si) being a composition of randomly oriented crystallites is a convenient object for EPR studies of the interaction of mobile defects with boundary grains, voids, and other imperfections. If the average distance between these imperfections is compared with the size of the disordered region induced by ion implantation, it will be possible to expect a noticeable change of the defect accumulation kinetics.

The fraction of mobile defects escaping from the disordered region and interacting with imperfections increases as the ion mass decreases [1]. So the largest effect of as-grown imperfections in radiation defect accumulation should be expected in light-ion-implanted semiconductors.

Taking this into account the aim of the present work was the study of the defect accumulation kinetics in light-ion-implanted poly-Si by the EPR technique.

2. Experimental

Silicon polycrystalline films were obtained by thermal decomposition of silane in an H_2 and N_2 atmosphere on 300 μm single-crystal silicon plates or on the same plates coated with a thermal SiO_2 film 0.4 μm thick. The films 1 μm thick were deposited at a rate of approximately 10^{-3} μms^{-1} at 700 °C (H_2 atmosphere) or 800 °C (N_2 atmosphere). The samples were implanted at room temperature with doses of 15 keV H^+, 500 keV H_2^+, 1.2 MeV He^+, and 60 keV N^+ and Ar^+ ions.

The density of VV-centres N_{vv} (cm^{-2}) [1, 2] was measured in the X-band of an EPR spectrometer at 77 K. In our case of a dual cavity, the sample under investigation was placed in one of the two sample holes and the reference sample in other hole.

The relative error of ΔN_{vv} (cm^{-2})[1]) estimations was ±5 %with the exception of the first point of the dose dependence measured in the initial (unannealed) poly-Si. The latter was estimated with an accuracy of 25%, because of the noticeable signal of VV-centres in the initial samples [7].

We have controlled the film structure by electron microscopy and electron diffraction techniques. Using the formula $\bar{\varepsilon} = kL\lambda/B$ [8] we have estimated the effective grain size $\bar{\varepsilon}$. Here L is the device constant, B the full width at half maximum (FWHM) obtained from the photometric curves of the electron diffraction patterns, k a coefficient estimated from the reference films (we have $k = 6 \times 10^{-4}$ μm). We have checked by electron microscopy the average grain size, estimated from FWHM. The dark field patterns have given approximately the same average grain size.

Some poly-Si films have been annealed at 850 to 1150 °C in a dry nitrogen atmosphere before irradiation.

3. Results

1. There are two stages in the dose dependence of the density of VV-centres in the poly-Si implanted by H^+, H_2^+, and He^+ ions. The first stage extends to the dose 3×10^{15} cm^{-2} and the second one lies in the dose interval 5×10^{15} to 10^{16} cm^{-2} (Fig. 1a, b).

2. In the poly-Si annealed before implantation the second stage appears at higher doses than in unannealed samples. In addition ΔN_{vv} decreases as the annealing temperature increases. The exception is only the heavy-dose irradiated sample annealed at 850 °C (Fig. 1b). The second stage appears at higher doses in the annealed poly-Si.

---

1) ΔN_{vv} is the difference between densities of VV-centres after and before ion irradiation.

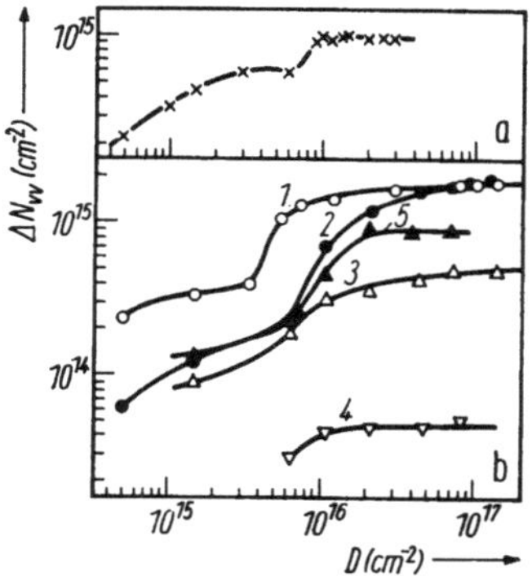

Fig. 1. The number of VV-centres in poly-Si versus dose of a) 15 keV H^+ ions, b) 500 keV H_2^+ ions, and b) 1.2 MeV He^+ ions (curve 5). The annealings of the films before irradiation are (1) $T_a = 20$, (2) 850, (3) 1000, (4) 1150 °C

The average grain size $\bar{\varepsilon}$ grows from 90 Å in the initial samples to 250 Å in the annealed ones at 850 °C. Further annealing up to 1150 °C does not lead to a remarkable growth.

3. When the density of VV-centres was saturated at high doses the poly-Si layers were not completely amorphized by H^+ and H_2^+ irradiation in contrast to the samples implanted with N^+ and Ar^+ ions where we get completely amorphous layers. Fig. 2 shows that H_2^+ ion implantation leaves the poly-Si layer structure unaltered. The average grain size did not change within the limits of accuracy.

The concentrations VV-centres were (2 to 4) $\times 10^{19}$ cm^{-3} and 2×10^{20} cm^{-3}, respectively, in light-ion(H^+, H_2^+)[2]) and medium-ion(N^+, Ar^+)-implanted poly-Si at doses saturating the density of VV-centres.

4. The amorphization dose D_a of N^+ ion-implanted poly-Si (Fig. 3) is less than that of the Si single crystal by a factor 100. In addition D_a has increased for the samples of poly-Si annealed before implantation (Fig. 3, curve 2). When the annealing temperature was above 1000 °C a difference of D_a between the single crystal and poly-Si was not observed.

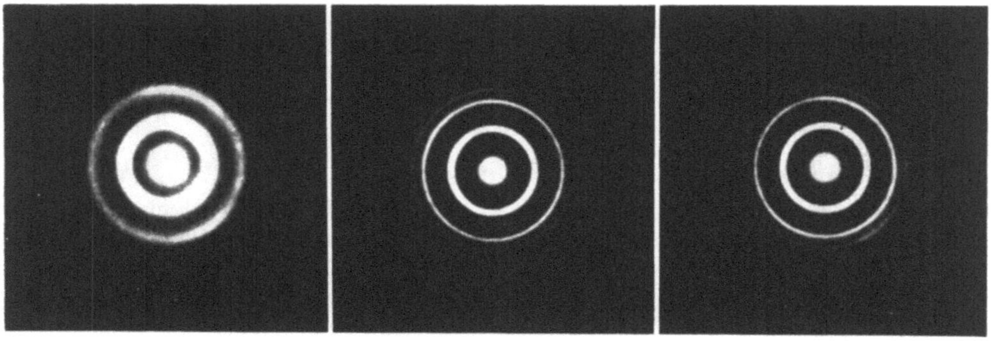

a b c

Fig. 2. The electron diffraction pattern of poly-Si. a) Initial sample; b) annealed at 850 °C; c) annealed at 850 °C and irradiated by 500 keV H_2^+ ions with the dose $D = 5 \times 10^{16}$ cm^{-2}

[2]) The H^+ ion implantation in amorphous Si is known to [9] lead to the interaction of hydrogen and VV-centres. As a result of this interaction the microvoids become non-paramagnetic. However, this effect vanishes in the present experiments due to the low dose H^+ ion implantation. We have measured the density of VV-centres in Si implanted with He^+ ions in the same dose range and have not observed a higher density of this defect (Fig. 1b).

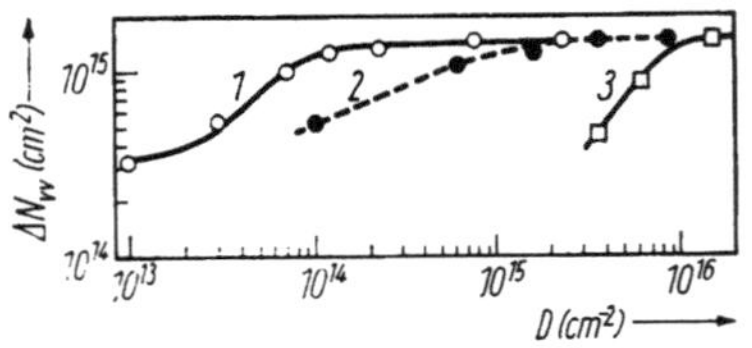

Fig. 3. The number of VV-centres versus the dose of 60 keV N^+ ions. (1) Initial poly-Si; (2) poly-Si annealed at 900 °C; (3) Si single crystal

All these quantitative results refer to the Si–SiO_2–poly-Si structure, obtained by film deposition at the substrate temperature of 700 °C in an H_2 atmosphere. Poly-Si films fabricated under other conditions described above have given qualitatively the same results.

4. Discussion

1. To convert a single crystal to the amorphous state by irradiation, a spatial separation of vacancies and interstitial atoms is required [1, 2]. It provides the accumulation of defects to a critical concentration n_c when the crystal-to-amorphous transition takes place. In poly-Si the main reason for the spatial separation and accumulation to n_c of vacancies and interstitial atoms appears to be related to the intrinsic stresses produced, for example, by the grain boundaries, voids, and impurity precipitations. The suggestion on the important role of the stresses in the defect accumulation up to amorphization of the implanted sample regions led us to explain the obtained results. Thus, the H_2^+ ion implantation of the Si single crystal did not lead to the appearence of VV-centres in contrast to the data on poly-Si in the studied dose interval.

As the annealing temperature of poly-Si increases two basic effects take place:

1) The number of feasible sinks of mobile defects diminishes due to the growth of the average grain size.

2) The intrinsic stresses are leveled off.

Simple estimations show that the density of VV-centres in annealed poly-Si goes down more quickly than the total surface of the grains. It gives us one of the arguments for the important role of the stresses in the defect accumulation. This conclusion is supported by other results. Thus, we have not reached complete amorphization of H^+ and H_2^+ ion-implanted layers of poly-Si, in spite of the fact that the concentration n_d of the displaced atoms for H^+ ions is greater than for N^+ and Ar^+ ions implanted in poly-Si (see Table 1). The concentration n_d is given by

$$n_d = (0.8\nu_a/2E_d)\,(D_0/R_p n_0)\,,$$

where ν_a is the nuclear stopping power, E_d the threshold energy which is equal to 20 eV in Si [10], D_0 the dose in the point where the density of VV-centres becomes

Table 1

The parameters of ν_a, D_0, and n_d in ion-implanted poly-Si and Si single crystal

substrate	ion	E (keV)	ν_a (keV)	D_0 (cm^{-2})	R_p (Å)	n_d
poly-Si	H^+	15	4.4	10^{16}	2300	0.77
poly-Si	N^+	60	22.5	10^{14}	1530	0.06
poly-Si	Ar^+	60	47.4	$< 3 \times 10^{13}$	640	0.09
single crystal of Si	N^+	60	22.5	10^{16}	1530	5.5
single crystal of Si	Ar^+	60	47.4	2×10^{14}	640	0.6

saturated (at N^+ and Ar^+ ion implantation $D_0 = D_a$), n_0 the number of atoms per cm^3. The value $0.8\nu_a/2E_d$ gives the number of atoms displaced by a single ion [11].

Firstly the simple defect accumulation (divacancies, tetravacancies) takes place in the stressed regions at ion implantation. When the concentrations of these defects reach the critical values, the crystalline-to-amorphous transition starts and we have a growth of the density of VV-centres (stage II of the dose dependence). In the first stage VV-centres appear to arise from the growth of the amorphous regions due to the flow of vacancies (or interstitial atoms) to these regions [12] existing in the initial poly-Si. Annealing of the sinks leads to a decrease of the role of the first stage (Fig. 1b).

2. In considering the fate of the Frenkel pairs in ion-implanted single crystal, three basic channels must be taken into account: 1) the annihilation of vacancies and interstitial atoms, 2) the migration of pairs from the implanted layer to the undamaged crystal, 3) the association of Frenkel pairs to stable complexes in the implanted layer. The layer amorphization is due to the third channel. If the contribution of the first and the second channels falls down, D_a decreases. This is the situation which appears to be realized in the N^+ ion-implanted initial poly-Si samples (Fig. 3). Here D_a is less than the amorphization dose in the Si single crystal by a factor 100. So there are a lot of imperfections in the poly-Si able to trap either vacancies or interstitial atoms. According to the theoretical estimations the defect concentration n_d required for amorphization is 0.02 to 0.04 [13]. The defect migration and annihilation were not taken into account in these estimations. n_d is close to the estimated value in the initial N^+ and Ar^+ ion-implanted poly-Si (see Table 1). After the high temperature annealing D_a is two orders of magnitude higher than in the initial samples. In the present case annihilation of Frenkel pairs is naturally suggested to be dominant, otherwise the number of displacement atoms will be sufficient to amorphize the layer with thickness 100 times larger than the ion projected range, i.e. 10 μm at N^+ ion implantation. Taking into account that the poly-Si films were 1 μm thick and the amorphous layer was spread to 0.1 μm in depth, we can make the conclusion that 99% of the Frenkel pairs are annihilated in poly-Si annealed at $T_a \geqq 1000$ °C. The same situation appears to be realized in Si single crystals.

5. Summary

The results obtained allow to propose that strained regions in the crystal (grain boundaries, voids, and other imperfections) play a dominant role in the crystal-to-amorphous transition at ion implantation. These regions contribute largely to the spatial separation of vacancies and interstitial atoms and their trapping in the vicinity of a great variety of imperfections up to the critical concentration.

The crystal-to-amorphous transition kinetics of the poly-Si sample containing strained regions was found to have two stages. The first stage has been caused by the growth of amorphous regions existing in the initial crystal, while the second is the accumulation of simple defects (divacancies, tetravacancies) in the stressed regions at ion implantation.

The annealing of poly-Si before implantation leads to the reduction of the role of the first stage because the intrinsic stresses are leveled off.

The annihilation of vacancies and interstitial atoms as well as the migration of pairs from the implanted layer to the undamaged crystal give a small contribution to the processes of defect accumulation at N^+ and Ar^+ ion implantation leading to a pronounced decrease in the amorphization dose D_a (approximately by two orders of magnitude). The annihilation processes dominate after annealing treatments at $T_a \geqq 1000$ °C.

Acknowledgements

The authors are indebted to Prof. L. S. Smirnov and Prof. L. N. Aleksandrov for fruitful discussions.

References

[1] L. S. Smirnov (Ed.), Physical Processes in Irradiated Semiconductors, Izd. Nauka, Novosibirsk 1977.
[2] A. V. Dvurechenskii and L. S. Smirnov, Radiat. Eff. **37**, 173 (1978).
[3] V. N. Gashtold, N. N. Gerasimenko, A. V. Dvurechenskii, and L. S. Smirnov, Fiz. Tekh. Poluprov. **9**, 835 (1975).
[4] A. I. Gerasimov, E. I. Zorin, P. V. Pavlov, and D. I. Tetelbaum, phys. stat. sol. (a) **12**, 679 (1972).
[5] I. A. Abroyan, A. I. Titov, and A. V. Khlebalkin, Fiz. Tekh. Poluprov. **11**, 1204 (1977).
[6] S. I. Romanov and L. S. Smirnov, Fiz. Tekh. Poluprov. **10**, 876 (1976).
[7] A. V. Dvurechenskii, N. N. Gerasimenko, and L. P. Potapova, Thin Solid Films **52**, 329 (1978).
[8] L. A. Yukova and M. A. Gurevich, The Electronography of Surface Layers and Films of Semiconductor Materials, Metallurgiya, Moscow 1971.
[9] B. I. Vikhrev, N. N. Gerasimenko, A. V. Dvurechenskii, and L. S. Smirnov, Fiz. Tekh. Poluprov. **8**, 1345 (1974).
[10] N. N. Gerasimenko, A. V. Dvurechenskii, V. I. Panov, and L. S. Smirnov, Fiz. Tekh. Poluprov. **5**, 1644 (1971); **6**, 991 (1972).
[11] J. W. Mayer, L. Eriksson, and J. A. Davis, Ion Implantation in Semiconductors: Si and Ge, Academic Press, New York/London 1970.
[12] N. N. Gerasimenko, A. V. Dvurechenskii, and G. P. Lebedev, Fiz. Tekh. Poluprov. **7**, 2297 (1973).
[13] M. L. Swanson, I. R. Parsons, and C. W. Hoelke, Radiat. Eff. **9**, 249 (1971).

(Received March 26, 1979)

phys. stat. sol. (a) 54, 463 (1979)

Subject classification: 1.5 and 3; 22.1.2

Institute of Physics of Semiconductors,
Siberian Department of Academy of Sciences of the USSR (a) and
Arifov Institute of Electronics, Uzbek Academy of Sciences (b), Tashkent

The Mechanism of Silicon Epitaxial Layer Growth from Ion-Molecular Beams

By

L. N. ALEKSANDROV (a), A. S. LUTOVICH (b) and E. D. BELORUSETS (b)

Energies of different stages of silicon homoepitaxial growth from ion-molecular beam (IMB) in vacuum: oxide layer destruction, physical absorption and chemisorption, surface diffusion etc. are estimated. Based on the discussion of the values obtained conclusions on the possible mechanism of the ion beam influence on nucleation and epitaxial layer growth are made. In the places of ion collision with crystallizing surface, point defects and local areas of atom excitation are formed which become centres of nucleation. Ions colliding with the centres of growth destroy the volume nuclei providing the conditions of two-dimensional covering. Ion beam makes the energetic additional feeding of diffusion and phase transition processes.

Проведена оценка по энергиям различных стадий процесса гомоэпитаксиального роста кремния из ионно-молекулярного пучка (ИМП) в вакууме: разрушения окисного слоя, физической абсорбции и хемосорбции, поверхностной диффузии и др. На основании обсуждения полученных величин сделаны выводы о возможном механизме действия ионного пучка на зарождение и рост эпитаксиальных слоев. В местах соударение иона с кристаллизующейся поверхностью образуются точечные дефекты и локальныя области возбуждения атомов, которые становятся центрами зарождения. Ионы, сталкиваясь с центрами роста разрушают объемные зародыши, обеспечивая условия двумерного азрастания. Ионный пучок производит энергетическую подпитку процессов диффузии и фазового перехода.

1. Introduction

Lately the interest in the activated crystallization of the epitaxial layers has been increased. The investigations of the crystal growth in the case of external effects contribute to a better understanding of elementary acts of this process and an elucidation of the conditions of its optimization. The methods of the epitaxial growth activation can be the excitation of condensed atoms, irradiation of the substrate by light, application of electric and magnetic fields, ion implantation and electron bombardment [1 to 6].

One of the most perspective methods of activated crystallization is, apparently, the method of growing from the ion-molecular beam (IMB) [7 to 11]. An ion beam of average energy (1 to 15 keV) effects directly the surface and the near surface area of some atomic layers. With this method it is possible to find the changes of the nucleation process and the conditions for the further growth, and also structural, electrical, and other properties of the films.

When growing germanium homoepitaxial films it was found that the growth "islands" being formed under the bombardment by ions with energies of 500 to 1000 eV are crushing and the surface of the epitaxial layer becomes smoother and this simplifies the further process of growth [8].

2. Experimental

Works [7, 9, 11] are dedicated to the low temperature epitaxial crystallization of silicon from IMB. With the help of this method the homoepitaxial silicon layers were grown on substrates oriented in (111) planes at 800 °C [7], in a vacuum of 5×10^{-7} Torr the ionization degree was about 1% and the accelerating field was about 15 keV. In [9, 11] with a vacuum of 5×10^{-6} Torr and deposition temperature of 800 to 700 °C the accelerating field of 600 V ensured the formation of a polycrystalline film, a voltage increase improved the film structure and blocks were integrated. With a field of 2000 V point electron-diffraction photographs were observed and at 2500 V the layers were epitaxial. On further energy increase the silicon layers became polycrystalline (Fig. 1). With the results of the above-mentioned works it is possible to assume the existence of some certain energy ranges and the concentration of the ion component of the beam for the formation of single crystal layers.

In [11] for the analysis of the density and energetic composition of the ion beam a 127° cylindrical condenser of Juz-Rozhanski type was used (Fig. 2, (5)). The analyzed beam had a rectangular cross-section of 1×4 mm² formed by the input diaphragm of the condenser. A Faraday cylinder was used to collect the ions (Fig. 2, (4)). The measurement of the analyzed ion current was carried out by an U-1-2 type dc electrometric amplifier which ensured a good agreement of the output analyzer impedance with the input impedance of the galvanometer, sufficiently decreasing the distortion of the investigated signal.

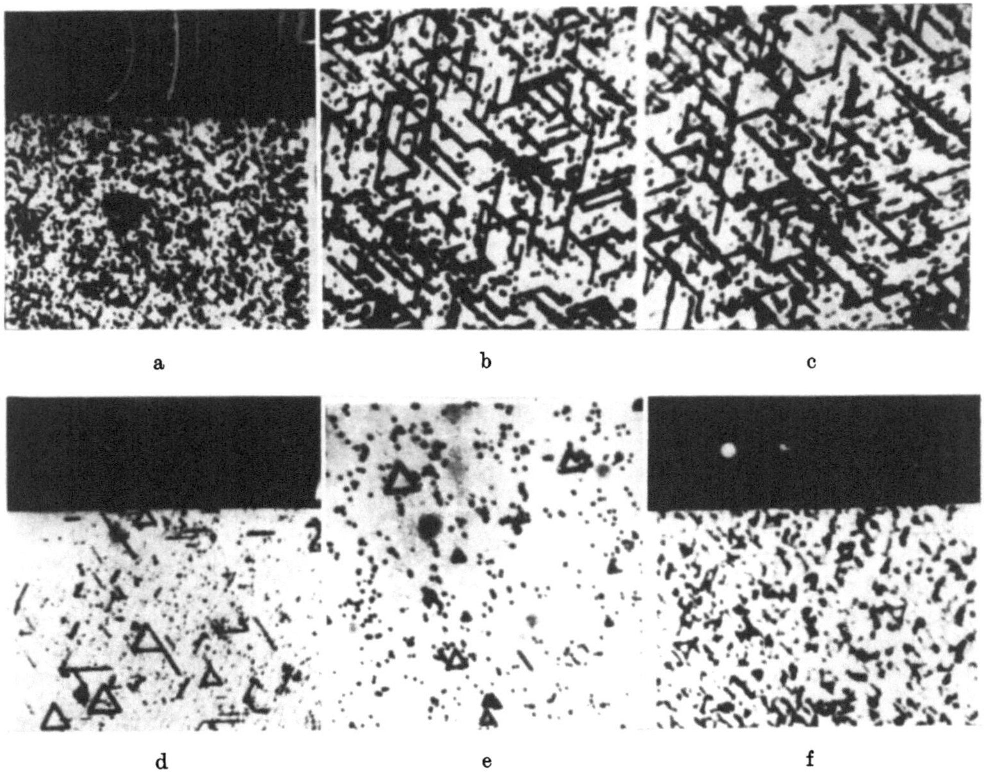

Fig. 1. The change of silicon homoepitaxial layer structure with the electric field accelerating the ions. a) $E = 600$, b) 1000, c) 1500, d) 2000, e) 2500, f) 3000 V (320 ×)

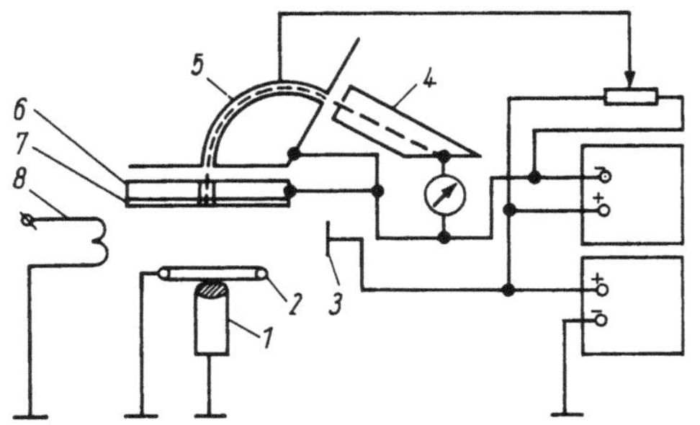

Fig. 2. The experimental installation scheme for the investigation of the process of epitaxial deposition from an ion-molecular beam and the investigation of the energetic spectrum of the ion beam. 1 Source of the evaporated substance, 2 inductor of HF generator, 3 anode, 4 Faraday cylinder, 5 Juz-Rozhanski analyzer, 6 support with a hole, 7 substrate, 8 cathode

The value of the analyzed ion current at the collector was 10^{-8} A corresponding to the density of the total ion current to the substrate 2.5×10^{-7} A/cm² with the square cross-section of the analyzed beam in the cylindrical condenser of 4 mm².

The spectra of the silicon ion energy distribution for the accelerating potentials 600, 800, 1000 V are characterized by the fact that with increasing accelerating potential the maximum peak of the ion density is approaching the value of the applied voltage.

With increasing accelerating potential there is a displacement of the equipotential surfaces of the electric field in the working volume in such a way that the majority of ions is coming through the accelerating potential difference approaching the value of the applied potential. In this case there is a decrease of ions with smaller energies with the simultaneous decrease of the total number of ions at the expense of the defocusing of the ionizing electron beam.

The elucidation of the epitaxial growth mechanism from the combined ion-molecular beam the determination of the energetic balance of the process, the establishment of the ways of non-thermal stimulation, and the discovery of the optimum growth conditions are of great interest.

3. Calculation of Atomic Energy

In present paper we have made an attempt to evaluate energetically the elementary process of silicon crystallization from IMB in vacuum on the (111) surface, that was conventionally technologically treated, and to explain the reasons of epitaxial growth stimulation and the decrease of epitaxial temperature in a certain interval of accelerating electric field. First of all it should be noted that this is only the first approximation and, therefore, in the calculations some simplifying assumptions were made. The epitaxial growth process from IMB is schematically presented in Fig. 3.

The epitaxial temperature of silicon under the usual conditions is 1200 to 1250 °C. But this temperature is not a threshold; as it was shown above it can be decreased in the case of IMB down to 800 °C and less [9 to 11].

For the estimation of the energy balance of the film deposition process we relate it to 1 cm² of the substrate surface and to the time of the single layer formation which in our conditions corresponds to 1 s. In the approximation of thermal equilibrium of adatoms with the substrate the adatom energy can be considered as consisting of three components, i.e. the energy of thermal motion of the substrate atoms (E_T), the energy carried by neutral atoms and molecules of the beam (E_K), and the energy of the accelerated ions in the beam (E_I)

$$E_A = E_T + E_K + E_I \,. \tag{1}$$

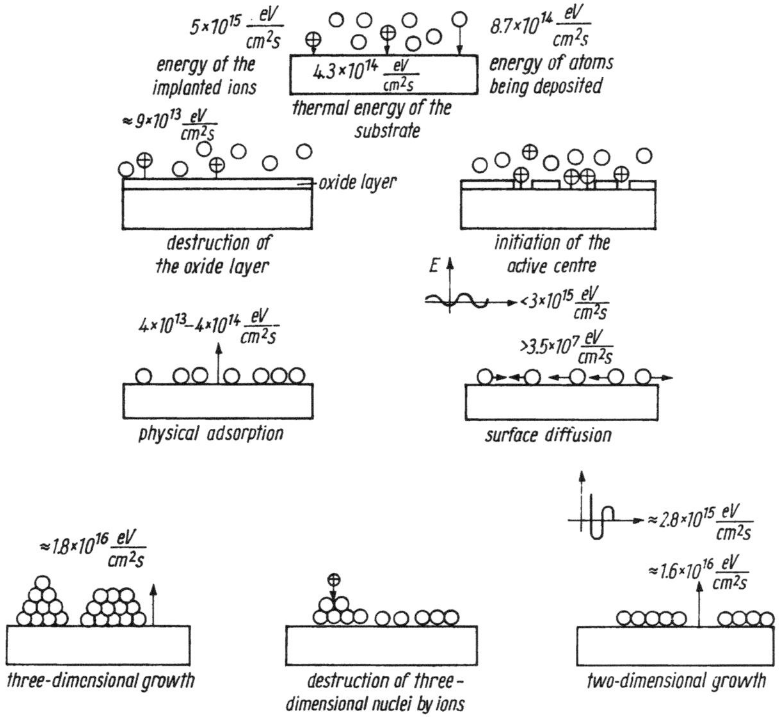

Fig. 3. Successive stages of epitaxial growth from an ion-molecular beam and typical energy values

The energy of substrate atoms at temperature T_S is determined as

$$E_T = 3n_0kT_S\,, \tag{2}$$

where n_0 is the reticular density of atoms at the surface $(1.5 \times 10^{15}\ \text{cm}^{-2}$ for $(111))$. k is the Boltzmann constant $(1.38 \times 10^{-23}\ \text{J/K})$.

The kinetic energy of atoms and molecules in the beam is determined by the source temperature T_I and the beam intensity

$$E_K = \tfrac{3}{2}\,kT_IF\,. \tag{3}$$

The beam intensity can be determined from the film growth rate $V\,(V = (F - F_\infty)/n_r a_0)$ or from (4) from the pressure P and the temperature T_0 of the vapour

$$F = \beta n \sqrt{\frac{kT_0}{2\pi m}}\,, \tag{4}$$

where $n = P/kT$ is the density, β is the adhesion coefficient, or from the known degree of the beam ionization α and the density of the ion current J formed by the given beam

$$J = F\alpha q\,, \qquad F = \frac{J}{\alpha q}. \tag{5}$$

Here F_∞ is the equilibrium current, q is the elementary charge $(1.6 \times 10^{-19}\ \text{C})$, a_0 is the interplane distance in the crystal lattice $(3.3 \times 10^{-8}\ \text{cm})$.

The energy of ions in the beam accelerated by the voltage U is given by

$$E_{\mathrm{I}} = JU = F\alpha qU\,. \tag{6}$$

Under the conditions of the experiment [11] at $T = 800$ and 1200 °C the values of E_{I} are 4.02 and 5.52×10^{14} eV/cm², respectively. The value of the current J is 2.5×10^{-7} A/cm², and at $U = 2500$ V $E_{\mathrm{I}} = 3.88 \times 10^{15}$ eV/cm² s.

If $\alpha = 0.01$ and $T_{\mathrm{I}} = 2600$ °C (the boiling temperature of silicon), $E_{\mathrm{K}} = 6.0 \times 10^{13}$ eV/s cm².

The corresponding intensity of the flow is $F = 1.6 \times 10^{14}$ at/cm² s which provides the film growth rate $V = 0.35$ Å/s, i.e. ≈ 0.1 monolayer per second. At $F \approx 10^{15}$ at/cm² s — $E_{\mathrm{K}} \approx 10^{14}$ eV/cm² s.

Thus, even for currents of small density and small degree of ionization, E_{I} is one order larger than E_{T} and two orders larger than E_{K}.

The decrease of adatom energy with decreasing substrate temperature during the growth is easily compensated by the ion current energy. As it is seen from (6) with a large fraction of the ion component (up to $\approx 1\%$) an accelerating field of some volts is sufficient. With an energy of some eV the ions excite surface reactions during the growth, but do not form new defects [13].

4. Discussion

Usually the substrate surface in insufficiently high vacuum is covered with an oxide layer, therefore, only the initial period of growth is essentially different from the following one. As it was assumed distortions in the oxide film are the centres of oriented nucleation in the clean silicon surface and in this case the development of the expitaxial layer follows the mechanism of three-dimensional growth [12].

It is clear that the more destructions in the oxide film take place, the more intense will be nucleation and growth. The cleaning of the substrate surface from oxides can be achieved by the ion action in the IMB method. Consequently in the initial moment of the deposition a portion of the ion energy is expended to damage the oxide film (9×10^{13} eV/cm² or more) and to increase the active parts of the surface which can be the centres of nucleation.

Apparently, in this case the process of epitaxial growth can be divided into two stages: the first initial stage with mainly three-dimensional growth and the second stage with mainly two-dimensional growth. The second stage begins at the moment of the full covering of the substrate [14], but in ultra-high vacuum it can start at once [15].

The atoms of the deposited substance approaching the substrate are adsorbed on it. Here physical adsorption is meant. The binding energy of the adsorbed atom with the solid surface is 0.01 to 0.1 eV/at [16]. The current of the adsorption released energy in our case is 4×10^{13} to 4×10^{14} eV/cm² s.

The state of the physical adsorption of atoms differs from the state of chemisorption by an activation barrier of 2.8×10^{15} eV/cm² s. [17] The energy released in the phase transition of silicon atoms from the vapour into the solid state for the first growth stage (three-dimensional growth) is 1.8×10^{16} eV/cm² s and for the second growth stage (two-dimensional growth) 1.6×10^{16} eV/cm² s.

The intermediate stage between adsorption and chemisorption is the surface diffusion of atoms. In the calculations of the diffusion activation energy we made the simplifying approximation that the atoms which get to the surface are moving on it separately of each other and not in clusters. In reality there can be surface diffusion of separate atoms and also of their associates. The activation energy of cluster movements can be essentially smaller than the activation energy for the diffusion of separate atoms [18]. Therefore, the activation energy of diffusion will be less than 3×10^{15} eV per cm^2 s [19].

From the energetic estimations made in the present work it follows that the ion beam produces an additional energetic feeding of the elementary processes of epitaxial growth. The phase transition of the atoms from vapour into solid state and surface diffusion are activated processes. The activation energy of the phase transition is 2.8×10^{15} eV/cm^2 s, the activation energy of surface diffusion is 3×10^{15} eV/cm^2 s which by the order of magnitude corresponds to the energy carried by the ion beam 5×10^{15} eV/cm^2 s.

Thus it is assumed that the ion beam stimulates the transition through the energy barriers in the elementary processes of crystallization.

The epitaxial temperature decrease during the growth by IMB can be explained by the increase of thermal energy of atoms at the expense of the energy carried by ions and also by the cleaning of the growth surface from silicon oxide and gas impurities by the ions. It is clear that this process requires some optimum conditions for its performing. Actually if the ion beam is fixed then with insufficient accelerating electric field the growth surface would be badly cleaned and the obtained epitaxial layers would have a lot of defects or they would be even polycrystalline. With excessively large electric field the layer structure can be bad as a result of radiation damages.

The given numerical values for the energy of destruction of the oxide layer and formation of three- and then two-dimensional growth centres are related to the conditions of this experiment. The improvement of vacuum in the system resulting in a decrease of the necessary temperature of epitaxial growth of a silicon film [14, 15] reduces the required contribution of the ion flow and the required accelerating voltage. In the given numerical estimations we assumed: the surface energy of the crystal–vapour interface is 1500 erg/cm^2, that of the substrate–nucleus interface decreases from 1100 to 500 erg/cm^2 in the process of surface cleaning. The angle of the contact between the three-dimensional nucleus and the substrate is 75°, the supersaturation as the ratio of silicon vapour pressures is ≈ 100; the 100×100 Å^2 windows with the density of 10^{10} cm^{-2} are formed on the 20 Å thick oxide film at its destruction, the energy of destruction of SiO_2 is ≈ 4.5 eV/mol.

The obtained epitaxial silicon films had a resistivity up to 500 Ω cm on substrates with $\varrho \approx 10^{-2}$ Ω cm.

5. Conclusion

The use of the accelerating electric field and electronic ionization of a molecular silicon beam enabled us to vary the contribution of the ion beam component and the total energy of adatoms in epitaxial film deposition.

The mechanism of the influence of the ion beam on nucleation and epitaxial growth of Si layers is described on the basis of calculations of the energy of individual processes.

The ion beam destroys the oxide film that covers the substrates, which favours the nucleation. In the places of collisions of ions with the growth surface structure defects and local areas of atom excitation are formed, they can become the nucleation centres. This also favours the intensity of nucleation.

Ions colliding with the growth centres destroy the volume nuclei and smooth the surface, i.e., the transformation of the volume nuclei into two-dimensional ones takes place. The ion beam also produces a stimulation of surface diffusion which favours the nucleation initiation, coalescence, of existing nuclei, and ordering of the crystal structure.

A decrease of the substrate temperature for the epitaxial layer growth from IMB can be explained by an increase of the atomic thermal energy at the expense of the energy carried by the ions and by cleaning the growth surface from silicon oxide and gas impurities by ions.

References

[1] V. I. PETROSYAN, Vichislitelnie sistemi, 15.ed. Nauka, Novosibirsk 1965 (p. 47).
[2] V. I. PETROSYAN, E. I. DAGMAN, and C. E. MITKOVSKII, in: Protsessi rosta monokristallicheskikh sloev poluprovodnikov, Part I, Nauka, Nivisibirsk 1968 (p. 588).
[3] SH. LUBI, YA. ČERVENAK, and YA. SCHILER, Fiz. tverd. Tela **12**, 1298 (1970).
[4] M. KUMAGAWA, H. SUMAMI, T. TEZASAKI, and I. NISHIZAWA, Japan J. appl. Phys. **7**, 1332 (1968).
[5] K. L. ČOPRA, in: Rost kristallov, Vol. 8, Nauka, Moskva 1968.
[6] R. N. TOMAS and M. KH. FRANKOMBE, Pryamoe preobrazovanie teplovoi energii v elektricheskuyu i toplivnie elementi **8**, 31 (1970).
[7] T. ITOH and T. NAKAMURA, Radiat. Effects, **9**, 1 (1971).
[8] B. G. ANOKHIN, A. E. LIKHTMAN, L. N. NEMIRIVSKII, and L. S. SEIDMAN, Elektronnaya tekhnika, Ser. 2, poluprovodnikovie pribori, 4 (54) p. 38, (1970).
[9] A. U. ARIFOV, A. S. LUTOVICH, and K. F. KLIMENKO, XIV. All-Union Conf. Emission Electronics. Tezisi dokl. FAN, Tashkent 1970 (p. 110).
[10] V. P. BELEVSKII and V. M. KOPTENKO, XIV. All-Union Conf. Emission Electronics, Tezisi dokl. FAN, Tashkent 1970 (p. 104).
[11] A. S. LUTOVICH and K. F. KLIMENKO, V. Symp. Semiconductor Crystal and Film Growth, Tez. dokl., Novosibirsk 1978 (p. 10).
[12] L. N. ALEKSANDROV, Kinetika obrasovaniya i struktura tverdikh sloev, Nauka, Novosibirsk 1972.
[13] L. N. ALEKSANDROV, Tez. dokl. 5. Symp. Processes of Semiconductor Crystal and Film Growth, Novosibirsk 1978 (p. 5).
[14] B. E. WATTS, R. R. BRADLEY, and B. A. JOYCE, Phil. Mag. **8**, 1163 (1968).
[15] L. N. ALEKSANDROV and R. N. LOVYAGIN, Thin Solid Films **20**, 1 (1974).
[16] F. F. VOLKENSTEIN, Elektronnaya teoriya kataliza na poluprovodnikakh, Fizmatgiz, Moskva 1960.
[17] F. F. VOLKENSTEIN, red. Elektronnie yavleniya v adsorbtsii i katalize na poluprovodnikakh, Lektsii na mezhdunarodnom simposiume, Mir, Moskva 1969.
[18] L. I. TRUSOV, IV. All-Union Meeting on Crystal Growth; Soderzhanie dokladov part I, Izd. AN Armyanskoi SSR, Tsakhkasor 1973 (p. 90).
[19] B. I. BOLTAKS, Diffuzyia i tochechnie defekti v poluprovodnika, Nauka, Leningrad 1972.

(Received March 14, 1979)

phys. stat. sol. (a) **54**, 471 (1979)

Subject classification: 18.4; 19; 22.5

Institute of Crystallography, Academy of Sciences of the USSR, Moscow

Spin Echo Spectra of ^{57}Fe and ^{19}F Nuclei in a Canted Antiferromagnet FeF_3

By

A. V. Zalesskii, V. V. Vanchikov, V. G. Krivenko, and A. N. Ivashchenko

The variation of the local fields $\boldsymbol{H}_l$ at ^{57}Fe and ^{19}F nuclei on temperature and external static field is studied for FeF_3 crystals by a spin echo technique in the magnetically ordered state. It is shown that F atoms in FeF_3 become none-quivalent with respect to the contributions to $\boldsymbol{H}_l$ at ^{19}F, the magnitude and the direction of $\boldsymbol{H}_l$ at ^{19}F are not the same for F atoms in the unit cell of FeF_3. The local fields at ^{19}F are proportional to the magnetization of the iron sublattice but do not coincide with the directions of $\boldsymbol{H}_l$ at ^{57}Fe nuclei.

Методом спинового эха изучена зависимость локальных полей $\boldsymbol{H}_l$ на ядрах ^{57}Fe и ^{19}F от температуры и внешнего постоянного поля для кристаллов FeF_3 в магнитоупорядоченном состоянии. Показано, что атомы фтора в FeF_3 становятся неэквивалентными по отношению ко вкладам в $\boldsymbol{H}_l$ на ядрах ^{19}F, величина и направление $\boldsymbol{H}_l$ на ядрах ^{19}F неодинаковы для атомов фтора в элементарной ячейке FeF_3. Локальные поля на ядрах ^{19}F пропорциональны намагниченности железной подрешетки, но не совпадают с направлениями $\boldsymbol{H}_l$ на ядрах ^{57}Fe.

1. Introduction

FeF_3 crystals at $T < T_N \approx 362$ K are canted antiferromagnets with weak ferromagnetism and belong to the $R\bar{3}c$ space group, which is typical for other trigonal "easy-plane" antiferromagnets (α-Fe_2O_3, $FeBO_3$).

The structure of FeF_3 was studied in [1]. The bimolecular rhombohedral unit cell — which is equal to the magnetic cell — may be represented as two unimolecular FeF_3 pseudo-cubes joined along the [111] direction (Fig. 1). The Fe atoms occupy the corners of the pseudo-cubes and F atoms are displaced from the mid-points of the edges with the bonding angle Fe–F–Fe equal to 153°. Fe and F atoms form regularly spaced parallel alternate planes perpendicular to the [111] direction, the arrangement of displaced fluorine atoms being almost intermediate between hexagonal and cubic close-packing. These displacements, as schematically shown in Fig. 1, are in opposite senses in adjacent pseudo-cubes, so that the true rhombohedral cell becomes bimolecular.

The antiferromagnetism in FeF_3 was established by neutron diffraction [2]. The magnetic moments of Fe^{3+} ions are ordered almost antiferromagnetically with the axis of antiferromagnetism perpendicular to [111]. A small canting of moments in the basal (111) plane produces a weak net moment normal to the direction of the spins and to [111]. The existence of weak ferromagnetism was proved by direct magnetic measurements [3]. The in-plane (111) anisotropy field H_a is very small ($H_a <$ < 0.01 Oe at room temperature [4]).

Local magnetic fields at ^{57}Fe nuclei were measured by the Mössbauer effect [5, 6] mainly to study the temperature dependence of the iron sublattice magnetization near the Néel temperature. The only observation of NMR in FeF_3 was reported in a short note by Petrov et al. [7]. A wide (55 to 80 MHz) complex spin echo spectrum was observed at 4.2 and 77 K. The spectrum was assigned to ^{19}F nuclei. The ^{57}Fe NMR signal was found only at room temperature.

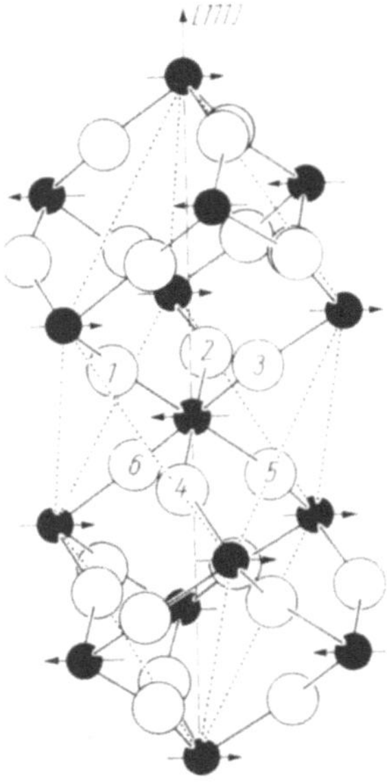

Fig. 1. Crystallographic and magnetic structure of FeF_3. Six F atoms inside the rhombohedral unit cell Fe_2F_6 (dotted lines) are shown. ○ F, ● Fe

We made further investigations of NMR in these interesting transparent magnetic crystals in order to obtain more detailed information about the local fields at ^{57}Fe and ^{19}F nuclei in a wide temperature range.

A commercial frequency-variable pulsed spectrometer was used to study randomly oriented small (0.2 to 0.5 mm) green hydrothermal FeF_3 crystals.

Preliminary results were reported at the XX. Ampère Congress, Tallin 1978. The data concerning ^{57}Fe in FeF_3 are also published by two of us in [8].

2. Main Results and Discussion

Fig. 2 gives an example of the spin echo spectrum for FeF_3 crystals at 77 K. The ^{19}F spectrum consists of the superposition of two broad peaks: a low-frequency peak and a high-frequency peak (the frequencies corresponding to these peaks hereafter are designated as $\nu(^{19}F)_{min}$ and $\nu(^{19}F)_{max}$). The values of local fields H_l at ^{19}F nuclei corresponding to $\nu(^{19}F)_{min}$ and $\nu(^{19}F)_{max}$ at 4.2 K are 15.25 and 19.5 kOe, respectively.

The approximate intensity ratio of the two superimposed line components $\nu(^{19}F)_{min}$ to $\nu(^{19}F)_{max}$ is about two. No correction for frequency has been made since it would not change considerably the estimation of this ratio.

A narrow ^{57}Fe NMR line was found at higher (with respect to the ^{19}F spectrum) frequency (see Fig. 2). The resonance frequency $\nu(^{57}Fe)$ for ^{57}Fe nuclei at 4.2 K is 84.94 MHz. This corresponds to the local field $H_l = 616.4$ kOe, which is near to Mössbauer effect value of 618.1 kOe given for 4.2 K in [5].

Both ^{57}Fe and ^{19}F signals were traced from 4.2 up to 340 K. At all temperatures it was possible to distinguish two broad peaks in the ^{19}F spectrum and a narrow single peak for ^{57}Fe nuclei. The temperature dependences of $\nu(^{57}Fe)$, $\nu(^{19}F)_{min}$, and $\nu(^{19}F)_{max}$ are shown in Fig. 3a.

The knowledge of the NMR frequencies in a wide temperature range made it possible to compare the dependence of the normalized frequency $\nu(T)/\nu(4.2\text{ K})$ on reduced

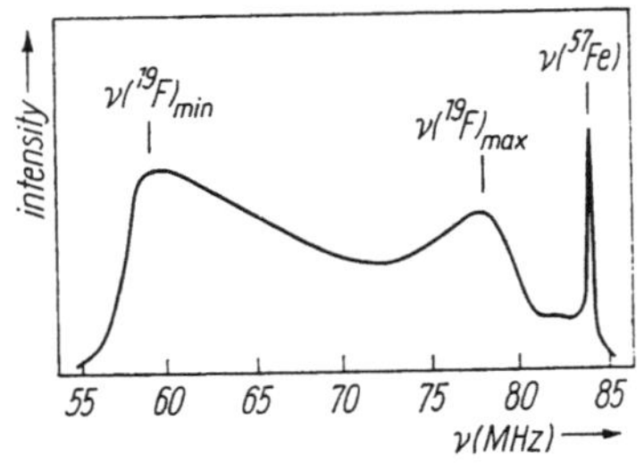

Fig. 2. Spin echo spectrum in FeF_3 at 77 K

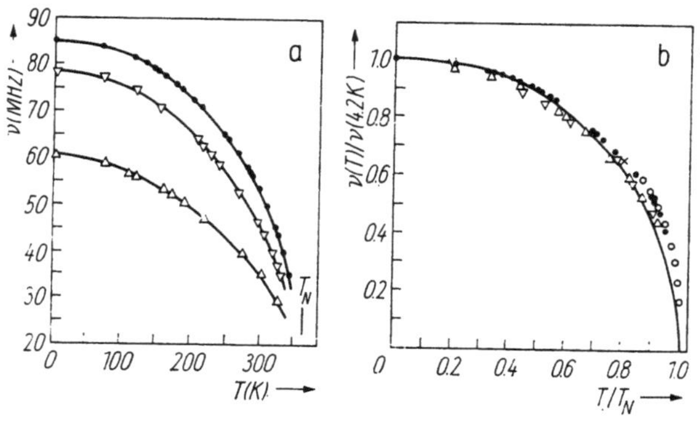

Fig. 3. a) Temperature dependence of resonance frequencies $\nu(^{57}Fe)$ (●), $\nu(^{19}F)_{min}$ (△), and $\nu(^{19}F)_{max}$ (▽). The sizes of circles and triangles correspond to measurement errors. b) The same NMR frequencies normalized to 4.2 K as a function of the reduced temperature T/T_N. Data taken from [7, 5] are shown by crosses and open circles, respectively. The solid line is the theoretical curve calculated from the molecular field approximation for $S = \frac{5}{2}$

temperature T/T_N with the dependence of the normalized magnetization for the iron sublattice versus T/T_N, calculated from the molecular field approximation for the spin $S = \frac{5}{2}$ (Fe^{3+} ions). More precisely it can be done for ^{57}Fe nuclei (black circles in Fig. 3b) because of the higher accuracy in the determination of the NMR frequency. Mössbauer effect data for the region of $0.85 < T/T_N < 0.99$ [5] and the room temperature value of $\nu(^{57}Fe)$ from [7] have been included in Fig. 3b for completeness. Deviations from the theoretical molecular field curve do not exceed those usually found for the temperature dependence of sublattice magnetization in other canted "easy-plane" antiferromagnets, α-Fe_2O_3 [9] and $FeBO_3$ [10].

The experimental values of $\nu(^{19}F)_{max}$ and $\nu(^{19}F)_{min}$ normalized to 4.2 K are also shown in Fig. 3b. From Fig. 3 a and b it follows that $\nu(^{19}F)_{min}$ and $\nu(^{19}F)_{max}$ decrease with increasing T proportionally to the change of $\nu(^{57}Fe)$. This proves that the local fields H_l at ^{19}F nuclei are entirely due to the magnetization of the iron sublattice.

The spin-lattice and spin-spin relaxation times T_1 and T_2 were measured both for ^{57}Fe and ^{19}F nuclei at different temperatures. The measurements of T_1 were made by means of observing the recovery of the longitudinal component of the nuclear spin magnetization from saturation using a four-pulse technique. T_2 was measured by the usual method of registration of the spin echo amplitude as a function of the time interval between two pulses. Pulse durations in all experiments were about 2 μs.

The spectral distribution of T_1 and T_2 for ^{19}F at 77 K is shown in Fig. 4. The statistical error of T_1 measurements is shown in Fig. 4 by the error bars. The T_2 measurements can be made more precisely (with accuracy about 10%), which fits into the size of the circles in Fig. 4.

Comparing Fig. 4 and 2 one can see an obvious correlation between the spin echo amplitude and T_1: the minima of T_1 correspond to the position of the intensity peaks at $\nu(^{19}F)_{min}$ and $\nu(^{19}F)_{max}$. T_1 and T_2 for ^{19}F are always shorter than for ^{57}Fe by a factor 5 to 10. The relaxation data for ^{57}Fe at different temperatures are shown in Table 1.

The ^{57}Fe and ^{19}F spectra were studied in the external static field $\boldsymbol{H}_0$. The direction of $\boldsymbol{H}_0$ was chosen perpendicular to the radio-frequency field. This geometry is the most

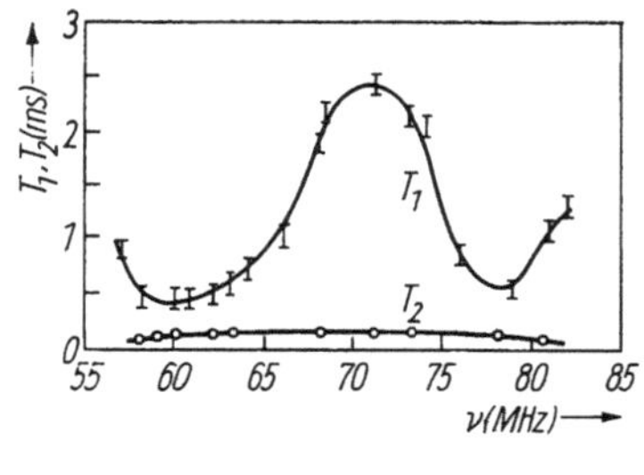

Fig. 4. Spectral distribution of relaxation times T_1 and T_2 for ^{19}F nuclei in FeF_3 at 77 K

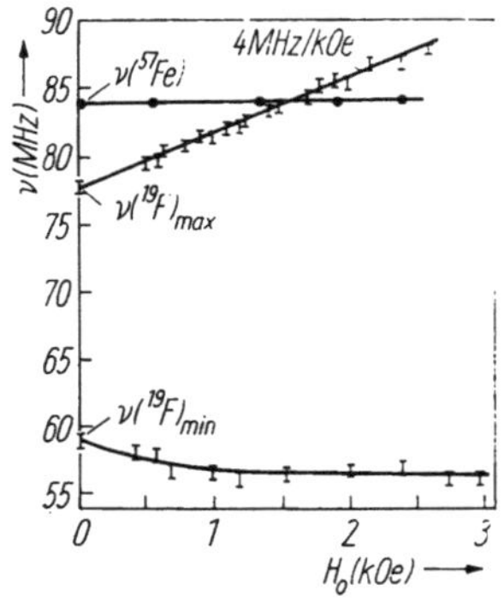

Fig. 5. Shift of $\nu(^{19}F)_{min}$, $\nu(^{19}F)_{max}$, and $\nu(^{57}Fe)$ in the external field H_0 at 77 K

favourable to observe NMR signal in the presence of $\boldsymbol{H}_0$ because it gives the maximum value of the enhancement factor η. The possibility to observe spectra of the same kind in $H_0 = 0$ as well as in $H_0 \neq 0$ indicates that the complex lineshape for ^{19}F nuclei is not due to domain wall effects, since in the fields H_0 used domain walls no longer exist. The amplitude of the echo decreases according to the relation $\eta \sim (H_a + H_0)^{-1}$ typical for intra-domain signals. In fields higher than approximately 3 kOe the signals become unobservable.

Table 1
Relaxation times for ^{57}Fe in FeF_3

T (K)	T_1 (ms)	T_2 (μs)
77	15 to 20	1000 ± 50
220	1.5 to 2.5	280 ± 15
295	≈0.3	80 ± 5

Fig. 5 shows the position of the peaks $\nu(^{57}Fe)$, $\nu(^{19}F)_{min}$, and $\nu(^{19}F)_{max}$ as a function of H_0 at 77 K. The frequency of the ^{57}Fe signal remains unchanged, while the peaks in the ^{19}F spectrum are broadened and shifted.

^{57}Fe NMR data for FeF_3 can be easily understood on the basis of our knowledge of magnetic and structural properties of similar canted antiferromagnets.

All iron atoms in FeF_3 occupy one type of crystallographic site and form two equivalent magnetic sublattices. The values of H_l and $\nu(^{57}Fe)$ at all ^{57}Fe nuclei are the same and proportional to the sublattice magnetization at a given temperature. The direction of $\boldsymbol{H}_l$ is collinear with the direction of Fe^{3+} spins. Because of a very small in-plane anisotropy, a comparatively weak external field $\boldsymbol{H}_0$ is needed to turn the Fe^{3+} spins and $\boldsymbol{H}_l$ at the ^{57}Fe nuclei preferably perpendicular (and the net moment parallel) to $\boldsymbol{H}_0$ in the "easy" (111) planes of different crystals. As a consequence, the NMR frequency $\nu(^{57}Fe)$ remains unchanged in $\boldsymbol{H}_0$.

As far as ^{19}F NMR data are concerned they are less perspicuous. Two main points should be explained:

(i) the shape of the ^{19}F spectrum which is spread over a wide frequency range and consists of two asymmetrical broad peaks:

(ii) the shift of the frequencies $\nu(^{19}F)_{min}$ and $\nu(^{19}F)_{max}$ in $\boldsymbol{H}_0$.

All F atoms are crystallographically equivalent and form a single sublattice, but they become non-equivalent with respect to the contributions to $\boldsymbol{H}_l$ from Fe atoms. $\boldsymbol{H}_l$ at ^{19}F nuclei may be represented as a vector sum of the dipole field $\boldsymbol{H}_d$ from the surrounding Fe moments and the transferred hyperfine field $\boldsymbol{H}_{hf}$:

$$\boldsymbol{H}_l = \boldsymbol{H}_d + \boldsymbol{H}_{hf} . \tag{1}$$

While the dipolar contribution can be calculated, there is still no adequate method for the calculation of the value and the direction of $\boldsymbol{H}_{hf}$ due to a spin-density transfer. That is why we restricted ourselves to calculating only $\boldsymbol{H}_d$ in (1) using the well-known expression

$$\boldsymbol{H}_d = \sum_i \left[\frac{3\boldsymbol{r}_i(\boldsymbol{\mu}_i \cdot \boldsymbol{r}_i)}{r_i^5} - \frac{\boldsymbol{\mu}_i}{r_i^3}\right], \tag{2}$$

where $\boldsymbol{\mu}_i$ is the magnetic moment of the Fe^{3+} ion ($5\mu_B$, $T = 0$) and $\boldsymbol{r}_i$ the radius vector from a given ^{19}F nucleus to the i-th Fe atom.

The dipole sum was calculated by means of a computer taking ten coordination spheres. The canting of Fe^{3+} moments was neglected. The calculation showed that the predominant contribution to $\boldsymbol{H}_d$ came from the nearest two Fe^{3+} moments ($\approx 90\%$ of H_d) and eight second Fe^{3+} neighbours ($\approx 7\%$).

Fig. 6 illustrates the calculated dependence of H_d (absolute values) for six fluorine atoms in the unit cell of 2 FeF_3 (shown by the numbers in Fig. 1) on the direction of Fe^{3+} spins in the (111) plane. The angle φ defines the direction of the axis of antiferromagnetic ordering of $\boldsymbol{\mu}$ with respect to one of the two-fold axes in the basal plane. When spins are directed in accordance with the symmetry elements of the crystal ($\varphi = 0°$, 60°, ... or 30°, 90°, ...) — this is the case when $H_0 = 0$ — there exist two values of H_d which are distributed among six F atoms according to the ratio 1:2. One can expect in this case the existence of two lines in the ^{19}F spectrum with the same intensity ratio. This agrees with the experimental ratio of $(^{19}F)_{min}$ to $(^{19}F)_{max}$ peak intensities in Fig. 2. However, the distance between peaks $\nu(^{19}F)_{max} - \nu(^{19}F)_{min} = 17$ MHz (4.15 kOe) at 4.2 K is larger than the calculated difference in H_d (1.2 kOe, 4.8 MHz), hence there should exist a considerable anisotropic contribution of $\boldsymbol{H}_{hf}$ to $\boldsymbol{H}_l$.

It should be emphasized that the width of the ^{19}F spectrum becomes less surprising if one takes into account the difference in the gyromagnetic ratios of ^{57}Fe (0.1378 MHz per kOe) and ^{19}F (4.011 MHz/kOe) nuclei. For example, the whole spectrum would have 0.6 MHz width if the gyromagnetic ratio for ^{19}F were the same as for ^{57}Fe.

Analysis of the dipolar contribution shows that besides the periodicity in absolute values of H_d for ^{19} F, there also exists a certain distribution of directions of $\boldsymbol{H}_d$ over six fluorine atoms. A similar situation may be expected for the $\boldsymbol{H}_{hf}$ contribution. As a result the directions of the total local field $\boldsymbol{H}_l$ at ^{19}F nuclei become different from those at ^{57}Fe nuclei. This is the reason why the ^{19}F spectrum exhibits a complicated deformation in $\boldsymbol{H}_0$, while the position of the ^{57}Fe signal remains unchanged. There is also

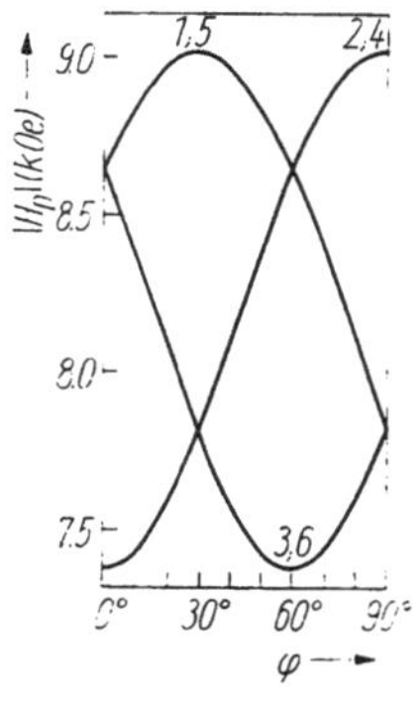

Fig. 6. Calculated variation of the dipole field H_d for six F atoms in the unit cell as a function of the direction of Fe^{3+} spin ordering in the (111) plane. The numbers correspond to the F ions shown in Fig. 1

an additional difficulty in the interpretation of the evolution of the ^{19}F spectrum in $\boldsymbol{H}_0$ because of random orientations of separate crystals with respect to $\boldsymbol{H}_0$: the angle φ in $H_0 \neq 0$ takes all possible values, $0 \leqq \varphi \leqq 2\pi$, which should cause the redistribution of NMR frequencies and intensities in the spectrum. The single-crystal measurements could help to avoid this complication.

From the results shown in Fig. 5 it follows that the local fields at ^{19}F nuclei, responsible for the $\nu(^{19}F)_{max}$ peak, have almost the same direction as $\boldsymbol{H}_0$ (i.e. perpendicular to Fe^{3+} moments and $\boldsymbol{H}_1$ at ^{57}Fe nuclei), because the $\nu(^{19}F)_{max}$ shift is about 4 MHz/kOe which is equal to the gyromagnetic ratio for ^{19}F. The shift of the low-frequency edge of the spectrum is more complicated.

3. Conclusion

FeF_3 is a very interesting object to study NMR because of the possibility to observe NMR spectra both on nuclei of magnetic and non-magnetic ions in a magnetically ordered state. This is of particular interest in connection with studying the origin of local fields at nuclei of non-magnetic ions such as F^-.

It has been shown in this paper that H_1 at ^{19}F is about 25 to 30% of the value of H_1 at ^{57}Fe, and consists of dipolar and hyperfine transferred contributions almost of the same order of magnitude. These contributions are anisotropic: the magnitude and the direction of $\boldsymbol{H}_1$ at ^{19}F depend on the mutual orientation of Fe^{3+} spins and the direction of Fe–F–Fe bondings, therefore, the total local field at ^{19}F becomes distributed in a certain way among all F atoms in the unit cell of FeF_3. The anisotropy of H_1 is large enough (4.25 kOe at 4.2 K) compared to the relatively low absolute values of H_1 at ^{19}F (15.25 to 19.5 kOe). As a result of these factors the observed NMR lineshape is rather complicated, and in the absence of the external field it consists of the superposition of two lines with the intensity ratio 1:2. Owing to the higher gyromagnetic ratio for ^{19}F nuclei the ^{19}F spectrum is very broad compared to the ^{57}Fe line.

It is worth to emphasize that the directions of $\boldsymbol{H}_1$ at the ^{19}F nuclei of non-magnetic F atoms do not coincide with the direction of the local field at ^{57}Fe, for some fluorine atoms $\boldsymbol{H}_1$ is almost perpendicular to the Fe^{3+} spins. This is rather a unique situation for magnetic materials.

Acknowledgement

The authors would like to thank Dr. B. Ya. Kotyuzhanskii for his help in the computer calculation of dipole sums.

References

[1] M. A. Hepworth, K. H. Jack, R. D. Peacock, and G. J. Westland, Acta cryst. **10**, 63 (1957).
[2] E. O. Wollan, H. R. Child, W. C. Koehler, and M. K. Wilkinson, Phys. Rev. **112**, 1132 (1958).
[3] L. M. Levinson, J. Phys. Chem. Solids **29**, 1331 (1968).
[4] R. Wolfe, A. J. Kurtzig, and R. C. LeCrow, J. appl. Phys. **41**, 1218 (1970).
[5] G. K. Wertheim, H. J. Guggenheim, and D. N. E. Buchanan, Phys. Rev. **169**, 465 (1968).
[6] U. Bertelsen, J. M. Knudsen, and H. Krogh, phys. stat. sol. **22**, 59 (1967).
[7] M. P. Petrov, A. P. Paugurt, and G. A. Smolenskii, Zh. eksper. teor. Fiz., Pisma **15**, 305 (1972).
[8] V. V. Vanchikov and A. N. Ivashchenko, Kristallografiya **24**, 169 (1979).
[9] K. Ono and A. Ito, J. Phys. Soc. Japan **17**, 1012 (1962).
[10] M. P. Petrov, G. A. Smolenskii, A. P. Paugurt, S. A. Kizhaev, and M. K. Chizhov, Fiz. tverd. Tela **14**, 109 (1972).

(Received April 17, 1979)

phys. stat. sol. (a) 54, 477 (1979)

Subject classification: 18.2.1; 21.1

Laboratoire d'Electricité, Université "Al. I. Cuza", Iaşi[1])

Propriétés d'indépendance des rémanences magnétiques hystérétiques et anhystérétiques dans les couches électrolytiques de Co–Ni–P

Par

C. PĂPUŞOI et P. APOSTOL

On rappelle quelques propriétés magnétiques remarquables des couches électrolytiques de Co–Ni–P, (16% Ni, 1% P et d'épaisseurs allant de 0,05 à 1,6 μm), mises en relief expérimentalement: notamment l'autonomie et l'additivité des aimantations rémanentes anhystérétiques partielles. On montre que cette propriété est plus générale car, d'une part les aimantations en question sont additives géométriquement et d'autre part, les rémanences magnétiques anhystérétiques provoquées par des champs magnétisants faibles et les rémanences magnétiques hystérétiques d'intensité comparable, peuvent coexister *indépendamment* dans un même échantillon. Finalement les faits observés sont interprétés quantitativement, dans leurs moindres détails, au moyen des diagrammes de Preisach associées aux couches de Co–Ni–P en essais.

Some remarkable magnetic properties of electrodeposited films of Co–Ni–P (16% Ni, 1% P, having thickness 0.05 to 1.6 μm) are presented, namely, an autonomy of partial anhysteretic remanent magnetizations (ARM) and an experimentally revealed additivity law. It is found that these properties are more general: on the one hand the ARMs are geometrically additive and, on the other hand, the ARM produced by small fields and the magnetic hysteretic remanences of comparable intensity can co-exist independently in the same sample. Finally, the observed facts are quantitatively interpred, in the minutest detail, using the experimental Preisach diagram for the Co–Ni–P films in question.

1. Introduction

Au départ il convient de préciser la terminologie dont on fait usage dans ce mémoire. Il existe deux procédés pour communiquer à une substance ferromagnétique, une aimantation rémanente, à une température constante:

a) On peut faire agir un champ magnétique H constant, puis le supprimer; le moment magnétique rémanent, que les matériaux en question (couches de Co–Ni–P) acquièrent ainsi, obéit à une variation hystérétique en fonction du champ magnétisant H, d'où la terminologie que nous venons d'adopter, "moment rémanent hystérétique" (abrégé M.R.H., ou A.R.H. pour l'aimantation rémanente hystérétique).

Selon Néel [2] c'est une aimantation rémanente "isotherme" (A.R.I.), mais nous avons évité cette terminologie pour distinguer une autre espèce d'aimantation rémanente, elle aussi isotherme, qu'on obtient par le procédé suivant:

b) On peut aussi, tout en maintenant la température constante, faire agir simultanément un champ magnétique constant H et un champ magnétique alternatif dont on fait décroître l'intensité d'une valeur $\tilde{H}_s$, suffisante pour réaliser la saturation magnétique, à zéro. Le moment magnétique rémanent qu'on mesure après suppression du champ magnétisant H, obéit à une variation anhystérétique en fonction de H, d'où la terminologie usuelle, "moment rémanent anhystérétique" (abrégé M.R.A., ou A.R.A. pour l'aimantation rémanente anhystérétique). Nous désignerons ce M.R.A. par $m(\tilde{H}_s; H; 0)$ pour rappeller ainsi les circonstances de son acquisition, et nous le

[1]) Iaşi—6600, Romania.

qualifierons de "total" pour indiquer que le champ magnétisant H a agi pendant la décroissance du champ alternatif, dans tout l'intervalle $(\tilde{H}_s, 0)$.

Récemment, l'un de nous a montré [1] que les M.R.A. que les couches électrolytiques de Co–Ni–P (16% Ni, 1% P) acquièrent au cours de la décroissance du champ alternatif dans des intervalles $(\tilde{H}_k, \tilde{H}_j)$ arbitraires, en présence d'un champ magnétisant H faible, gardent une sorte d'autonomie dans leurs intervalles d'acquisition; ces M.R.A. "partiels" $m(\tilde{H}_k; H; \tilde{H}_j)$ obéissaient à une loi quantitative d'additivité d'une simplicité tout à fait inattendue: Disons seulement que la somme des M.R.A. partiels, acquis dans des intervalles $(\tilde{H}_k, \tilde{H}_j)$ consécutifs depuis $\tilde{H}_s$ à zéro, dans le même champ magnétisant H, est égale au M.R.A. total $m(\tilde{H}_s; H; 0)$.

S'aidant d'essais nouveaux, qui sont d'ailleurs à l'origine de ce mémoire, nous allons montrer que cette propriété d'indépendance des moments magnétiques des porteurs de l'A.R.A. dans les couches de Co–Ni–P en question, est plus générale: D'une part, les M.R.A. partiels sont additifs *géométriquement* et, d.'autre part un M.R.A. provoqué par un champ H faible et un M.R.H. de module comparable et de n'importe quelle direction par rapport au M.R.A. (même en sens inverse), peuvent coexister *indépendamment* dans le même échantillon.

D'abord nous allons exposer les résultats obtenus, pour tenter ensuite une explication des faits observés.

2. Résultats expérimentaux

Les essais ont porté sur des couches électrolytiques de Co–Ni–P d'une même composition (83% Co, 16% Ni, 1% P), déposées sur des cylindres en cuivre (longueur 65 mm, diamètre 8 mm) et d'épaisseurs différentes allant de 0,1 à 1,8 μm. Nous laissons de coté tout ce qui a trait aux méthodes et appareils de mesure (magnétomètres), un peu particuliers étant données les aimantations très faibles à mesurer.

Les courbes de la Fig. 1 résument une de nos nombreuses séries d'essais; ceux-ci portant dans ce cas sur un échantillon de couche de Co–Ni–P d'épaisseur 1,6 μm, ont été conduits de la façon suivante:

1. L'échantillon, initialement désaimanté, est soumis à l'action d'un champ magnétique constant faible $H = 400$ A/m et d'un champ magnétique alternatif

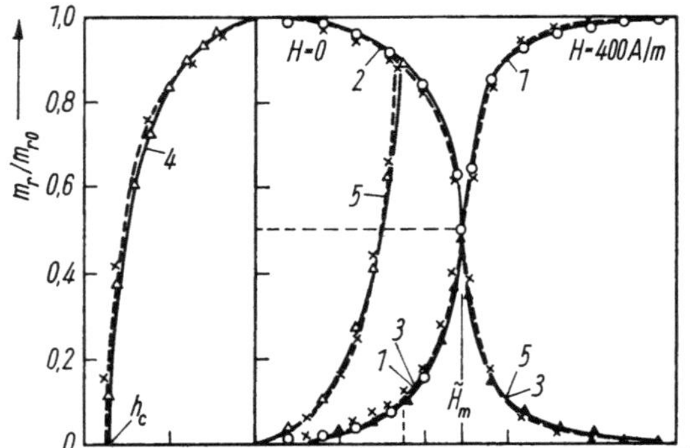

Fig. 1. Echantillon de couche de Co–Ni–P, 1,6 μm épaisseur. Courbes expérimentales (trait plein) et théoriques (trait pointillé) relatives à: (1) l'aimantation rémanente anhystérétique progressive dans le champ constant $H = 400$ A/m; (2) la désaimantation partielle par action de champs alternatifs croissants de zéro à $\tilde{H}_m$; (3) la réaimantation de l'échantillon désaimanté apparemment par opposition d'une A.R.A. égale, suivie de la désaimantation complète en champs alternatifs croissants; (4) la désaimantation apparente par action de champs constants h provoquant des M.R.H. en sens inverse du M.R.A. initial; (5) la réaimantation par disparition du M.R.H. suivie de la désaimantation par disparition du M.R.A. en champs alternatifs croissants de zéro à $\tilde{H}_s$

décroissant d'une valeur $\tilde{H}$ ($< \tilde{H}_s$) à zéro. Après suppression du champ magnétisant H, on mesure au magnétomètre (en champ nul) le M.R.A. $m(\tilde{H}; H; 0)$. Dans des essais successifs on fait croître $\tilde{H}$ jusqu'à ce que, pour $\tilde{H} = \tilde{H}_s$, l'échantillon acquiert le M.R.A. total $m(\tilde{H}_s; H; 0)$. La Fig. 1 donne la courbe d'aimantation rémanente anhystérétique progressive (courbe 1 en trait plein, passant par les cercles blancs); cette courbe a été déterminée en reportant, en fonction de $\tilde{H}$, les valeurs des M.R.A. ainsi acquis, $m(\tilde{H}; H; 0)$ ($= m_r$), réduites à la valeur du M.R.A. total, $m(\tilde{H}_s; H; 0)$ ($= m_{r0}$), soit $m(\tilde{H}; H; 0)/m(\tilde{H}_s; H; 0) = f(\tilde{H})$ ($= m_r/m_{r0}$).

2. L'échantillon, portant maintenant le M.R.A. total $m(\tilde{H}_s; H; 0)$, est soumis à un champ alternatif dont on fait décroître l'intensité à zéro depuis une valeur $\tilde{H}$, en l'absence de tout champ constant ($H = 0$). On mesure le moment magnétique restant que nous désignerons par $m(\tilde{H}_s; H; 0)\ \tilde{H}$. On fait croître $\tilde{H}$, dans des essais successifs, jusqu'à ce que (pour une certaine valeur $\tilde{H} = \tilde{H}_m$) le moment magnétique restant soit exactement moitié du M.R.A. total; on a ainsi $m(\tilde{H}_s; H; 0)\ \tilde{H}_m = m(\tilde{H}_s; H; 0)/2$. La courbe 2 (en trait plein, passant par les cercles blancs; Fig. 1) a été déterminée avec les valeurs réduites du moment magnétique restant, correspondant aux champs $\tilde{H}$ croissants de zéro à $\tilde{H}_m$, soit $m(\tilde{H}_s; H; 0)\ \tilde{H}/m(\tilde{H}_s; H; 0) = f(\tilde{H})$ ($= m_r/m_{r0}$). Finalement, avec l'échantillon doué d'un moment magnétique rémanent qui est la moitié du M.R.A. initial, $m(\tilde{H}_s; H; 0)\ \tilde{H}_m$, on fait agir simultanément le champ magnétisant $-H$ ($= 400$ A/m) en sens inverse du M.R.A. initial, et un champ magnétique alternatif décroissant à zéro depuis la valeur $\tilde{H}_m = 2{,}45 \times 10^4$ A/m. On observe, après suppression du champ H, une aimantation nulle, que nous désignerons par $m(\tilde{H}_s; +H; \tilde{H}_m; -H; 0) = 0$ pour rappeler ainsi ces circonstances.

3. Avec l'échantillon désaimanté comme on l'a vu, $m(\tilde{H}_s; +H; \tilde{H}_m; -H; 0) = 0$, on fait ensuite l'expérience suivante, qui peut paraître paradoxale: On applique un champ magnétique alternatif que l'on fait décroître à zéro depuis une valeur $\tilde{H}$ donnée ($\tilde{H} < \tilde{H}_m$), tout champ constant à l'emplacement de l'échantillon étant soigneusement compensé ($H = 0$). Mesurant ensuite l'aimantation rémanente, on constate l'apparition d'un moment magnétique dirigé dans le sens du M.R.A. initial; désignons-le par $m(\tilde{H}_s; +H; \tilde{H}_m; -H; 0)\ \tilde{H}$. Les essais sont renouvelés pour des champs $\tilde{H}$ croissant successivement jusqu'à ce que $\tilde{H} = \tilde{H}_s$. Reportant, en fonction de $\tilde{H}$, les valeurs réduites de ce moment magnétique rémanent, on obtient la courbe (3) (en trait plein, passant par des triangles noirs; Fig. 1). On observe que le moment magnétique croît d'abord, pour atteindre exactement la valeur initiale du M.R.A. (moitié du M.R.A. total) quand $\tilde{H} = \tilde{H}_m$, ensuite diminue pour $\tilde{H} > \tilde{H}_m$ et disparaît complètement quand $\tilde{H} = \tilde{H}_s$ (8×10^4 A/m). Faut-il ajouter que l'application du champ magnétisant $+H$, dans le sens initial du M.R.A. total, alors que le champ alternatif décroissait à zéro depuis la valeur $\tilde{H}_m$, reconstituerait exactement le M.R.A. total initial, $m(\tilde{H}_s; H; 0)$.

Il s'ensuit qu'en réalité, l'échantillon était apparemment désaimanté et qu'il portait véritablement deux aimantations autonomes mais opposées: $m(\tilde{H}_s; +H; \tilde{H}_m)$, reste du M.R.A. total initial, et $-m(\tilde{H}_m; -H; 0)$ acquise au cours de la décroissance du champ alternatif dans l'intervalle ($\tilde{H}_m$, 0) en présence du champ $-H$. Alors, ces deux aimantations rémanentes anhystérétiques partielles sont additives géométriquement.

Tout se passe comme si l'on avait, dans le même échantillon, deux ensembles *indépendants* de porteurs d'aimantation rémanente: l'un responsable de la moitié de l'A.R.A. totale dirigée dans le sens du champ magnétisant $+H$ et insensible à tout champ alternatif d'intensité allant de zéro à $\tilde{H}_m$, et l'autre responsable de la moitié de l'A.R.A. dirigée dans le sens du champ $-H$ (opposé à l'A.R.A. initiale), qui disparaît complètement quand le champ alternatif croît de zéro à $\tilde{H}_m$.

Quelques essais analogues à ceux qui viennent d'être décrits, ont été faits en provoquant des rémanences magnétiques hystérétiques en sens opposé au M.R.A. initial. Ceux-ci ont été conduits de la façon suivante:

4. L'échantillon, portant initialement le M.R.A. total $m(\tilde{H}_s; H; 0)$ acquis dans le même champ magnétisant faible, $H = 400$ A/m, est soumis à l'action d'un champ magnétique constant $\boldsymbol{h}$ dirigé en sens inverse du M.R.A.; après suppression de ce champ $\boldsymbol{h}$, on mesure au magnétomètre le moment magnétique restant que nous désignerons par $m(\tilde{H}_s; H; 0)\, h$, On répète ces opérations pour des champs h_1, h_2, ... successivement croissants (en valeur absolue) jusqu'à ce qu'on amène le moment magnétique total de l'échantillon à être nul; ainsi, pour un certain champ, h_c, on aura $m(\tilde{H}_s; H; 0)\, h_c = 0$. L'allure de cette désaimantation, par action de champs constants h en sens inverse de l'A.R.A. initiale, est indiquée par la courbe 4 en traits pleins (Fig. 1). Cette courbe 4 a été déterminée, en reportant en fonction de h, les valeurs réduites du moment magnétique restant, soit $m(\tilde{H}_s; H; 0)\, h/m(\tilde{H}_s; H; 0) = f(h)$ $(= m_r/m_{r0})$.

5. Après avoir désaimanté ainsi l'échantillon (avec $m(\tilde{H}_s; H; 0)\, h_c = 0$ pour $h_c = -1{,}75 \times 10^4$ A/m), on fait agir un champ magnétique alternatif, décroissant d'une intensité maximum $\tilde{H}$, en l'absence de tout champ magnétique constant. On désigne par $m\,|(\tilde{H}_s; H; 0)\, h_c\,|H$ le moment magnétique qu'on mesure dès que le champ alternatif H est ramené à zéro. En renouvelant ces opérations pour des champs $\tilde{H}$ successivement croissants, et en reportant en fonction de $\tilde{H}$ les valeurs réduites des moments magnétiques mesurés, on obtient la courbe 5 en trait plein (Fig. 1), soit $m\,|(\tilde{H}_s; H; 0)\, h_c|\, \tilde{H}/m(\tilde{H}_s; H; 0) = f(\tilde{H})$ $(= m_r/m_{r0})$. On observe alors un fait assez paradoxal à première vue: une aimantation rémanente apparaît et croît dans le sens du M.R.A. initial, pour atteindre un maximum de près de 0,9 de la valeur du M.R.A. initial quand le champ H est de l'ordre de grandeur du champ constant h_c (ayant rendu nulle l'aimantation globale de l'échantillon), ensuite décroît, pour disparaître complètement quand $\tilde{H}$ s'approache de H_s.

Il en ressort donc, qu'en réalité, l'aimantation n'était pas nulle mais l'échantillon portait deux aimantations, en quelque sorte indépendantes: le M.R.A. initial, plus ou moins altéré par le champ h_c, et un M.R.H., opposé, dû au champ h_c; le premier stable pour des champs alternatifs allant jusqu'à $|\tilde{H}| = h_c$, le deuxième disparaissant complètement pour ces champs. De tels faits, qui ne sont pas sans analogie avec ceux qui viennent d'être établis dans le cas des A.R.A. opposées, nous paraissent nécessiter deux ensembles indépendants de porteurs d'aimantation rémanente, dans le même échantillon: l'un responsable du M.R.H. et l'autre de la plus grande part du M.R.A. initial.

6. On reprend enfin l'expérience qui vient d'être décrite au paragraphe 2, pour compléter la courbe de désaimantation de l'échantillon, par action de champs alternatifs $\tilde{H}$ croissants, jusqu'à disparition complète du M.R.A. initial, $m(\tilde{H}_s; H; 0)$. Avec les valeurs des moments magnétiques $m(\tilde{H}_s; H; 0)\, \tilde{H}$ qui subsistent après chaque retour de $\tilde{H}$ à zéro (en l'absence de tout champ constant), on obtient la courbe

$m(\tilde{H}_s; H; 0)\ \tilde{H}/m(H_s; H; 0) = f(\tilde{H})$; celle-ci, indiquant la disparition progressive du M.R.A. initial, s'identifie effectivement avec la courbe 2 pour les champs allant de zéro à $\tilde{H}_m$ et avec la courbe 3 pour les champs allant de $\tilde{H}_m$ à $\tilde{H}_s$.

On observe que cette désaimantation obéit à une loi simple, correspondant à une indépendance des M.R.A. partiels, acquis au cours de la décroissance du champ alternatif dans des intervalles arbitraires de valeurs ($\tilde{H}_k$, $\tilde{H}_j$), et à une sorte de réversibilité: le moment magnétique acquis par franchissement d'un intervalle $\tilde{H}_k - \tilde{H}_j$ à la décroissance du champ alternatif, disparaît dans le même intervalle exactement, quand le champ alternatif croît de $\tilde{H}_j$ à $\tilde{H}_k$. Les courbes 1, d'aimantation rémanente anhystérétique progressive, et 2, 3, de désaimantation, sont exactement complémentaires et symétriques par rapport à une droite parallèle à l'axe des $\tilde{H}$, passant par l'ordonnée $m_r/m_{r0} = \frac{1}{2}$, traduisant en fait la loi d'additivité [1] à laquelle obéissent les M.R.A. partiels.

Finalement faut-il préciser que tout ce qui vient d'être dit ici des propriétés d'indépendance des M.R.A. partiels opposés l'un à l'autre, d'une part, et des M.R.H. opposés au M.R.A., d'autre part, n'est valable que dans le cas des M.R.A. acquis en champs magnétisants H faibles seulement.

3. Interprétation des résultats

L'ensemble des faits ainsi acquis conduit à penser que les couches de Co–Ni–P en question possèdent les propriétés magnétiques d'un ensemble de particules, magnétiques monodomaines, à champ coercitif H_c étalé sur une large échelle de valeurs; ce fait est d'ailleurs suggéré par le spectre large des valeurs du champ alternatif qui est nécessaire pour détruire complètement une A.R.A. totale. Il est à supposer, d'autre part, qu'entre ces particules (que ce soient des cristallites [3] ou des agrégats laméllaires [4] à comportement de monodomaines magnétiques) s'exèrcent des interactions magnétostatiques appréciables, vu la susceptibilité rémanente anhystérétique de ces couches, qui est finie dans les champs magnétisants faibles [5]. On peut représenter ces interactions dans un ensemble de particules monodomaines, par la distribution d'un champ magnétique d'interaction H_i. Selon Néel [6], l'effet de ce champ se traduit par le déplacement du cycle d'hystérésis élémentaire de chaque particule, suivant l'axe des champs, d'une quantité H_i, sans que la largeur du cycle ($2H_c$) soit modifiée. Alors, les particules de l'ensemble, associé à une couche de Co–Ni–P, doivent comporter des cycles élémentaires d'hystérésis asymétriques, à champs critiques différents, $H_+ = H_c - H_i$, $H_- = -H_c - H_i$. On obtient le diagramme de Preisach pour cet ensemble, en faisant correspondre à chaque particule un point du plan H_+H_-. La distribution du moment magnétique rémanent, suivant les valeurs des H_+ et H_-, est donnée par la fonction tridimensionnelle de Preisach, $m(H_+, H_-)$. L'un de nous a montré [7] que les fonctions de Preisach associées aux couches de Co–Ni–P en question, sont statistiquement stables; c'est-à-dire que le champ d'interaction H_i est, statistiquement, indépendant de l'état d'aimantation rémanente de l'échantillon et que, par conséquent, la répartition, des particules dans le diagramme de Preisach est, pratiquement, indépendante de l'intensité du champ magnétique extérieur.

Le diagramme de Preisach (Fig. 2), déterminé expérimentalement [7] pour l'échantillon en essais (1,6 μm épaisseur), donne la répartition numérique du moment magnétique rémanent; celle-ci est limitée au quatrième quadrant du plan H_+H_-, où sont représentées, par leurs points correspondants (H_+, H_-), toutes les particules de l'ensemble associé. Après désaimantation par action de champs magnétiques alternatifs décroissants, le moment magnétique de chaque particule restera bloqué dans le sens du champ critique le plus faible; alors l'ensemble *désaimanté* comporte des particules

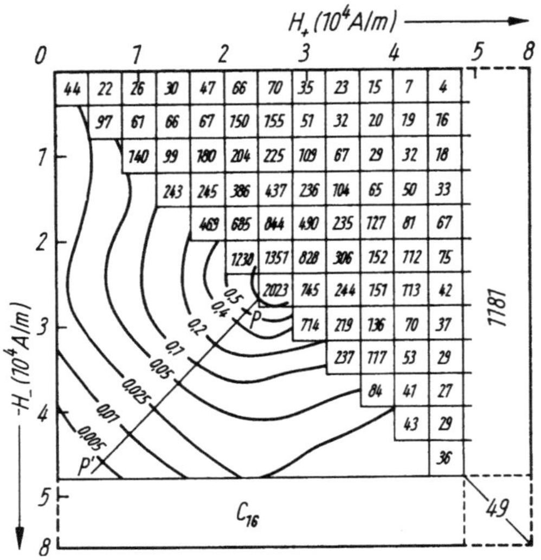

Fig. 2. Echantillon de couche de Co–Ni–P, 1,6 µm épaisseur; diagramme de Preisach expérimental. Les valeurs figurées dans chaque rectangle élémentaire, multipliées par 10^{-7}, représentent le moment magnétique associé, en Am^2

"polarisées", mais leur moment magnétique résultant doit être nul, vu la symétrie de cette distribution par rapport à la diagonale $H_+ + H_- = 0$ (Fig. 3a).

Nous avons intégré la fonction de Preisach $m(H_+, H_-)$ associée à l'échantillon en essais, en accord avec les processus d'aimantation rémanente qui viennent d'être décrits précédemment (Section 2), en vue de déterminer les courbes correspondantes, prédites au moyen du diagramme de la Fig. 2; celles-ci, reportées par des traits interrompus dans la Fig. 1, ont été déterminées de la façon suivante:

1. La courbe 1 est relative à l'aimantation rémanente anhystérétique progressive, $m(\tilde{H}; H; 0) = f(\tilde{H})$, pour $H = 400$ A/m. Le M.R.A. correspondant à un point d'abscisse $\tilde{H}_1$ de cette courbe, est donné par l'intégrale numérique de la fonction $m(H_+, H_-)$ correspondant à la région du diagramme de Preisach, limitée par les droites $H_+ = -H_-$, $H_+ + H_- = 2H$ (= 800 A/m), $H_- = 0$ et $H_+ = \tilde{H}_1$ (Fig. 3b). Les valeurs obtenues, en donnant au H_+ des valeurs $\tilde{H}_1, \tilde{H}_2, \ldots$ croissantes successivement jusqu'à $H_+ = \tilde{H}_s$, ont été réduites à la valeur du M.R.A. total, $m(\tilde{H}_s; H; 0)$, et reportées en fonction de $\tilde{H}$ dans la Fig. 1 (courbe 1 en traits interrompus). Le M.R.A. total, dû au champ H, est donné par la somme des moments magnétiques des particules dont les points figuratifs sur le diagramme de Preisach sont situés dans la bande diagonale (hachurée), limitée par les droites $H_+ = -H_-$ et $H_+ + H_- = 2H$.

2. Maintenant nous allons simuler l'attaque du M.R.A. total par des champs alternatifs d'intensité progressivement croissante, jusqu'à la valeur $\tilde{H}_m$ pour laquelle le moment magnétique restant $m(\tilde{H}_s; H; 0)\, \tilde{H}_m$ est moitié du M.R.A. total. Cette désaimantation partielle, correspondant à la courbe 2 expérimentale, $m(\tilde{H}_s; H; 0)\, \tilde{H}/m(\tilde{H}_s; H; 0) = f(\tilde{H})$ (Fig. 1), découle du diagramme de Preisach comme l'on voit dans la Fig. 3c: Le moment magnétique restant, $m(\tilde{H}_s; H; 0)\, \tilde{H}_1$, correspondant à un point d'abscisse H_1 de cette courbe 2, est donné par l'intégrale numérique de la fonction de Preisach correspondant à la région du diagramme qui est limitée par les droites $H_+ + H_- = 0$, $H_+ = \tilde{H}_s$, $H_+ + H_- = 2H$ et $H_+ = \tilde{H}_1$ (l'aire PQRS, Fig. 3c). Avec les valeurs réduites du moment magnétique restant (obtenues en donnant au H_+ des valeurs croissantes $\tilde{H}_1, \tilde{H}_2, \ldots, \tilde{H}_m$), reportées en fonc-

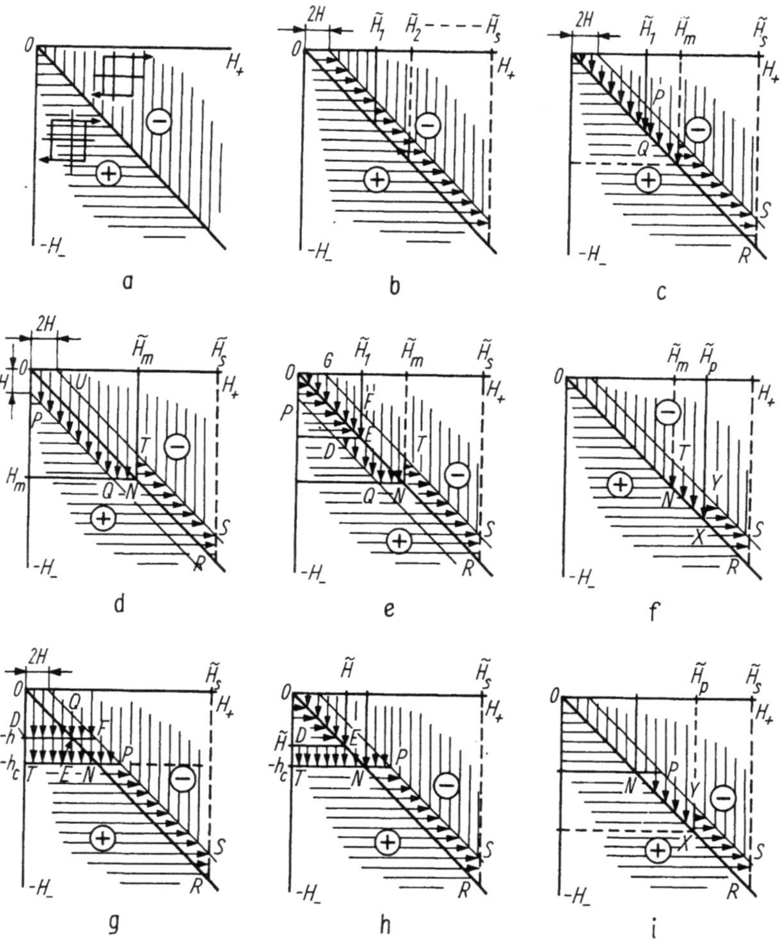

Fig. 3. Simulation au moyen du diagramme de Preisach, des processus d'aimantation auxquels se rapportent les courbes de la Fig. 1: a) la distribution des particules "polarisées" de l'ensemble désaimanté; b) l'aimantation rémanente anhystérétique dans le champ constant H (courbe 1); c) la désaimantation partielle (courbe 2); d) désaimantation apparente par superposition de deux M.R.A. égaux et opposés; e) réaimantation par disparition progressive du M.R.A. acquis dans l'intervalle $(0, \tilde{H}_m)$ (courbe 3 ascendante); f) disparition progressive en champs $\tilde{H} > \tilde{H}_m$ du M.R.A. acquis dans l'intervalle $(\tilde{H}_m, \tilde{H}_s)$ (courbe 3 descendante); g) désaimantation apparente par superposition d'un M.R.H. en sens opposé au M.R.A. initial (courbe 4); h) réaimantation par disparition progressive du M.R.H. en champs alternatifs d'intensité croissant jusqu'à $|\tilde{H}| = |h_c|$ (courbe 5 ascendante); i) disparition du M.R.A. en champs alternatifs croissants de $|\tilde{H}_c|$ à $\tilde{H}_s$

tion de H dans la Fig. 1, on a déterminé la courbe 2 (traits interrompus), prédite au moyen du diagramme, pour cette désaimantation partielle.

La désaimantation apparente de l'echantillon, par application du champ $-H$ (en sens inverse du M.R.A. initial) et du champ alternatif décroissant à zéro depuis la valeur H_m, est simulée au moyen du diagramme de Preisach, de la façon indiquée dans la Fig. 3d; le moment magnétique de la distribution est ramené à zéro, vu la symétrie du diagramme. On a ainsi $m(\tilde{H}_s; +H; \tilde{H}_m; -H; 0) = 0 = m(\tilde{H}_s; H; 0)\ \tilde{H}_m -$

— $m(\tilde{H}_m; -H; 0)$. Effectivement, le moment $m(H_m; -H; 0)$, donné par l'intégrale de la fonction de Preisach limitée à l'aire OPQN (Fig. 3d), correspond aux particules à champ coercitif compris dans l'intervalle $(0, \tilde{H}_m)$; ce moment magnétique est égal et opposé au moment $m(\tilde{H}_m; +H; 0)$ — correspondant à l'aire ONTU — qui était dû aux particules à champ coercitif compris dans le même intervalle $(0, \tilde{H}_m)$. D'autre part, il est opposé et égal au moment $m(\tilde{H}_s; H; 0)\ \tilde{H}_m\ (= m(\tilde{H}_s; H; \tilde{H}_m))$ — correspondant à l'aire NRST — qui est dû aux particules à champ coercitif compris dans l'intervalle $(\tilde{H}_m, \tilde{H}_s)$.

3. Enfin il sera facile d'interpréter le fait qui parut paradoxal pour un échantillon désaimanté: la réaimantation de celui-ci (courbe 3 expérimentale) par action de champs alternatifs croissants (jusqu'à l'acquisition du M.R.A. initial, c'est-à-dire moitié du M.R.A. total, dans le champ $\tilde{H}_m$), suivie de la disparition progressive de ce moment magnétique, pour $\tilde{H} > \tilde{H}_m$.

Le processus d'aimantation qui se rattache à la partie ascendante de cette courbe 3 (correspondant aux champs étalés entre zéro et $\tilde{H}_m$), est simulé au moyen du diagramme de Preisach dans la Fig. 3e. L'action d'un champ alternatif $\tilde{H}_1$ $(< \tilde{H}_m)$ fait disparaître effectivement la fraction du M.R.A. que l'échantillon avait acquis au cours de la décroissance du même champ $\tilde{H}_1$ en présence du champ constant $-H$ (moment magnétique donné par l'intégrale correspondant à l'aire (OPDE), respectivement $m(\tilde{H}_1; -H; 0)$). L'ensemble des particules possède ainsi une aimantation rémanente qui est donnée par la somme géométrique de deux moments magnétiques: l'un, $m(\tilde{H}_m; -H; \tilde{H}_1)$, correspondant aux particules de l'aire DQNE (polarité (—)), et l'autre, $m(\tilde{H}_s; H; \tilde{H}_m)$, correspondant à l'aire NRST (polarité (+)). Le moment résultant, (en polarité (+)) est opposé et égal exactement au moment magnétique disparu par l'effet du champ $\tilde{H}_1$, ou autrement dit, ce moment est exactement $m(\tilde{H}_1; H; 0)$ que l'échantillon, préalablement désaimanté, avait acquis au cours de la décroissance à zéro du champ $\tilde{H}_1$, en présence du champ constant $+H$; celui-ci correspond à l'aire OEFG (Fig. 3e). La courbe 3 ascendante, en pointillé (Fig. 1), a été déterminée avec les valeurs réduites de ce moment magnétique résultant, relevées sur le diagramme de la Fig. 2, pour des champs $\tilde{H}_1$, $\tilde{H}_2$, ... croissants jusqu'à $\tilde{H}_m$.

La disparition progressive du M.R.A., par action de champs alternatifs d'intensité croissante de $\tilde{H}_m$ à $\tilde{H}_s$ (partie descendante de la courbe 3), est simulée dans la Fig. 3f. L'application du champ alternatif $\tilde{H}_p > \tilde{H}_m$ fait disparaître le moment magnétique $m(\tilde{H}_p; H; \tilde{H}_m)$ (correspondant à l'aire NXYT) et laisse inaltéré le moment magnétique restant, $m(\tilde{H}_s; H; \tilde{H}_p)$, lié à l'aire XRSY (Fig. 3f); celui-ci disparaîtra à son tour quand le champ alternatif croissant franchira l'intervalle $(\tilde{H}_p, \tilde{H}_s)$.

4. La désaimantation apparente de l'échantillon (initialement porteur de l'A.R.A. totale), par action de champs constants $\boldsymbol{h}$ opposés à l'A.R.A. (courbe 4 expérimentale, Fig. 1), s'interprète, au moyen du diagramme de Preisach, de la façon indiquée dans la Fig. 3g. L'A.R.A. initiale, correspondant au champ magnétisant $+H$, est due aux particules aimantées positivement, qui sont situées dans la bande diagonale ORSQ. Par application et suppresions d'un champ $-h$ $(< h_c)$, toutes les particules représentées en-dessus de la droite $H_- = -h$ s'aimantent négativement. Alors le moment magnétique restant, $m(\tilde{H}_s; H; 0)\ h$, correspondant au point d'abscisse $-h$ de la courbe 4, est donné par la somme géométrique de deux aimantations: l'une due aux particules aimantées *positivement*, dans le champ $+H$, par le mécanisme de l'A.R.A.

(particules appartenant à l'aire ERSF), et l'autre, due aux particules aimantées *négativement*, dans le champ $-h$, par le mécanisme de l'A.R.H. (l'aire ODE). On s'aperçoit facilement que l'effet du champ h est double: tout se passe comme si ce champ, d'une part provoquait une A.R.H. comme il le ferait sur l'échantillon dépourvu d'aimantation (l'aire ODE), d'autre part réduisait partiellement l'A.R.A. initiale (l'aire OEFQ). Ce deuxième effet sera moins important à mesure que le champ $+H$ ayant provoqué l'A.R.A. est plus faible.

Pour $h = h_c$, les deux moments magnétiques opposés, le M.R.H. dû au champ h_c (l'aire OTN) et le reste du M.R.A. (l'aire NRSP), sont égaux et ainsi, le moment magnétique résultant est nul. La courbe 4 (en pointillé dans la Fig. 1) a été obtenue avec les valeurs réduites du moment magnétique résultant, donné par l'intégrale numérique de la distribution de Preisach $m(H_+, H_-)$ pour les aires correspondantes aux M.R.H. et M.R.A. opposés.

5. Nous allons tenter maintenant d'expliquer les faits résumés par la courbe 5 expérimentale (Fig. 1): la réaimantation suivie de la désaimantation complète de l'échantillon (qui initialement était apparemment désaimanté), par action de champs magnétiques alternatifs croissants. Par l'effet d'un champ alternatif $|\tilde{H}| < |h_c|$, l'aimantation ("négative") due aux particules appartenant à l'aire ODE correspondante (Fig. 3h) disparaît, et le moment magnétique résultant sera donné par l'intégrale numérique de la fonction de Preisach correspondant aux aires DENT (reste du M.R.H. négatif) et NRST (M.R.A. positif, un peu altéré par le champ h_c). Cette résultante, que nous avons notée $m \mid (\tilde{H}_s; H; 0)\, h_c \mid \tilde{H}$, doit augmenter jusqu'à ce que $|\tilde{H}| = |h_c|$; la courbe 5 ascendante (traits pointillés), correspondant en fait à la disparition progressive du M.R.H. provoqué par le champ h_c (l'aire OTN de la Fig. 3h), a été déterminée avec les valeurs réduites de ce moment magnétique résultant, relevées sur le diagramme de Preisach pour l'échantillon en question. Le moment magnétique maximum qu'on peut réaliser au cours de cette désaimantation, $m \mid (\tilde{H}_s; H; 0)\, h_c \mid \tilde{H}_c$, représente la plus grande part du M.R.A. initial. En effet, vu la configuration de la distribution (Fig. 2), pour les champs H très faibles, ce M.R.A. maximum — qui est dû aux particules appartenant à l'aire NRSP (Fig. 3h) — est très voisin de la valeur du M.R.A. total, ou autrement dit, la fraction du M.R.A. total, détruite par le champ h_c, est relativement petite (moins de 10% du M.R.A. total, dans le cas considéré ici).

Enfin, la partie descendante de cette courbe 5 correspond à la disparition progressive du M.R.A. restant, par l'effet des champs alternatifs d'intensité $|\tilde{H}| > |h_c|$. Le moment magnétique restant, correspondant à un champ alternatif $\tilde{H}_p\ (> \tilde{H}_c)$, $m \mid (\tilde{H}_s; H; 0)\, h_c \mid \tilde{H}_p$, sera donné par l'intégrale numérique de la fonction de Preisach pour l'aire XRSY (Fig. 3i). Ce moment restant disparaît complètement quand $\tilde{H} = \tilde{H}_s$, comme l'indique la courbe 5 descendante (en traits pointillés dans la Fig. 1); celle-ci, déterminée avec les valeurs réduites de ce moment magnétique, s'identifie dans cette région (où $|\tilde{H}| > |h_c|$) avec la courbe de disparition progressive du M.R.A. total en champs alternatifs croissants.

4. Conclusion

L'accord excellent entre les résultats des mesures directes d'aimantation rémanente (courbes en trait plein) et ceux prédits au moyen de la fonction de Preisach associée aux couches de Co–Ni–P en essais (courbes en trait pointillé), concernant une propriété magnétique tellement remarquable comme l'est celle d'indépendance des M.R.A. partiels d'une part, et des M.R.H. opposés aux M.R.A. d'autre part, nous permet d'en tirer les conclusions suivantes:

a) D'abord ces résultats confirment la stabilité statistique de la fonction de distribution de Preisach, associée aux couches de Co–Ni–P. Effectivement, on a pu prédire avec une précision remarquable, des faits concernant les propriétes des rémanences magnétiques purement *anhystérétiques,* au moyen d'une distribution de particules (associée à la couche de Co–Ni–P) qui a été déterminée par utilisation exclusive des courbes d'aimantation rémanente *hystérétique.*

b) Il en ressort que l'indépendance des M.R.A. partiels de même que celle des M.R.A. vis-à-vis du M.R.H. est une propriété attribuable non seulement aux particules monodomaines *indépendantes* (pour lesquelles $H_i = 0$), mais aussi aux particules en *interaction* magnétostatique (pour lesquelles $H_i \leqq H$). Cependant, cette indépendance, et particulièrement l'autonomie des M.R.A. partiels dans leurs intervalles d'acquisition, serait possible à condition que le champ d'interaction H_i soit indépendant de l'état magnétique de l'ensemble des particules. En effet, cette condition est satisfaite dans le cas des couches de Co–Ni–P en question, vu la stabilité statistique des fonctions de Preisach associées. Par ailleurs nous avons montré que cette stabilité est assurée pour des champs extérieurs assez forts: Par example, le M.R.A. dû aux particules à champ d'interaction faible ($H_i \leqq H = 4 \times 10^2$ A/m) est insensible à l'action d'un champ extérieur constant relativement fort, $h_c = 175 \times 10^2$ A/m (sauf une petite fraction due aux particules à champ coercitif $H_c \leqq 175 \times 10^2$ A/m).

c) La symétrie des courbes d'aimantation (1) et de désaimantation (2 et 3) traduit en fait la loi d'additivité des M.R.A. partiels, mise en relief expérimentalement [1]. On comprend maintenant que cette additivité est *géométrique,* vu l'autonomie *réelle* des M.R.A. partiels dans leurs intervalles d'acquisition.

d) Cette propriété d'indépendance des moments magnétiques des porteurs de l'aimantation rémanente dans les couches de Co–Ni–P, de même que la propriété d'exacte sommation géométrique de ces moments magnétiques, confèrent aux couches de Co–Ni–P des qualités remarquables de bons matériax uutilisables pour l'enregistrement magnétique.

Bibliographie

[1] P. Apostol, An. Șt. Univ. "Al. I. Cuza" Iași (Ib) **21**, 45 (1975).
[2] L. Néel, Ann. Géophys. (Paris) **5**, 99 (1949).
[3] G. Bate et D. E. Speliotis, J. appl. Phys. **34**, 1073 (1963).
[4] J. S. Sallo et K. H. Olsen, J. appl. Phys. **32**, 203S (1961).
[5] P. Apostol, An. Șt. Univ. "Al. I. Cuza", Iași (Ib) **23**, 37 (1977).
[6] L. Néel, Appl. sci. Res. (Netherlands) **B4**, 18 (1954).
[7] P. Apostol, An. Șt. Univ. "Al. I. Cuza", Iași (Ib) **21**, 19 (1975).

(Received March 22, 1979)

phys. stat. sol. (a) 54, 487 (1979)

Subject classification: 14.3.1; 22.4.2

Zentralinstitut für Elektronenphysik der Akademie der Wissenschaften der DDR, Berlin (a) and A. F. Ioffe Physico-Technical Institute, Academy of Sciences of the USSR, Leningrad (b)

Hopping Conductivity of Undoped ZnSe Thin Films

By

R. RENTZSCH[1]) (a), I. S. SHLIMAK (b), and H. BERGER (a)

Results on the temperature dependence of the dc conductivity at low and high electric field strengths, and the frequency dependence of the ac conductivity are interpreted as hopping conductivity. Near room temperature the low field conductivity is determined by nearest-neighbour hopping, and at lower temperatures by variable range hopping conductivity which is given by $\sigma \sim \exp(-(T_0/T)^b)$, $b = (1+x)/(1+d+x) \approx 0.6$ to 0.8. The high field limit of the current density yields $j \sim \exp(-(F_0/F)^c)$. The obtained correspondence $c \approx b$ can be taken as an evidence of the hopping conduction mechanism.

Die Ergebnisse der Temperaturabhängigkeit der Gleichstromleitfähigkeit bei kleinen und großen elektrischen Feldstärken sowie der Frequenzabhängigkeit der Wechselstromleitfähigkeit werden als Hoppingleitfähigkeit gedeutet. Bei Temperaturen nahe Raumtemperatur wird die Leitfähigkeit bei kleinen Feldern durch Hopping über nächste Nachbarn und bei tieferen Temperaturen durch variable range hopping der Form $\sigma \sim \exp(-(T_0/T)^b)$, $b = (1 + x)/(1 + d + x) \approx 0{,}6$ bis 0,8 bestimmt. Im Grenzfall hohen Feldes ergibt sich für die Stromdichte $j \sim \exp(-(F_0/F)^c)$. Die beobachtete Übereinstimmung von $c \approx b$ wird als ein Beweis für den Hoppingleitungsmechanismus genommen.

1. Introduction

In recent years there is an increasing interest in heavily distorted crystals like heavily doped compensated (h.d.c.) crystals. The results of this research show that some essential electronic properties are similar to those of amorphous semiconductors [1]. Recent papers concerning conductivity measurements of amorphous films [2] as well as of h.d.c. crystals of II–VI semiconducting compounds [3] have used the concept of hopping conductivity. The reason is that carrier transport at low temperatures is determined by the existence of a potential relief due to lattice distortions or nonhomogeneities in amorphous semiconductors and fluctuations of the concentration of charged impurities in h.d.c. semiconductors. From this point of view it seems interesting to prove whether or not the conductivity of not intentionally doped ZnSe films at low temperature is a hopping one, because this substance has a strong tendency to self-compensation, and in films the concentration of defects is in general relatively high.

2. Experimental

The ZnSe films evaporated in vacuum onto quartz glass substrates were polycrystalline and of cubic structure. At the substrate temperature of 300 °C and the deposition rate of 10 Å/s the mean crystallite size was 50 to 100 nm depending on film thickness. The conductivity of these films at room temperature is very low ($\sigma = 10^{-11}\ \Omega^{-1}\ \mathrm{cm}^{-1}$). After recrystallization of the films both the structure and the conductivity were essentially modified:

---

[1]) Rudower Chaussee 5, DDR-1199 Berlin, DDR.

Fig. 1. Typical electron micrographs of polycrystalline ZnSe films. a) As-deposited film; b) after recrystallization at 700 °C

(i) The mean crystallite size increases to 0.1 to 4 μm depending on the recrystallization temperature between 500 and 700 °C [4] (Fig. 1). At a thickness of about 1 μm the film has single-crystalline areas from the substrate up to the free surface.

(ii) The conductivity of the films increases by about seven to ten orders of magnitude and is isotropic. Therefore, the conductivity of recrystallized films is determined by the bulk properties of the crystallites rather than by the grain boundaries. For the contact separation of 8 to 20 μm used, in the current direction there are only few grain boundaries, whose influence will be neglected.

3. Results

All results presented in this paper deal with recrystallized films. Fig. 2 shows the temperature dependence of the dc conductivity of films after recrystallization at various temperatures T_a. At a fixed temperature the conductivity increases with the annealing temperature. A temperature independent activation energy exists only in

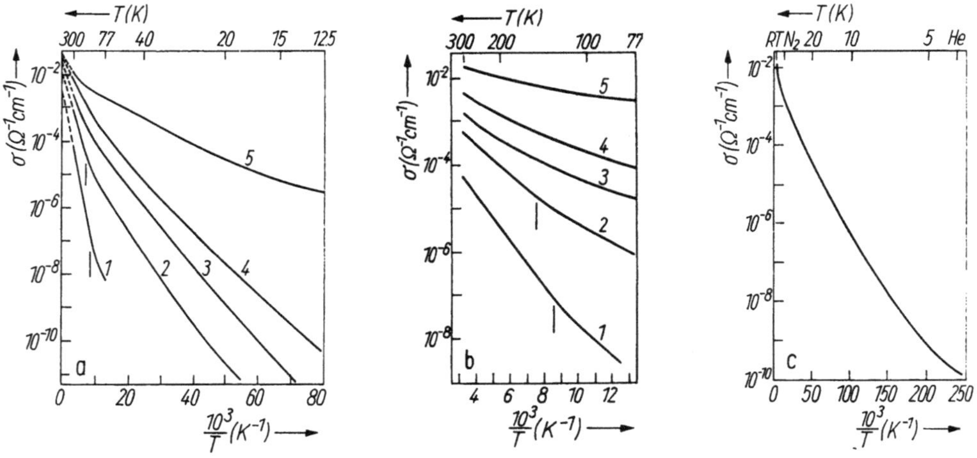

Fig. 2. a) Temperature dependence of the dc conductivity after annealing at temperature T_a (°C): (1) 500, (2) 550, (3) 600, (4) 650, (5) 700 (the lower limit of a temperature independent activation energy is indicated by a mark); b) dc conductivity in the high-temperature region 77 to 300 K; c) temperature dependence of curve 5 at $T = 4.2$ to 300 K

Table 1

No.	T_a (°C)	$\sigma(0)$ (Ω^{-1} cm^{-1})	ε_3^{RT} (meV)	$b = \frac{1+x}{1+d+x}$	T_0 (K)	c
1	500	2.0×10^{-3}	104	1	1.18×10^3	
2	550	1.4×10^{-2}	86	0.86	5.27×10^2	
3	600	2.0×10^{-2}	57	0.80	6.82×10^2	
4	650	3.3×10^{-2}	49	0.70	1.09×10^3	
5	700	4.0×10^{-2}	23	0.68	3.14×10^1	0.75 — 1.0

a narrow range below room temperature. This is clearly evident for curves 1 and 2 in Fig. 2. The magnitude of these activation energies ε_3 decreases with increasing annealing temperature (Table 1).

In Fig. 3 the ac conductivity at 77 K is plotted. From the ω-dependent part we get a dependence $\sigma \sim \omega^n$ with n between 0.6 and 0.8. There is a tendency that the lower the dc conductivity (indicated on the ordinate in Fig. 3), the lower is the frequency at which the conductivity starts to increase.

Typical dependences of the current density on electric field strength are shown in Fig. 4. The measurements were made in a helium bath at 2 to 4 K. In the high-field limit the curves tend to a uniform one, which is a result of a vanishing temperature dependence.

Starting from the lowest temperature we found a shift of the limit of the ohmic range to lower field strength with increasing temperature (marks in Fig. 4) contrary to the expected behaviour. Passing the λ-point of liquid helium ($T_\lambda = 2.17$ K) we found further a discontinuity of the critical field strength at which non-ohmic behaviour is found. These facts indicate some heating to take place in the sample at $T > T_\lambda$ Therefore, we will essentially restrict our discussion to the behaviour at $T \leqq T$.

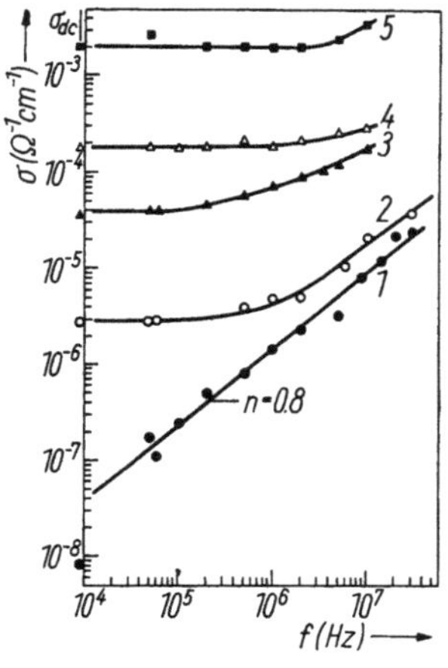

Fig. 3

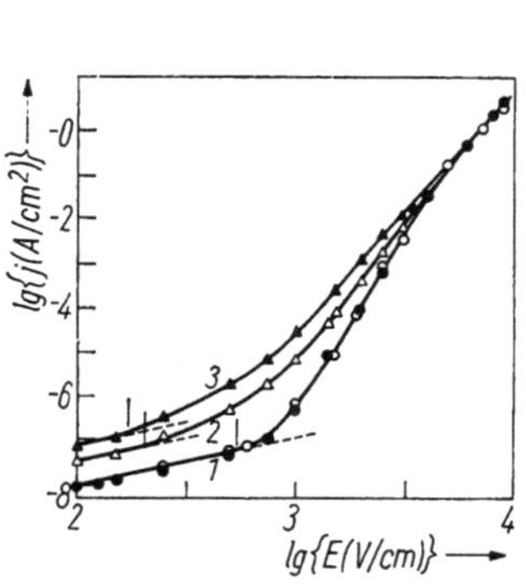

Fig. 4

Fig. 3. ac conductivity at 77 K after annealing at temperature T_a (°C): (1) 500, (2) 550, (3) 600, (4) 650, (5) 700

Fig. 4. Dependence of the current density on the electric field strength in the film after recrystallization at 700 °C for various temperatures T (K): (1) ● 2.1, ○ 2.3, (2) 3.5, (3) 4.2

where overheating is inhibited due to the better thermal conductivity of superfluid helium.

4. Discussion

We assume that the conductivity below room temperature is a hopping one. An important evidence for the hopping mechanism are the ac conductivity results at 77 K. They show a frequency dependence $\sigma(\omega) \sim \omega^n$ with n between 0.6 and 0.8 which is typical for hopping conductivity [5, 11]. The dc conductivity results in low and high electric fields must be also interpreted in this way. At high temperatures a temperature independent activation energy can be determined either by electron excitation from localized states into the conduction band (σ_1-conductivity) or by nearest-neighbour hopping (σ_3-conductivity). In favour of the latter process speaks, firstly, that in no case a Hall mobility is found and, secondly, that the conductivity intercept $\sigma(0)$ in the expression $\sigma = \sigma(0) \exp(-\varepsilon/kT)$ at $1/T = 0$ is in the range from 3×10^{-3} to $4 \times 10^{-2}\ \Omega^{-1}\ \text{cm}^{-1}$ which is more than four orders of magnitude smaller than the minimum metallic conductivity in extended states being expected to be about $800\ \Omega^{-1}\ \text{cm}^{-1}$ for fourfold co-ordination [5]. Thus we assume that the activation energy reflects the distance of the Fermi level from the percolation level in the impurity band. The experimentally found decrease of ε_3 and increase of $\sigma(0)$ with increasing annealing temperature T_a corresponds to a lowering of the compensation in h.d.c. crystals [6]. Therefore, our hypothesis is that annealing lowers the compensation.

For all samples the low-temperature conductivity shows decreasing activation energy at temperature lowering (sliding activation energy). This can be due to a hopping process over states always closer to the Fermi level (variable range hopping) as well as due to correlation effects in hopping conduction, i.e. multi-electron transitions [7].

The higher the density of states at the Fermi level is, the higher is the temperature at which variable range hopping can compete with nearest-neighbour hopping. Indeed, as can be seen in Fig. 2, the transition to sliding activation energy occurs at higher temperature for samples with smaller ε_3, i.e. with the Fermi level closer to the peak of the density of states of the impurity band. The generalized expression for the variable range hopping conductivity, derived in [8, 9], is

$$\sigma = \sigma(0) \exp\left[-\left(\frac{T_0}{T}\right)^b\right] \tag{1}$$

with $b = (1 + x)/(1 + d + x)$ and the density of states near the Fermi level $g(E) = N_0 |E|^x$ where d is the dimension of the random network in the percolation theory. For getting a high accuracy in the determination of b we used the analysis given in [10]. From the usual $\lg \sigma$ versus T^{-1} plot we get $\varepsilon(T)$ by means of graphical differentiation. In the case of a temperature dependence of the conductivity according to (1), $\lg(\varepsilon(T)/kT)$ should be a linear function of $\lg T$ with slope b.

A corresponding analysis yields that in the measured temperature interval a behaviour according to (1) with b between 0.68 and 1.0 is observed (Fig. 5). These values for b are clearly larger than for Mott's law ($b = \frac{1}{4}$) where the b-values decrease with increasing T_a (Table 1). In our opinion, this means that hopping does not take place very near the Fermi energy but via a level only somewhat lower than the percolation level.[2])

[2]) The other possibility of interpreting the low-temperature conductivity is correlated hopping [7]. Up to now, it cannot be taken into account, because we have no information on the impurity concentration and compensation.

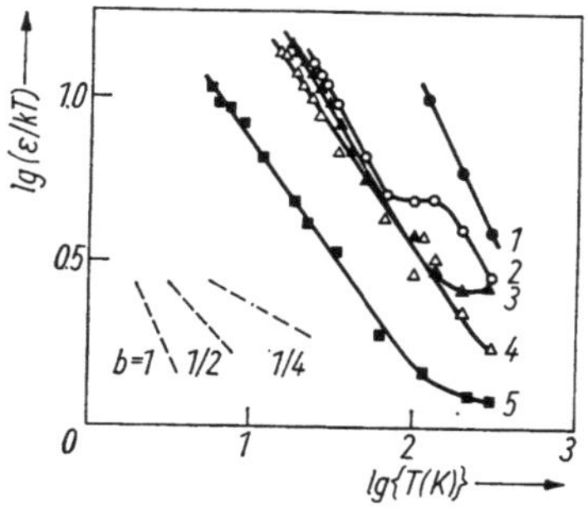

Fig. 5

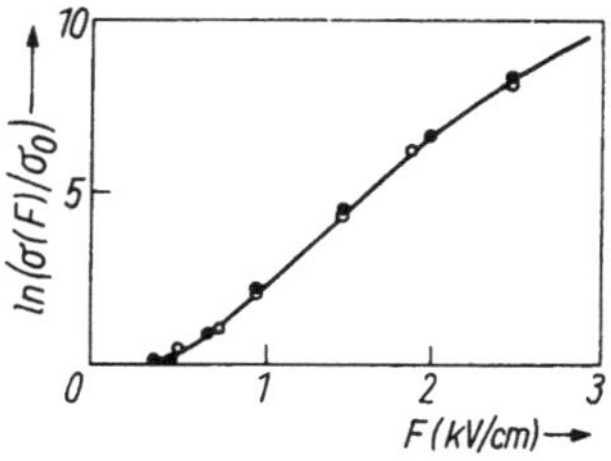

Fig. 6

Fig. 5. Plot of the expression lg (ε/kT) versus lg T from curves 1 to 5 in Fig. 2

Fig. 6. ln $(\sigma(F)/\sigma_0)$ versus F of curve 1 in Fig. 4 at medium electric field strength

The results of the current–voltage characteristics in the non-ohmic range at 2.1 K are also typical for hopping conductivity. At a "medium" field strength $F \leqq kT/ea \approx \approx 2 \times 10^3$ V/cm we obtained a dependence which is comparable with theoretical calculations of the field-enhanced thermal hopping given in [13 to 15],

$$\sigma(F) = \sigma_0 \exp\left(\frac{Fer}{kT}\right), \tag{2}$$

where r is the hopping distance (Fig. 6). This dependence is in accordance with results in other heavily doped compensated materials [12]. From the slope of the lg $(\sigma(F)/\sigma_0)$ versus F curve we estimated $r \approx 80$ Å. Taking into account that the Bohr radius a of a shallow hydrogen-like impurity in ZnSe is about 28 Å [16], we get $r/a \approx 3$ which is a magnitude typical for this parameter. On the other hand, from the hopping distance the calculated mean concentration of localized states $N \geqq r^{-3}$ is of the order of magnitude of 10^{18} cm^{-3}, which seems to be a reasonable concentration for intrinsic impurities in evaporated films of II–VI compounds [17].

At "high" field strength $F > 2 \times 10^3$ V/cm, up to breakdown we get for the current density

$$j(E) \sim \exp\left[-\left(\frac{F_0}{F}\right)^c\right]. \tag{3}$$

The exponent c is obtained from an analysis which is analogous to that given before for the determination of b in the ohmic range. By graphical differentiation of lg j versus F^{-1} we get $\Delta(F)$ and from the shape of a lg $(\Delta(F)/F)$ versus lg F plot we find c between 0.7 and 1 and F_0 between 4×10^4 and 1×10^5 V/cm (Fig. 7). The c-value coincides with that of b in (1) (see, for comparison, Table 1, No. 5). The coincidence of b and c is not accidental. Indeed in the case of variable range hopping in a three-dimensional random network for $g(\mu_F) = \text{const}$ it follows that $\sigma \sim \exp(-T^{-1/4})$ and that the current–voltage characteristic in the high-field limit is given by [18] $j \sim \sim \exp(-F^{-1/4})$. If the density of states near the Fermi level is $g(E) = N_0 |E|^x$, variable range hopping takes place at a concentration of localized states

$$N \sim 2 \int_0^{|E|} \tilde{E}^x \, d\tilde{E} = \frac{2}{x+1} E^{x+1}. \tag{4}$$

The mean hopping distance can be estimated by

$$r \approx N^{-1/3} \sim E^{-(x+1)/3}. \tag{5}$$

Hopping conductivity without thermal activation means that the drop of the potential energy of the electron over the hopping distance in the electric field direction exceeds

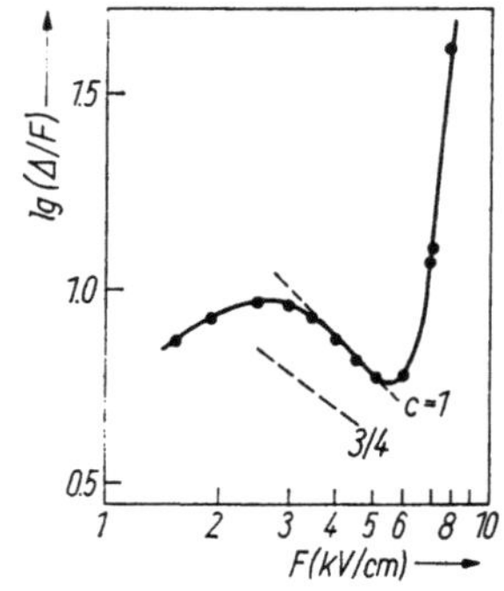

Fig. 7. Determination of c in (3) from the slope of lg (Δ/F) versus lg F

the width of the energy range in which hopping takes place [18], i.e.

$$E \approx eFr\,. \tag{6}$$

By inserting (6) into (5) the hopping distance can be rewritten as

$$r \sim F^{-(1+x)/(4+x)}\,. \tag{7}$$

Taking into account that $j \sim \exp(-r/a)$ we get finally

$$j \sim \exp\left[-\left(\frac{F_0}{F}\right)\right]^{(1+x)/(4+x)}. \tag{8}$$

From a comparison of (8) with (3) we find that

$$c = \frac{1+x}{4+x} \approx b\,.$$

The coincidence of b and c therefore underlines the connection of the charge transport in low and high electric fields and evidences once more the concept of hopping conduction in thin films of ZnSe.

References

[1] S. M. Ryvkin and I. S. Shlimak, phys. stat. sol. (a) **16**, 515 (1973).
[2] P. K. Lim and D. E. Brodie, Canad. J. Phys. **55**, 1512, 1641 (1977).
[3] N. V. Agrinskaya and E. D. Krymova, Fiz. Tekh. Poluprov. **6**, 1783 (1972).
N. V. Agrinskaya and O. A. Matveev, Fiz. Tekh. Poluprov. **11**, 116 (1977).
[4] D. Steinberg, Dissertation, Humboldt-Universität, Berlin 1974.
[5] N. F. Mott and E. A. Davis, Electronic Processes in Non-Crystalline Materials, Clarendon Press, Oxford 1971.
[6] H. Fritzsche and K. Lark-Horovitz, Phys. Rev. **113**, 999 (1959).
E. A. Davis and W. D. Compton, Phys. Rev. **140**, A2183 (1965).
[7] M. L. Knotek and M. Pollak, Phys. Rev. B **9**, 664 (1974).
[8] M. Pollak, J. non-crystall. Solids **11**, 1 (1972).
[9] E. M. Hamilton, Phil. Mag. **26**, 1043 (1972).
[10] A. G. Zabrodskii, Fiz. Tekh. Poluprov. **11**, 595 (1977).
[11] M. Pollak and T. H. Geballe, Phys. Rev. **133**, A1742 (1961).
M. Pollak, Phys. Rev. **138**, A1822 (1965).
[12] A. G. Zabrodskii, A. N. Ionov, and I. S. Shlimak, Fiz. Tekh. Poluprov. **8**, 503 (1974).
[13] D. Redfield, Adv. Phys. **24**, 463 (1975).
R. M. Hill, Phil. Mag. **24**, 1307 (1971).
[14] B. I. Shklovskii, Fiz. Tekh. Poluprov. **10**, 1440 (1976).
[15] M. Pollak and I. Riess, J. Phys. C **9**, 2339 (1976).
[16] G. E. Hite, D. T. F. Marple, M. Aven, and B. Segall, Phys. Rev. **156**, 850 (1967).
[17] R. Rentzsch and H. Berger, Thin Solid Films **37**, 235 (1976).
[18] B. I. Shklovskii, Fiz. Tekh. Poluprov. **6**, 2335 (1972).

(Received May 8, 1979)

phys. stat. sol. (a) 54, 493 (1979)

Subject classification: 14.3.3

Department of Electrical Engineering, University of Illinois, Urbana[1]) *(a)*
and Department of Electrical Engineering and Systems Science, Michigan State University, E. Lansing[2]) *(b)*

Wave Interpretation of *nl* Product in Gunn Instability

By

HANG H. LAI (a), BONG HO (b), and G. E. STILLMAN (a)

A novel approach deriving the threshold value for a Gunn device to exhibit instability is given. A two-stream model simulating the electrons in the lower and upper valleys travelling with different speeds is used in the analysis. The solution obtained from the transport equation indicates that whenever the $\sqrt{n}\, l$ product exceeds $3.41 \times 10^8\ \text{cm}^{-1/2}$, a domain in the device is formed; this value is compatible to the result of others [1]. A critical *nl* product of the order of $10^{11}\ \text{cm}^{-2}$ for the device as a whole to exhibit amplifier mode is also derived. The result of this analysis can be extended to describe other interactions such as the TRAPATT mode of an IMPATT diode.

Es wird ein neues Verfahren zur Ableitung der Schwellenwerte für das Auftreten einer Gunnbauelementinstabilität angegeben. Ein Zweiströmemodell, das die Elektronen in den unteren und oberen Tälern, die unterschiedliche Geschwindigkeiten besitzen, simuliert, wird für die Analyse benutzt. Die aus der Transportgleichung erhaltene Lösung zeigt, daß immer, wenn das $\sqrt{n}\, l$-Produkt den Wert $3{,}41 \times 10^8\ \text{cm}^{-1/2}$ übersteigt, eine Domäne im Bauelement gebildet wird; dieser Wert ist mit den Ergebnissen anderer Autoren kompatibel. Ein kritisches Produkt *nl* der Größenordnung $10^{11}\ \text{cm}^{-2}$, für das das Bauelement als ganzes Verstärkungsverhalten zeigt, wird ebenfalls abgeleitet. Das Ergebnis dieser Analyse läßt sich auch zur Beschreibung anderer Wechselwirkungen, wie die TRAPATT-Mode einer IMPATT-Diode, verwenden.

1. Introduction

It is generally recognized that microwave devices exhibit instability when their *nl* product exceeds a critical value of the order of $10^{12}/\text{cm}^2$ [2, 3]. Previously, the explanation of this critical value has been based on the negative resistance (or conductance) derived using Poisson's equation and the continuity equation for the charged carriers in the Gunn device and the boundary conditions of the device. In order to obtain the simple relation mentioned above, it was necessary to make several approximations. In the following, we derive a $\sqrt{nl}$ (or nl^2) relationship which described when a domain in a Gunn device is formed, and obtain a result similar to that of McCumber and Chynoweth [2]. In this analysis, a solution of wave interaction of the carriers in the solids is obtained from the transport equation combined with other basic equations. Two charged particle streams, with different drift velocities simulating the upper and lower valley electrons in Gunn devices, are closely coupled together. A value corresponding to the convective instability is obtained from the dispersion relationship for the propagation constant in a domain. A positive real part for the propagation constant is observed whenever the sample $\sqrt{nl}$ product exceeds a critical value. This positive real part propagation constant corresponds to a gain factor for the domain as it travels to the output end of the device (see Fig. 3).

[1]) Urbana, Illinois 61801, USA.

[2]) E. Lansing, Michigan 48824, USA.

2. Analysis

The equations of motion for the charged particles in the upper and lower valleys in a Gunn device with temperature and collision effects are [4]

$$\frac{d\boldsymbol{v}_a}{dt} = -\frac{e}{m_a^*}\boldsymbol{E} - \nu_a\boldsymbol{v}_a - \frac{v_{Ta}^2}{n_{0a}}\nabla n_a\,, \tag{1a}$$

$$\frac{d\boldsymbol{v}_b}{dt} = -\frac{e}{m_b^*}\boldsymbol{E} - \nu_b\boldsymbol{v}_b - \frac{v_{Tb}^2}{n_{0b}}\nabla n_b\,, \tag{1b}$$

where the subscript "a" or "b" indicates values in question are in valley a or b, respectively, and

$$\left.\begin{aligned}
\boldsymbol{v}_a &= \boldsymbol{v}_{0a} + \boldsymbol{v}_{1a} = \text{(dc + ac) components}\,,\\
m_a^* &= \text{carrier effective mass in valley a,}\\
\boldsymbol{E} &= \text{electric field}\,,\\
\nu_a &= \text{collision frequency in valley a,}\\
n_a(\boldsymbol{r}) &= n_{0a}(\boldsymbol{r}) + n_{1a}(\boldsymbol{r}) = \text{(dc + ac) components of}\\
&\quad\ \text{carrier concentration}\,,\\
v_{Ta} &= \text{carrier thermal velocity in valley a}\,,\\
\omega_a^2 &= \frac{\bar{n}_{0a}e^2}{\varepsilon m_a^*}\,, \qquad k_{Ta} = \frac{v_{Ta}}{v_{0a}}\,, \qquad \Omega_n = \frac{\omega}{\omega_a}\,,
\end{aligned}\right\} \tag{2}$$

wehre $\bar{n}_{0a}$ is the average value of $n_{0a}(\boldsymbol{r})$ over the entire active lengh.
A similar set of parameters for b also applies when the subscript a is replaced by b. Without loss of generality, let us assume that valley a resembles the lower valley and valley b stands for the upper valley in a Gunn device. Then Poisson's equation and other basic equations describing the system are

$$\nabla\cdot\boldsymbol{E} = -\frac{e}{\varepsilon}(n_a + n_b)\,, \tag{3}$$

$$\nabla\cdot\boldsymbol{J}_a - j\omega e n_a = 0\,, \tag{4}$$

$$\boldsymbol{J}_a = -e(n_{0a} + n_{1a})(\boldsymbol{v}_{0a} + \boldsymbol{v}_{1a}) \approx -e(n_{0a}\boldsymbol{v}_{1a} + n_{1a}\boldsymbol{v}_{0a})\,, \tag{5}$$

where a first-order perturbation approximation was made in (5) and a time harmonic factor $e^{j\omega t}$ was implicitly included in all the ac components.

Combining equations (1) through (5) and considering a one-dimensional case only, it can be shown that

$$(1 - k_{Ta}^2)\frac{\partial^2 n_a}{\partial z^2} + \left(\frac{\nu_a}{v_{0a}} + 2j\frac{\omega}{v_{0a}}\right)\frac{\partial n_a}{\partial z} + \left(\frac{\omega_a^2}{v_{0a}^2} - \frac{\omega^2}{v_{0a}^2} + j\frac{\omega\nu_a}{v_{0a}^2}\right)n_a - \frac{\omega_a^2}{v_{0a}^2}n_b = 0 \tag{6}$$

and

$$(1 - k_{Tb}^2)\frac{\partial^2 n_b}{\partial z^2} + \left(\frac{\nu_b}{v_{0b}} + 2j\frac{\omega}{v_{0b}}\right)\frac{\partial n_b}{\partial z} + \left(\frac{\omega_b^2}{v_{0b}^2} - \frac{\omega^2}{v_{0b}^2} + j\frac{\omega\nu_b}{v_{0b}^2}\right)n_b - \frac{\omega_b^2}{v_{0b}^2}n_a = 0\,. \tag{7}$$

Equations (6) and (7) describe the electron concentration in the upper and lower valleys as a function of position in a Gunn device, and thus relate to the carrier bunching, or domain formation in the space charge region in the device. On the other hand, since (6) and (7) represent electrons travelling (due to dc applied voltage) with different effective masses, they represent equations of motion with two streams moving relative to each other. Due to the interaction of these streams, energy is transferred from one stream to another, as well as from dc to ac values, etc. [4].

Our object is to solve (6) and (7) simultaneously and to determine the carrier concentration at the output port of the device for a given value at the input. To do that, let us assume the solutions are of the form $N_i e^{\Gamma z}$ (i = a or b), where N_i are constants. Substituting this into (6) and (7), we have

$$(1 - k_{\mathrm{Ta}}^2)\,\Gamma^2 N_{\mathrm{a}} + \left(\frac{v_{\mathrm{a}}}{v_{0\mathrm{a}}} + 2j\,\frac{\omega}{v_{0\mathrm{a}}}\right)\Gamma N_{\mathrm{a}} + \left(\frac{\omega_{\mathrm{a}}^2}{v_{0\mathrm{a}}^2} - \frac{\omega^2}{v_{0\mathrm{a}}^2} + j\,\frac{\omega v_{\mathrm{a}}}{v_{0\mathrm{a}}^2}\right)N_{\mathrm{a}} - {} - \frac{\omega_{\mathrm{a}}^2}{v_{0\mathrm{a}}^2}\,N_{\mathrm{b}} = 0\,, \tag{8}$$

$$(1 - k_{\mathrm{Tb}}^2)\,\Gamma^2 N_{\mathrm{b}} + \left(\frac{v_{\mathrm{b}}}{v_{0\mathrm{b}}} + 2j\,\frac{\omega}{v_{0\mathrm{b}}}\right)\Gamma N_{\mathrm{b}} + \left(\frac{\omega_{\mathrm{b}}^2}{v_{0\mathrm{b}}^2} - \frac{\omega^2}{v_{0\mathrm{b}}^2} + j\,\frac{\omega v_{\mathrm{b}}}{v_{0\mathrm{b}}^2}\right)N_{\mathrm{b}} - {} - \frac{\omega_{\mathrm{b}}^2}{v_{0\mathrm{b}}^2}\,N_{\mathrm{a}} = 0\,. \tag{9}$$

For a nontrivial solution for N_{a} and N_{b}, the following secular equation is obtained:

$$\begin{aligned}
&(1 - k_{\mathrm{Ta}}^2)(1 - k_{\mathrm{Tb}}^2)\,\Gamma'^4 + \left[(1 - k_{\mathrm{Ta}}^2)\left(\frac{v_{\mathrm{b}}v_{0\mathrm{a}}}{v_{0\mathrm{b}}\omega_{\mathrm{a}}} + 2j\,\frac{\omega v_{0\mathrm{a}}}{v_{0\mathrm{b}}\omega_{\mathrm{a}}}\right) + (1 - k_{\mathrm{Tb}}^2)\,\times\right.\\
&\left.\times\left(\frac{v_{\mathrm{a}}}{\omega_{\mathrm{a}}} + 2j\,\frac{\omega}{\omega_{\mathrm{a}}}\right)\right]\Gamma'^3 + \left[(1 - k_{\mathrm{Ta}}^2)\left(\frac{\omega_{\mathrm{b}}^2 v_{0\mathrm{a}}^2}{v_{0\mathrm{b}}^2\omega_{\mathrm{a}}^2} - \frac{\omega^2 v_{0\mathrm{a}}^2}{v_{0\mathrm{b}}^2\omega_{\mathrm{a}}^2} + j\,\frac{\omega v_{\mathrm{b}} v_{0\mathrm{a}}^2}{v_{0\mathrm{b}}^2\omega_{2}^{\mathrm{a}}}\right) + \right.\\
&\left.+ (1 - k_{\mathrm{Tb}}^2)\left(1 - \frac{\omega^2}{\omega_{\mathrm{a}}^2} + j\,\frac{\omega v_{\mathrm{a}}}{\omega_{\mathrm{a}}^2}\right) + \left(\frac{v_{\mathrm{a}}}{\omega_{\mathrm{a}}} + 2j\,\frac{\omega}{\omega_{\mathrm{a}}}\right)\left(\frac{v_{\mathrm{b}}v_{0\mathrm{a}}}{v_{0\mathrm{b}}\omega_{\mathrm{a}}} + 2j\,\frac{\omega v_{0\mathrm{a}}}{v_{0\mathrm{b}}\omega_{\mathrm{a}}}\right)\right]\times\\
&\times \Gamma'^2 + \left[\left(1 - \frac{\omega^2}{\omega_{\mathrm{a}}^2} + j\,\frac{\omega v_{\mathrm{a}}}{\omega_{\mathrm{a}}^2}\right)\left(\frac{v_{\mathrm{b}}v_{0\mathrm{a}}}{v_{0\mathrm{b}}\omega_{\mathrm{a}}} + 2j\,\frac{\omega v_{0\mathrm{a}}}{v_{0\mathrm{b}}\omega_{\mathrm{a}}}\right) + \left(\frac{v_{\mathrm{a}}}{\omega_{\mathrm{a}}} + 2j\,\frac{\omega}{\omega_{\mathrm{a}}}\right)\times\right.\\
&\left.\times\left(\frac{\omega_{\mathrm{b}}^2 v_{0\mathrm{a}}^2}{v_{0\mathrm{b}}^2\omega_{\mathrm{a}}^2} - \frac{\omega^2 v_{0\mathrm{a}}^2}{v_{0\mathrm{b}}^2\omega_{\mathrm{a}}^2} + j\,\frac{\omega_{\mathrm{b}}^2 v_{0\mathrm{a}}^2}{v_{0\mathrm{b}}^2\omega_{\mathrm{a}}^2}\right)\right]\Gamma' + \left[\left(1 - \frac{\omega^2}{\omega_{\mathrm{a}}^2} + j\,\frac{\omega v_{\mathrm{a}}}{\omega_{\mathrm{a}}^2}\right)\times\right.\\
&\left.\times\left(\frac{\omega_{\mathrm{b}}^2 v_{0\mathrm{a}}^2}{v_{0\mathrm{b}}^2\omega_{\mathrm{a}}^2} - \frac{\omega^2 v_{0\mathrm{a}}^2}{v_{0\mathrm{b}}^2\omega_{\mathrm{a}}^2} + j\,\frac{\omega v_{\mathrm{b}} v_{0\mathrm{a}}^2}{v_{0\mathrm{b}}^2\omega_{\mathrm{a}}^2}\right) - \frac{\omega_{\mathrm{b}}^2 v_{0\mathrm{a}}^2}{\omega_{\mathrm{a}}^2 v_{0\mathrm{b}}^2}\right] = 0\,,
\end{aligned} \tag{10}$$

where

$$\Gamma' = \frac{\Gamma v_{0\mathrm{a}}}{\omega_{\mathrm{a}}}, \qquad \Gamma = \Gamma_{\mathrm{R}} + j\Gamma_{\mathrm{I}}\,. \tag{11}$$

The solution to (10) will give us the dispersion relationship between the propagation constant Γ (or Γ') and the operating frequency ω. Since equation (10) is an algebraic equation of 4th order, one can analytically find the solution without making any approximations [5].

Here, instead of using the relation of negative resistance, we shall look for the real parts of the roots of (10) that are positive. These positive real roots correspond to the number of charged particles in a domain as the domain travels toward the output before finally being collected.

Let us consider a simple case in which the temperature and collision effects are neglected for both streams. A computer solution for this particular case is shown in Fig. 1 and 2. In Fig. 1, the real part of the normalized propagation constant Γ' is plotted against the operating frequency with appropriate normalization for five cases, with each case a different relative drift velocity. Parameters used for these

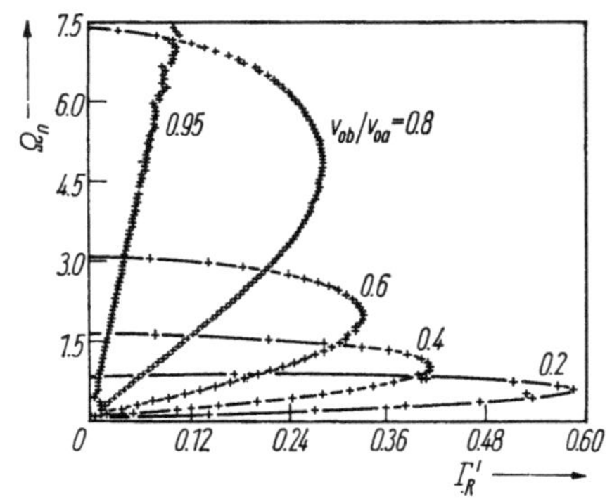

Fig. 1. Real parts of the propagation constant (Γ_R') are plotted against the normalized frequency (Ω_n) for five different relative velocities between carriers in the lower and upper valleys of a Gunn diode

plots are intended to simulate the data for GaAs Gunn devices [6], i.e., the effective mass for the electrons in the lower valley was $0.065m_0$, while that for the upper valley was $0.85m_0$. The drift velocity (v_{0a}) for a typical applied voltage was chosen to be 10^7 cm/s, the relative dielectric constant for this device was chosen to be 13. Fig. 2 is a similar plot for the same set of data except that the imaginary part of the propagation constant Γ' is plotted against the normalized operating frequency Ω_n. It is observed from these figures that the real part of Γ' increases as ω increases for each relative velocity and reaches a peak value at a certain value of Ω_n and then decreases to zero (i.e. no gain). This thus sets an upper limit for the operating frequency. The fact that the peak values of Γ_R' in Fig. 1 are larger for a smaller value of (v_{0b}/v_{0a}) can be explained by the observation that the energy transfer is larger when the difference in energy (or in velocity) between streams is larger. However, this high gain advantage is offset by a smaller range of operating frequency ω, within which Γ_R' exists. Although the peak value of Γ_R' is small for values (v_{0b}/v_{0a}) approaching unity, the range of the operating frequency for these cases is rather wide.

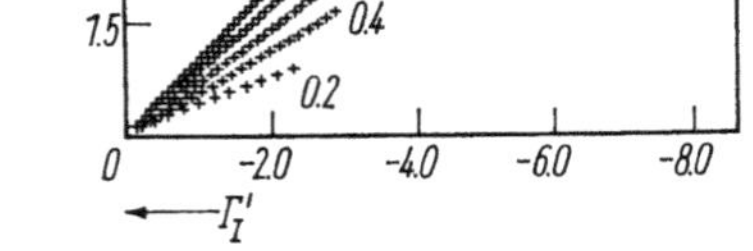

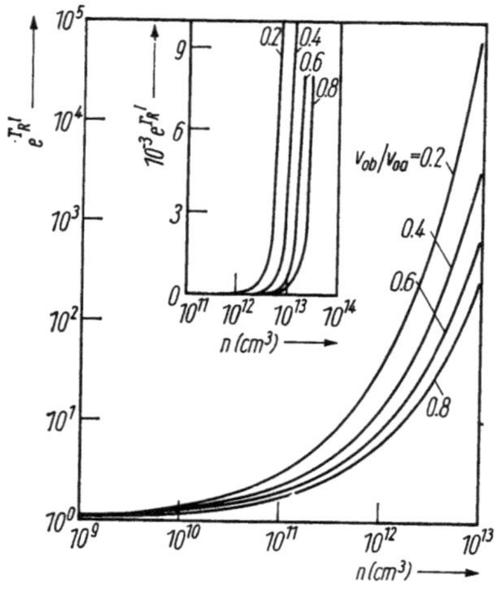

Fig. 2 Fig. 3

Fig. 2. Imaginary parts of the propagation constant (Γ_I') corresponding to the values shown in Fig. 1

Fig. 3. Growth factor $e^{\Gamma_R' l}$ is plotted against the carrier concentration n. All parameters used in this figure are the same as in Fig. 1, and Γ_R' was chosen to be the largest value for each curve in Fig. 1. The inset in this figure is a semi-lg plot as opposed to the lg–lg plot, and $l = 10^{-3}$ cm

From Fig. 2, it is noted that the group velocity (the slope) of the curve is nearly constant for each curve and that it is larger for a larger value of (v_{0b}/v_{0a}).

It is of interest to consider the growth factor $e^{\Gamma_R l}$ for a Gunn device of typical length, say 10 μm, as a function of carrier concentration. A result is shown in Fig. 3. In this figure, the growth factor is plotted against the carrier concentration with (v_{0b}/v_{0a}) as parameter. The values of Γ_R were chosen to be the corresponding maximum values for each curve in Fig. 1, and all other parameters were chosen to be the same as in Fig. 1 and 2. It is seen from this figure that the growth factor rises rather sharply when the carrier concentration reaches about $10^{13}/\text{cm}^3$, in this case, the nl product is $10^{13} \times 10^{-3} = 10^{10}/\text{cm}^2$.

Let us consider now the condition under which a domain is formed. Referring to Fig. 1, for a reasonable value of (v_{0b}/v_{0a}), say 0.6, the normalized cut-off frequency Ω_n is about 3 or lower, i.e.

$$\frac{\omega}{\omega_a} \leqq 3\,, \tag{12}$$

where

$$\omega = \frac{v_{0a} 2\pi}{l} \tag{13}$$

and l is the active length in the device.

With all the parameters used in Fig. 1 substituted into (12), we have

$$\sqrt{n}\; l \geqq 3.411 \times 10^3/\text{cm}^{1/2}\,. \tag{14}$$

Equation (14) gives a threshold value for the space charge in the device to form a domain. This result compares favorably with that of others [6]. Although the value given in (14) will insure the creation of a domain in the device, it does not guarantee that the domain will be sustained, grow, and propagate toward the output port. Thus it is interesting to consider the condition that the device as a whole exhibits an amplifier mode. This will involve the boundary conditions imposed on the device. Using the solution obtained from (10) and assuming that the cross-sectional area of the device is A, it can be shown [4] that the input admittance for the device is

$$Y = \frac{\varepsilon A \omega \left[1 - \dfrac{\omega_a^2}{\omega(\omega + j\nu_a)} - \dfrac{\omega_b^2}{\omega(\omega + j\nu_b)}\right]}{1 + \Gamma l - e^{\Gamma l}}\,, \tag{15}$$

where we have assumed that there is only one root from (10) with a positive real part.

Again, let us make the assumption that $\omega_a \gg \omega_b$ and the collision and temperature effects are small. It can be shown that the real part of the admittance is

$$Y_R = \frac{\varepsilon A \omega\, (1 - \omega_a^2/\omega^2)\, (\Gamma_I - \Gamma_I\, e^{\Gamma_R l} \cos \Gamma_I l + \Gamma_R\, e^{\Gamma_R l} \sin \Gamma_I l)}{(1 + \Gamma_R l - e^{\Gamma_R l} \cos \Gamma_I l)^2 + (\Gamma_I l - e^{\Gamma_R l} \sin \Gamma_I l)^2}\,. \tag{16}$$

In order to have amplification, the input admittance of the device must exhibit a negative value, i.e., equation (16) must have a negative value. As can be seen from (16), for a fixed value of l, equation (16) oscillates between positive and negative values for different values of carrier concentration.

Using the parameters considered in the previous example, we shall find the carrier concentrations that satisfy (10), and also that the first value of n exceeds the condition imposed by (12) such that (16) is negative. The results are shown in Table 1. In this table, first-minimum carrier concentrations that exceed the value dictated by (10) for three fixed values of l are tabulated, and the corresponding nl products are listed It is seen from the table that our result agrees very well with that of others. The n

Table 1

Carrier concentrations derived from equation (16) such that the input admittance is negative for three different fixed-length devices

l (10^{-4} cm)	1	10	100
n (cm^{-3})	2.0×10^{15}	7.0×10^{13}	7.0×10^{11}
nl (cm^{-2})	2.0×10^{11}	7.0×10^{10}	0.7×10^{10}

product values are slightly lower than those of others. This deviation may come from the fact that the data we used were idealized, i.e. no collision and no temperature effects, or it may be due to the fact that the difference in drift velocities between valleys was too large; consequently the cut-off values in Ω_n are small in this case. If the temperature and collision effects were included, the cut-off in values would have been larger in certain cases.

3. Conclusion

A simple approach deriving the threshold value for a $\sqrt{n}\, l$ product for which a domain is formed in a Gunn device and an nl condition on the device to exhibit amplification mode was derived in this paper (equation (14)). This approach employed the transport equation along with other basic equations to set up the dispersion relationship for the propagation constant of the domain in a Gunn device. A two-stream scheme used in this paper simulated the carriers from two energy levels travelling related to each other. The result of an idealized example (no collision or temperature effects) showed that both the $\sqrt{n}\, l$ and nl critical values for the domain to form and grow agreed well with that of others [1, 2].

Although the results we discussed here were derived from an idealized case, the approach applies to any general case. With the thermal and collision effects included in the analysis, the early computer solution shows that the thermal and collision effects reduce the gain value and in certain cases these effects may induce instability even beyond the normal cut-off operating frequency, which agrees with other discoveries [4]. The result of this work indicates that wave interaction analysis can be utilized to describe other solid-state phenomena such as the TRAPATT mode of the IMPATT diode, which basically is the wave interaction between drifting charge particle stream and a moving plasma. It is believed that this wave interaction technique described here gives more information for the energy exchange mechanism and the electron dynamic behavior inside the device.

References

[1] P. J. BULMAN, G. S. HOBSON, and B. C. TAYLOR, Transferred Electron Devices, Academic Press, Inc., 1972.

[2] D. E. MCCUMBER and A. G. CHYNOWETH, IEEE Trans. Electron. Devices **13**, 4 (1966).

[3] I. B. BOTT and W. FAWCETT, The Gunn Effect in Gallium Arsenide, in: Advances in Microwaves, Vol. 3, Ed. L. YOUNG, Academic Press, Inc., 1968.

[4] M. C. STEELE and B. VURAL, Wave Interaction in Solid State Plasmas, McGraw-Hill Publ. Co., 1969.

[5] H. H. LAI and J. D. DYSON, The Determination of the Direction of Arrival of an Interference Field, RRL Rep., Dept. of El. Engng., Univ. Illinois, Urbana, Dec. 1977.

[6] G. E. STILLMAN, C. M. WOLFE, and J. O. DIMMOCK, Far-Infrared Photoconductivity in High Purity GaAs, in: Semiconductors and Semimetals, Vol. 12, Ed. R. K. WILLARDSON and A. C. BEER, Academic Press, Inc., 1977.

(Received February 26, 1979)

phys. stat. sol. (a) **54**, 499 (1979)

Subject classification: 1.3 and 1.4; 21.6

Rijksuniversitair Centrum Antwerpen[1]) *(a) and Wihuri Physical Laboratory, University of Turku*[2]) *(b)*

High Resolution Electron Microscopic and Electron Diffraction Study of the Au–Mg System

I. The Monoclinic $Au_{15}Mg_4$ Phase[3])

By

M. VAN SANDE (a), G. VAN TENDELOO (a), S. AMELINCKX[4]) (a), and P. AIRO (b)

During a systematic study of the gold-rich side of the Au–Mg system a new phase with approximate composition $Au_{15}Mg_4$ is discovered using a combination of electron diffraction and high resolution electron microscopy. The new monoclinic structure can be described as a two-dimensional long period superstructure of the DO_{22} phase, the antiphase boundaries forming an angle of 115.35°. It is stable in the temperature range between 400 and 280 °C. At lower temperatures it evolves into a two-dimensional long period antiphase boundary structure presumably of the $Au_{3+}Zn$-type.

Während einer systematischen Untersuchung der goldreichen Seite des Au–Mg-Systems wird eine neue Phase mit der näherungsweisen Zusammensetzung $Au_{15}Mg_4$ mittels einer Kombination von Elektronenbeugung und hochauflösender Elektronenmikroskopie gefunden. Die neue monokline Struktur läßt sich als zweidimensionale langperiodische Superstruktur der DO_{22}-Phase beschreiben, wobei die Antiphasengrenzen einen Winkel von 115,35° bilden. Sie ist im Temperaturbereich zwischen 400 und 280 °C stabil. Bei niedrigeren Temperaturen entwickelt sie sich zu einer zweidimensionalen langperiodischen Antiphasengrenzenstruktur vom $Au_{3+}Zn$-Typ.

1. Introduction

Gold–magnesium alloys have been studied for years already, but most of the attention has been devoted to alloys with composition around Au_3Mg [1 to 3]. For these alloys different polytypes (such as 7R, 4H, and 10H) of the $M = 2$ structure (DO_{23} structure) are found using X-ray or electron diffraction [1, 3]. Moreover around 23 at% Mg a complex hexagonal phase has been identified [2]. A two-dimensional long period antiphase boundary (LPAPB) structure has first been suggested [3] and recently definitely identified [4], based only on electron diffraction observations.

In a series of recent papers devoted to the Au–Mn system [5 to 9] we have demonstrated how the combination of high resolution electron microscopy and diffraction constitutes a powerful tool to solve difficult structural problems related with interface modulated structures. Difficulties in the interpretation of X-ray data often result from the presence of a large number of orientation and translation variants in alloys having structures of low symmetry. In electron diffraction it is usually possible to obtain single domain diffraction patterns; this simplifies the problem considerably. Using the technique introduced in [10, 11], it is then feasible in many cases to derive the superstructure mainly from geometrical features of the diffraction pattern, provided the basic structure is known.

1) Groenenborgerlaan 171, B-2020 Antwerpen, België.

2) Vesilinnante 5, SF-2500 Turku 50, Finland.

3) Work performed under the auspices of the association R.U.C.A.–S.C.K., Mol, with financial support of the I.I.K.W.

4) Also at S.C.K./C.E.N., Boeretang, B-2400 MOL, België.

In this paper we shall apply the same methods to structural problems in the Au–Mg system, which is notorious for exhibiting intricate diffraction patterns due to a large number of variants. In particular we investigate in detail the structure of an alloy with a nominal composition in the vicinity of Au_4Mg. As it turns out, this alloy offers a good example of the complications that can result from the presence of structural modulations occurring at different structural levels.

Anticipating the results we shall show that the alloy forms a one-dimensional very long period superstructure of a two-dimensional periodic APB structure of the DO_{22} structure, which in turn is a superstructure based on the f.c.c. structure. We shall therefore in our discussion distinguish between the primary long period superstructure (two-dimensional), and the secondary long period superstructure (one-dimensional), the DO_{22} structure being considered as the basic structure.

2. Experimental

2.1 Preparation of material

The alloys with nominal composition Au–18 at% Mg where prepared at Wihuri Physical Laboratory in Turku, Finland, by melting together the constituent elements in an induction furnace in alumina crucibles enclosed in a stainless steel chamber, under a continuous flow of argon. The composition was checked by X-ray analysis. The ingots were cold-rolled to ± 200 μm thick foils and afterwards annealed at 700 °C to remove the induced dislocations.

2.2 Specimen preparation

Specimens for electron microscopic observations were prepared by punching 3 mm discs out of the foils. They were electro-polished to transparency using a double jet technique with different electrolytes. The first method is described in [12] and the second one in [13].

2.3 Thermal treatment

Material used in the present investigation has been first heat-treated at 700 °C and subsequently annealed at 320 °C for several days in order to grow larger domains of the new phase.

It has been found that this phase is stable in the temperature range between approximately 280 and 400 °C. According to Terasaki and Watanabe [14] who independently discovered this unidentified phase and called it the "X-phase", it occurs in the range between 18 and 22 at% Mg [14].

Quenching from higher temperatures reveals small domains of a one- and two-dimensional long period superstructure, while annealing at lower temperatures reveals a $Au_{3+}Zn$-type two-dimensional superstructure [4]. The relation between these different phases and the evolution from one into the other will be treated in a subsequent paper.

3. Observations

3.1 The diffraction pattern

3.1.1 Single domain diffraction pattern

The diffraction patterns are usually complicated by the presence of several orientation variants. However, in well annealed specimens the domain size is sufficiently large so as to allow the selection of a single domain and to obtain the pattern reproduced in Fig. 1.

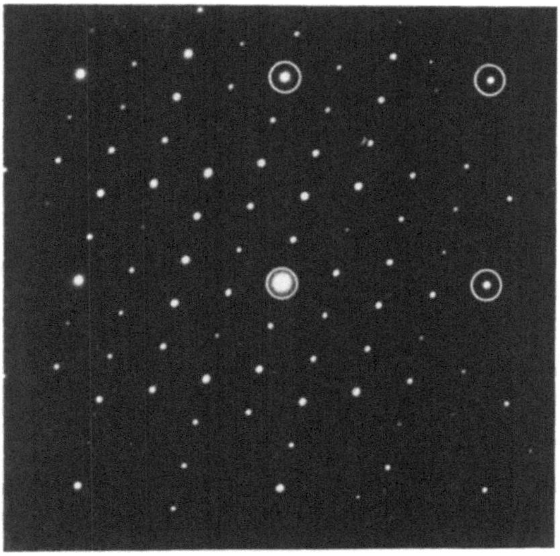

Fig. 1. Single domain [010]* diffraction pattern of Au–18 at% Mg annealed for several days at 320 °C

The most intense spots are clearly those typical of a cube section of the reciprocal lattice of an f.c.c. crystal. Tilting experiments confirm that the structure is based on the f.c.c. structure.

In this cube section the superlattice spots form a lattice with a parallelogram-shaped unit mesh with base vectors $\boldsymbol{B}_1'$ and $\boldsymbol{B}_3'$. In the direction perpendicular to this section there is no evidence for superlattice spots suggesting that the unit cell has a large mesh size parallel to the considered cube section, but it is only one f.c.c. unit cell thick in the direction perpendicular to this section, i.e. $\boldsymbol{B}_2 = \boldsymbol{b}_2$.

It is interesting to note that the most intense superlattice spots are those closest to the positions marked by crosses in Fig. 2. These crosses, together with the basic f.c.c. spots, form a pattern which is characteristic of the DO_{22} structure.

We can therefore express $\boldsymbol{B}_1$, $\boldsymbol{B}_2$, and $\boldsymbol{B}_3'$ in terms of the reciprocal base vectors $\boldsymbol{b}_1$, $\boldsymbol{b}_2$, and $\boldsymbol{b}_3$ of the DO_{22} structure, with its tetragonal axis parallel with a cube direction as indicated in Fig. 2. The positions marked by crosses can then be indexed as 002, 202, 101, and 103 with respect to the DO_{22} reciprocal lattice.

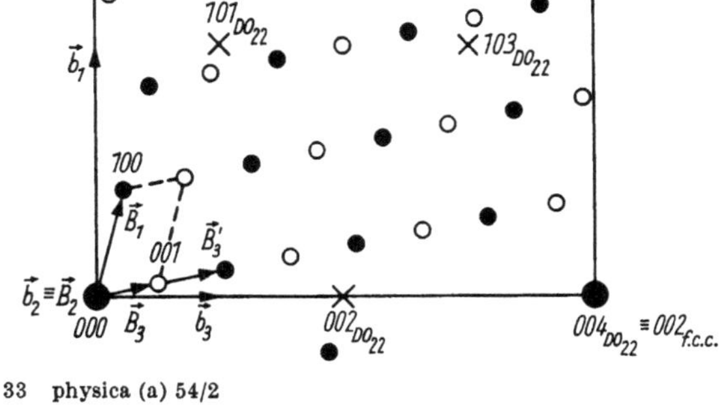

Fig. 2. Schematic single domain diffraction pattern: DO_{22} reflections are indicated by crosses, extinctions by open circles. The reciprocal unit cell is outlined

The most densely packed rows of visible superlattice spots are along the $[201]_{DO_{22}}$ direction; their separation corresponds to an interplanar spacing of 10.2 Å. Tilting experiments reveal that there are systematic extinctions: all rows with l = odd, marked by open dots, are in fact absent. The true reciprocal base vectors for the monoclinic cell outlined in Fig. 2 are therefore $\boldsymbol{B}_3 = \frac{1}{2}\boldsymbol{B}_3'$ and $\boldsymbol{B}_1$.

The relation between $\boldsymbol{B}_1$, $\boldsymbol{B}_2$, and $\boldsymbol{B}_3$ and $\boldsymbol{b}_1$, $\boldsymbol{b}_2$, and $\boldsymbol{b}_3$ is thus

$$\begin{pmatrix} \boldsymbol{B}_1 \\ \boldsymbol{B}_2 \\ \boldsymbol{B}_3 \end{pmatrix}_{\mathrm{m}} = \begin{pmatrix} \frac{8}{19} & 0 & \frac{4}{19} \\ 0 & 1 & 0 \\ \frac{1}{19} & 0 & \frac{10}{19} \end{pmatrix} \begin{pmatrix} \boldsymbol{b}_1 \\ \boldsymbol{b}_2 \\ \boldsymbol{b}_3 \end{pmatrix}_{\mathrm{DO}_{22}} \tag{1}$$

Related to the base vectors $\boldsymbol{b}_1'$, $\boldsymbol{b}_2'$, and $\boldsymbol{b}_3'$ of the f.c.c. the base vectors are

$$\begin{pmatrix} \boldsymbol{B}_1 \\ \boldsymbol{B}_2 \\ \boldsymbol{B}_3 \end{pmatrix}_{\mathrm{m}} = \begin{pmatrix} \frac{8}{19} & 0 & \frac{2}{19} \\ 0 & 1 & 0 \\ \frac{1}{19} & 0 & \frac{5}{19} \end{pmatrix} \begin{pmatrix} \boldsymbol{b}_1' \\ \boldsymbol{b}_2' \\ \boldsymbol{b}_3' \end{pmatrix}_{\mathrm{f.c.c.}} \tag{2}$$

The base vectors in direct space as derived from $\boldsymbol{B}_1$, $\boldsymbol{B}_2$, and $\boldsymbol{B}_3$ are then

$$\begin{pmatrix} \boldsymbol{A}_1 \\ \boldsymbol{A}_2 \\ \boldsymbol{A}_3 \end{pmatrix}_{\mathrm{m}} = \begin{pmatrix} \frac{5}{2} & 0 & -\frac{1}{4} \\ 0 & 1 & 0 \\ -1 & 0 & 2 \end{pmatrix} \begin{pmatrix} \boldsymbol{a}_1 \\ \boldsymbol{a}_2 \\ \boldsymbol{a}_3 \end{pmatrix}_{\mathrm{DO}_{22}} \tag{3}$$

and

$$\begin{pmatrix} \boldsymbol{A}_1 \\ \boldsymbol{A}_2 \\ \boldsymbol{A}_3 \end{pmatrix}_{\mathrm{m}} = \begin{pmatrix} \frac{5}{2} & 0 & -\frac{1}{2} \\ 0 & 1 & 0 \\ -1 & 0 & 4 \end{pmatrix} \begin{pmatrix} \boldsymbol{a}_1' \\ \boldsymbol{a}_2' \\ \boldsymbol{a}_3' \end{pmatrix}_{\mathrm{f.c.c.}} \tag{4}$$

where $\boldsymbol{a}_1$, $\boldsymbol{a}_2$, and $\boldsymbol{a}_3$ and $\boldsymbol{a}_1'$, $\boldsymbol{a}_2'$, and $\boldsymbol{a}_3'$ are the base vectors of the DO_{22} lattice and of the f.c.c. lattice, respectively.

Assuming for a moment that we had been unaware of the systematic extinctions along lines l = odd, we would find the unit cell to be only half this size, i.e., the base vector along the $\boldsymbol{A}_3$-direction would then be $\frac{1}{2}\boldsymbol{A}_3$. However, this is not a possible symmetry translation in an f.c.c. lattice.

Based on these considerations only, we find that the base vector must be doubled and hence become $\boldsymbol{A}_3$, introducing automatically the extinctions along l = odd. These extinctions are also compatible with the presence of a glide mirror plane parallel with (010) and a glide vector $\frac{1}{2}\boldsymbol{A}_3$.

3.1.2 Multidomain diffraction patterns

The complicated multidomain patterns can all be interpreted as superpositions of single domain patterns related by symmetry operations. The multidomain diffraction pattern of Fig. 3a can, for instance, be decomposed in four single domain diffraction patterns of the type shown in Fig. 1, in the manner represented schematically in Fig. 3b. Anticipating the discussion of the domain structure, we shall find that the

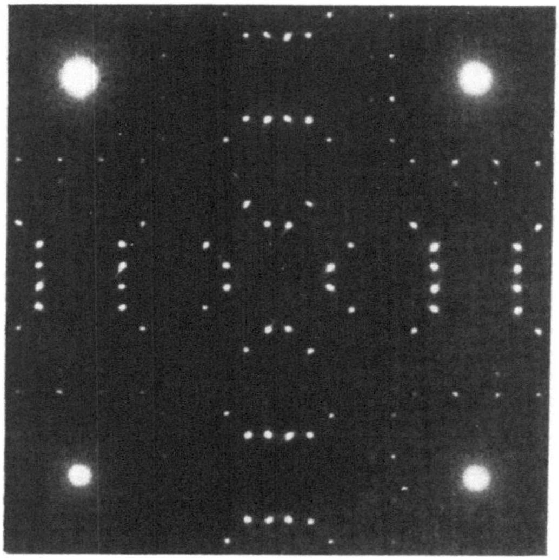

a

Fig. 3. a) [010]* diffraction pattern of four coaxial variants of Au–Mg. b) Schematic representation of the diffraction pattern in Fig. 3a. Reflections belonging to different variants are marked differently. The symmetry elements relating different variants are indicated: △ E, ■ C_4^1, ▲ m_1, □ m_2

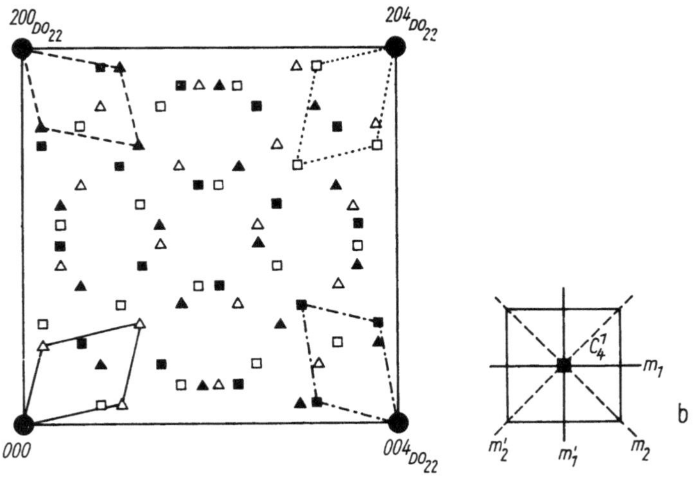

pattern of Fig. 3a, b is due to four variants with common twofold axis parallel with the electron beam. In any given area there is a tendency to form co-axial variants only, i.e., the short axis (A_2-axis) is preferentially parallel in neighbouring domains.

3.2 Direct resolution

3.2.1 The primary structure

It was shown previously that relevant structural images can be obtained in the case of ordered alloys by collecting in the objective aperture only superlattice spots, rejecting the spots due to the basic structure [5 to 9]. The same technique was used here. The images are rather more focus- and thickness-sensitive than in the previously studied alloys Au_5Mn_2 and Au_4Mn. We could nevertheless obtain images which, as we will see, reveal the structure as determined from the diffraction pattern; such images are reproduced in Fig. 4 and 5. The periodicity of the structure can clearly be recognized, and subunit cell detail is visible as well in Fig. 4 as in Fig. 5. Fig. 6 is the resolution image of a region containing several coaxial variants.

3.2.2 The secondary structure

The widely spaced quasi-periodic antiphase boundaries that give rise to the secondary structure, can clearly be recognized in Fig. 7a. They are oriented along homologeous directions in different domains of the primary structure. They may join in groups of four producing perfect material.

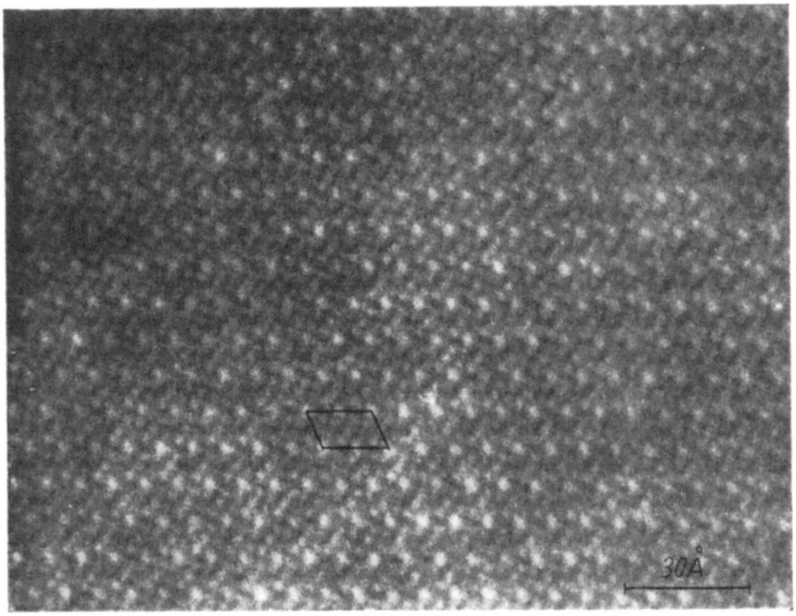

Fig. 4. [010]* high resolution dark-field micrograph obtained by centering the objective diaphragm around $(110)_{f.c.c.}$ and taking a diaphragm size as large as possible, but excluding the basic f.c.c. reflections

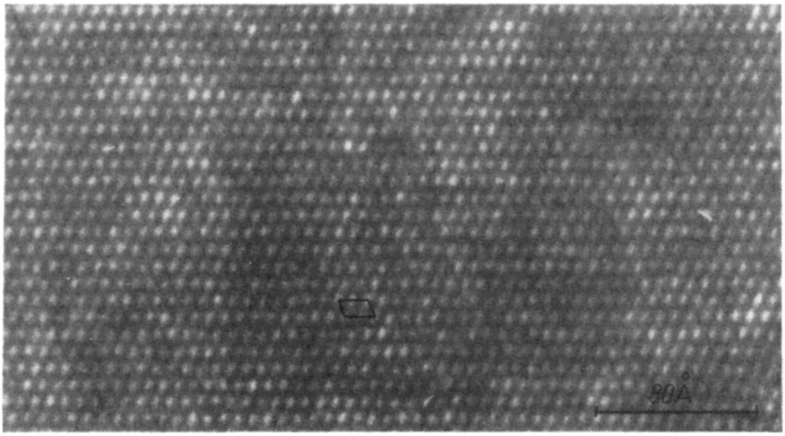

a

Fig. 5

b

Fig. 5. High resolution dark-field micrographs taken under the same conditions as in Fig. 4, at a lower and at a higher magnification

Fig. 6. Dark-field micrograph taken under the same conditions as in Fig. 4, revealing different coaxial variants

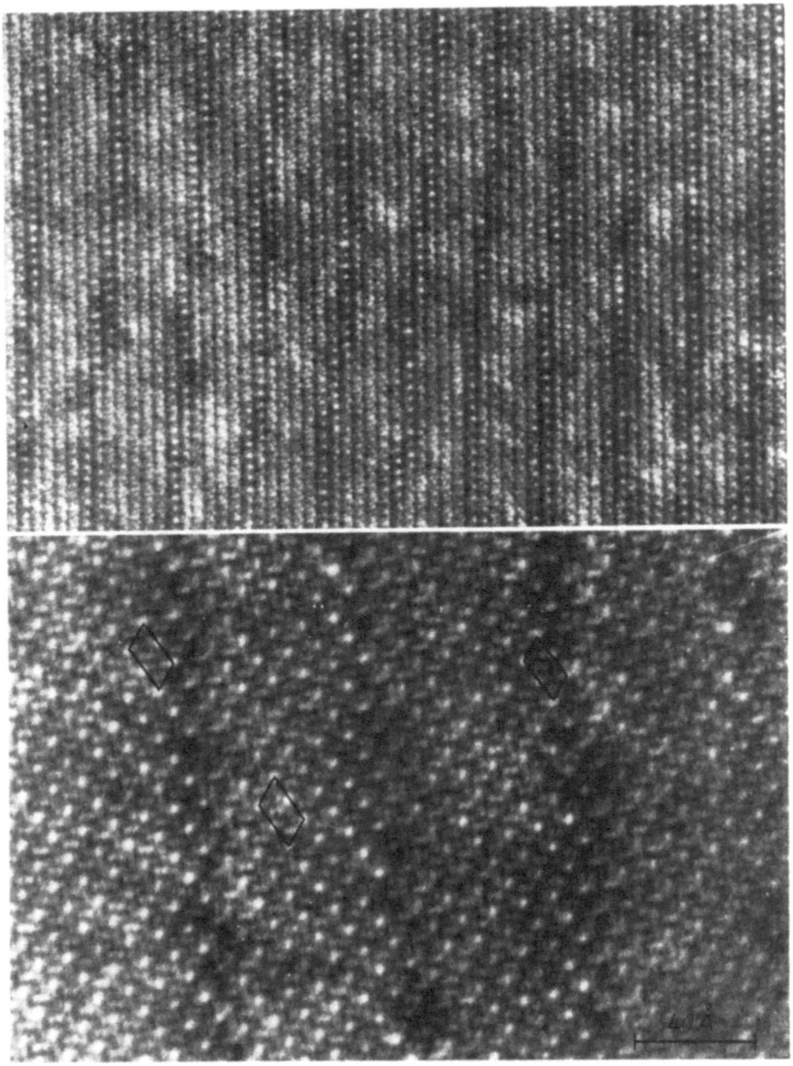

a

b

Fig. 7. a) Dark-field micrograph showing jogging quasi-periodic APB's in the $Au_{15}Mg_4$ structure (1 700 000 ×). b) Higher magnification of three parallel APB's, allowing to deduce the projection of the displacement vector

In Fig. 7 b three successive boundaries of the sequence are shown at high resolution. The off-sets of the structure images produced by the presence of the APB's can clearly be observed; they are the same for all boundaries in the sequence. They are used further below to derive a model (Section 4.2).

4. Interpretation

4.1 The primary structure

From Fig. 1 it could be concluded that the configuration of superlattice spots can be described as consisting of groups of equidistant spots along rows parallel with the

$[201]_{DO_{22}}$ direction, and associated with spots of the DO_{22} structure. These groups of spots are shifted with respect to the positions of the DO_{22} spots marked by crosses in Fig. 2. The fractional shift for the rows $l = 2$ and $l = 4$ is $\pm\frac{1}{2}$. However, the rows with $l =$ odd are systematically absent. Although the DO_{22} spots are also not there, we can nevertheless deduce the fractional shift of the absent rows with respect to the absent 103 and $101_{DO_{22}}$ spots because the geometry is completely defined by spots which are present. The fractional shift for the rows $l =$ odd is $\pm\frac{1}{4}$, as deduced from Fig. 1 and 2.

The most intense superstructure spots clearly occur in the vicinity of the DO_{22} spot positions. All geometrical features strongly suggest that the structure can be derived from the DO_{22} structure by the periodic introduction of APB's parallel with $(201)_{DO_{22}}$ planes and with a DV $\boldsymbol{R}_1 = \frac{1}{4}[20\bar{1}]_{DO_{22}}$, causing the correct fractional shift, as is evident from Table 1.

Table 1

$\boldsymbol{g}$ (DO_{22})	calculated $\boldsymbol{g}\cdot\boldsymbol{R}_1$	observed (mod 1)
103	$-\frac{1}{4}$	$\pm\frac{1}{4}$
101	$\frac{1}{4}$	$\pm\frac{1}{4}$
002	$-\frac{1}{2}$	$\pm\frac{1}{2}$
202	$\frac{1}{2}$	$\pm\frac{1}{2}$

This result can also be derived more directly. It is sufficient to assume that the structure is a LPAPB structure of the DO_{22} structure with the APB perpendicular to the $[201]_{DO_{22}}$ direction and with a displacement vector of the type $[u0w]_{DO_{22}}$. The fractional displacements then immediately lead to the relations of Table 2.

Table 2

DO_{22} spot	relation	
103	$u + 3w = \pm\frac{1}{4} \pmod 1$	(5)
101	$u + w = \pm\frac{1}{4} \pmod 1$	(6)
002	$2w = \pm\frac{1}{2} \pmod 1$	(7)
202	$2u + 2w = \pm\frac{1}{2} \pmod 1$	(8)

From (7) we deduce that $w = \pm\frac{1}{4} \pmod 1$; from (6) then follows $u = 0, \pm\frac{1}{2} \pmod 1$. The values so obtained are also consistent with relations (5) and (8). The possible displacement vectors are thus $\pm\frac{1}{4}[20\bar{1}]$ or $\pm\frac{1}{4}[001]$. Only the first one is consistent with the underlying f.c.c. or DO_{22} structure since the second one does not connect atomic sites in the f.c.c. structure.

As a result of the choice of the monoclinic unit cell (Fig. 8a) one of its sidefaces is parallel with the plane of the APB's. The unit cell contains two subcells, indicated also in Fig. 8. The second subcell must be derived from the first one by the requirement that the monoclinic unit cell should be c-face-centered, in agreement with the observed systematic extinctions. Also a glide mirror plane parallel with (010) and with a glide vector $\frac{1}{2}\boldsymbol{A}_3$ should be present. The primitive unit cell would in fact be triclinic.

We can now fill the first subcell with a small island of DO_{22} structure, in the correct orientation as suggested by the most intense superlattice spots in the diffraction

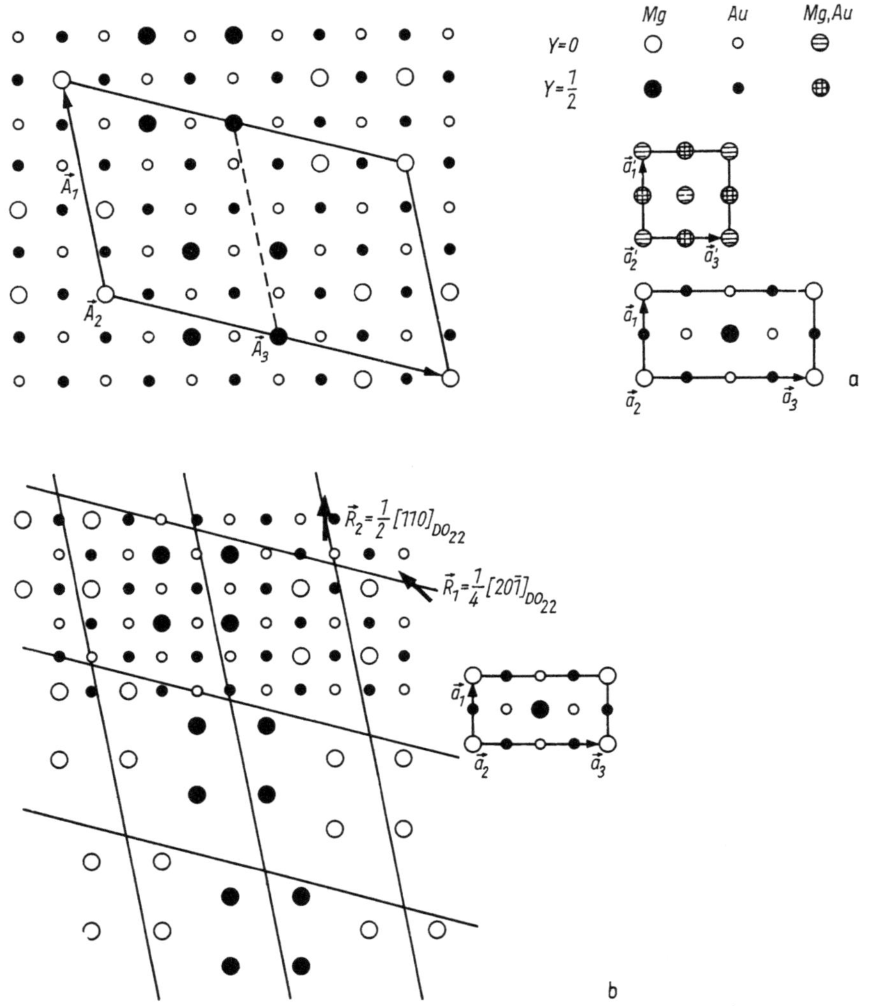

Fig. 8. a) Tentative structure of $Au_{15}Mg_4$ derived from diffraction evidence as well as from high resolution observations. b) Configuration of the two families of APB's to be introduced in the DO_{22} structure in order to generate the $Au_{15}Mg_4$ structure

pattern. Repetition of the pattern according to the translation symmetry of the lattice leads to the structure of Fig. 8a.

This structure can be described as being derived from the DO_{22} structure by the periodic introduction of APB's along two different planes (Fig. 8b):

(i) APB's parallel with the $(201)_{DO_{22}} \equiv (401)_{f.c.c.}$ planes with a DV $\boldsymbol{R} = \frac{1}{4}[20\bar{1}]_{DO_{22}}$ and a spacing of 10.2 Å;

(ii) APB's parallel with the $(1010)_{DO_{22}} \equiv (105)_{f.c.c.}$ planes with a DV $\boldsymbol{R} = \frac{1}{2}[110]_{DO_{22}}$ and a spacing of 8.2 Å.

The second family of interfaces automatically results from the presence of the first one and from the translation symmetry.

The structure can alternatively be described as consisting of parallel columns of single unit cells of Cu_3Au structure, placed according to the pattern defined by the translation symmetry and imbedded in a gold matrix (Fig. 8a). The columns occur at two different levels. The ideal composition corresponds to $Au_{3.75}Mg$ or $Au_{15}Mg_4$.

The trial structure is thus a two-dimensional periodic antiphase boundary derivative of the DO_{22} structure, which is itself a superstructure of the f.c.c. structure.

4.1.1 The primary structure: diffraction evidence

For a number of superlattice reflections the extinction distances were calculated assuming the trial structure represented in Fig. 8a. The results are summarized in Table 3. The same data are represented schematically in Fig. 9 where the size of the dots is a measure for the relative intensities of the corresponding diffraction spots. It is clear that the correspondence between the intensities, calculated on the basis of the kinematical theory, and the observed values is quite satisfactory (Fig. 1).

It is therefore felt that the structure is consistent with the diffraction evidence. We shall see below that also the structural images support the proposed model.

Table 3

reflection	extinction distance (in Å)
100	2955
200	1035
300	2169
400	7064
500	14517
002	8838
004	966
006	6736
008	4863
102	21491
104	2292
106	2682
202	1165
204	1370
206	1614
302	2549
304	32119
306	5485
402	1375
404	15891
406	287

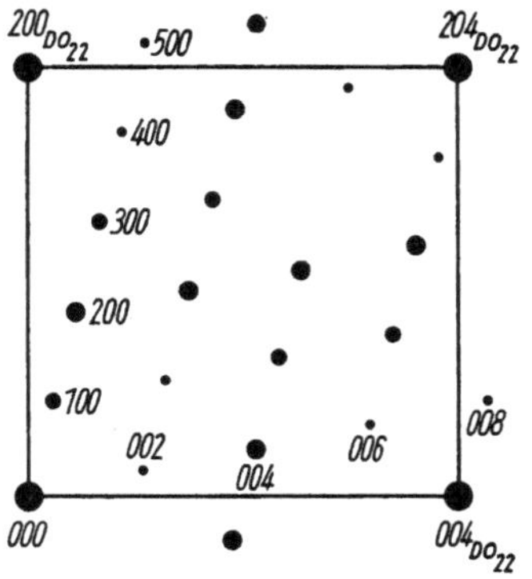

Fig. 9. Calculated [010]* diffraction pattern assuming the structure of Fig. 8

4.1.2 Structure images

Structure images are made by collecting superstructure spots only in the selector aperture. In this way information on the superstructure is obtained, whereas information on the basic f.c.c. structure is rejected intentionally.

Images obtained in this way exhibit striking white dots forming a pattern having the configuration and scale of the columns of Cu_3Au structure (Fig. 5a). Under certain conditions the dots acquire a square shape, and in certain images these squares can even be resolved into four dots (Fig 5b). It should be noted that in the proposed structure the squares are separated by different distances along the two families of APB's. This difference in spacing can be observed in the images of Fig. 5b.

4.2 The secondary structure

It is evident from the diffraction pattern (Fig. 10) that in some regions a further one-dimensional very long period superstructure is present. It can be derived from the primary structure by the introduction of quasi-periodic antiphase boundaries with a spacing of about 130 Å and in planes of the type (101) as referred to the DO_{22} structure. The displacement vector can again be derived from the fractional displacement of the super-superstructure spots with respect to the spots due to the primary superstructure. It can, however, more directly be derived from the structure images (Fig. 7), where the APB's cause a visible shift in the structure images of the primary structure. One finds for the projection of the DV: $\boldsymbol{R} = \frac{1}{2}[\bar{1}01]$. A model for one APB of this secondary structure is represented in Fig. 11.

Fig. 10. [010] diffraction pattern showing spot splitting due to the introduction of quasi-periodic APB's in the primary structure

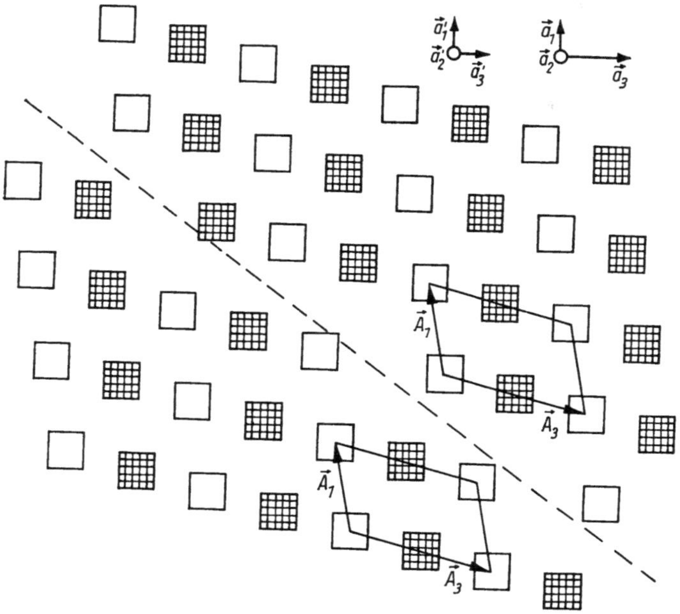

Fig. 11. Schematic model for quasi-periodic APB's in the primary structure

4.3 Structural variants of the primary structure

4.3.1 Orientation variants

The monoclinic phase $Au_{15}Mg_4$ is clearly a superstructure of the f.c.c. disordered cell. The point groups of the structures are $G = 4/m\ \bar{3}\ 2/m$ of order $p = 48$ for the f.c.c. structure and $H = 2/m$ of order $q = 4$ for the ordered monoclinic structure, respectively. The number of orientation variants is $n = p/q = 12$ [15]. This situation is identical to Au_5Mn_2, which has been discussed previously [5]. Along any of the three cube directions four coaxial variants, having their A_2-axis in common, are to be expected. Evidence for the presence of such variants is found from the diffraction pattern (Fig. 3a) as well as from the direct image (Fig. 6).

Between these four coaxial variants three different interfaces are possible:

1. parallel twins, having their A_2-axis parallel, but rotated about this axis over 90°, one with respect to the other;

2. antiparallel twins having antiparallel A_2-axes and a $(100)_{f.c.c.}$ plane parallel with A_2 as a mirror plane;

3. antiparallel twins having a $(110)_{f.c.c.}$ plane parallel with A_2 as a mirror plane.

Between non-coaxial variants two more interfaces (perpendicular twins) can be expected:

4. The two variants are related by a (110) mirror plane not parallel with A_2 (or a rotation over 180° about a twofold axis of the f.c.c. structure not perpendicular to A_2).

5 The two variants are related by a rotation over 120° about a threefold axis of the f.c.c. structure.

Like in Au_5Mn_2 large areas have a common A_2-axis, and only coaxial variants occur over a relatively large area. This is probably related to the fact that the DO_{22} structure is the basic structure for both these structures and also occurs in the initial stages of the ordering process.

4.3.2 *Translation variants*

The number of translation variants equals the volume of the primitive ordered unit cell divided by the volume of the primitive unit cell of the disordered phase [15]. In the present case it follows from Section 3.1.1. that the volume of the twofold monoclinic cell of the superstructure is equal to 19/2 times the volume of the f.c.c. cell, which is itself fourfold. The number of translation variants is then 19, resulting in 18 different antiphase boundaries. Not all of them will be equally favourable of course, and in reality only a limited number of different APB's will occur.

5. Conclusion

In gold–magnesium alloys with nominal composition around Au–18 at% Mg different phases occur. A new phase with an ideal composition of $Au_{3.75}Mg$ was isolated and its structure determined, as well from the diffraction pattern as from the high resolution images. The structure is a good example of the presence of structure modulation by interfaces at different structural levels; it is a two-dimensional long period superstructure of the DO_{22} structure, which is itself a superstructure of the f.c.c. structure. Moreover, certain parts of this two-dimensional superstructure contain a very long period one-dimensional superstructure.

References

[1] K. Burkhardt and K. Schubert, Z. Metallk. **56**, 864 (1965).
[2] K. Burkhardt, K. Schubert, R. S. Toth, and H. Sato, Acta cryst. **B24**, 137 (1968).
[3] H. Sato and R. S. Toth, J. Phys. Chem. Solids **29**, 2015 (1968).
[4] P. Airo, phys. stat. sol. (a) **47**, K107 (1978).
[5] G. van Tendeloo, R. Wolf, D. van Dyck, and S. Amelinckx, phys. stat. sol. (a) **47**, 105 (1978).
[6] G. van Tendeloo and S. Amelinckx, phys. stat. sol. (a) **49**, 337 (1978).
[7] G. van Tendeloo, R. de Ridder, and S. Amelinckx, phys. stat. sol. (a) **49**, 655 (1978).
[8] G. van Tendeloo, R. Wolf, J. van Landuyt, and S. Amelinckx, phys. stat. sol. (a) **47**, 539 (1978).
[9] R. Wolf, G. van Tendeloo, J. van Landuyt, and S. Amelinckx, phys. stat. sol. (a) **48**, 39 (1978).
[10] J. van Landuyt, R. de Ridder, R. Gevers, and S. Amelinckx, Mater. Res. Bull. **5**, 353 (1970).
[11] R. de Ridder, J. van Landuyt, and S. Amelinckx, phys. stat. sol. (a) **9**, 551 (1972).
[12] R. Wolf, Micron **8**, 17 (1977).
[13] J. Tenko, V. Lehtimäki, and V. Hovi, J. Phys. E **11**, 510 (1978).
[14] O. Terasaki and D. Watanabe, Proc. Internat. Conf. Modulated Structures, Hawai 1979 (to be published).
[15] G. van Tendeloo and S. Amelinckx, Acta cryst. **A30**, 431 (1974).

(Received April 23, 1979)

phys. stat. sol. (a) **54**, 513 (1979)

Subject classification: 3 and 11; 22.1.2

California Institute of Technology, Pasadena[1])

Regrowth Behaviour of Ge Implanted ⟨100⟩ Si

By

P. Revesz[2]) and J. W. Mayer

Regrowth kinetics of amorphous Si layers formed by implanting ^{75}Ge ions to concentrations of about 0.5 at % into ⟨100⟩ Si are obtained by using 1.5 MeV ^{4}He backscattering and channeling measurements. The growth of epitaxial Si is linear in time with an activation energy that coincides with that obtained on ^{28}Si implanted-amorphous Si. The Ge atoms occupy substitutional sites in the regrown layer.

Кинетика рекристаллизации аморфного кремния, созданного ионным внедрением германия в концентрациях вплоть до 0,5 процентов в подложку кремния с ориентацией ⟨100⟩ была получена методом обратного рассеяния ионов гелия с энергией 1,5 MeV, а также методом, основанном на эффекте каналирования. Рост эпитаксиального кремния является линейной функцией во времени, а энергия активации этого процесса совпадает с тем, что было получено для аморфного кремния, созданного ионным внедрением ^{28}Si. Ионы германия после отжига занимают замещающее положение в решетке кремния.

1. Introduction

It is now well-established that implanted-amorphous silicon layers can grow epitaxially on the underlying crystal substrate in the solid-phase at temperatures around one-half the melting point in K [1, 2]. Use of implantation techniques allows the formation of the amorphous layer with both controlled impurity-concentrations and a contamination-free amorphous to crystal interface.

Epitaxial regrowth on ⟨100⟩ Si proceeds linearly with time with few defects in contrast to the non-uniform growth rates and high defect concentrations found with ⟨111⟩ oriented Si [3, 4]. For this reason, most studies of regrowth have been made on ⟨100⟩ oriented Si.

The presence of impurities at concentrations greater than $10^{20}/cm^3$ in ⟨100⟩ Si has a pronounced influence on the growth rate. Conventional dopants such as B, P, or As lead to an increase in growth rate [5 to 7] whereas complex-forming impurities such as O, N, and C lead to a decrease in growth rate [8]. One approach to evaluate the impurity dependence was to investigate ion size effects by implanting noble gases (Ne, Ar, and Kr) which would act as inert impurities. It was found [9, 10] that the presence of gas atoms slows down the growth rate. However, the presence of gas bubbles and microtwins led to the speculation [9] that the gas bubbles act both as a sink for vacancies, hence reducing the growth rate, and as nucleation center for microtwins.

The present work investigates the influence of Ge, which should act as an inert impurity, but with a larger ionic radius than Si. Concentrations of Ge around 0.5 at% were used since this concentration produced about an order of magnitude or greater change in growth rate for the other impurities investigated.

---

[1]) Pasadena, California 91125, USA.

[2]) Present address: Central Research Institute for Physics, Budapest, Hungary.

2. Experiment

Wafers of ⟨100⟩ orientation from Czochralski grown silicon with 2—10 Ωcm resistivity were used for the experiments. The wafers were chemically-mechanically polished on one side in a manner suitable for device fabrication. Then the wafers were implanted with $^{75}Ge^+$ ions to a concentration of about $2.5 \times 10^{20}/cm^3$ over depths of about 2000 Å. The implantation energies were 50, 100, 150, 200, and 250 keV with doses 3×10^{14} cm^{-2}, 6×10^{14} cm^{-2}, 8×10^{14} cm^{-2}, 10^{15} cm^{-2}, and 2×10^{15} cm^{-2}, respectively. The substrate temperature during implantation was kept at LN_2 temperature to avoid beam heating effects. Upon cleaving the wafers, small samples were annealed in a vacuum of $\approx 7 \times 10^{-7}$ Torr at temperatures of 500, 525, 550, and 575 °C for different times.

3. Results and Discussion

Backscattering spectra taken with ⟨100⟩ aligned beam of as-implanted and annealed samples at 550 °C are shown in Fig. 1. The leading edge of the silicon signal at about 0.85 MeV corresponds to scattering of the ^{4}He particles from silicon at the surface of the sample. The backscattering yield from the implanted region is the same as for random incidence of the He beam, indicating this layer is amorphous. The steps at lower energies (0.7 to 0.8 MeV) in the spectra correspond to scattering of the particles from silicon atoms at the amorphous–crystalline interface.

With standard values of stopping cross-section and the assumption of bulk density for the amorphous layer the thickness of the amorphous silicon layer was calculated in Å [11]. The narrowing of the silicon signal from the amorphous layer and the decreasing of the dechanneling tails at the low energy part of the spectra with increasing annealing time show the amorphous silicon layer recorders epitaxially on the underlying crystal substrate. Measuring the thicknesses of the amorphous layers for different annealing time and temperature, we determined the regrowth kinetics of the amorphous layers. As it is shown in Fig. 2, the regrowth of the amorphous layer is a linear function of the annealing time. The accuracy of the absolute thickness measurement is mainly given by the uncertainty of the stopping cross-sections, which is between 5 and 10 percent. The regrowth rate versus reciprocal temperature is plotted in Fig. 3. Values of regrowth rate agree closely with measurements made by Csepregi et al. [12]. This shows that the presence of 0.5 at% of Ge has a negligible effect on the growth kinetics of amorphous-implanted Si.

Channeling measurements can also be used to study the lattice location of the Ge atoms in the regrown layer. The signals in the spectra in Fig. 1 at about 1.2 MeV

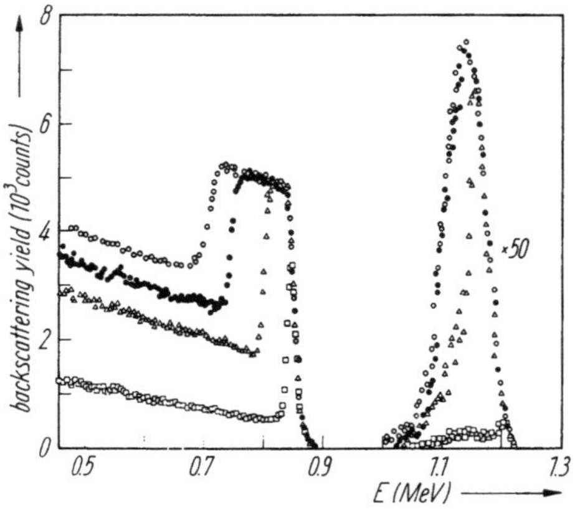

Fig. 1. Aligned ⟨100⟩ spectra for 1.5 MeV $^4He^+$ backscattered from Si samples implanted with Ge. Several spectra for isothermal annealing at 550 °C are shown. ○ as-implanted, ● 10 min, △ 20 min, □ 30 min

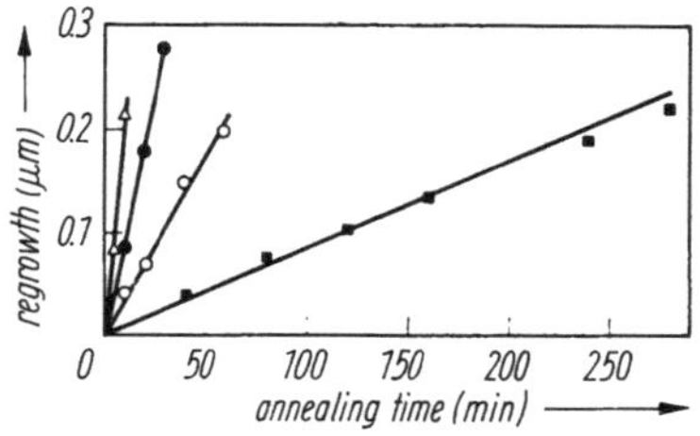

Fig. 2

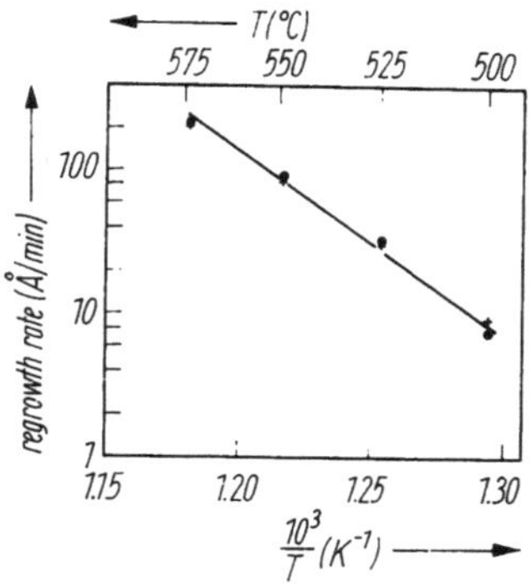

Fig. 3

Fig. 2. Epitaxial regrowth on ⟨100⟩ Si as a function of annealing time for temperatures of ■ 500, ○ 525, ● 550, and △ 575 °C

Fig. 3. Regrowth rate versus reciprocal temperature for ⟨100⟩ Si implanted with Ge to a concentration of 0.5 at%. The crosses represent data on Si implanted ⟨100⟩ Si obtained by Csepregi et al. [12], ● this work

correspond to the aligned yield of ^{4}He particles backscattered from Ge ions. The spectrum taken for the sample annealed for 30 min at 550 °C, where the amorphous layer has nearly completely regrown, shows that the aligned yield from the Ge atoms has decreased an order of magnitude. This along with similar results for ⟨111⟩ aligned yields indicates that the Ge atoms lie predominantly on substitutional lattice sites aligned with the underlying axial direction. Angular scans on fully annealed samples showed that the minimum yield, χ_{min}, and axial half width, $\Psi_{1/2}$, of the Ge signal coincides with that of the Si signal. This result confirms that the Ge is substitutional in the regrown layer.

The depth profile of Ge atoms in Si are shown in Fig. 4. Channeling measurements do not allow assignment of lattice locations of Ge in the amorphous layer and, hence, the aligned and random incidence (not shown) yields from Ge coincide in the as-implanted sample. As the amorphous/crystal interface moves toward the surface as a result of thermal anneal, there is a decrease in the Ge yield at depths corresponding to the interface position and beyond. This is expected due to the high substitutional nature of the Ge in the regrown layer. Random spectra show that there is no redistribution of the Ge due to diffusion processes.

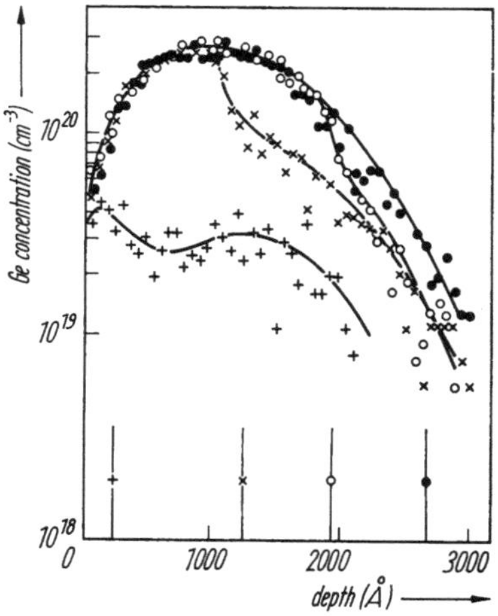

Fig. 4. Depth profile of Ge in silicon obtained from the ⟨100⟩ aligned backscattering spectra for isothermal annealing at 550 °C. ● as-implanted, ○ 10 min, × 20 min, + 30 min. Position of crystal Si/amorphous-Si interface is indicated by vertical lines

4. Conclusion

Channeling effect measurements show that the presence of 0.5 atomic percent of Ge has no effect on the kinetics of the epitaxial regrowth of amorphous Si on ⟨100⟩ Si. This indicates that the impurity dependence of growth kinetics is not simply associated with ion size effects. This is in agreement with the previous observation that As and P (at similar concentrations) cause the same [5] increase in growth rates although their ion sizes differ.

Acknowledgements

We are grateful to T. W. Sigmon for providing us the Ge implanted silicon wafers. The work was supported by NSF and the Institute of Cultural Relations, Budapest, Hungary (P. Revesz).

References

[1] S. S. Lau and W. F. van der Weg, in: Thin Films — Interdiffusion and Reactions, Chap. 12, Eds. J. M. Poate, K. N., Tu, and J. W. Mayer, Wiley Interscience, New York 1978.
[2] T. Tokuyama, M. Miyao, and N. Yoshihiro, Japan J. appl. Phys. 17, 1301 (1978).
[3] L. Csepregi, J. W. Mayer, and T. W. Sigmon, Appl. Phys. Letters 29, 92 (1976).
[4] A. Golanski, A. Fiderkiewicz, H. Rzewuski, M. Lefeld-Sosnowska, J. Gronkowski, R. Grötzschel, U. Kreissig, and H. Bartsch, phys. stat. sol. (a) 38, 139 (1976).
[5] L. Csepregi, E. F. Kennedy, T. J. Gallagher, J. W. Mayer, and T. W. Sigmon, J. appl. Phys. 48, 4234 (1977).
[6] Y. Ohmura, T. Inoue, and Y. Yamamoto, J. appl. Phys. 49, 3597 (1978).
[7] H. Nishi, T. Sakurai, and T. Furuya, J. Electrochem. Soc. 125, 461 (1978).
[8] E. F. Kennedy, L. Csepregi, J. W. Mayer, and T. W. Sigmon, J. appl. Phys. 48, 4241 (1977).
[9] P. Revesz, M. Wittmer, J. Roth, and J. W. Mayer, J. appl. Phys. 49, 5199 (1978).
[10] M. Wittmer, J. Roth, P. Revesz, and J. W. Mayer, J. appl. Phys. 49, 5207 (1978).
[11] W. K. Chu, J. W. Mayer, and M-A. Nicolet, Backscattering Spectrometry, Academic Press, New York 1978.
[12] L. Csepregi, E. F. Kennedy, J. W. Mayer, and T. W. Sigmon, J. appl. Phys. 49, 3906 (1978).

(Received April 9, 1979)

phys. stat. sol. (a) 54, 517 (1979)

Subject classification: 13.4 and 14.3.4; 22

Department of Physics, Semiconductors, and Microelectronics, Voronezh State University

The Energy Spectrum of Discrete Centres in the High Inhomogeneous Electric Field of the Space Charge Region of a Semiconductor

By

V. F. Synorov, V. D. Linnik, and B. I. Sysoev

A lowering of the barrier surrounding the centre in electrically inhomogeneous materials is analyzed. It is shown that the activation energy decrease at the space charge region may substantially differ from the Frenkel effect. A possibility of the occurrence of the space-charge-enhanced emission is pointed out in the case when the carrier release probability does not depend practically on the electric field strength, but is determined by its derivative. A simple approximation formula for the computation of the decrease in the activation energy of the Coulombic attractive and neutral centre in an inhomogeneous electric field is offered.

Проведен анализ понижения барьера окружающего центр, в электрически неоднородных материалах. Показано, что уменьшение энергии активации в поле объемного заряда может существенно отличаться от закона Френкеля. Отмечена возможность существования эмиссии, облегченной объемным зарядом, когда вероятность выброса носителя практически не зависит от напряженности электрического поля, а определяется ее производной. Предложена простая аппроксимирующая формула расчета уменьшения энергии активации кулоновского притягивающего и нейтрального центра в неоднородном электрическом поле.

1. Introduction

The effect of a high electric field on the activation energy of a discrete Coulomb attractive centre has been first studied by Frenkel [1] who showed that the decrease of the activation energy ΔW for an attractive centre is proportional to $\sqrt{E}$ (E electric field strength). The model given in [1] has been more accurately defined by Gubanov [2] who introduced the Coulomb screening potential of interaction between a carrier and an impurity centre and obtained the value of ΔW for both the charged centre when $r_0 \to \infty$ (r_0 characteristic screening length determined by the type of the centre and the parameters of the material) as well as for the neutral centre when r_0 is of the order of an interatomic distance.

It has been assumed in both papers [1, 2] that the electric field does not depend on the coordinate, i.e. is homogeneous. This is well justified for isotropic materials with a sufficiently low charge carrier concentration (for instance, homogeneous dielectrics and high-resistivity semiconductors with a wide band gap).

It is well known, however, that in the overwhelming majority of cases electric fields in semiconductors are essentially inhomogeneous, the degree of this inhomogeneity being determined by the space charge density and reaching considerable values. The basic reasons accounting for the occurrence of space charge regions (SCR), responsible for the electric field inhomogeneity, are interphase boundaries, p–n junctions, statistical fluctuations, ion impurity concentrations, etc.

A thermally stimulated ionization of discrete trapping centres in the electric field of the semiconductor space charge region has been analyzed in [3]. It has been shown that there exists a continuous series of activation energy (decreasing from W_{max} to 0) in the semiconductor SCR, the potential barrier lowering being maximum close to the semiconductor surface, $\Delta W = \Delta W_{max}$ (where the electric field strength is maximum), and decreases monotonously with distance.

It was supposed in [3] that the lowering of the potential barrier surrounding the centre was only due to the value of the electric field strength at a given point of SCR i.e. the electric field in the neighbourhood of the centre was, as previously, regarded constant and the coordinate dependence of the activation energy decrease was unambiguously related to the coordinate dependence only of the electric field strength value.

Apparently, these assumptions are valid for electric fields varying smoothly enough, and need a more precise definition when the electric field strength throughout the characteristic width of the potential well of the localized state undergoes essential changes.

The Bohr radius a_B may be used as the Coulomb centre potential well width. The degree of the electric field inhomogeneity is simply characterized by the screening length $L_D \equiv |E/(dE/dx)|$. Quantitative estimations show that these values become commensurable when $n_0 = 1.1 \times 10^{19} \frac{T}{l^4 \varepsilon} \left(\frac{m^*}{m_0}\right)^2$ cm^{-3} (where n_0 is the total concentration of charge carriers participating in the electric field screening, l the number of the Bohr orbit, ε the relative permittivity of the semiconductor, T the absolute temperature, m^*/m_0 the ratio of the charge carrier effective mass and the mass of a free electron). It can be seen that at distances of the order of the first Bohr orbit the electric field appears to be inhomogeneous even at concentrations of the order of 10^{17} cm^{-3}, ($l = 1$, $m^*/m_0 = 0.1$, $\varepsilon = 10$, $T = 100$ K). The inhomogeneous distribution of the electric field is most clearly defined for the excited states of the impurity centre ($l > 1$) at low temperatures, in materials with high permittivity and small effective mass of charge carriers.

Therefore, a detailed consideration of the process of thermal ionization of charged localized centres in electrically inhomogeneous materials seems to be of interest.

2. Theoretical Analysis

We shall treat the thermal emission of carriers from a ground state of the potential well into one of the allowed bands, irrespective of the mechanism of the process.

The account of the cascade mechanism of the process leads only to a change of the temperature–field dependence of the pre-exponential factor in the expression for the charge release probability [4].

In order to describe the interaction of a carrier (an electron, to be more exact) with a localized centre, it is convenient to use a potential of the type

$$U = -\frac{a}{|r|^n}, \tag{1}$$

where a and n are constants dependent on the type of the centre and the parameters of the semiconductor, $|r|$ is the distance between the carrier and the potential well centre (Fig. 1a, b). If $a = a_1 \equiv q^2/4\pi\varepsilon\varepsilon_0$ and $n = n_1 \equiv 1$, (1) corresponds to the electron potential energy in the field of a charged centre. If $n = n_2 \equiv 4$ and $a = a_2 \equiv \equiv \alpha/2\,(q/4\pi\varepsilon\varepsilon_0)^2$, (1) describes the interaction of the electron with a neutral centre in dipole approximation [5]. Here q is the charge of the electron, $\varepsilon\varepsilon_0$ the absolute permittivity of the semiconductor, α the impurity centre polarizability. In the general case the total potential energy of a carrier in an inhomogenous electric field of SCR is equal

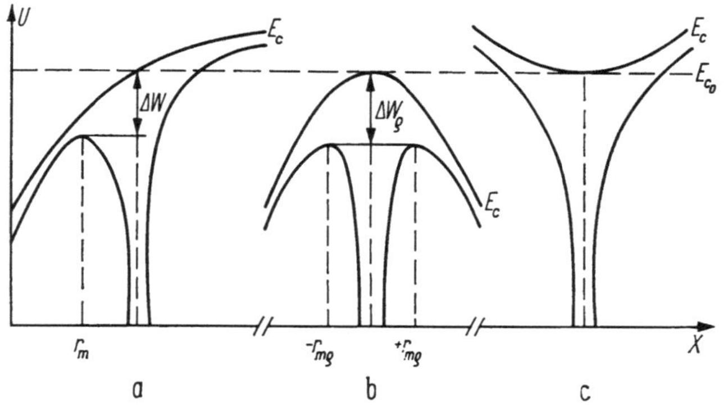

Fig. 1. The energy diagram of charged localized centre in an inhomogeneous electric field. a) $E > 0$, $\varrho < 0$; b) $E = 0$, $\varrho < 0$; c) $E = 0$, $\varrho > 0$; E_{c0} is the unperturbed conduction band edge

to

$$U(x, r) = -\frac{a}{|r|^n} + q \int_0^r E(x + r')\,\mathrm{d}r' , \tag{2}$$

where x is the impurity centre coordinate (Fig. 1a).

Bearing in mind that $\Delta W(x) = |U(x, r_\mathrm{m})|$ by means of series expansion of the electric field strength in the neighbourhood of the point x and integrating by terms we obtain

$$\Delta W(x) = \left| -\frac{a}{|r_\mathrm{m}|^n} + q \sum_{k=0}^{\infty} \frac{\partial^k E(x)}{\partial x^k} \frac{r_\mathrm{m}^{k+1}}{(k+1)!} \right| , \tag{3}$$

where $|r_\mathrm{m}|$ is the distance between the centre and the potential barrier maximum (Fig. 1a, b), which can be found from $\partial U/\partial r|_{r=r_\mathrm{m}} = 0$:

$$\frac{an}{|r_\mathrm{m}|^n r_\mathrm{m}} - q \sum_{k=0}^{\infty} \frac{\partial^k E(x)}{\partial x^k} \frac{r_\mathrm{m}^k}{k!} = 0 . \tag{4}$$

The system of equations (3) and (4) enables one to evaluate the activation energy decrease at the semiconductor SCR for the general case. It is clear from (3) and (4) that in order to calculate the activation energy decrease one requires the values of the derivatives $\partial^k E(x)/\partial x^k$, which within a certain degree of accuracy can be found by means of successive differentiation of the Poisson equation for the electric field strength $\partial E/\partial x = \varrho/\varepsilon\varepsilon_0$ where ϱ is the total density of all space charges forming the inhomogeneous potential profile in the semiconductor.

In general the analysis of (4) and (5) is rather complicated, but for investigating the effect of the electric field inhomogeneity on the process of impurity centre thermal ionization it is sufficient to confine oneself to the first two terms in the sums, since their contribution to the barrier lowering is maximum due to the fast absolute series convergence in (3) and (4). It is equivalent to the investigation of the barrier lowering effect caused by a linear inhomogeneous electric field $(\varrho(x) = \varrho = \text{const})$.

This approximation is mostly justified for the Schottky depletion layers and is quite applicable for the SCR with slowly varying $\varrho(x)$ when the space charge density in the neighbourhood of the trapping centre can be considered constant.

The following algebraic equation is, therefore, obtained from (4):

$$q\varrho\, |r_\mathrm{m}|^{n+2} - qE\varepsilon\varepsilon_0\, |r_\mathrm{m}|^n r_\mathrm{m} + an\varepsilon\varepsilon_0 = 0 . \tag{5}$$

The effective energy decrease in a linear inhomogeneous electric field, equation (3), transforms into

$$\Delta W = \left| -\frac{a}{|r_m|^n} + qEr_m + \frac{q\varrho}{2\varepsilon\varepsilon_0} r_m^2 \right| . \tag{6}$$

The first two terms in (6) look absolutely analogous to the formula describing the barrier lowering by a homogeneous electric field [1, 2], in our case, however, r_m is estimated with regard to the space charge influence ($\varrho \neq 0$) from (5). The third term in (6) characterizes a direct SCR contribution to the decrease of the impurity centre activation energy.

In slightly inhomogeneous electric fields $\left(|\varrho| \ll \varepsilon\varepsilon_0 |E| (qE/an)^{1/(n+1)}\right)$ from (5) and (6) we get

$$\Delta W_E = (n + 1)\, a^{1/(n+1)} \left(\frac{qE}{n}\right)^{n/(n+1)} . \tag{7}$$

It should be noted that (7) transforms into the Frenkel effect $\Delta W \sim \sqrt{E}$ when $a = a_1$ and $n = n_1$ for the charged centre, and into the dependence $\Delta W \sim E^{0.8}$ closely related to the empirical Poole law [6] ($\Delta W \sim E$) for a neutral centre, when $a = a_2$ and $n = n_2$.

In the extreme case of low electric fields when $|E| \ll (|\varrho|/\varepsilon\varepsilon_0)^{(n+1)/(n+2)} (an/q)^{1/(n+2)}$ (such a situation is realized in the vicinity of the potential profile extrema Fig. 1 b, c), from (5) and (6) we have

$$|r_{m\varrho}| = \sqrt{\frac{an\varepsilon\varepsilon_0}{-q\varrho}} , \tag{8}$$

$$\Delta W_\varrho = \left(\frac{n+2}{2}\right) \frac{a}{|r_{m\varrho}|^n} = \frac{(n+2)}{2}\, a^{2/(n+2)} (-q\varrho)^{n/(n+2)} . \tag{9}$$

A consequence of (8) and (9) is the existence of the space-charge-enhanced thermal ionization of the impurity centre near the extrema, i.e. in spite of the fact that the electric field strength is equal to or approaches zero, the barrier lowering is different from zero and is determined by the space charge density. In particular, for the impurity centre interacting with the electron the above-mentioned effect of the space-charge-caused decrease of the activation energy is possible only when the space charge density is negative (Fig. 1 b). As distinct from the homogeneous-field-enhanced ionization [1, 2], the space-charge-enhanced thermal emission near the potential profile maxima may occur in positive ($r_m > 0$) as well as in the negative direction of the x-axis ($r_m < 0$). In case of a positive space charge ($\varrho > 0$, Fig. 1 c) the opposite effect of the activation energy increase should be expected, i.e. space-charge-hindered carrier emission.

3. Results and Discussion

The limiting cases mentioned above allowed analytical solutions of the problem of the lowering of the potential barrier surrounding the centre. The activation energy decrease in the whole range of E- and ϱ-values in a linear inhomogeneous field can be estimated by numerical calculations of (5) and (6).

The results of numerical calculations according to (5) to (9) are given in Fig. 2. It can be seen that in a slightly inhomogeneous field the curves (5) and (6) practically coincide with the straight lines (1) and (2), i.e. the barrier lowering in this region is caused primarily by the electric field strength. The monotonous increase of ΔW with the growth of the space charge density is due to a considerable contribution of the electric field inhomogeneity to the decrease of the effective activation energy. With

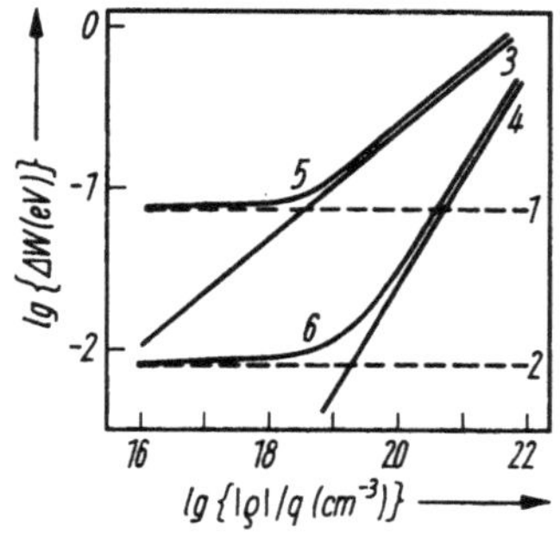

Fig. 2. The dependence of the effective activation energy decrease on the degree of the electric field inhomogeneity ($\varepsilon = 10$). (1) The charged centre, $E = 10^5$ V/cm, $\varrho = 0$, $n = 1$, formula (7); (2) the neutral centre, $E = 10^5$ V/cm, $\varrho = 0$, $n = 4$. $\alpha = 4.67 \times 10^{-37}$ Fm^2, formula (7); (3) the charged centre, $E = 0$, $n = 1$, formula (9); (4) the neutral centre, $E = 0$, $n = 4$, $\alpha = 4.67 \times 10^{-37}$ Fm^2, formula (9); (5) the charged centre, $E = 10^5$ V/cm, $n = 1$, equations (5), (6); (6) the neutral centre $E = 10^5$ V/cm, $n = 4$, $\alpha = 4.67 \times 10^{-37}$ Fm^2, equations (5), (6)

further increasing ϱ the curves (5) and (6) asymptotically approach the straight lines (3) and (4), i.e. the barrier lowering does not depend on the electric field strength (due to $|E| \ll [|\varrho|/\varepsilon\varepsilon_0]^{(n+1)/(n+2)} [an/q]^{1/(n+2)}$ but is determined only by the degree of its inhomogeneity.

It should be pointed out that as it follows from (8) and (9) the space-charge-caused decrease of the activation energy for the charged centre is porportional to $|\varrho|^{1/3}$ and for the neutral centre $\Delta W_\varrho \sim |\varrho|^{2/3}$ (Fig. 2, straight lines 3 and 4).

The analysis of numerical calculations according to (5) to (9) showed that in the ranges of electric field strength 10^2 to 10^6 V/cm and space charge density 10^{16} to 10^{22} cm^{-3} the activation energy decrease can be described by the following approximate formula:

$$\Delta W = \sqrt[k]{\Delta W_E^k + \Delta W_\varrho^k}\,, \tag{10}$$

where $k = 2.88 - 0.383n$, ($n \leqq 4$), ΔW_E and ΔW_ϱ are found from formulae (5) and (9), respectively, the approximation errors not exceeding 2.5%.

Thus, analyzing the high-field effect in electrically inhomogeneous materials we should take into account the contribution of the electric field inhomogeneity to the decrease of the potential barriers surrounding the centre.

The specific character of the interaction between discrete charged localized centres and the inhomogeneous potential profile during the experiment may cause the registration of a quasi-continuous energy band of localized states.

References

[1] Y. I. Frenkel, Zh. eksper. teor. Fiz. **2**, 1292 (1938).
[2] A. I. Gubanov, Zh. tekh. Fiz. **24**, 308 (1954).
[3] B. I. Sysoev, V. D. Linnik, and V. F. Synorov, Fiz. Tekh. Poluprov **12**, 1639 (1978).
[4] V. M. Abakumov, L. N. Kreshuk, and I. N. Yassievitch, Zh. eksper. teor. Fiz. **74**, 1019 (1978).
[5] M. Lax, Phys. Rev. **119**, 1502 (1960).
[6] Poole, Phil. Mag. **27**, 58 (1914).

(Received April 16, 1979)

phys. stat. sol. (a) 54, 523 (1979)

Subject classification: 14.3.3; 16; 22.2.1

Institute of Semiconductor, Academy of Sciences of the Ukrainian SSR, Kiev

Photosensitivity Losses in Ideal and Real Schottky Barriers

By

N. L. DMITRUK and O. YU. BORKOVSKAYA

Various sources of recombination losses of photocurrents of the Schottky diodes are analysed as depending upon the semiconductor doping, the life time of minority carriers, the barrier height, and the interfacial layer between the metal and semiconductor. A general expression for the photocurrent is obtained within the scope of a diffusion theory approach with suitable boundary conditions. Recombination losses of photocurrent of Au–GaAs photodiodes are studied experimentally by measuring their spectral and field dependences.

Проанализированы различные источники рекомбинационных потерь фототока диодов Шоттки в зависимости от легирования материала, времени жизни неосновных носителей тока, высоты барьера и зазора между металлом и полупроводником. Получено общее выражение для фототока в приближении диффузионной теории при подходящих граничных условиях. Экспериментально изучены рекомбинационные потери в фотодиодах Au–GaAs посредством спектральных и полевых характеристик.

1. Introduction

At present there is no unique description of the photoelectric properties of the metal–semiconductor contact with Schottky barrier, being one of the most perspective transformers of solar energy into electric one, moreover the premises of the phenomenological theory of such photoconverters are not always founded enough. Therefore, the mechanisms of photocurrent and photosensitivity losses in Schottky barriers are analyzed in this work depending upon the incident light spectrum, bias voltage on the barrier, semiconductor parameters, and dielectric gap between metal and semiconductor.

The mechanisms of photocurrent in ideal Schottky and real barriers with dielectric gap are considerably different. To obtain the formula for the ideal barrier short circuit photocurrent, being conditioned by the hole component (in the n-type semiconductor)

$$i_{ph} = qI\left(1 - \frac{e^{-kl}}{1 + kL_p}\right) \tag{1}$$

provided that $L \gg L_p$. (I is the light intensity, k the light absorption coefficient in the semiconductor, L_p the diffusion length of the minority current carriers, L the thickness of the sample, l the thickness of the space charge region.) Gärtner used [1] the rigid boundary condition of an infinite "recombination" velocity at the boundary of space charge region (SCR) and semiconductor bulk $x = l$, i.e. the full hole concentration there is $p(l) = 0$. Such strong hole exclusion leads to the concentration excess at the boundary of $\Delta p(l) = -p_0$.

2. The Photocurrent of the Ideal Barrier

To take into account the recombination losses of photocurrent in SCR and on the surface of the Schottky barrier it is necessary to take into consideration the electron component of photocurrent, calculating the photocurrent of the ideal barrier as the difference of hole and electron flows:

$$i_{ph} = -q\left(j_p(0) - j_n(0)\right), \tag{2}$$

To calculate the electron component of photocurrent it is necessary to set suitable boundary conditions for the concentrations and flows. The zero concentration conditions for electrons (majority carriers) on the semiconductor surface and on the boundary of SCR and bulk and for holes on the surface [2, 3] were usually used,

$$\Delta n(0) = \Delta n(l) = 0\,, \qquad \Delta p(0) = 0\,, \tag{3a}$$

and the matching conditions of the solutions on the boundary $x = l$ are

$$\Delta p_{sc}(l) = \Delta p_v(l)\,, \qquad j_{psc}(l) = j_{pv}(l)\,. \tag{3b}$$

The zero conditions on the metal–semiconductor boundary are quite suitable for the calculation of the photocurrent in the absence of a gap as they are a particular case of the known Shockley theory and reflect the strong exchange of the electrons with the metal. At the same time the condition $\Delta n(l) = 0$ cannot be well founded since the detailed calculation of the behaviour of the excess concentrations $\Delta n(x)$ and $\Delta p(x)$ showed that $\Delta n(x)$ passes through a maximum in the environment of this point [4].

Calculating the ideal barrier photocurrent by formula (2) we used the matching condition for electrons in the form (3b) instead of (3a) and also the following condition on the surface:

$$D_n \left.\frac{d\Delta n}{dx}\right|_{x=0} = (s_n + v_n + \mu_n E_{max}) \cdot \Delta n(0) \tag{4}$$

and obtained [5] the expression

$$i_{ph} = qI \left[1 - \frac{e^{-kl}}{1 + kL_p} - \frac{\int\limits_0^l (1 - e^{-kx})\, e^{\psi(x)}\, dx + L_n(1 - e^{-kl}/(1 + kL_n))}{\int\limits_0^l e^{\psi(x)}\, dx + L_n + \dfrac{D_n}{s_n + v_n}\, e^{\psi(0)}}\right], \tag{5}$$

where L_n is the diffusion length of electrons, $\psi(x) = q^2 N_d (x - l)^2 / 2\varepsilon\varepsilon_0 k_B T$ — the electrostatic potential in the SCR of the semiconductor which is assumed to be invariable under illumination (i.e. short circuit regime, no accumulation of non-equilibrium carriers in SCR). Expression (5) transfers into that obtained in [6] at substitution $L_n = 0$. The calculation by formula (5) shows that the surface recombination S_n and the direct electron emission into the metal v_n lead to a photosensitivity decay in the short-wave region of the spectrum and the diffusion of electrons from the bulk ($L_n \neq 0$) enhances the decay. The effect is most important in slightly doped material (where the maximum field E_m is weaker) and for direct bias voltage on the diode (the power phototransformers work in these conditions).

But in expression (5) the recombination in SCR was not taken into consideration. Though it is not as intensive as in the bulk, it can significantly decrease the photosensitivity of the Schottky diodes. As it follows from the analysis according to statistics of Shockley-Read recombination [7], the current of recombination losses in SCR can be approximately presented as

$$j_{pr} \approx 2\, \frac{\Delta p(l)}{\tau_p(x_m)} \left(\frac{c_n N_d}{c_p\, \Delta p(l)}\right)^{1/2} \left|\frac{d\psi}{dx}\right|^{-1}_{x=x_m}, \tag{6}$$

where c_n and c_p are the cross-sections of electron and hole capture by the recombination centres, $x = x_m$ is the position of the maximum of the function of the recombination losses calculated by the formula $\psi(x_m) = \ln\,(c_n N_d / c_p\, \Delta p(l))^{1/2}$, $\tau_p(x_m)$ is the lifetime of the minority current carriers in the point $x = x_m$. As it follows from (6) at increasing photon energy ($\hbar\omega$) the photosensitivity decay starts earlier than the losses due to surface recombination become considerable. Moreover, the spectrum of losses $j_{pr}(\hbar\omega)$ also has a maximum which increases under forward voltage and approaches the surface. The value of these losses in a typical Au–GaAs diode with $N_d = 10^{16}$ cm^{-3}, $L_p = 10^{-4}$ cm, $I = 10^{12}$ cm^{-2} s^{-1} without an injecting voltage is equal to about 15% at maximum.

3. The Photocurrent of the Real Barrier

Above in the calculation of the main component of photocurrent (hole current) we considered the ideal diodes, which possess no intermediate insulating layer between the metal and the semiconductor. Another situation takes place in real diodes with such a layer. This situation was earlier considered in [8], where for the extreme case of tunnel MOS structures the photocurrent was calculated according to the formula: $i_{ph} = A\,\Delta p(0)$ assuming an accumulation of minority carriers at the interlayer and a very weak perturbance of their distribution within SCR by the photocurrent. In the same work the recombination in SCR and surface recombination were taken into account for the rough approximation of a linear (but not of parabolic) potential distribution. The qualitative analysis of carrier accumulation in Schottky photodiodes carried out in [9] has shown that because of an increasing number of carrier collisions within the boundary potential well, the tunnel current across the dielectric interlayer into the metal increases and for transparency coefficients $D_p^* \gg 2s_p/v_T$ where s_p and v_T are surface recombination and thermal velocity of minority carriers the barrier photosensitivity is practically not weakened due to the existence of the interlayer. Moreover a detailed calculation of the photocurrent under such conditions [5, 7] has shown that its spectral characteristics are close to those of the above discussed case of a nearly ideal barrier. However, for real barriers other boundary conditions are preferable, which represent a particular case of the general boundary conditions given by Pekar for the metal–semiconductor contact [10]

$$j_n(0) = -(s_n + v_n)\,\Delta n(0)\,, \qquad j_p(0) = -(s_p + v_p)\,\Delta p(0)\,, \tag{7}$$

where s_p and s_n are surface recombination velocities for holes and electrons, respectively, v_n and v_p their emission rates into the metal through the interlayer depending on the transparency coefficient $D^*_{n,p}(T)$: $v_{n,p} = (v_{T\,n,p}/4)\,D_{n,p}(T)$. It should be noted that in the stationary case the equality $s_n\,\Delta n(0) = s_p\,\Delta p(0)$ holds. The final expression for the short-circuit photocurrent (assuming that recombination in SCR has a negligible effect) has the form

$$i_{ph} = qI\,\frac{1 - e^{-kl}/(1 + kL_p) + I_2}{1 + I_1} + qj_n(0)\left[1 + \frac{s_n D_n\,e^{-\varphi(0)}}{s_p(s_n + v_n)\,L_p(1 + I_1)}\right], \tag{8}$$

$$I_1 = \frac{1}{L_p}\int_0^l e^{-\varphi(x)}\,dx\,, \qquad I_2 = \frac{1}{L_p}\int_0^l (1 - e^{-kx})\,e^{-\varphi(x)}\,dx\,,$$

where the expression for $j_n(0)$ coincides with the last term of (5). A comparison of (5) with (1) shows that even neglecting the electron contribution $j_n(0)$ the photocurrent for a real diode is lower than that for an ideal one. Only for $I_1, I_2 \ll 1$ their magnitudes and the spectral characteristics of the photocurrent coincide with each other. The corrections which have the order of l/L_p, are remarkable for illumination of the semiconductor with light from the edge of the fundamental absorption for the case of low doping level and small diffusion length L_p. For example in Au–GaAs diodes with

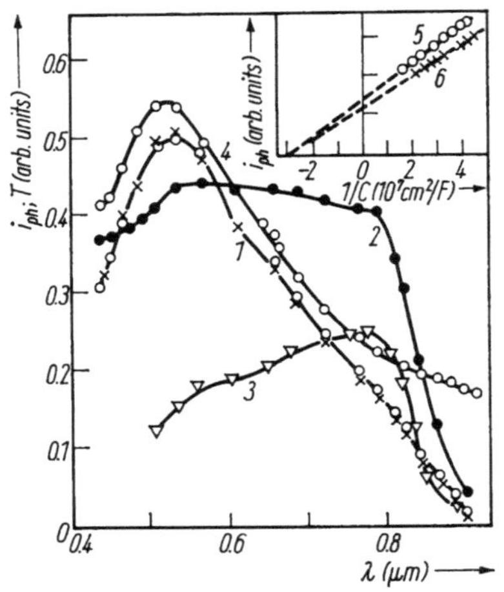

Fig. 1. $i_{ph}(\lambda)$ dependence of Au–GaAs Schottky diodes (1) $N_d = 4.7 \times 10^{16}$ cm^{-3}, $n = 1.06$ (○), $n = 1.6$ (×); (2) and (3) $N_d = 1.4 \times 10^{14}$ cm^{-3}, $n = 1.08$ (2), $n = 1.45$ (3); (4) Optical transmittance of Au film on a quartz substrate. The inset shows the i_{ph} $(1/C)$ dependence at $\lambda = 0.865$ μm, $N_d = 4.7 \times 10^{16}$ cm^{-3}, $n = 1.06$ (5), and $n = 1.6$ (6)

$L_p = 1.4 \times 10^{-5}$ cm and $N_d = 1.3 \times 10^{16}$ cm^{-3} the correction for the photocurrent ($k = 2 \times 10^3$ cm^{-1}) achieves about 10%, whereas for $k = 10^5$ cm^{-1} only 1.5%. For the forward biases occurring in phototransformers under load the correction increases strongly.

4. Experimental Results and Discussion

To check the above-mentioned spectral, field, and current–voltage (C–U) characteristics (at 1 MHz) were measured experimentally at room temperature on Au–GaAs Schottky diodes, performed by vacuum evaporation on weakly heated GaAs surfaces.

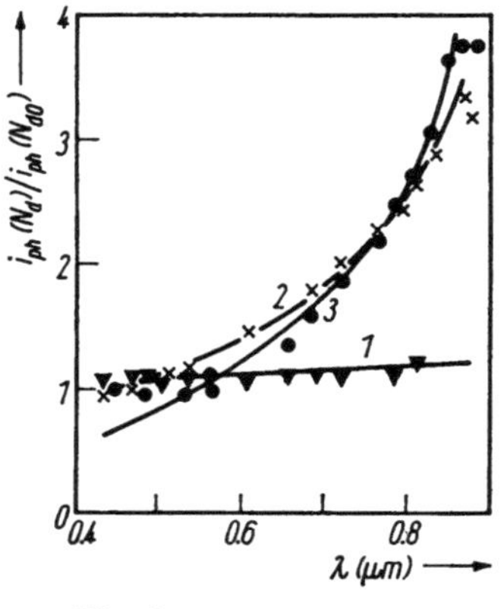

Fig. 2

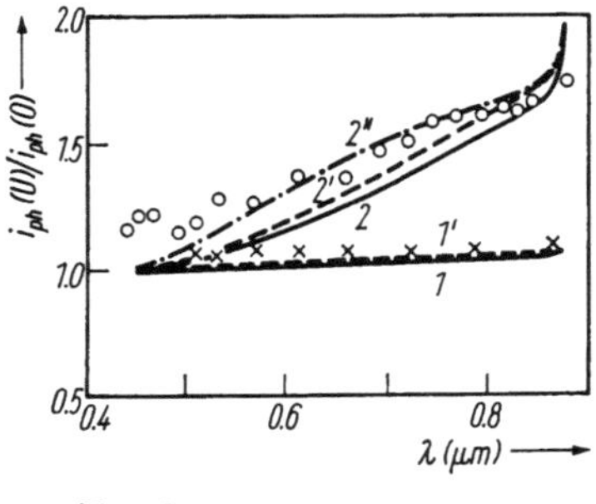

Fig. 3

Fig. 2. Dependence of $i_{ph}(N_d)/i_{ph}(N_{d0})$ on λ at $N_{d0} = 2.5 \times 10^{17}$ cm^{-3}, $L_{p0} = 0.3$ μm: (1) $N_d = 4.7 \times 10^{16}$ cm^{-3}, $L_p = 0.29$ μm; (2) 4.8×10^{16} cm^{-3}, 8 μm; (3) 1.4×10^{14} cm^{-3}, 6.8 μm. Points are measured values, solid curves represent dependences calculated according to formula (5)

Fig. 3. Dependence of $i_{ph}(U)/i_{ph}(0)$ on λ at $U = -4$ V before (1), (1′) and after (2), (2′), (2″) irradiation of Au–GaAs diodes with 4 MeV electrons. (Before irradiation $N_d = 2.8 \times 10^{16}$ cm^{-3}, $L_p = 3.5$ μm, after it $N_d = 1.3 \times 10^{16}$ cm^{-3}, $L_p = 0.14$ μm.) Points are measured values, solid curves represent results of the calculation according to formula (1); dashed lines calculated according to formula (8) and dash-dotted line calculated according to formula (8) with the recombination in SCR being taken into account according to formula (6)

The experimental conditions could be varied 1. by varying the interval between etching and gold evaporation it was possible to vary the thickness of the insulating layer, which was evaluated from the nonideality parameter of forward voltage–current characteristics (n); 2. by using materials with different doping by shallow donors N_d it was possible to distinguish the contributions of the surface recombination channel and of the direct transfer of electrons to the metal; 3. by irradiation with fast electrons up to doses $\Phi = 2 \times 10^{15}\,\text{cm}^{-2}$ ($E = 4$ MeV) the diffusion length could be varied, which was determined by the $i_{ph}(1/C)$ dependence cut-off at the axis $1/C$ in the region of the fundamental absorption edge.

Fig. 1 shows the spectral characteristics of four diodes, based on slightly doped ($N_d = 1.4 \times 10^{14}\,\text{cm}^{-3}$) and heavily doped ($N_d = 4.7 \times 10^{16}\,\text{cm}^{-3}$) GaAs with two value of the nonideality parameter ($n = 1.07$ and 1.6). It is seen that the characteristics $i_{ph}(1/C)$ of heavily doped GaAs do not depend on the nonideality parameter, but in slightly doped GaAs in the case of large n, the short-wave decay is sharper. These differences are explained by differences in the diffusion lengths of the material (L_p) and by the effect of surface recombination velocity being remarkable for $N_d \leqq 2 \times 10^{16}\,\text{cm}^{-3}$. Fig. 2 presents the photocurrent ratio for the cases of slight and heavy doping and the theoretical dependences according to (5). It is seen that indeed in the material with $N_d = 1.4 \times 10^{14}\,\text{cm}^{-3}$ the surface recombination velocity plays a substantial role and can be evaluated by matching the experimental and theoretical curves. As a result $D_n/(s_n + v_n) \approx 10^{-6}$ cm.

The $i_{ph}(\lambda)$ dependences for materials with different L_p for the long-wave region are usually described by (1), which allows to determine the value of L_p before and after irradiating the diodes with fast electrons. The relative spectra $(i_{ph}U)/i_{ph}(0)$ convenient for a comparison with theory are shown in Fig. 3. It is seen that a better agreement with (8) can be achieved for the material with low L_p, the recombination in SCR being taken into account according to (6).

Thus the theoretical analysis of photocurrent allows to explain principal regularities of spectral and field characteristics of Au–GaAs Schottky barriers.

References

[1] W. W. GÄRTNER, Phys. Rev. **116**, 84 (1959).

[2] A. M. VASILYEV and A. L. LANDSMAN, Semiconductor photoconverters, Soviet Radio, Moscow 1971 (in Russian).

[3] S. S. LI, F. A. LINDHOLM, and C. T. WANG, J. appl. Phys. **43**, 4123 (1972).

[4] N. L. DMITRUK and V. A. ZUEV, Ukr. fiz. Zh. **11**, 1176 (1966).

[5] N. L. DMITRUK and O. YU. BORKOVSKAYA, Ukr. fiz. Zh. **24**, 524 (1979).

[6] A. A. GUTKIN and V. B. SEDOV, Fiz. Tekh. Poluprov **9**, 1761 (1975).

[7] N. L. DMITRUK and O. YU. BORKOVSKAYA, Mikroelektronika **8**, 64 (1979).

[8] N. L. DMITRUK and A. K. TERESHENKO, Elektronnaya tekhnika, Ser. 2, N 4, 68 (1973)..

[9] A. A. GUTKIN and V. B. SEDOV, Fiz. Tekh. Poluprov. **10**, 1589 (1976).

[10] S. I. PEKAR, Izv. Akad. Nauk SSSR, Ser. Fiz., N4–5, 422 (1941).

(Received April 9, 1979)

phys. stat. sol. (a) 54, 529 (1979)

Subject classification: 5; 18.1; 22.8.2

Institut de Physique B5, Université de Liège[1])

A Mössbauer Study of the Effects of the Substitution of Fe by Cr in $SrFe_{12}O_{19}$

By

P. M. RAO[2]), A. GÉRARD, and F. GRANDJEAN

The study of the preferential substitution of Fe by Cr in the series $SrFe_{12-x}Cr_xO_{19}$ allows an unambiguous attribution of the hyperfine parameters of the Mössbauer spectra to the five crystallographic sites of the structure. The evolution of the crystalline parameters and of the magnetic properties as a function of Cr content is also described.

En étudiant la substitution préférentielle du Fe par le Cr dans la série $SrFe_{12-x}Cr_xO_{19}$, nous avons pu attribuer sans ambiguité les paramètres hyperfins des spectres Mössbauer aux cinq sites cristallographiques de la structure. Nous décrivons aussi l'évolution des paramètres cristallins et des propriétés magnétiques en fonction du contenu en Cr.

1. Introduction

Because of their remarkable magnetic properties, hexagonal ferrites have given rise to several studies of fundamental or applied scientific interest. Indeed, both $BaFe_{12}O_{19}$ and $SrFe_{12}O_{19}$ have large coercive forces and remanences. Recently, $SrFe_{12}O_{19}$ has been found superior to $BaFe_{12}O_{19}$ with a remanent magnetization within 20% of the theoretical value while maintaining the coercive force above 3000 G [1].

The Curie temperature of $SrFe_{12}O_{19}$ amounts to (723 ± 5) K [2].

The crystalline structure of the M-type hexagonal ferrites, as $SrFe_{12}O_{19}$, may be described as a superposition of hexagonal blocks (R) containing Sr^{2+} ions and spinel blocks (S). The iron ions are distributed among five lattice sites different in coordination and spin orientation [2, 3] as described in Table 1.

Table 1

The five iron sites in $SrFe_{12}O_{19}$, their coordination, point symmetry, population, spin orientation, and block situation

site	coordination	point symmetry	population	spin	block
12k	octahedral	m	6	up	S–R
$4f_1$	tetrahedral	3m	2	down	S
$4f_2$	octahedral	3m	2	down	R
2a	octahedral	$\bar{3}$m	1	up	S
2b	trigonal or bipyramidal	$\bar{6}$m2	1	up	R

[1]) B-4000 Sart-Tilman par Liège I, Belgium.

[2]) Present address: Department of Material and Metallurgical Engineering, Stevens Institue of Technology, Hoboken, New Jersey 07030, USA.

Several authors have analyzed the room temperature Mössbauer spectrum of $SrFe_{12}O_{19}$ [1, 2, 4 to 7]. In spite of differences in resolution, they all found the same general shape. But the attribution of the different components of the spectrum (four or five sextets) is very difficult, because of the lines superposition. More realistic decompositions have been given by Van Wieringen and Rensen [4] and Vogel and Evans [1]. This latter one will be discussed in comparison with our results and conclusions.

In order to make this attribution more easily, we have substituted part of the Fe^{3+} ions by Cr^{3+} ions. Indeed, the Cr^{3+} ion usually does not occupy tetrahedral sites, thus we may hope to substitute ten of the twelve Fe^{3+} ions by Cr^{3+} ions. We have prepared and studied the series $SrFe_{12-x}Cr_xO_{19}$ with $x = 0, 3, 6, 8, 9, 10$.

The compositions with $x = 2$ and 6 have been studied by neutron diffraction [8]. The authors conclude that Cr^{3+} ions are equally distributed among sites 2a and 12k and among sites 2a, 12k and $4f_2$, respectively.

The substitution of Fe by Cr not only decreases the number of different sites for Fe ions but also lowers the Curie temperature. The compound $SrCr_{3.2}Fe_{8.8}O_{19}$ has a Curie temperature amounting to (580 ± 4) K [9]. For higher x-values, we shall see that the compounds are paramagnetic at lower temperatures. Thus, paramagnetic spectra may be recorded at reasonably high temperatures.

2. Experimental

The samples were prepared from equimolal quantities of the carbonate $SrCO_3$ and of the oxides Fe_2O_3 and Cr_2O_3, in powders. They were crushed in an agate mortar, put in an alumina crucible, and heated up to 950 °C for several days and then between 1200 and 1300 °C for at least 1 d.

X-ray spectra show the lines of the $SrFe_{12}O_{19}$ structure. Traces of Fe_2O_3 could not be found neither in the X-ray spectra nor in the Mössbauer spectra. The crystalline parameters of the series $SrFe_{12-x}Cr_xO_{19}$ were determined from the X-ray spectra and are plotted in Fig. 1 as a function of x. In comparison, the results obtained by Bertaut et al. [8] for the $BaFe_{12-x}Cr_xO_{19}$ series and the compounds $SrFe_{12-x}Cr_xO_{19}$ with $x = 2$ and 6 are reproduced. The good agreement between both series is a proof of the good quality of our samples. The accuracy of our X-ray measurements is about ±0.01 Å on the a-axis and ±0.05 Å on the c-axis.

The Mössbauer spectrometer is of constant acceleration type, with a 10 mC source of ^{57}Co diffused in a Rh matrix. The high-temperature spectra have been recorded in a vacuum furnace built in the laboratory, the temperature being measured and controlled with an accuracy of 0.025 K by a Cr–Al thermocouple. The temperature scale of the furnace has been calibrated against the temperature dependence of the hyperfine field of Fe_2O_3.

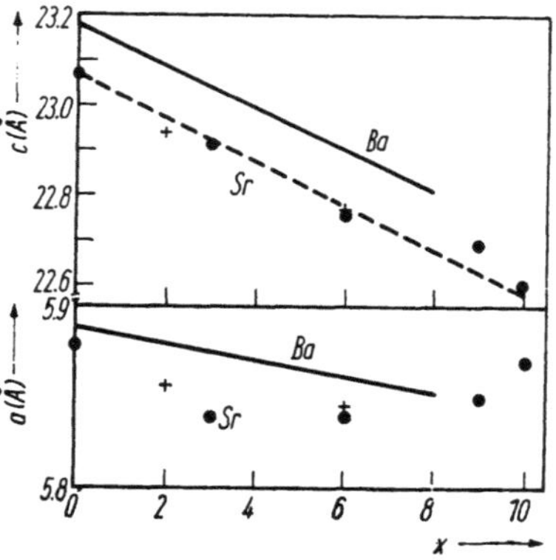

Fig. 1. Crystalline parameters of $Sr(Ba)Fe_{12-x}Cr_xO_{12}$ series as a function of Cr content. For Sr: ● our results, + [8]; for Ba: [8]

3. Mössbauer Spectra

3.1 Paramagnetic spectra

Fig. 2 displays the paramagnetic spectra of the series $SrFe_{12-x}Cr_xO_{19}$, with $x = 10, 9, 8$ at room temperature and with $x = 6, 3$ at high temperature. All the spectra were least-squares fitted by a sum of several symmetrical doublets and the parameters are given in Table 2.

In $SrFe_2Cr_{10}O_{19}$, the iron atoms are expected to occupy only the $4f_1$ sites. The spectrum (Fig. 2a) shows one main intense (97%) doublet (No. 2 in Table 2) and a second one (No. 1) with a very low intensity (3%). The main doublet is associated with iron on $4f_1$ sites and the other one with iron on 2b sites which is well known to present a large quadrupole splitting [1] because of its strongly distorted environment. The linewidths of both doublets (0.24 and 0.34 mm s^{-1}) do not imply any further decomposition of the spectrum and are smaller than those given for the magnetic spectrum of $SrFe_{12}O_{19}$ by Vogel and Evans [1].

The spectrum of $SrFe_3Cr_9O_{19}$ (Fig. 2, curve b) clearly shows the doublet associated with site 2b (No. 1). Its intensity amounts to 17% and thus, iron occupies first $4f_1$ sites and then 2b. The central part of Fig. 2, curve b was decomposed into three doublets, whose linewidths show no more possible decomposition. Doublet No. 2 is easily associated with $4f_1$ sites by comparison with the spectrum of $SrFe_2Cr_{10}O_{19}$. Up to now, it is impossible to attribute the other two doublets (No. 3 and 4) to sites 12k, $4f_2$, or 2a.

The same analysis was done for $x = 8, 6, 3$. For $x = 8$, the intensity of 2b site is abnormaly low and this is probably due to a lower first firing temperature of the compound. The X-ray spectrum confirms a sample of low quality but as it is not determinant for our discussion, we did not try to improve it. As expected, the relative intensities of 2b and $4f_1$ sites decrease as x decreases. Furthermore, the relative intensity of sites $4f_1$ is in good agreement with theoretical values for the different x-values, if this site is preferentially occupied by Fe. This attribution gives to the tetrahedral sites (No. 2) an isomer shift greater than those observed for the octahedral sites (No. 3 and 4). This is in opposition to the attribution done by Vogel and Evans

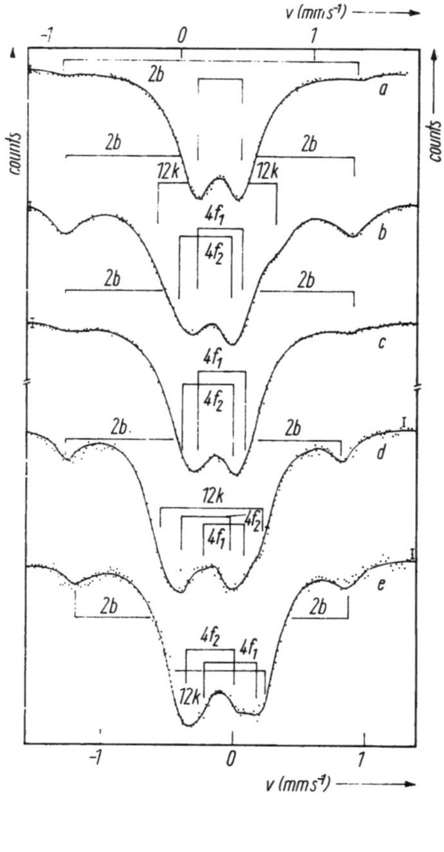

Fig. 2. Mössbauer spectra of $SrFe_{12-x}Cr_xO_{19}$ series (a) $x = 10$, (b) 9, (c) 8 at $T = 293$ K, the velocity scale for (a), (b), and (c) is on the top. (d) $x = 6$, $T = 870$ K; (e) $x = 3$, $T = 800$ K. The velocity scale for (d) and (e) is on the bottom. For (a) to (e), the solid lines are the results of least-squares fits with symmetrical doublets as discussed in the text

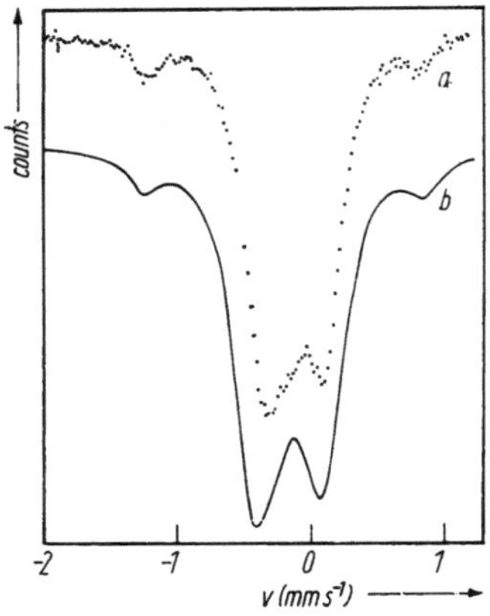

Fig. 3. (a) Mössbauer spectrum of $SrFe_{12}O_{19}$ at 870 K. (b) Mössbauer spectrum computed from hyperfine parameters given in Table 2

[1]. We have no definite explanation for this unexpected large isomer shift on a tetrahedral site. The intensity of the largest quadrupole splitting (No. 3) rapidly varies with Cr substitution, thus it could be attributed to 12k site which is known to be first occupied by Cr [8] and which must have the fastest population variation. This is in agreement with the magnetic spectrum analysis.

The paramagnetic spectrum of $SrFe_{12}O_{19}$ at 870 K (Fig. 3, curve a) is very similar to the spectrum observed by Kreber et al. [10] for $BaFe_{12}O_{19}$ at 870 K. Since this spectrum is the result of the superposition of five doublets, it is extremely difficult to decompose it into several doublets with a classical least-squares procedure. Thus, we followed another way.

We synthesized the spectra by superimposing doublets. We deduced the hyperfine parameters of these doublets from the results obtained on the $SrFe_{12-x}Cr_xO_{19}$ series, assumed a variation with temperature amounting to about 10%, between room temperature and 870 K, associated the hyperfine parameters with the sites as given in Table 2, and obtained the spectrum of Fig. 3, curve b rather similar to the experimental one. In order to check the values given by Vogel and Evans [1], we computed a spectrum with these values and concluded that the high-velocity line was broader than the low-velocity line which is in contradiction with the experiment. Furthermore, the quadrupole splitting of 2b site amounting to 2.6 mm s^{-1} is obviously too large.

Our attribution of the hyperfine parameters to 12k, $4f_2$, and 2a sites, if it gives a spectrum in good agreement with the experimental one, is still rather arbitrary. The analysis of the magnetic spectrum will confirm it.

3.2 Magnetic spectrum of $SrFe_{12}O_{19}$

Fig. 4, curve a displays the magnetic spectrum of $SrF_{12}O_{19}$ at 300 K. It is nearly identical to the spectrum of sample No. 57 shown by Vogel and Evans [1]. Because of the superposition of five sextets, it is very difficult to fit the spectrum by means of the least-squares method.

In Fig. 4, curve b a spectrum is computed from the hyperfine parameters given in Table 3, in agreement with the results obtained on the paramagnetic spectra of $SrFe_{12-x}Cr_xO_{19}$ series. The point symmetry for $4f_2$, $4f_1$, 2a, and 2b sites implies that the asymmetry parameter η equals zero and that the electric field gradient tensor principal axis is parallel to the c-axis and thus parallel to the easy magnetization axis. Consequently, the angle θ between these two directions is equal to zero. For 12k site, the point symmetry allows θ to be different from zero and a value of 30° is not excluded. In first approximation, η is taken equal to zero for this site. The linewidth is equal to 0.32 mm s^{-1} which is the mean value observed in the paramagnetic spectra (Table 2). The intensity ratios of the five Zeeman patterns are taken equal to the five sites population ratios of the stoichiometric compound.

It is clear that the calculated spectrum accounts very well for the experimental one. The slight differences which subsist may be due to deviations from the theoretical

Table 2

Mössbauer parameters of the paramagnetic spectra of the series $SrFe_{12-x}Cr_xO_{19}$

x	T (K)	No. 1 site 2b δ	ε	Γ	I	No. 2 site $4f_1$ δ	ε	Γ	I	No .3 site 12k δ	ε	Γ	I	No. 4 site $4f_2$ δ	ε	Γ	I	No. 5 site 2a δ	ε	Γ	I
10	295	0.20	2.25	0.24	3	0.23	0.34	0.34	97												
9	295	0.17	2.18	0.30	17	0.25	0.33	0.32	40	0.23	0.90	0.24	9	0.13	0.38	0.32	34				
	467	0.03	2.10	0.30	16	0.11	0.36	0.30	51	0.06	0.87	0.14	6	−0.04	0.38	0.26	27				
8	295	0.16	2.20	0.30	4	0.28	0.36	0.32	48					0.16	0.38	0.34	48				
6	489	0.03	2.15	0.24	11	0.14	0.37	0.30	33	0.08	0.77	0.24	20	0.01	0.39	0.28	36				
	870	−0.18	2.08	0.24	10	−0.03	0.32	0.20	10	−0.12	0.77	0.26	27	−0.17	0.37	0.31	52				
3	800	−0.15	2.06	0.28	9	0.01	0.40	0.26	24	−0.07	0.68	0.32	28	−0.14	0.37	0.30	39				
0*)	870	−0.20	2.10	0.32	8	−0.10	0.27	0.32	17	−0.15	0.55	0.32	50	−0.20	0.33	0.32	17	−0.32	0.25	0.32	8

δ is the isomer shift relative to Rh, ε the quadrupole splitting, and Γ the linewidth (all in mm s^{-1}); I the relative intensity in % (line area).
*) Values used to calculate the spectrum of Fig. 3, curve b.

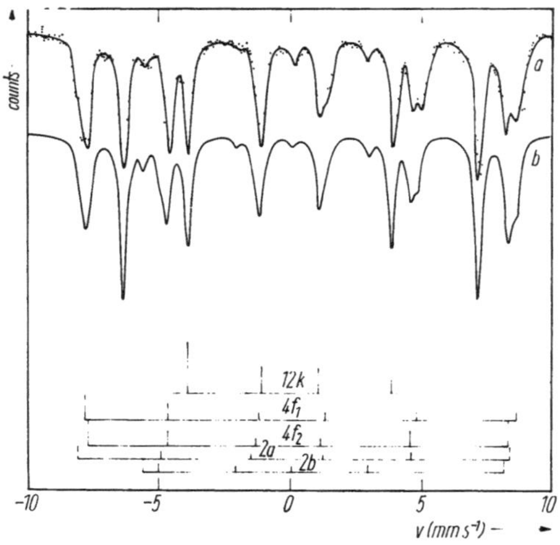

Fig. 4. (a) Mössbauer spectrum of $SrFe_{12}O_{19}$ at 300 K. The solid line is not a last-squares fit. (b) Computed spectrum with the hyperfine parameters given in Table 3

populations of the different sites. The experimental intensities of the 12k site, especially the low- and high-velocity lines show that either this site is not fully occupied or there is saturation. A trial to compute a spectrum with a 10% deficit in the 12k population did not fully account for the experimental intensities and both phenomena are probably coexisting.

The attribution of the hyperfine parameters to site 12k is unambiguous since it is the most intense one and is in agreement with the conclusions drawn from the paramagnetic spectra. Consequently, the remaining sites $4f_2$ and 2a have very similar hyperfine parameters. We see the presence of site 2a in the left wing of the lines at -8 and -4.5 mm s^{-1} and in the right wing of the line at 8.5 mm s^{-1}. Thus, we do the attribution given in Table 3 and Fig. 4.

Table 3
Hyperfine parameters of $SrFe_{12}O_{19}$ at 300 K

site	intensity (%)	H (kOe)	ε (mms^{-1})	δ/Rh (mms^{-1})	θ
12k	50	415	0.70	0.25	30°
$4f_2$	17	493	0.38	0.20	0°
$4f_1$	17	507	0.30	0.30	0°
2a	8	508	0.28	0.08	0°
2b	8	425	2.30	0.20	0°

Compared to the attribution done by Vogel and Evans [1], the quadrupole splitting of 12k is higher, but an angle of 30° is introduced, sites $4f_1$ and $4f_2$ are reversed, the quadrupole splitting of 2b is lower, the hyperfine fields of all sites are very similar.

3.3 *Magnetic spectra of $SrFe_{12-x}Cr_xO_{19}$*

Punctual informations about the magnetic behaviour of the series $SrFe_{12-x}Cr_xO_{19}$ were obtained.

The compound $SrFe_9Cr_3O_{19}$ is magnetically ordered at room temperature. Its Mössbauer spectrum is in agreement with a decreasing population of Fe on 12k sites

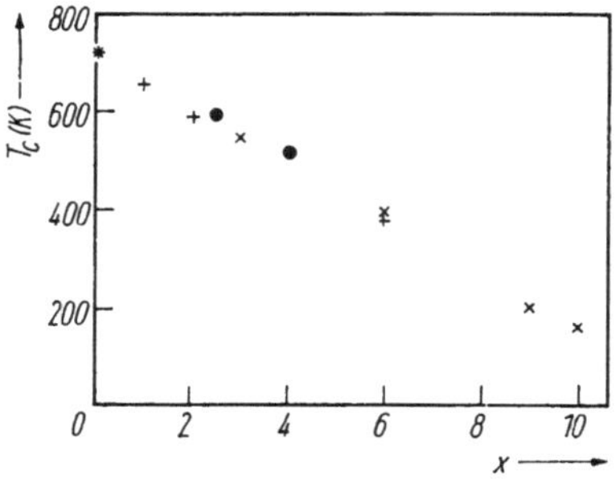

Fig. 5. Curie temperature of $BaFe_{12-x}Al_xO_{19}$ (●), $SrFe_{12-x}Ga_xO_{19}$ (+) (from [2] and [5]), and $SrFe_{12-x}Cr_xO_{19}$ (×) (this work) as a function of x

compared to $SrFe_{12}O_{19}$ and in agreement with the paramagnetic data (Table 2). The Curie temperature was roughly determined as equal to (550 ± 25) K in very good agreement with the value of 580 K for $x = 3.2$ [9]. $SrFe_6Cr_6O_{19}$ has a Curie temperature around (400 ± 10) K. $SrFe_3Cr_9O_{19}$ is magnetically ordered below (220 ± 10) K. The spectrum at 77 K is compatible with a very low iron population on 12k sites.

$SrFe_2Cr_{10}O_{19}$ has a Curie temperature amounting to (175 ± 20) K and a hyperfine magnetic field on $4f_1$ site amounting to about 440 kOe at 77 K. The broadness of the lines at 77 K could not be explained without knowledge of the magnetic structure of Cr ions.

In Fig. 5 we have plotted as a function of x the Curie temperatures given in [2] and [5] for $BaFe_{12-x}Al_xO_{19}$ and $SrFe_{12-x}Ga_xO_{19}$, and determined in this work for $SrFe_{12-x}Cr_xO_{19}$. The agreement between the three series and the linear variation are remarkable. Surprisingly the magnetic Cr ions have the same effect as the non-magnetic Al and Ga ions. Thus, the magnetic interaction between Fe ions seems to determine the magnetic properties of these compounds.

4. Conclusions

By studying the Mössbauer spectra of the solid solutions $SrFe_{12-x}Cr_xO_{19}$, we have been able to associate the hyperfine parameters with the five crystallographic sites of the structure. Indeed, as generally accepted, Cr atoms do not occupy the tetrahedral sites, this directly allows the attribution of $4f_1$ site. The 2b site has a very large quadrupole splitting because of its strongly distorted environment and is therefore automatically attributed. The evolution of the intensity of a doublet (No. 3) in the paramagnetic spectra with increasing Cr content lets us assume that it corresponds to 12k site. Thus, from the analysis of the paramagnetic spectra, three sites are attributed. From the magnetic spectrum of $SrFe_{12}O_{19}$, 12k site is unambiguously attributed because of its intensity, confirming the assumption made in the analysis of paramagnetic spectra. Consequently 2a and $4f_2$ sites have very similar parameters, but 2a sites may be separated in the highly resolved magnetic spectrum. In conclusion, the five sites are finally attributed.

Acknowledgement

One of the authours (P.M.R.) is extremely thankful to Prof. J. Duchesne for providing him with a postdectoral fellowship to carry out this work.

References

[1] R. H. Vogel and B. J. Evans, J. appl. Phys. 49, 1570 (1978).
[2] G. Albanese, M. Carbucicchio, and A. Deriu, phys. stat. sol. (a) 23, 351 (1974).
[3] V. Adelsköld, Ark. Kemi, Mineralog. Geol., Ser. A 12, No. 29, 1 (1938).

[4] J. S. VAN WIERINGEN and J. G. RENSEN, Z. angew. Phys. **21**, 69 (1966).
[5] G. ALBANESE, G. ASTI, and P. BATTI, Nuovo Cimento **54**, 339 (1968).
[6] T. A. KHIMICH, V. F. BELOV, M. N. SHIPKO, and E. V. KORNEEV, Soviet Phys. — J. exper. theor. Phys. **30** 217 (1970).
[7] V. F. BELOV, T. A. KHIMICH, M. N. SHIPKO, I. S. ZHEDULEV, E. V. KORNEEV, and N. S. OVANESYAN, Soviet Phys. — J. exper. theor. Phys. **37**, 1089 (1973).
[8] E. F. BERTAUT, A. DESCHAMPS, R. PAUTHENET, and S. PICKART, J. Phys. Radium **20**, 404 (1959).
[9] K. P. BELOV, YU. V. JERGIN, L. I. KOROLEVA, R. Z. LEVITIN, and A. V. PEDKO, phys. stat. sol. **12**, 219 (1965).
[10] E. KREBER, U. GONSER, A. TRAUTWEIN, and F. E. HARRIS, Proc. V. Internat. Conf. Mössbauer Spectroscopy, Part 1, Nuclear Information Centre, Bratislava 1973 (p. 130).

(Received May 10, 1979)

phys. stat. sol. (a) 54, 537 (1979)

Subject classification: 7; 1.4; 21

Reactor Research Centre, Kalpakkam[1])

Theory of Ultrasonic Scattering by Grains in Polycrystals

By

R. RANGANATHAN and S. DATTAGUPTA

A theory of ultrasonic scattering by a polycrystal, valid for a wide range of ratios of wavelength and grain-size, is presented. The present treatment concerns the study of wave equations with random elements to take account of the fact that the elastic constants of the crystal change values abruptly at grain boundaries. The solutions of the wave equations, which are valid for weak inhomogeneities, are given in terms of the spatial correlation function of elastic moduli. Appropriate modelling of this correlation function yields closed expressions for the ultrasonic attenuation coefficient. These can be studied in the limiting cases of large and small wavelengths in order to make contact with existing theoretical results. Satisfactory agreement of the present theory with experimental data in Al is also obtained.

Es wird eine Theorie der Ultraschallstreuung durch einen Polykristall angegeben, die für einen weiten Bereich des Verhältnisses von Wellenlänge zur Korngrenze gültig ist. Die Methode umfaßt die Untersuchung der Wellengleichung mit statistischen Elementen um die Tatsache zu berücksichtigen, daß die elastischen Konstanten des Kristalls sich abrupt an den Korngrenzen ändern. Die Lösungen der Wellengleichungen, die für schwache Inhomogenitäten gültig sind, werden durch die räumlichen Korrelationsfunktionen der Elastizitätsmoduln ausgedrückt. Eine geeignete Modellierung dieser Korrelationsfunktion ergibt geschlossene Ausdrücke für den Ultraschalldämpfungskoeffizienten. Diese können im Grenzfall großer und kleiner Wellenlängen untersucht werden und mit existierenden theoretischen Ergebnissen in Zusammenhang gebracht werden. Befriedigende Übereinstimmung der dargestellten Theorie mit experimentellen Werten für Al wird ebenfalls erhalten.

1. Introduction

The scattering of ultrasonic waves offers a useful technique for investigating the grains in a polycrystalline medium. The method has important technological applications in testing for (i) flaws in forgings, (ii) improper fabrication of materials, etc. [1].

In general, any inhomogeneity in a material which differs in elastic modulus or density from the surrounding medium, scatters elastic waves. In a polycrystalline medium, the inhomogeneity arises from grains which are oriented differently with respect to each other. The theoretical calculation of ultrasonic scattering in a polycrystal is an involved one and progress has been made [1] only when the following conditions are satisfied:

(i) The grains are weakly anisotropic, i.e. the deviation of the modulus from its average value is much smaller than the modulus itself.

(ii) Preferred orientation is absent, i.e. the grains are all randomly oriented.

(iii) The grains are roughly spherical ("equiaxed").

(iv) The material has single-crystal grains of a single phase with no voids or inclusions. The latter condition implies that the elastic anisotropy of a grain is the major contributory factor to scattering, and density variations may be neglected.

Even after making the simplifying assumptions enumerated above, the theoretical study has been limited to two extreme cases in which the ratio of the wavelength of the ultrasound (λ) and the average grain diameter ($\bar{D}$) is $\gg 1$ or $\ll 1$. The expressions

[1]) Kalpakkam 603102, India.

for the attenuation coefficient α (in nepers per unit length) are then obtained as follows:

$$\alpha = Tf^4 S\,, \qquad \lambda \gg 2\pi \bar{D}\,, \tag{1}$$

$$\alpha = \bar{D} f^2 \Sigma\,, \qquad \lambda \ll 2\pi \bar{D}\,, \tag{2}$$

where f is the ultrasonic frequency, $\bar{D}$ and T are the average diameter and volume of the grain, respectively, and S and Σ are parameters of the material. Often in practice, however, one has to deal with a situation in which the ratio of λ and $\bar{D}$ is arbitrary, a possibility which the existing theories cannot handle [2]. Attempts to interpret experimental attenuation data in the intermediate region ($\lambda \sim \bar{D}$) are rather crude [1]. In this paper, we present a calculation of α which is valid for arbitrary values of $\lambda/\bar{D}$. Results given in (1) and (2) then follow as limiting cases of the present theory.

The mathematical discussion presented here is based on the same simplifying assumptions (i) to (iv) listed above. The calculation involves the solution of a standard set of four wave equations for the dilatational and shear components of the strain tensor [3]. However, the novel element of the present treatment is to regard the inhomogeneous terms in the above-mentioned wave equations as random. This is a consequence of the fact that the elastic modulus, which is constant within a grain, changes randomly at the grain boundary.

The application of the techniques to solve random differential equations [4] to the present set of wave equations in their complete generality is very complex. However, since the changes in the elastic constants from grain to grain are usually small, the equations can be solved in the Born approximation [4, 5]. Here also we follow a similar approach as has been elaborated by Chernov [6]. However, the Chernov method has to be extended to take account of the tensorial nature of the elastic properties which gives rise to an effect called mode conversion [1]. Mode conversion is especially important for $\lambda \gg \bar{D}$ when most of the scattered energy from an incident dilatational wave is carried off by shear waves.

We present in Section 2 the basic equations and their solutions. The attenuation coefficient α is expressed in terms of the correlation function of the spatially random elastic modulus. In Section 3, analytical expressions for α are given based on two assumed forms for the correlation function. The comparison of theory with experimental attenuation data in Al is presented in Section 3.3.

2. Wave Equations and Their Solutions

2.1 Basic equations

In setting up the wave equations, we follow entirely the notations used in [3]. As mentioned in Section 1, the grains are assumed to be randomly oriented, so that the bulk material may be regarded as elastically isotropic on the average, characterized by the two Lamé constants λ_0 and μ_0. The elastic constants of a grain, as encountered by an incident stress wave, are written as

$$\left.\begin{aligned} &\lambda = \lambda_0 + \lambda'\,, \qquad \mu = \mu_0 + \mu'\,, \\ &c = c_0 + c' = \lambda_0 + 2\mu_0 + \lambda' + 2\mu'\,, \end{aligned}\right\} \tag{3}$$

where the random variables λ' and μ' are treated as small perturbations (Born approximation). The dilatational component (Δ) and the shear components (φ_x, φ_y, and φ_z) can be similarly decomposed as

$$\Delta = \Delta^0 + \Delta'\,, \qquad \varphi_x = \varphi_x^0 + \varphi_x'\,, \quad \text{etc.}\,, \tag{4}$$

where the quantities with superscript zero are solutions of the wave equations with incident stress wave only; the primed quantities are the additional contributions (again assumed small) arising from the perturbation terms in (3). Introducing the propagation constants k and h for the dilatational and shear waves, respectively, as

$$k^2 = \frac{p^2}{V_1^2} = \frac{\varrho_0 p^2}{c_0}, \qquad h^2 = \frac{p^2}{V_s^2} = \frac{\varrho_0 p^2}{\mu_0} \tag{5}$$

with V_1 and V_s as the velocities of the dilatational and shear waves in a homogeneous medium of density ϱ_0, and p $(= 2\pi f)$, the angular frequency of the incident wave, Bhatia [3] derives

$$\left.\begin{aligned} \nabla^2 \Delta' + k^2 \Delta' &= \psi_1 , \\ \nabla^2 \varphi_x' + h^2 \varphi_x' &= \psi_2 , \\ \nabla^2 \varphi_y' + h^2 \varphi_y' &= \psi_3 , \\ \nabla^2 \varphi_z' + h^2 \varphi_z' &= \psi_4 . \end{aligned}\right\} \tag{6}$$

For a plane dilatational wave of amplitude A incident on the inhomogeneity in x-direction, the quantities ψ_1 to ψ_4 are given by

$$\psi_1 = -\frac{ikA}{c_0}\left[\frac{\partial^2 c'}{\partial x^2} + 2ik\frac{\partial c'}{\partial x} - k^2 c' - \frac{1}{2}\left(\frac{\partial^2 c'}{\partial y^2} + \frac{\partial^2 c'}{\partial z^2}\right)\right] e^{ikx} , \tag{7}$$

$$\psi_2 = 0 , \tag{8}$$

$$\psi_3 = -\frac{3ikA}{2\mu_0}\left(\frac{\partial^2 c'}{\partial z\, \partial x} + ik\frac{\partial c'}{\partial z}\right) e^{ikx} , \tag{9}$$

$$\psi_4 = \frac{3ikA}{2\mu_0}\left(\frac{\partial^2 c'}{\partial x\, \partial y} + ik\frac{\partial c'}{\partial y}\right) e^{ikx} . \tag{10}$$

In writing (7) to (10), we have neglected any possible variation in density from grain to grain (see also Section 1). We have also made use of the relation

$$c = \beta + \tfrac{4}{3}\mu , \tag{11}$$

where β is the bulk modulus, and the fact that β is independent of the orientation of a grain, i.e.

$$\beta' = 0 . \tag{12}$$

This enables us to express μ' entirely in terms of c' which is the single random variable that appears in the relevant equations.

In (6), we have equations for secondary scattered waves with ψ's as the source terms determined by the primary wave. The solutions for the waves scattered by a volume V can be written as

$$\Delta'(r) = -\frac{1}{4\pi}\int_V \frac{e^{ikr}}{r}\psi_1(\xi, \eta, \zeta)\, dV \quad (dV = d\xi\, d\eta\, d\zeta) , \tag{13}$$

$$\left.\begin{aligned} \varphi_x'(r) &= 0 , \\ \varphi_y'(r) &= -\frac{1}{4\pi}\int_V \frac{e^{ihr}}{r}\psi_3(\xi, \eta, \zeta)\, dV , \end{aligned}\right\} \tag{14}$$

and similarly for φ_z'. In (13) and (14) r is the distance between the source point (ξ, η, ζ) and the field point (x, y, z), i.e.,

$$r = [(x-\xi)^2 + (y-\eta)^2 + (z-\zeta)^2]^{1/2} . \tag{15}$$

2.2 The Chernov method

At this stage we depart from Bhatia [3] who straightaway evaluates the integrals in (13) and (14) by assuming the wavelength λ to be much larger than the dimension of the scattering volume. Instead, we have to consider correlations between the values of c' at different source points. Since the method adopted is a straightforward application of [6], we outline merely the crucial steps in the calculation.

We want to evaluate the integral in (13) over a cubic volume of side L such that $r \gg L$ but $L \gg r_0$ (Fig. 1), where r_0 is the correlation length associated with the random elastic modulus c'. From Fig. 1

$$r \sim R - \boldsymbol{n}_1 \cdot \boldsymbol{r}' , \tag{16}$$

where $\boldsymbol{n}_1$ is the unit vector in the scattering direction and $\boldsymbol{r}'(\xi, \eta, \zeta)$ defines the source point. We then have from (13)

$$\Delta'(r) \sim - \frac{e^{ikR}}{4\pi R} \int\limits_{V_1} e^{-ik\boldsymbol{n}_1 \cdot \boldsymbol{r}_1'} \psi_1(\boldsymbol{r}_1') \, dV_1 , \tag{17}$$

where from (7)

$$\psi_1(\boldsymbol{r}_1') = - \frac{ikA}{c_0} \left[\frac{\partial^2}{\partial \xi_1^2} + 2ik \frac{\partial}{\partial \xi_1} - k^2 - \frac{1}{2}\left(\frac{\partial^2}{\partial \eta_1^2} + \frac{\partial^2}{\partial \zeta_1^2}\right)\right] c'(\boldsymbol{r}_1') \, e^{ik\xi_1} . \tag{18}$$

The attenuation coefficient α can be shown to be related (see (28) below) to the statistical average of $|\Delta'|^2$, the average (denoted by $\langle ... \rangle$) being performed over the random variation of c'. The expression for $\langle |\Delta'|^2 \rangle$ can then be given in terms of the correlation function, $\langle c'(\boldsymbol{r}_1') \, c'(\boldsymbol{r}_2') \rangle$ by assuming that the operation $\langle ... \rangle$ can be interchanged with the integration and differentiation operations. This implies that the correlation function is spatially stationary. The expression for $\langle |\Delta'|^2 \rangle$ can be further simplified by introducing the relative coordinates $\boldsymbol{r}' = \boldsymbol{r}_1' - \boldsymbol{r}_2'$, and the centre of mass coordinates $\boldsymbol{r} = \frac{1}{2} (\boldsymbol{r}_1' + \boldsymbol{r}_2')$, and keeping in mind the fact that the correlation function $\langle c'(\boldsymbol{r}_1') \, c'(\boldsymbol{r}_2') \rangle$ depends only on $\boldsymbol{r}'$. Denoting the latter by $N(\boldsymbol{r}')$, we have relations like (cf. [6])

$$\frac{\partial N}{\partial \xi_1} = \frac{\partial N}{\partial \xi} , \qquad \frac{\partial N}{\partial \xi_2} = - \frac{\partial N}{\partial \xi} , \qquad \frac{\partial^2 N}{\partial \xi_1 \, \partial \xi_2} = - \frac{\partial^2 N}{\partial \xi^2} , \tag{19}$$

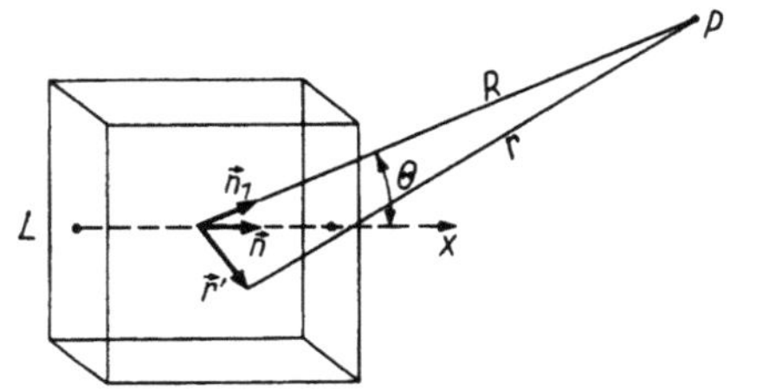

Fig. 1. The scattering configuration

etc.. We thus obtain

$$\langle|\Delta'|^2\rangle = \frac{k^2A^2V}{16\pi^2c_0^2R^2}\int\left\{\left[\frac{\partial^4}{\partial\xi^4}-\frac{\partial^2}{\partial\xi^2}\left(\frac{\partial^2}{\partial\eta^2}+\frac{\partial^2}{\partial\zeta^2}\right)+\frac{1}{4}\left(\frac{\partial^2}{\partial\eta^2}+\frac{\partial^2}{\partial\zeta^2}\right)^2\right]+\right.$$
$$+\,2ik\left[2\,\frac{\partial^3}{\partial\xi^3}-\frac{\partial}{\partial\xi}\left(\frac{\partial^2}{\partial\eta^2}+\frac{\partial^2}{\partial\zeta^2}\right)\right]+k^2\left(-6\,\frac{\partial^2}{\partial\xi^2}+\frac{\partial^2}{\partial\eta^2}+\frac{\partial^2}{\partial\zeta^2}\right)-$$
$$\left.-\,4ik^3\frac{\partial}{\partial\xi}+k^4\right\}N(\boldsymbol{r})\,\mathrm{e}^{i\boldsymbol{K}\cdot\boldsymbol{r}}\,\mathrm{d}V\,, \tag{20}$$

where V is the volume of the cube (Fig. 1), and $\boldsymbol{K}$ is defined by

$$\boldsymbol{K} = k(\boldsymbol{n} - \boldsymbol{n}_1)\;; \tag{21}$$

$\boldsymbol{n}$ being the unit vector in the direction of the incident wave (see Fig. 1). It is easy to note that

$$K = 2k\sin\theta/2\,, \tag{22}$$

where θ is the scattering angle.

The integrals containing derivatives of $N(\boldsymbol{r})$ can be evaluated by parts and by demanding that $N(\boldsymbol{r})$ and its derivatives vanish on the faces of the cube, the dimensions of which are much larger than the correlation length [6]. After some straightforward algebra, we obtain finally

$$\langle|\Delta'|^2\rangle = \frac{k^6A^2V}{64\pi^2c_0^2R^2}\,(9\cos^4\theta - 6\cos^2\theta + 1)\,N(\boldsymbol{K})\,, \tag{23}$$

where $N(\boldsymbol{K})$ is defined by

$$N(\boldsymbol{K}) = \int N(\boldsymbol{r})\,\mathrm{e}^{i\boldsymbol{K}\cdot\boldsymbol{r}}\,\mathrm{d}V\,. \tag{24}$$

In a similar manner, one can derive

$$\langle(|\varphi_y'|^2 + |\varphi_z'|^2)\rangle = \frac{9k^2h^4A^2V}{64\pi^2\mu_0^2R^2}\,(\cos^2\theta - \cos^4\theta)\,N(\boldsymbol{H})\,, \tag{25}$$

where

$$\left.\begin{aligned} N(\boldsymbol{H}) &= \int N(\boldsymbol{r})\,\mathrm{e}^{i\boldsymbol{H}\cdot\boldsymbol{r}}\,\mathrm{d}V\,,\\ \boldsymbol{H} &= k\boldsymbol{n} - h\boldsymbol{n}_1\,,\\ H^2 &= h^2 + k^2 - 2hk\cos\theta\,.\end{aligned}\right\} \tag{26}$$

In (23) to (26),

$$N(\boldsymbol{r}) = \langle c'(\boldsymbol{r})\,c'(0)\rangle\,. \tag{27}$$

2.3 The attenuation coefficient

Bhatia [3] has shown that for a plane progressive wave the energy $I\,\mathrm{d}S$ flowing per unit time across a surface of area $\mathrm{d}S$ is given by

$$I\,\mathrm{d}S = \tfrac{1}{2}\,[V_1c_0\,|\Delta'|^2 + V_s\mu_0(|\varphi_x'|^2 + |\varphi_y'|^2 + |\varphi_z'|^2)]\,\mathrm{d}S\,. \tag{28}$$

The attenuation coefficient α is defined as

$$\alpha = \frac{1}{VI_0}\int\mathrm{d}S\,\langle I\rangle\,, \tag{29}$$

where I_0 is the incident intensity and $\langle I\rangle$ is the average scattered intensity. Since for an incident dilatational wave

$$I_0 = \tfrac{1}{2}\, V_l c_0\, |\Delta_0|^2\,, \tag{30}$$

where

$$\left.\begin{aligned} |\Delta_0|^2 &= k^2 A^2 \\ \text{and} \qquad dS &= 2\pi R^2 \sin\theta\, d\theta \quad \text{(cf. Fig. 1)}\,, \end{aligned}\right\} \tag{31}$$

we have from (28) and (29)

$$\alpha = \frac{k^4}{32\pi c_0^2} \int\limits_0^\pi \sin\theta\, d\theta\, (9\cos^4\theta - 6\cos^2\theta + 1)\, N(\boldsymbol{K}) +$$

$$+ \frac{V_s}{V_l}\,\frac{9h^4}{32\pi\mu_0 c_0} \int\limits_0^\pi \sin\theta\, d\theta\, (\cos^2\theta - \cos^4\theta)\, N(\boldsymbol{H})\,. \tag{32}$$

3. Models for $N(\boldsymbol{r})$

3.1 Exponential correlation

For spherical grains, the correlation function $N(\boldsymbol{r})$ can be assumed to be isotropic. As mentioned earlier, the elastic modulus c' is constant within a grain. This means that the waves scattered from two points within a grain are strongly correlated. The probability that such a situation occurs for two points separated by a distance r, can be expressed as $\exp(-r/r_0)$ where r_0 is proportional to the average radius of the grains. We may therefore take

$$N(\boldsymbol{r}) = N(r) = \langle c'^2\rangle_0 \exp(-r/r_0)\,, \tag{33}$$

where $\langle\ldots\rangle_0$ indicates the average over all possible orientations of the grains in a sample.

From (33) one can easily obtain

$$N(\boldsymbol{K}) = N(K) = (1 + K^2 r_0^2)^{-1}\, 8\pi r_0^3 \langle c'^2\rangle_0 \tag{34}$$

and a similar expression for $N(\boldsymbol{H})$ with H replacing K.

Bearing in mind that the θ-dependences of K and H are given by (22) and (26), evaluation of α from (32) requires the knowledge of three integrals over θ. The final result can be written as

$$\alpha = \frac{r_0^3 k^4 \langle c'^2\rangle_0}{4c_0^2}(A_1 - A_2 + A_3) + \frac{9r_0^3 h^4 \langle c'^2\rangle_0}{4\mu_0 c_0}\,\frac{V_s}{V_l}(A_4 - A_5)\,, \tag{35}$$

where

$$A_1 = \frac{3}{2q^8(1+4q^2)}(12q^8 + 64q^6 + 64q^4 + 24q^2 + 3) -$$

$$- \frac{9}{8}\,\frac{(1+2q^2)^3}{q^{10}} \ln(1+4q^2)\,, \tag{36}$$

$$A_2 = \frac{6}{q^4(1+4q^2)}(2q^4 + 4q^2 + 1) - \frac{3(1+2q^2)}{2q^6}\ln(1+4q^2)\,, \tag{37}$$

$$A_3 = \frac{2}{(1+4q^2)}\,, \tag{38}$$

$$A_4 = \frac{1}{2q^2m^2}\left\{1 + \frac{[1 + (q^2 + m^2)]^2}{[1 + (m - q)^2]\,[1 + (m + q)^2]} - \right.$$
$$\left. - \frac{1 + (q^2 + m^2)}{2qm} \ln\left[\frac{1 + (q + m)^2}{1 + (m - q)^2}\right]\right\}, \tag{39}$$

$$A_5 = \frac{1}{8q^4m^4}\left\{\frac{4}{3} q^2m^2 + 3(1 + q^2 + m^2)^2 - \frac{(1 + q^2 + m^2)^4}{[1 + (m - q)^2]\,[1 + (m + q)^2]} - \right.$$
$$\left. - \frac{(1 + q^2 + m^2)^3}{qm} \ln\left[\frac{1 + (q + m)^2}{1 + (m - q)^2}\right]\right\}, \tag{40}$$

where

$$q = kr_0\,, \qquad m = hr_0\,. \tag{41}$$

On averaging over all possible orientations of the grains it can be shown that [9]

$$\langle (c')^2\rangle_0 = \frac{16}{525}\,\mu^2\,, \qquad \mu = (C_{11} - C_{12} - 2C_{44})\,, \tag{42}$$

where C_{ij} are the characteristic elastic constants.

We may now derive (1) and (2) of Section 1 as special cases of (35). In the limit $\lambda \gg \overline{D}$, $q^2 \ll 1$, $m^2 \ll 1$ it is easy to see from (34), that

$$N(K) \approx N(H) \approx 8\pi r_0^3 \langle (c')^2\rangle_0\,. \tag{43}$$

Substituting (43) in (32) and making use of (42) and (5), we get

$$\alpha = \tfrac{32}{7}\,\pi r_0^3 f^4 S\,, \tag{44}$$

where

$$S = \frac{8\pi^3\mu^2}{375\varrho_0^2 V_1^3}\left(\frac{2}{V_1^5} + \frac{3}{V_s^5}\right). \tag{45}$$

Equation (44) reduces to (1) if we identify the average volume of the grain as

$$T = \tfrac{32}{7}\,\pi r_0^3\,. \tag{46}$$

In this model, therefore, the mean radius of the grains is given by

$$a = (\tfrac{24}{7})^{1/3}\, r_0\,. \tag{47}$$

With this identification and the definition (45) of S, equation (44) agrees exactly with the expression of α for cubic crystals quoted by Papadakis [1] from results derived earlier [3, 7, 8].

It is to be noted that the result (44) in the so-called Rayleigh limit can be obtained independently by assuming for the correlation function

$$N(\boldsymbol{r}) = v_0 \langle (c')^2\rangle_0\, \delta(\boldsymbol{r})\,, \tag{48}$$

where the quantity v_0 has the dimension of a volume. Equation (48), when substituted into (32), yields again the result (44) provided we have

$$v_0 = 8\pi r_0^3\,. \tag{49}$$

The physical meaning of the above statement is obvious: when the wavelength of the ultrasound is much larger than the average grain size, the grains appear as point objects.

On the other hand, for $\lambda \ll \overline{D}$, $q^2 \gg 1$, $m^2 \gg 1$ we get from (36) to (40)

$$A_1 \approx \frac{9}{2q^2}, \qquad A_2 \approx \frac{3}{q^2}, \qquad A_3 \approx \frac{1}{2q^2}, \qquad A_4 \approx A_5 \approx 0\,. \tag{50}$$

Equation (35) then yields

$$\alpha \approx \frac{32\pi^2\mu^2}{525\varrho_0^2 V_1^6} r_0 f^2 \,, \tag{51}$$

having made use of (42) and (5). The expression for α is therefore of the form given in (2), in which

$$\bar{D} = (7/3)^{1/3}\, a \quad (\text{cf. } (47)) \,,$$
$$\Sigma = \frac{16\pi^2\mu^2}{525\varrho_0^2 V_1^6} \,. \tag{52}$$

Equation (51) agrees exactly with the form of α (for cubic crystals) derived from theories which neglect mode conversion (in this case, from dilatational to shear) [1]. The present analysis demonstrates explicitly how the contributions of the shear components, proportional to the terms A_4 and A_5, tend to become unimportant as the wavelength decreases.

For Al, we have [9]

$$\langle (c')^2 \rangle_0 / c_0^2 = 3 \times 10^{-4} \,,$$

$$V_1 = 6.326 \times 10^5 \text{ cm/s} \,, \qquad V_s = 3.13 \times 10^5 \text{ cm/s} \,.$$

Using these values, we present in Fig. 2 and 3 plots of α versus f on the basis of the calculated expression (35), and the results (44) and (51) in the low and high frequency regimes, for $2a = 0.013$ cm. The graphs indicate the ranges of validity of the approximate theories.

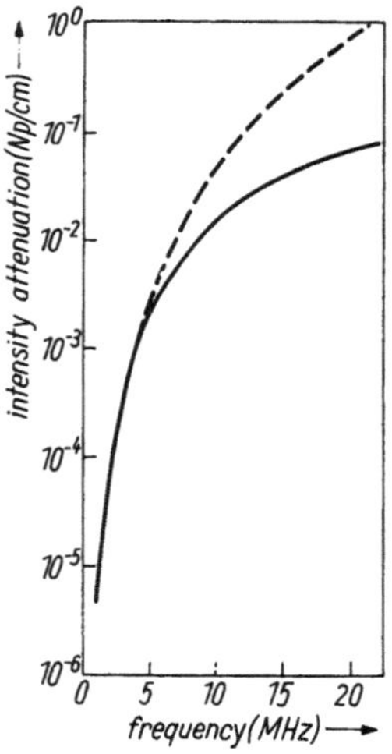

Fig. 2

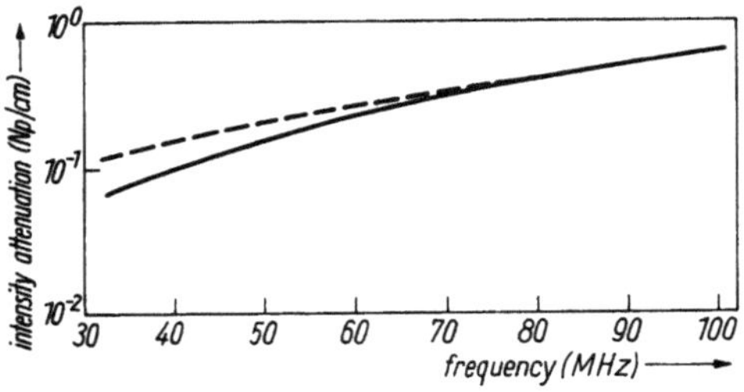

Fig. 3

Fig. 2. Comparison of α calculated on the basis of the low frequency form (44) (dashed line) and the more general result (35) (solid line). The Rayleigh limit is found to extend up to $f \approx 2$ MHz corresponding to $\lambda \approx 25\bar{D}$

Fig. 3. Comparison as in Fig. 2 but for the intermediate and high frequency regimes ($\lambda \ll \bar{D}$), using the limiting forms (51) (dashed line) and (35) (solid line). Equation (51) seems to become valid above $f \approx 70$ MHz corresponding to $\lambda \approx 0.7\bar{D}$

3.2 Gaussian correlation

Next we study another physically acceptable form of the correlation function which like the exponential correlation case, also yields tractable mathematical expressions for α. We assume

$$N(\boldsymbol{r}) = N(r) = \langle (c')^2 \rangle_0 \exp(-r^2/r_0^2) . \tag{53}$$

From (24) and (26), then

$$N(\boldsymbol{K}) = \langle (c')^2 \rangle_0 \left(\sqrt{\pi}\, r_0\right)^3 \exp(-q^2 \sin^2 \theta/2) , \tag{54}$$

$$N(\boldsymbol{H}) = \langle (c')^2 \rangle_0 \left(\sqrt{\pi}\, r_0\right)^3 \exp[-\tfrac{1}{4}(m^2 + q^2)] \exp[\tfrac{1}{2} mq \cos \theta] , \tag{55}$$

where q and m are given by (41). Substituting in (32) and carrying out the integrals over θ, we get

$$\alpha = \frac{\langle (c')^2 \rangle_0}{16 c_0^2} \sqrt{\pi}\, r_0^3 k^4 (B_1 - B_2 + B_3) + \frac{9 \langle (c')^2 \rangle_0}{16 \mu_0 c_0} \sqrt{\pi}\, r_0^3 h^4 \frac{V_s}{V_l} (B_4 - B_5) , \tag{56}$$

where

$$B_1 = \frac{9}{X^5}(X^4 \sinh X - 4X^3 \cosh X + 12X^2 \sinh X - 24X \cosh X + 24 \sinh X)\, e^{-X} , \tag{57}$$

$$B_2 = \frac{6}{X^3}(X^2 \sinh X - 2X \cosh X + 2 \sinh X)\, e^{-X} , \tag{58}$$

$$B_3 = \frac{1}{X}(\sinh X)\, e^{-X} , \tag{59}$$

$$B_4 = \frac{1}{Y^3}(Y^2 \sinh Y - 2Y \cosh Y + 2 \sinh Y)\, Z , \tag{60}$$

$$B_5 = \frac{1}{Y^5}(Y^4 \sinh Y - 4Y^3 \cosh Y + 12Y^2 \sinh Y - 24Y \cosh Y + 24 \sinh Y)\, Z , \tag{61}$$

in which

$$X = \tfrac{1}{2} q^2 , \qquad Y = \tfrac{1}{2} qm , \qquad Z = \exp[-\tfrac{1}{4}(q^2 + m^2)] . \tag{62}$$

As in the exponential correlation case, we can show that (56) reduces to (1) and (2) in the low and high frequency limits, respectively, where S and Σ are again given by

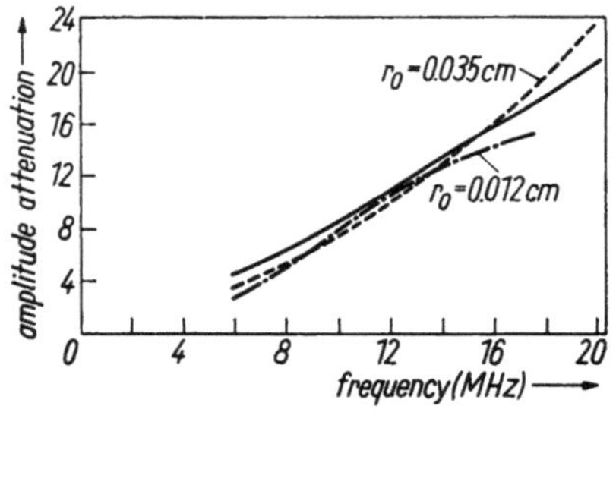

Fig. 4. Comparison between theory and experimental results for amplitude attenuation ($= \alpha/2$, in units of dB/30.48 cm) in Al with average grain diameter of 0.023 cm [9]. ——— data points; ---- based on exponential correlation; –·–·– based on Gaussian correlation

(45) and (52), while

$$T = \frac{4}{7} (\sqrt{\pi}\, r_0)^3 , \qquad \bar{D} = \left(\frac{7\pi}{3}\right)^{1/3} a . \tag{63}$$

The expressions for α which are derived above for an incident of dilatational wave, can be obtained easily also for the case of an incident shear wave.

3.3 Comparison with experiment on Al

In Fig. 4 we present experimental data on Al [9] for the amplitude attenuation ($= \alpha/2$) for $\bar{D} = 0.023$ cm along with the calculated points for the exponential and Gaussian correlation cases. Using the values as given in the last paragraph of Section 3.1 with the correlation length r_0 as a free parameter, least-squares fits with data up to 20 MHz yield, for the exponential case, $r_0 \approx 0.035$ cm, while for the Gaussian case, $r_0 \approx 0.012$ cm. Experimental data which are available up to 35 MHz do not agree so well with the theory beyond 20 MHz. This discrepancy may be attributed to the break-down of one or more of the four conditions listed in the second paragraph of Section 1, errors in the used values of V_l and V_s, wide spread in grain size, and uncertainty in $\langle (c')^2 \rangle_0 / c_0^2$ which can be as high as 20% [1]. Finally, it should be mentioned that the method of small perturbations as employed here is not strictly valid at very high frequencies [6]. An estimate, however, shows that this restriction in the case of Al (with $\bar{D} = 0.023$ cm) becomes prohibitive at a frequency as high as 250 MHz.

Acknowledgements

The authors are grateful to B. Purniah for his unflagging and skilful assistance in computations. One of the authors (S.D.) wishes to thank A. B. Bhatia for a very fruitful discussion.

References

[1] E. P. PAPADAKIS, in: Physical Acoustics, Vol. IV B, Ed. W. P. MASON, Academic Press, New York 1968 (p. 269).
[2] A. G. EVANS, B. R. TITTMAN, L. AHLBERG, B. T. KHURI-YUKUB, and G. S. KINO, J. appl. Phys. 49, 2669 (1978).
[3] A. B. BHATIA, J. Acoust. Soc. Amer. 31, 16 (1959).
[4] N. G. VAN KAMPEN, Physics Rep. 24, 171 (1976).
[5] C. L. PEKERIS, Phys. Rev. 71, 268 (1947).
[6] L. A. CHERNOV, Wave propagation in a Random Medium, Chap. IV, McGraw-Hill, New York 1960.
[7] E. M. LIFSHITZ and G. D. PARKHOMOVSKII, Zh. eksper. teor. Fiz. 20, 175 (1950).
[8] L. G. MERKULOV, Soviet. Phys. — J. techn. Phys. 1, 59 (1956).
[9] W. P. MASON and H. J. MCSKIMIN, J. appl. Phys. 19, 940 (1948).

(Received May 7, 1979)

phys. stat. sol. (a) 54, 547 (1979)

Subject classification: 1; 22.2.4

Laboratoire de Minéralogie-Cristallographie, associé au C.N.R.S., Université Pierre et Marie Curie, Paris[1]) *(a) and IBM Thomas J. Watson Research Center, Yorktown Heights*[2]) *(b)*

Boundary Conditions in the Numerical Integration of Takagi-Taupin Equations

Application to the Laue and Bragg Cases

By

Y. EPELBOIN (a, b) and P. RIGLET (a)

Boundary conditions in the numerical integration of the Takagi-Taupin equations are investigated in both Laue and Bragg reflections. They show how an incident spherical or plane wave may be simulated, and they discuss the accuracy of the calculation. In the Laue case the conclusion is that a spherical incident plane wave is well simulated lightening one elementary source, and a plane wave lightening more than 30 elementary sources. In the Bragg case it is shown that the accuracy of the numerical integration depends upon the distribution of sources and the thickness of the crystal.

On a étudié l'influence des conditions aux limites dans l'intégration numérique des équations de Takagi-Taupin pour la réflexion de Laue comme celle de Bragg. On montre comment on peut simuler de façon satisfaisante une onde incidente plane ou sphérique. Dans le cas de Laue il suffit d'un point source pour simuler une onde incidente sphérique; par contre il faut au moins une trentaine de points pour obtenir une onde incidente plane. Dans le cas de Bragg la précision de l'intégration numérique dépend de la distribution de sources élémentaires mais aussi de l'épaisseur du cristal.

1. Introduction

The integration of the Takagi-Taupin equations [1, 2] may not be performed in most cases by means of classical mathematics. It would be impossible to simulate the contrast of the images of various defects such as dislocations without a computer, and numerical algorithms had to be written to solve this problem [2 to 4]. These different algorithms differ by their accuracy and the time needed to integrate the equations by means of a computer, but a common difficulty remains which is how to take into account the boundary conditions on the surfaces of the crystal.

In conventional X-ray topography the incident wave falling upon a crystal may be considered as a spherical wave [5]; in the case of a multiple crystal spectrograph or in some synchrotron experiments the incident wave may be considered as a plane wave. It is thus necessary to consider both boundary conditions in the simulation of X-ray topographs. The aim of this paper is to discuss this point on the basis of the numerical algorithm written by Tournarie (cf. [3]). In the Laue case we will show the influence of the incident wave on the propagation of the wave fields in the plane of incidence and on the contrast of a dislocation. In the Bragg case we will consider the boundary conditions for a thin crystal and calculate the theoretical rocking curve with a good accuracy.

We will explain how it is possible to simulate either an incident spherical or plane wave very simply in the same computer routine.

[1]) 4 place Jussieu, 75230 Paris Cédex 05, France.

[2]) P.O. Box 218, Yorktown Heights, New York 10598, USA.

2. Boundary Conditions in the Takagi-Taupin Equations

Let us recall the well-known equations

$$\left.\begin{aligned} \frac{\partial}{\partial s_0} D_0(r) &= -i\pi K \chi_{\bar{h}} D_h(r)\,, \\ \frac{\partial}{\partial s_h} D_h(r) &= -i\pi K \chi_h D_0(r) + 2i\pi \left[K\beta_h - \frac{\partial}{\partial s_h} \boldsymbol{h} \cdot \boldsymbol{u}(r)\right] D_h(r)\,, \end{aligned}\right\} \qquad (1)$$

where χ_h and $\chi_{\bar{h}}$ are the Fourier coefficients of the dieelectric susceptibility. — We do not take into account the polarization factor C in this numerical calculation. — $\boldsymbol{u}(r)$ is the local deformation at point r inside the crystal and $\boldsymbol{h}$ the reciprocal lattice vector of the reflection; $\boldsymbol{s}_0$ and $\boldsymbol{s}_h$ are unit vectors along the refracted and reflected directions;

$$\beta_h = \frac{\boldsymbol{k}_h^2 - \boldsymbol{k}_0^2}{2\boldsymbol{K}^2}\,,$$

where $\boldsymbol{k}_0$ and $\boldsymbol{k}_h$ are the wave vectors inside the crystal and $\boldsymbol{K}$ the wave vector in the vacuum. The extremities of both wave vectors $\boldsymbol{k}_0$ and $\boldsymbol{k}_h$ are matched through the value of β_h. In the integration we will choose their extremities at point P_1 (Fig. 1), which is the intersection of the asymptote to the dispersion surface with the normal to the entrance surface of the crystal drawn from the extremity of vector $\boldsymbol{K}$ in the vacuum. The exact position of M is a function of the departure of the incident wave from the exact Bragg condition. This choice for the extremities of the wave vectors permits to write very easily the continuity of the tangential component through the surface of the crystal.

2.1 Boundary conditions on the entrance surface

In the Laue case without absorption the boundary conditions may be written as

$$\left.\begin{aligned} D_0 &= \frac{\partial}{\partial s_0} D_0 = 0\,, \\ D_h &= \frac{\partial}{\partial s_h} D_h = 0 \end{aligned}\right\} \qquad (2)$$

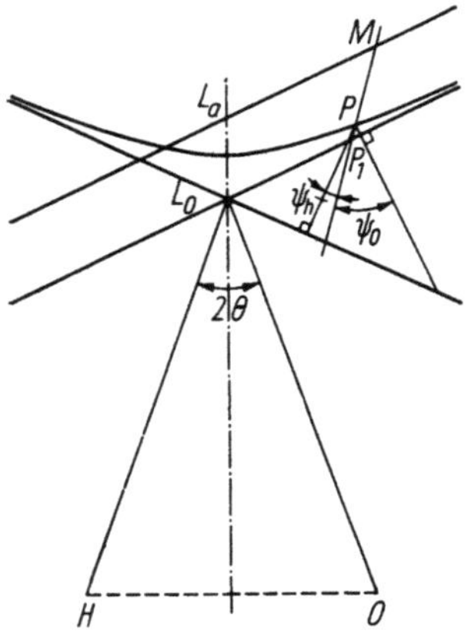

Fig. 1. Conditions of continuity along the entrance surface. For simplicity only branch 1 of the dispersion surface is shown

along $\mathbf{s}_0$ and as

$$\left.\begin{aligned} &D_0(\xi) = D_a(\xi)\,, \\ &D_h(\xi) = 0\,, \\ &\frac{\partial D_0}{\partial s_0} = 0\,, \\ &\frac{\partial D_h}{\partial s_h} = -\,i\pi K \chi_h D_a(\xi) \end{aligned}\right\} \tag{3}$$

on the entrance surface of the crystal.

In the Bragg case the boundary conditions are the same along the $\mathbf{s}_0$ direction, but differ along the surface since the characteristic $\mathbf{s}_h$ is not in the crystal [6]:

$$\left.\begin{aligned} &D_0(\xi) = D_a(\xi)\,, \\ &\frac{\partial D_h}{\partial s_h}(\xi) = -i\pi K \chi_h D_a(\xi) + 2i\pi K \beta_h' D_h(\xi)\,. \end{aligned}\right\} \tag{4}$$

$D_a(\xi)$ is the amplitude of the incident wave at point ξ along the surface. It may be shown that

$$\beta_h = -\eta_h \sqrt{\chi_h \chi_{\bar{h}}}\,\sqrt{\gamma_h/\gamma_0}\,,$$

where η is the usual deviation parameter.

The normal photoelectric absorption is taken into account in equations (1) by means of an imaginary part for the wave vectors [7]. Thus we must write $\boldsymbol{k}_0 = \boldsymbol{k}_{0r} + i\boldsymbol{k}_i$. It must be noticed that $\boldsymbol{k}_h$ possesses the same imaginary part since both wave vectors are linked by the reciprocal lattice vector $\boldsymbol{h}$:

$$\boldsymbol{k}_0 = \boldsymbol{k}_h + \boldsymbol{h}\,. \tag{5}$$

We may write $\boldsymbol{k}_0^2 = \boldsymbol{k}_{0r} + 2i\boldsymbol{k}_{0r}\cdot\boldsymbol{k}_i$, neglecting the second-order term. Since $\boldsymbol{k}_0^2 = \boldsymbol{K}^2(1+\chi_0)$, we write $2\boldsymbol{k}_{0r}\cdot\boldsymbol{k}_i = \boldsymbol{K}^2\chi_{0i}$.

The direction of $\boldsymbol{k}_i$ may be chosen arbitrarily, and it is convenient to follow Takagi and set it along the reflection plane. Then it is easy to establish

$$|\boldsymbol{k}_i| = -K\,\frac{\chi_{0i}}{2\cos\theta_B} = \frac{\mu_0}{4\pi\cos\theta_B}\,, \tag{6}$$

where μ is the photoelectric absorption.

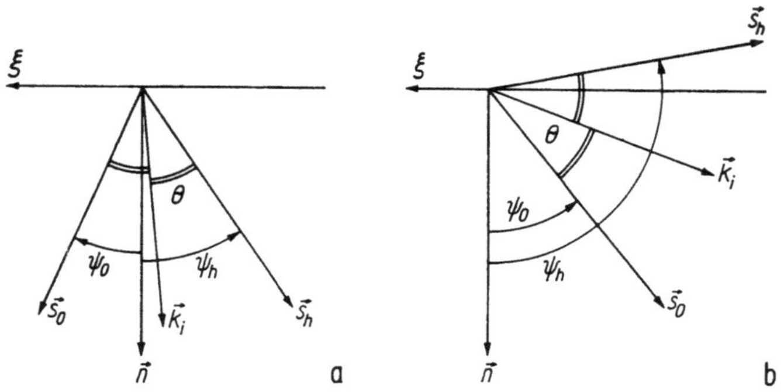

Fig. 2. Geometry of a) the Laue and b) Bragg reflection

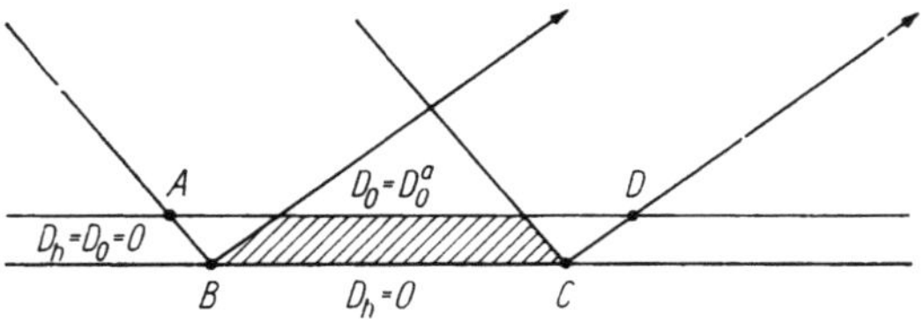

Fig. 3. Boundary conditions in the Bragg case

If we take k_i into account in boundary conditions (equations (3) and (4)), we must write

$$D_0(\xi) = D_a(\xi) \exp\left[\frac{\mu_0}{2\cos\theta_B}\sin(\psi_0+\theta_B)\,\xi\right], \tag{6}$$

where $\psi_0 = \sphericalangle(\boldsymbol{n}, \boldsymbol{s}_0)$ and θ_B is always taken as a positive angle (Fig. 2).

2.2 Boundary conditions on the exit surface

The intensity of diffracted waves outside the crystal is given by

$$I_h(\xi) = |D_h(\xi)|^2 \exp(-4\pi\boldsymbol{k}_i\cdot\boldsymbol{r})$$

or, taking into account equation (4)

$$I_h(\xi) = |D_h(\xi)|^2 \exp\left\{-\mu_0\left[\frac{t}{\gamma_0}+\frac{\sin(\psi_0+\theta_B)}{\cos\theta_B}\right]\right\} \tag{7}$$

in the Laue case and

$$I_h(\xi) = |D_h(\xi)|^2 \exp\left[-\mu_0\frac{\sin(\psi_0+\theta_B)}{\cos\theta_B}\right] \tag{8}$$

in the Bragg case. t is the thickness of the crystal and $\gamma_0 = \cos\psi_0$; ξ is a coordinate along the exit surface.

2.3 Boundary conditions for a thin crystal in the Bragg case

An additional boundary condition must be added in the Bragg case for a thin crystal on the back side of the crystal as suggested by Batterman and Hildebrandt [8]:

$$\left.\begin{aligned} &D_h(\xi) = 0\,,\\ &\frac{\partial D_0}{\partial s_0}(\xi) = 0\,,\\ &\frac{\partial D_h}{\partial s_h}(\xi) = -\,i\pi K\chi_h D_0(\xi). \end{aligned}\right\} \tag{9}$$

This means that all diffracted waves are back-scattered by the second surface of the crystal as shown in Fig. 3.

3. Simulation of the Boundary Conditions

We do not recall the principle of the integration of the Takagi-Taupin equations, which has already been explained [9].

The convenient choice of the extremity of the wave vectors permits to simulate an incident spherical wave by the data of one point source on the entrance surface of the crystal. An incident plane wave will be simulated by a distribution of point sources on the surface as shown in Fig. 4. If TRANSV is the distance between two following sources along the surface, their amplitudes will be linked by the relation

$$D_a(n+1) = D_a(n) \exp\left[\frac{\mu_0}{2\cos\theta_B} \sin(\psi_0 + \theta) \cdot \mathrm{TRANSV}\right], \tag{10}$$

$D_a(n)$ being the amplitude of the n-th source and $D_a(n+1)$ the amplitude of the $(n+1)$-th one.

It has been shown [10] that the data of one incident point source corresponds exactly to the usual assumption of an incident spherical wave when the step of integration is decreased to zero.

It is difficult to simulate an incident spherical wave which possesses a given lateral extension. The amplitude of the wave in the vacuum, on the entrance surface of the crystal, is given by [11]

$$D_a(\xi) = \frac{A}{4\pi r_0} \exp\left[-i\pi K \frac{\gamma_0^2}{r_0} \xi^2\right], \tag{11}$$

where r_0 is the distance from the point source to the crystal. Usually r_0 is of the order of 40 cm, and this phase factor oscillates very rapidly along the surface. Thus, the distribution of sources must be compact enough to take these oscillations into account. This means that the step of integration TRANSV must be small enough and difficulties may appear [9]: the step of integration would be so narrow that the computing time would be too long and numerical errors would not be negligible. This is the reason why an extended spherical wave will not be considered. This problem may be solved only by the use of a variable step algorithm.

On the other hand, the simulation of an incident plane wave is rather easy: lightening many points on the entrance surface of the crystal and following relation (10) is enough to obtain a satisfactory boundary condition. If the number of lightened points is N, the width of the source will be FENTE $= N \times$ TRANSV in the case of an incident plane wave and FENTE = TRANSV in the case of an incident spherical wave. This permits to calculate the true divergence of the source by the knowledge of the dispersion of the amplitudes of the wave vectors: $\Delta \boldsymbol{K} \cdot \Delta\xi = 1$. Thus the angular divergence of the distribution of sources is $\Delta\theta = 1/K\gamma_0\,\Delta\xi$.

Table 1

Relation between the number of elementary sources, the width of the simulated entrance slit, and the angular divergence of the incident wave

N	1	14	28	67
Δ (µm)	0.7	10	20	30
$\Delta\theta$	21.2″	1.48″	0.74″	0.29″

Table 1 gives the divergence of the incident wave as a function of the number of lightened points in the case of the 220 MoKα reflection in a symmetric Laue setting for silicon. It clearly shows that one incident point corresponds to a satisfactory angular divergence for the simulation of a section topograph. However, the step of integration would have to be decreased to simulate a traverse topograph [9]. A plane wave is simulated with quite a good accuracy, lightening about 30 points in the Laue case, and all along the field of the image in the Bragg case. Fig. 4a summarize both cases for the Laue diffraction and Fig. 4b for the Bragg case.

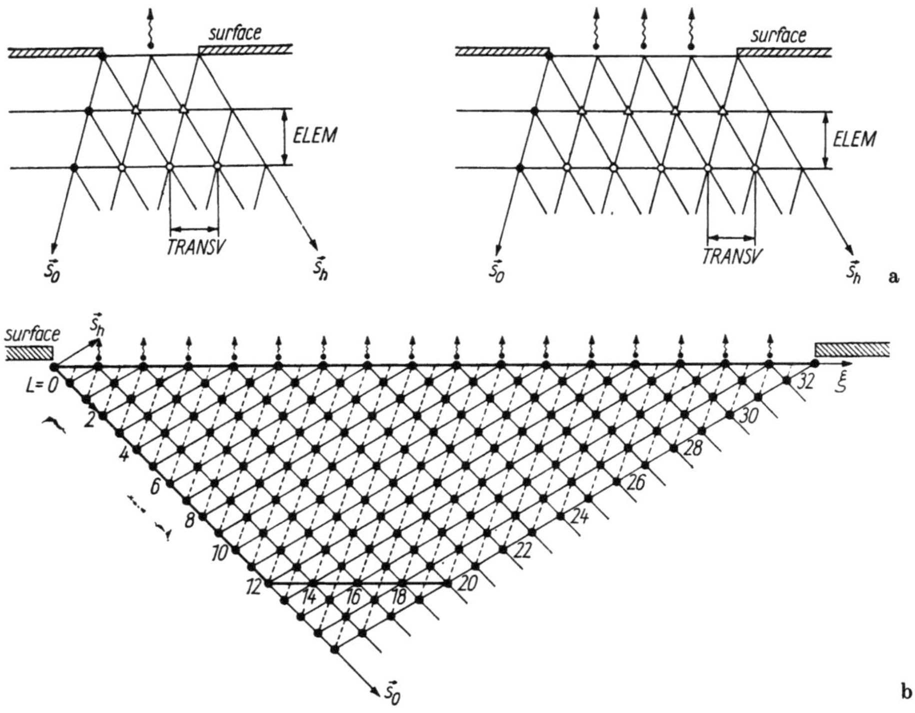

Fig. 4. Simulation of the boundary conditions on the entrance surface. a) Laue case: incident spherical and plane wave. b) Bragg case: incident plane wave. ● Unlighted point, △ calculated at the line n, ○ calculated at the line $(n + 1)$; points with wavy lines mean lighted points

4. Application to the Laue Case

To visualise the influence of the boundary conditions we have simulated the paths of the wave fields in a plane of incidence, lightening a variable number of sources on the entrance surface. We will then show how the contrast of a dislocation is also modified by these conditions.

4.1 Paths of the wave fields

The series of simulations of Fig. 5 corresponds to an incident wave at exact Bragg condition. The data correspond to MoKα 220 reflection for a symmetric setting. We have studied a silicon crystal 800 μm thick. In Fig. 5a one point source is lightened on the entrance surface, thus corresponding to an incident spherical wave. Hyperbolic fringes are clearly visible, which are characteristic of an incident spherical wave. When seven elementary sources are lightened on the entrance surface (Fig. 5b), these fringes remain visible. However, we may notice that the wave fields do not propagate along the edges of the Borrmann fan, since the divergence of the source is not big enough; to consider an entrance slit of this size (≈ 5 μm) and an incident spherical wave, it would be necessary to describe the incident wave following the scheme described by equation (11). All following simulations correspond to larger entrance slits; the representation of an incident plane wave is already satisfactory in Fig. 5e, where 27 elementary source are lightened (slit ≈ 20 μm large). According to Table 1 the divergence of the source is 0.74″. Fig. 5h corresponds to an entrance slit of 50 μm; 67 sources are lightened on the entrance surface and as the incident wave is at exact Bragg condition, wave fields

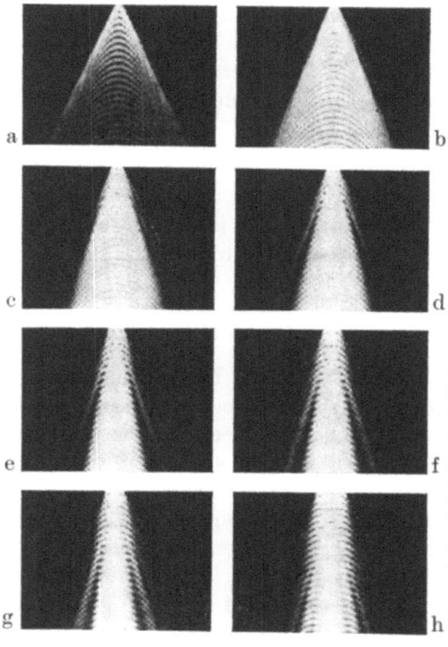

Fig. 5. Paths of the wave fields in the Laue case. Silicon — 800 μm, MoKα 220. a) One point source (slit 0.7 μm), b) 7 point sources (slit 5 μm), c) 14 point sources (slit 10 μm), d) 20 point sources (slit 15 μm), e) 27 point sources (slit 20 μm), f) 34 point sources (slit 25 μm), g) 40 point sources (slit 30 μm), h) 67 point sources (slit 50 μm)

propagate along the reflection planes. Pendellösung fringes are noticeable along the trajectory of the rays. This simulation corresponds to an incident plane wave of good quality. All simulations, from Fig. 5c to 5h show narrow beams which propagate near the edges of the Borrmann fan. They are due to a diffraction effect by the edges of the entrance slit, and would only disappear if an infinite lateral extension was considered.

4.2 Contrast of a dislocation

We have simulated the image of a dislocation to show how the boundary conditions influence the image. The dislocation of Burgers vector $\boldsymbol{b}=\frac{1}{2}[\bar{1}10]$ is inclined in the crystal. It has been previously described [12], and the diffraction conditions are the same as in the precedent simulations. Fig. 6a is the simulation with one elementary source; it is in very good agreement with the experiment. Fig. 6b corresponds to an entrance slit of 5 μm. As previously explained (Fig. 5b) the image vanishes along the edges of the simulations. Moreover it must be noticed that the direct image disappears; this is due to the poor divergence of the incident wave, which should be increased to take into account all the areas which contribute to this image. It must also be remembered that the present algorithm is not satisfactory for the study of the direct image [9]. Then Fig. 6c and 6d present the same image, using an incident wave of 10 and 100 μm,

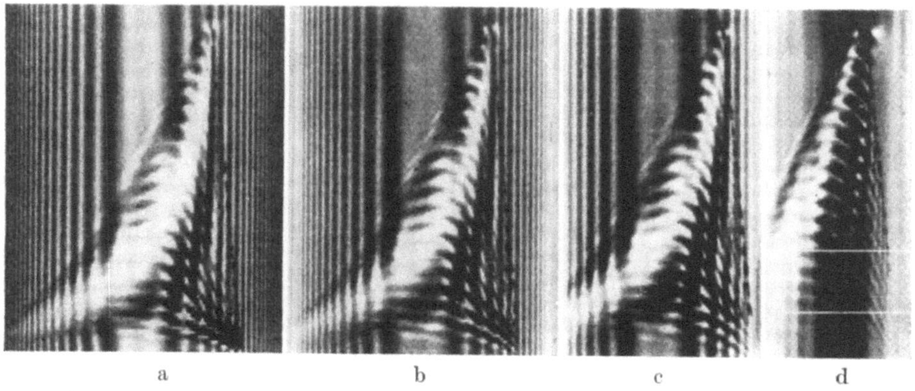

Fig. 6. Image of a dislocation. Diffraction conditions as in Fig. 5, geometry of the dislocation as in [12]. a) Slit 0.7, b) 5, c) 10, d) 100 μm

respectively. The direct image of the dislocation does not exist anymore and the width of the dynamical image is decreased.

These simulations demonstrate that the assumption of one elementary source on the entrance surface is the only simple way to simulate a section topography with satisfactory accuracy.

5. Application to the Bragg Case

We will show that the present choice for the boundary conditions is also accurate for the Bragg case and that an incident plane wave may be described by a distribution of sources along the surface of the crystal.

5.1 Rocking curve of the crystal

The integrated intensity and rocking curve of a perfect crystal has been calculated by Zachariasen [13] and Wilkins [14]. In the case of thin crystals, secondary peaks are visible on the wings (Fig. 7). They are due to interferences between the wave fields back-scattered along the second surface of the crystal and the ones scattered along the entrance surface. Fig. 7 presents the rocking curve of such a crystal: Curve a has been drawn using the analytical equations by Zachariasen and Wilkins, curve b has been obtained by a numerical integration of the Takagi-Taupin equations. It corresponds to an epitaxial layer of $Ge_{0.7}Al_{0.3}As$, 1.3 μm thick, on a GaAs substrate. The studied reflection is 400, with reflecting planes parallel to the surface. There is a good agreement between both curves, which proves that the boundary conditions used in the numerical calculations simulate an incident plane wave with good accuracy. However, if the distance between the elementary sources is not small enough, the height of the secondary peaks is decreased and their position is shifted (Fig. 8). As mentioned in Section 3 this is due to the sampling of the sources, which is not compact enough to take into account

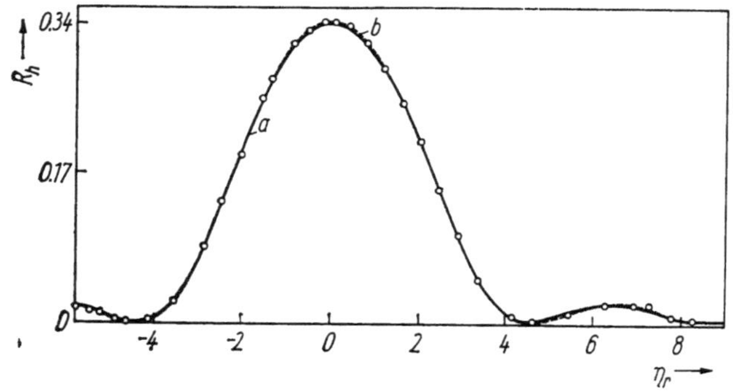

Fig. 7. Rocking curves as calculated by (a) Zachariasen-Wilkins (full line) and (b) Takagi-Taupin (dashed line). Epitaxial layer: 1.3 μm $Ga_{0.7}Al_{0.3}As$, reflections 400, $\lambda = 1.23783$ Å; symmetric setting, step of integration 0.2 μm

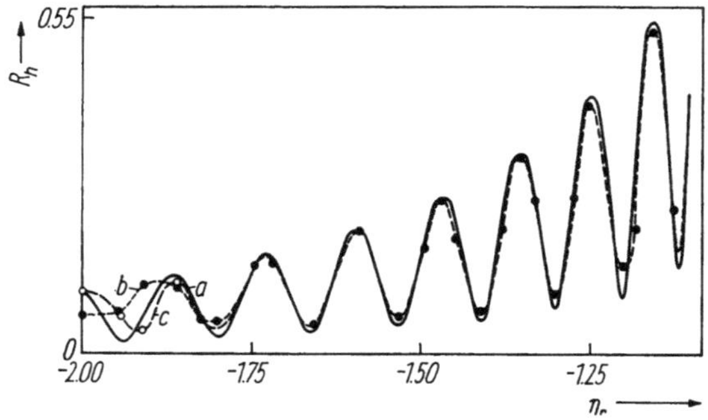

Fig. 8. Influence of the step of integration on the rocking curve. Epitaxial layer 30 μm $Ga_{0.7}Al_{0.3}As$, reflection 400, $\lambda = 1.54$ Å. Only the oscillations for $\eta < 0$ are shown. (a) Zachariasen-Wilkins (full line); (b) Takagi-Taupin, step of integration 1 μm (closed circles); (c) Takagi-Taupin, step 0.5 μm (open circles)

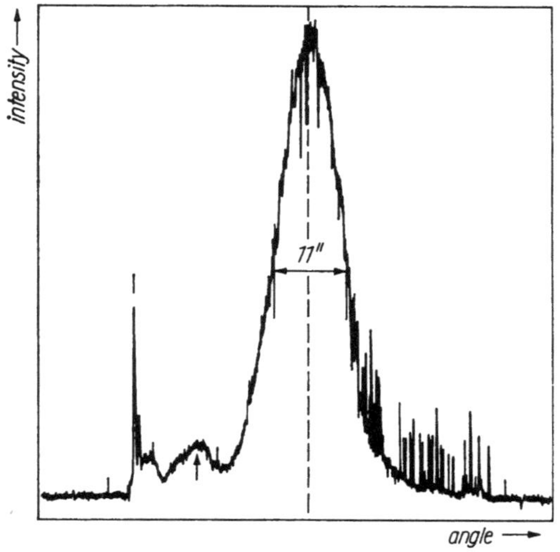

Fig. 9. Experimental rocking curve. Same conditions as in Fig. 7

the fast oscillations of the incident wave for large values of η. This would be the same in the Laue case [9]. The thinner the epitaxial layer, the larger are the secondary peaks; thus this phenomenon is more sensitive for thick layers were a small shift in the position of the peaks is more visible, as shown in curves b and c of Fig. 8. It would be visible in Fig. 7 for very large values of η only. This means that the choice of the step of integration must be done, for weak beam reflections, considering the thickness of the crystal, the numerical integration of the diffraction equations, and the departure from the exact Bragg condition.

5.2 *Experimental rocking curve of a crystal*

Epitaxial layers of $GaAl_{1-x}As_x$ on a GaAs substrate have been studied in reflection topography using a double-crystal spectrograph and the synchrotron facility at L.U.R.E.[3]) Results and experiment are presented in a different paper [15]. The incident beam falling on the crystal has a divergence of 0.29″ and thus may be considered as a good plane wave. The wavelength is $\lambda = 1.237\,83$ Å. The monochromator is described elsewhere [16]. As the crystal is curved by the epitaxial layer, the departure from exact Bragg condition is 4″ per millimetre of crystal through the beam. Thus it is possible to record the rocking curve of the epitaxial layer by a densitometric measurement of the density of grey through the topograph (Fig. 9). The wing on the left-hand side is in agreement with the theoretical one (Fig. 7) and is due to the finite thickness of the layer. It is not visible on the right-hand side due to the presence of dislocations.

6. Conclusions

The accuracy of the numerical integration of the Takagi-Taupin equations is very sensitive to the boundary conditions on the surfaces of the crystal.

In both Laue and Bragg reflections an incident plane wave may be simulated, lightening elementary sources along the surface in sufficient number. The sampling of these sources must be done taking into account all parameters of the calculation, i.e. the departure from the exact Bragg condition, the convergence of the numerical algorithm, and the thickness of the crystal.

---

[3]) Laboratoire pour l'Utilisation du Rayonnement Synchrotron, 92000 Orsay, France.

To simulate an incident spherical wave it is sufficient to lighten one elementary source on the surface, as shown in the Laue case. The simulation of a section topograph considering an incident spherical wave with a large lateral extension may be difficult since the sampling of the sources would be dependent on the oscillations of the incident wave. It is thus possible to simulate either an incident plane wave or an incident spherical wave by the data of the distance between the sources and the width of the incident beam falling upon the crystal.

Acknowledgement

One of us (Y. E.) gratefully acknowledges IBM for a fellowship which has made part of this work possible.

References

[1] S. Takagi, Acta cryst. **15**, 1131 (1962).
[2] D. Taupin, Bull. Soc. Franc. Minér. Crist. **87**, 469 (1964).
[3] A. Authier, C. Malgrange, and M. Tournarie, Acta cryst. **A24**, 126 (1968).
[4] P. V. Petrashen, Fiz. tverd. Tela **18**, 3729 (1976).
[5] N. Kato, J. Phys. Soc. Japan **18**, 12 (1963).
[6] T. Uragami, J. Phys. Soc. Japan **27**, 147 (1969).
[7] S. Takagi, J. Phys. Soc. Japan **26**, 1239 (1969).
[8] B. W. Batterman and G. Hildebrandt, Acta cryst. **A24**, 150 (1968).
[9] Y. Epelboin, Acta cryst. **A33**, 758 (1977).
[10] N. Kato, Acta cryst. **A32**, 453 (1976).
[11] A. Authier and D. Simon, Acta cryst. **A24**, 517 (1968).
[12] Y. Epelboin, J. appl. Cryst. **7**, 372 (1974).
[13] W. H. Zachariasen, Theory of X-Ray Diffraction in Crystals, Dover Publ., New York 1967.
[14] S. W. Wilkins, Acta cryst. **A34**, 343 (1978).
[15] J. F. Petroff, M. Sauvage, and P. Riglet, to be published.
[16] H. Hashizume, J. F. Petroff, M. Sauvage, and P. Riglet, to be published.

(Received May 7, 1979)

phys. stat. sol. (a) **54**, 557 (1979)

Subject classification: 14.2; 18; 21

Institut für Theoretische Physik und Reaktorphysik der Technischen Universität, Graz[1]) *(a)* *and Atominstitut der Österreichischen Universitäten, Wien*[2]) *(b)*

Dynamic Destruction of Superconductivity in Pb–n% Sb Alloys

By

E. SCHACHINGER (a), H. W. WEBER (b), and H. ZLÖBL (a)

The dynamic behaviour of the superconductive to normal transition of Pb–Sb alloys subjected to longitudinal external magnetic fields $H_{||}$ is investigated experimentally. The experimental data are compared with theoretical results obtained by Pippard and Lifshitz. In the case $H_{||} \gg H_c$ the experimental results are in good agreement with theoretical predictions. For $H_{||} \approx H_c$, however, an additional dependence of the transition time on the transport current is observed.

Das dynamische Verhalten des Übergangs supraleitend–normal von Pb–Sb-Legierungen, die longitudinalen äußeren Magnetfeldern $H_{||}$ ausgesetzt werden, wird experimentell untersucht. Die experimentellen Werte werden mit theoretischen von Pippard und Lifshitz erhaltenen Werten verglichen. Für den Fall $H_{||} \gg H_c$ befinden sich die experimentellen Ergebnisse in guter Übereinstimmung mit theoretischen Vorhersagen. Für $H_{||} \approx H_c$ wird jedoch eine zusätzliche Abhängigkeit der Übergangszeit vom Transportstrom beobachtet.

1. Introduction

Pippard [1] and Lifshitz [2] first gave a theoretical description of the phase transition in cylindrical samples placed into longitudinal magnetic fields. According to this theory the propagation of the interphase boundary is limited by electromagnetic damping which accompanies the spatial propagation of the magnetic field within the normal region of the sample (see Fig. 1).

If we assume as a zero-order approximation the field strength in the normal region to be equal to H_c (the thermodynamic critical field at the sample temperature), the magnetic flux Φ inside the normal region is given by

$$\Phi = \mu_0 \pi \big(a^2 - r^2(t)\big) H_c\,. \tag{1}$$

Here $r(t)$ denotes the actual position of the interphase boundary $\big(0 \leqq r(t) \leqq a\big)$ and a is the sample radius. The eddy current density $\boldsymbol{g}$ is then given by

$$|\boldsymbol{g}| = \mu_0 \sigma H_c \frac{r(t)}{a} \frac{\mathrm{d}r(t)}{\mathrm{d}t}\,, \tag{2}$$

where σ denotes the electric conductivity. If the longitudinal magnetic field at the sample surface $H_{||} = H_c(1 + p)$ and $p = (H_{||} - H_c)/H_c$ we derive from the first Maxwell equation

$$-\int\limits_{H_c}^{H_c(1+p)} \mathrm{d}H = -pH_c = \int\limits_{r(t)}^{a} g\,\mathrm{d}r' = \int\limits_{r(t)}^{a} \mu_0 \sigma H(r') \frac{r(t)}{r'} \frac{\mathrm{d}r(t)}{\mathrm{d}t}\,\mathrm{d}r' \tag{3}$$

with $H(r') = H_c$ according to our first approximation.

[1]) A-8010 Graz, Österreich.
[2]) A-1020 Wien, Österreich.

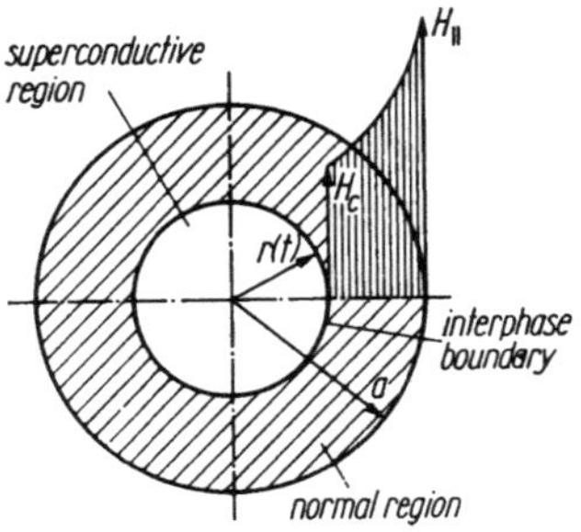

Fig. 1. Schematic representation of the phase transition in a cylindrical sample placed into a longitudinal external magnetic field $H_{||}$

Furthermore, we use the boundary condition, that at $t = 0 \to r(t) = a$. From (3) we derive an equation for the time t which is needed by the interphase boundary to reach the position $r(t)$:

$$t = \frac{\mu_0 \sigma}{p} \left[\frac{1}{4} \left(a^2 - r(t)^2\right) - r(t)^2 \ln \frac{a}{r(t)} \right]. \tag{4}$$

The transition is finished at the time t_T, where the boundary has reached the axis of the cylindrical sample $\left(r(t) = 0\right)$. This condition leads to an equation for the transition time:

$$t_T = \frac{\mu_0 \sigma a^2}{4} \frac{1}{p}. \tag{5}$$

Using the fact, however, that at the surface of the sample the magnetic field $H(r' = a)$ is equal to $H_{||}$ instead of H_c, we compute the following modified equation for the transition time:

$$t_T = \frac{\mu_0 \sigma a^2}{4} \frac{(1 + p)}{p}. \tag{6}$$

This equation was reported first by Faber [3], the results of this electrodynamic theory were extended by Keller [7] lateron.

In his extensive work Faber [3 to 6] showed that good agreement between experimental and theoretical results was achieved for the case $H_{||} \gg H_c$. For $H_{||} \approx H_c$, however, differences between experiment and theory were noted. Ittner [8] pointed out that thermodynamic effects must be considered in this case.

Previously the voltage drop at the surface of the sample induced by the movement of the interphase boundary was used in all experiments to determine the appearance of the normal state. This voltage may be computed using the following equation:

$$E(a) = \mu_0 H_c \frac{r(t)}{a} \frac{\mathrm{d}r(t)}{\mathrm{d}t}, \tag{7}$$

where $E(a)$ denotes the electric field strength at the sample surface. This voltage drop disappears when the boundary reaches the axis of the sample. However, because of the slow motion of the boundary in the vicinity of the axis this voltage drop is very small resulting in problems with an accurate determination of the transition time.

In order to avoid this problem a new experimental method was designed [9]. Some details of this technique are reported in Section 2. In Section 3 we present results on several type I superconductors which were aimed in particular to verify the influence of the electric conductivity on the transition time as predicted by (5) and (6).

2. Experimental

The basic idea of our experimental method consists of determining the normal state by the appearance of an ohmic voltage drop produced by a small transport current I through the sample. However, since this current produces an additional field $H_\perp = I/2\pi a$ which is oriented perpendicular to the applied longitudinal field $H_{||}$, it is necessary to measure the critical field curve $H_c = f(H_{||}, H_\perp)$ for each sample first. In the case of type I superconductors this curve is given simply by

$$H_c = \sqrt{H_{||}^2 + H_\perp^2}\,. \tag{8}$$

Using this relation for each transport current a critical longitudinal field $H_{||c}$ is obtained, where the field starts to penetrate from the sample surface ($t = 0$). After the transition time t_T an ohmic voltage drop appears, which is much larger (at least by one order of magnitude) than the induced voltage present in the time interval $0 \leqq t \leqq t_T$ and used in previous experiments. Therefore, the determination of the transition time requires two experimental steps: measurement of $H_c = f(H_{||}, H_\perp)$ and hence $H_{||c}$, and measurement of the ohmic voltage drop.

2.1 The critical curve $H_c = f(H_{||}, H_\perp)$

To measure the critical curve an experimental equipment which is shown schematically in Fig. 2 was used. The transport current through the sample is supplied by the 220 V, 50 Hz grid via the silicon controled rectifier (SCR) and the serial resistor R_N. The SCR operates as a switch and is triggered by an electronic control system. Since only one trigger pulse is used, the current flow through the sample is restricted to one single current pulse in order to avoid damages of the sample. The external field H is supplied by a Nb–25% Zr coil of about 25 cm length and 6 mm diameter. The current through this coil which is proportional to the field H is regulated by the serial resistor R_1. Wires of 1 mm diameter were used as samples and centred carefully inside the coil.

Using the oscilloscope (OSC) the voltage across the sample and the voltage drop along the resistor R_N, which is a measure of the transport current, are recorded simultaneously by photographs triggered by the electronic control system. The critical current — and hence the critical value of $H_\perp$ — is determined by the first appearance of an ohmic voltage drop at the sample surface for fixed $H_{||}$.

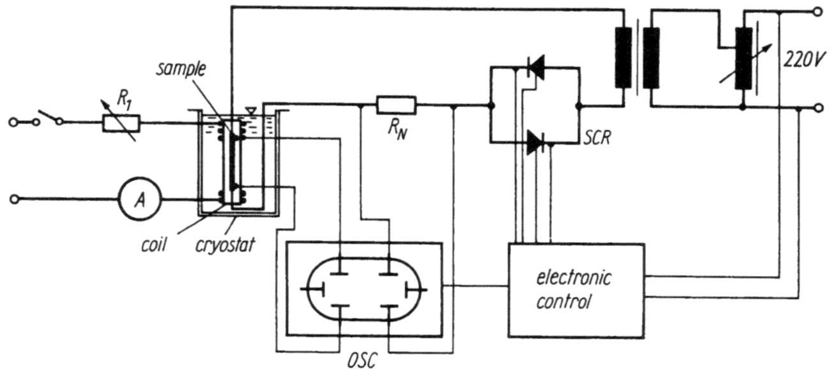

Fig. 2. Experimental equipment for critical curve measurements

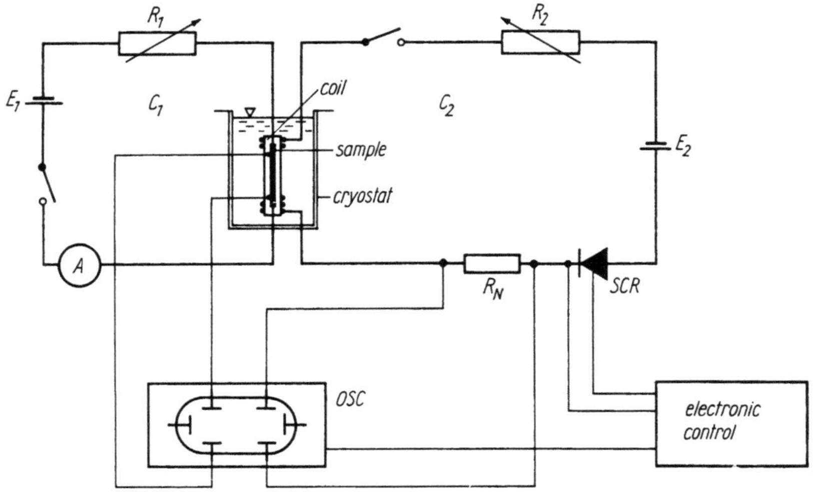

Fig. 3. Experimental equipment for transition time measurements

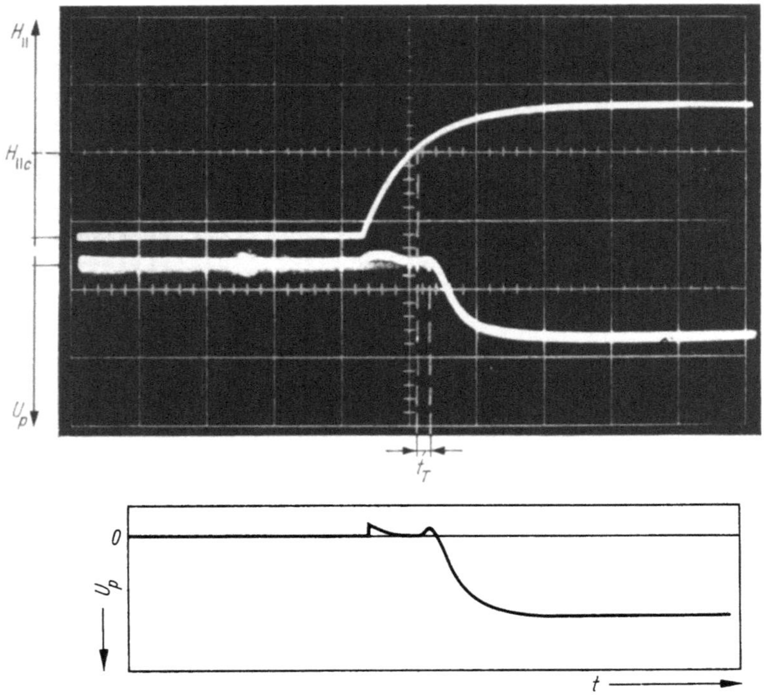

Fig. 4. Typical oscillograph display showing a transition time experiment. $H_{||}$ external longitudinal field, U_p voltage across the sample, t_T transition time

2.2 Measurement of the transition time

The experimental equipment used for the measurements of the transition time is shown schematically in Fig. 3. A slightly modified equipment has been used for measurements on Nb and Nb alloys under the influence of short magnetic field pulses previously [10].

A constant transport current is applied to the sample by the circuit C_1 (serial resistor R_1, battery E_1, and ammeter A). The current through the field coil is now regulated by the serial resistor R_2 and switched by the SCR, which is triggered by the electronic control system. After the completion of one measurement the current through the coil is switched off by the switch (SW); then the SCR is ready for the next trigger pulse (circuit C_2).

In this case the electronic control system triggers the oscilloscope to record the current through the coil ($I \sim H_{||}$, R_N) and the voltage drop along the sample (U_p) by photographs. A typical display together with a schematic drawing of the voltages along the sample is shown in Fig. 4. First, upon applying the external field a small pick-up voltage in the measuring circuit is observed, which decays rapidly. Then the critical external field $H_{||c}$ for a given transport current, which was determined from the critical curve of the sample as described in Section 2.1, is reached and marked in the figure. At this particular field, where the transition starts, small induced voltages may be observed in the case of small transport currents. When the transition is finished, a sharp ohmic voltage drop is observed, which is also marked in Fig. 4. The resulting time interval is the transition time t_T.

3. Experimental Results

First static magnetization curves were measured at 4.2 K for all the samples listed in Table 1. The corresponding results are shown in Fig. 5 and clearly display type I behaviour of the samples. This result was confirmed by the experiments on the critical curves shown in Fig. 6. In this figure relation (8) is indicated by a dashed quarter circuit, which is followed closely by the experimental data. Moreover, in the case of small transport currents the samples show surface superconductivity as indicated by

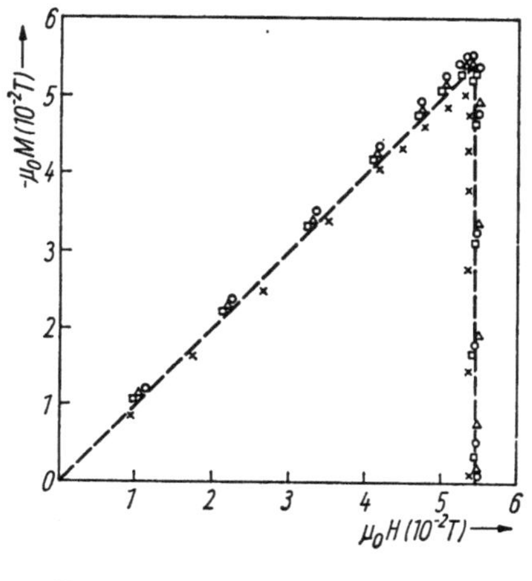

Fig. 5

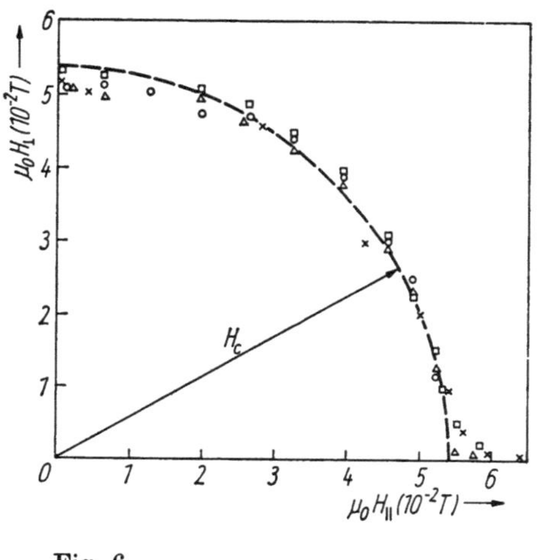

Fig. 6

Fig. 5. Magnetization curves of the samples ($T = 4.2$ K). × Pb (99.99%), ○ Pb–0.185%Sb △ Pb–0.4%Sb, □ Pb–1.0%Sb

Fig. 6. Critical curves of the samples (symbols as in Fig. 5)

the tails on the right-hand side of the diagram. In agreement with the predictions of the Ginzburg-Landau theory (e.g. [11]) this effect occurs only in longitudinal fields, i.e. in our geometry, when the disturbance caused by the transport current and the corresponding field $H_\perp$ is very small.

Furthermore, measurements of the transition temperature and the residual resistivity ratio were made. Whereas, no significant influence of the alloy composition on T_c could be detected, the resistivity ratio decreases in a systematic way (Table 1).

Table 1
List of samples used

sample	residual resistivity ratio	T_c (K) ($\pm$0.02 K)
Pb (99.99%)	1050	7.10
Pb–0.185% Sb	84.7	7.10
Pb–0.4% Sb	64.0	7.12
Pb–1.0% Sb	56.0	7.10

Finally, in Fig. 7 the experimental results on the transition times are plotted as a function of the external field $H_\parallel$ for different transport currents I. The theoretical results of the electrodynamic theory (see (6)) calculated with the appropriate conductivities σ according to Table 1 are shown as dashed lines. It will be noted, that good agreement between theory and experiment is found for $H_\parallel \gg H_c$. In particular the increase of the transition time with increasing conductivity predicted by (6) is followed closely by the experimental data (Fig. 8). As remarked by Ittner [8] previously, the validity of the electrodynamic theory extends to increasingly lower fields ($H_\parallel \to H_c$) with increasing conductivity in agreement with our results.

For fields $H_\parallel \approx H_c$, where the electrodynamic theory fails, a marked increase of the transition times is observed, which is caused by additional thermodynamic effects.

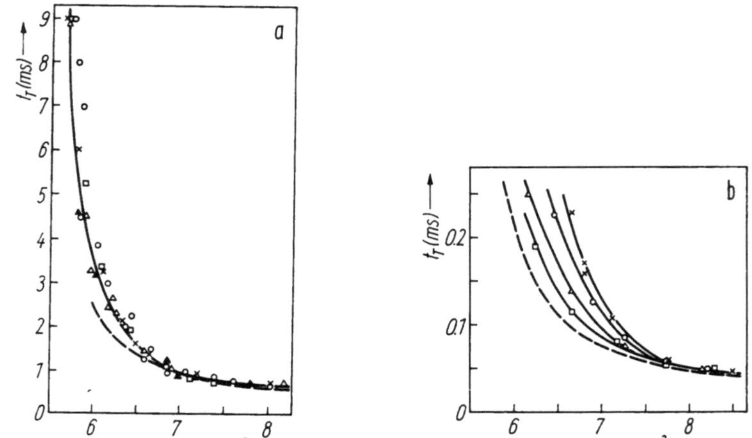

Fig. 7. a) Transition time t_T as a function of the external field and the transport current I_s. a) Sample Pb(99.99%), 1 mm diameter. ▲ I_s = 11, × 9, ○ 8, △ 6, □ 5 A; - - - - theory. b) Sample Pb–1.0% Sb, 1 mm diameter, □ I_s = 18, △ 15, ○ 12.5, × 10 A; - - - - theory

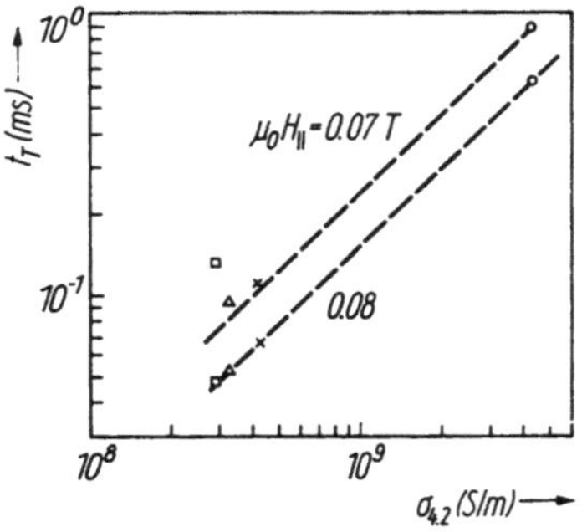

Fig. 8. Comparison of theoretical predictions by equations (5) and (6) with the transition times measured for the various samples. The transport current for all the samples I_s was 10 A. ○ Pb (99.99%), × Pb–0.185% Sb, △ Pb–0.4% Sb, □ Pb–1.0% Sb; ---- theory

Considering the results by Ittner [8], who calculated these effects for hollow cylinders, where a plane geometry may be used, qualitative agreement with respect to the increase of t_T is noted. However, in order to arrive at quantitative results an appropriate modification of the theory using the present cylindrical geometry has to be made [12].

As a final remark, we wish to point out, that all samples except for the highest purity one show a marked dependence of the transition times on the transport current. Since this dependence occurs only in the "thermodynamic" limit and becomes more pronounced with decreasing conductivity an appropriate modification of present thermodynamic theories to account for this dependence should be interesting.

References

[1] A. B. Pippard, Phil. Mag. **41**, 243 (1950).
[2] E. M. Lifshitz, Zh. eksper. teor. Fiz. 9, 834 (1950).
[3] T. E. Faber, Proc. Roy. Soc. **A214**, 392 (1952).
[4] T. E. Faber, Proc. Roy. Soc. **A219**, 75 (1953).
[5] T. E. Faber, Proc. Roy. Soc. **A223**, 174 (1954).
[6] T. E. Faber, Proc. Roy. Soc. **A231**, 353 (1955).
[7] J. B. Keller, Phys. Rev. **111**, 1497 (1958).
[8] W. B. Ittner, Phys. Rev. **111**, 1483 (1958).
[9] E. Schachinger, Ber. Inst. theor. Phys. Reaktorphys., ITPR–78007.
[10] J. Gerhold and E. Schachinger, Cryogenics **17**, 269 (1977).
[11] D. Saint-James, G. Sarma, and E. J. Thomas, Type II Superconductivity, Pergamon Press, 1969.
[12] E. Schachinger, to be published.

(Received April 27, 1979)

phys. stat. sol. (a) 54, 565 (1979)

Subject classification: 2 and 5; 18.2; 21.1.1

Laboratoire de Structure Electronique des Solides, ERA n⁰ 100, Université Louis Pasteur, Strasbourg[1])

Mössbauer Effect and Short-Range Order in Fe–P–B and Fe–P–C Amorphous Alloys[2])

By

A. AMAMOU

The hyperfine field of some amorphous iron-based alloys (Fe–P–B, $Fe_{80}B_{20}$, and $Fe_{75}P_{15}C_{10}$) obtained by "splat cooling" is investigated. The isomer shift can be interpreted either in terms of the electron transfer from the metalloid to iron "d"-bands, or in terms of the change in the "s"-electron density. The experimental spectra are fitted according to two procedures, by assuming, respectively, a continuous and discrete distribution of hyperfine fields. These methods yield distributions which have approximately the same shape and the same average hyperfine field. Such a shape can be related to local environments which are similar, on an average, to the crystalline counterpart. The average hyperfine field values and the observed magnetic moment suggest that the distribution around the nucleus is constant for the alloys investigated. On the other hand, it is similar to those of various crystalline compounds of transition metals and metalloids.

Nous présentons une étude de champ hyperfin dans des alliages amorphes à base de fer (Fe–P–B, $Fe_{80}B_{20}$ et $Fe_{75}P_{15}C_{10}$) obtenus par trempe ultra-rapide. Le déplacement isomérique peut être interprété en termes de transfert de charge du métalloïde à la bande "d" du fer ou bien en termes de diminution de la densité électronique "s". Les spectres expérimentaux peuvent être ajustés à des spectres calculés en supposant une distribution $P(H)$ soit continue, soit discrète de champs hyperfins; bien que différant dans le détail, l'allure des $P(H)$ obtenues dans les deux cas est semblable. Ces distributions de champs hyperfines peuvent être reliées aux divers environnements locaux des atomes de fer qui seraient, en moyenne, semblables à ceux existant dans le composé cristallin équivalent. La valeur moyenne du champ hyperfin comparée au moment magnétique suggère que la distribution des moments autour du noyau de fer d'une part est constante dans les alliages étudiés, d'autre part est sensiblement la même que dans de nombreux composés métal de transition–métalloïde.

1. Introduction

In this paper we present some results of Mössbauer absorption experiments on Fe–P–B and Fe–P–C amorphous systems, and we discuss the interpretation which can be deduced from various fits of the experimental data. The investigated alloys prepared by splat-cooling were $Fe_{79}P_{21-c}B_c$ ($5\ at\% \leqq c \leqq 16\ at\%$), $Fe_{94-c}P_6B_c$ ($11\ at\% \leqq c \leqq 19\ at\%$), $Fe_{80}B_{20}$, and $Fe_{75}P_{15}C_{10}$. Structural studies on amorphous alloys [1, 2, 18] show that the local environment of a given atom is not completely random, but that the coordination number and the nature of the near neighbours are well defined on an average.

Magnetic investigations in crystalline systems have shown that the magnetic state of an atom is sensitive to the number and nature of its neighbours [3, 5, 29, 30]. Therefore such studies are useful to approach the short-range order (SRO) in the amorphous state. A suitable way to investigate local properties of atoms are NMR and Mössbauer spectroscopy.

---

[1]) 4, rue Blaise Pascal, 67000 Strasbourg, France.

[2]) Part of this work was carried out in California Institute of Technology, Pasadena, California 91125, and sponsored by ERDA Contract No. AT(04-3)-822.

Previous hyperfine field studies on amorphous alloys have concluded a very broad distribution with an appreciable probability near zero field. Therefore, it has been suggested [6] that the resistivity minimum found in many amorphous ferromagnetic alloys is due to Kondo scattering by the weak magnetic moments. An alternative explanation is given by Harris et al. [7] who relate such a minimum to low-energy structural excitations. Then it would be interesting to determine whether or not there is any evidence of an appreciable probability distribution at zero field.

2. Summary of Previous Studies on the Investigated Alloys

Previous studies [8] on the Fe–P–B system (75 at% $\leqq c_{Fe} \leqq$ 83 at%) show that these alloys are ferromagnetic with the Curie temperature T_C raningg between 570 and 670 K and the average magnetic moments $\bar{\mu}$ per iron atom between 1.90 and 2.05μ_B. The magnetic properties show in particular that in $Fe_{79}P_{21-c}B_c$ the variations of T_C and $\bar{\mu}$ with c are comparable to those observed in crystalline $Fe_{75}P_{25-c}B_c$. Such alloys exhibit a crystallographic transition at $c_0 \approx 10$ at% B; for $c < c_0$ the compound has an ε-structure, and for $c > c_0$ the compound has an ε_1-structure. Therefore, it has been suggested that in the amorphous state a similar transition involving the SRO occurs. A first account of our Mössbauer studies on $Fe_{79}P_{21-c}B_c$, in agreement with the magnetic ones, has been presented [9]. NMR studies on $Fe_{79}B_{19}Ga_2$ showed a broad hyperfine field distribution (HFD) on boron nuclei centred at 27 kOe [10].

The Fe–P–C system has been extensively studied by measurements of structure, specific heat [11], magnetization, and Mössbauer spectroscopy [4]. The magnetic properties show that the system is ferromagnetic with the same moment per iron atom as in Fe_3P (1.8μ_B); the HFD on phosphorus deduced from NMR measurements is centred at 27 kOe [12].

Mössbauer studies on $Fe_{80}B_{20}$ (2605 Meltglass, Allied Chemical) have been performed by several authors [13 to 17]. Although the experimental results have been fitted in different ways, the obtained HFD's are qualitatively the same, i.e., a broad HFD centred at around 260 kOe at 300 K and the similarity between the crystalline and amorphous state have been pointed out [13, 17].

3. Experimental Results and Qualitative Analysis

Amorphous foils of Fe–P–B, $Fe_{80}B_{20}$, and Fe–P–C were obtained and checked in the same way as previously reported [18, 8]. We have also investigated an Allied Chemical $Fe_{80}B_{20}$ (2605 Meltglass). The Curie transition is very sharp (1 to 3 K [8]), and the resistivity at high temperature, measured for some samples, exhibits an irreversible crystallization process. Mössbauer measurements were performed at room temperature with a standard spectrometer operating in a constant acceleration mode [19]; the source was Co^{57} in copper, and the absorber was the as-quenched amorphous foil of about 40 μm thickness. The absorption pattern was recorded on a multichannel analyser, and the average data provide 256 absorption values per spectrum.

In this section we only discuss the qualitative features of the experimental data. In a first approximation for all specimens the intensity of absorption $I(v)$ as a function of velocity v is roughly symmetrical, although some small dissymmetries can be noticed when the spectra are considered in detail (Fig. 1). This approximative symmetry shows that on an average (i) there is no large quadrupole splitting and (ii) the isomer shift is uniform. The hyperfine field deduced from the 6 peak minima corresponds to a value of about 240 kOe for Fe–P–C and of 250 to 270 kOe for Fe–P–B. A very rough evaluation from the outer lines of the width at the half intensity gives about 100 kOe for $Fe_{80}B_{20}$ and 150 kOe for Fe–P–C. The fact that there is no apparent quadrupole splitting can be explained in the same way as for Fe–Si amorphous alloys

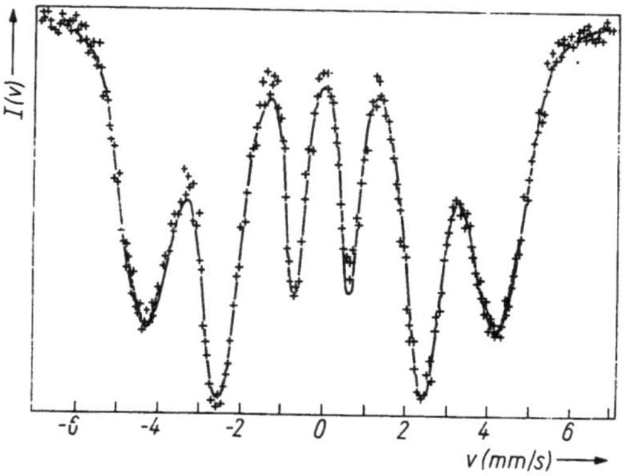

Fig. 1. $Fe_{83}P_6B_{11}$: Mössbauer spectrum and fits obtained with a continuous HFD. I_s = = 0.10 mm/s, $R = 2.4$, $\gamma = 4$ (full line) and $\gamma = 0.3$ (dashed line)

[20], i.e., the values of the angle θ between the axis of the field gradient and the hyperfine field axis are randomly distributed.

From the symmetry axis of the spectra we can define the isomer shift I_s related to pure iron (Table 1). I_s is likely an average value over the iron atoms. The isomer shift in iron compounds was originally calibrated in terms of a donor model [22]; a more complex mechanism of charge transfer has been proposed by Fatseas [21] in order to account for the experimental results in iron metallic compounds.

In $Fe_{79}P_{21-c}B_c$ the isomer shift decreases from 0.22 to 0.10 mm/s (the uncertainty is about 0.03 mm/s), when the boron content increases from 5 to 16 at%. Such a variation is similar to that observed in crystalline $Fe_{75}P_{25-x}B_x$ [23], if we define the average I_s $(= \sum_i I_{si}J_i$ where I_{si} and J_i are the isomer shift and the relative intensity of the i-th site, respectively).

These isomer shift results can be interpreted either in terms of a donor model or by direct effects of changes in "s"-electron density with composition; both mechanisms can be involved. In Fe–P–B amorphous alloys the electron transfer from the metalloid to the iron "d"-band has been estimated from magnetization studies as about 1.4 electrons for boron and 1.6 for phosphorus [24]; being given the usual values of the change of I_s per transferred electron ($\leqq$ 0.25 mm/s per 3d electron [13]), in a donor model the contribution of d-change can be roughly estimated as a maximum value of 0.1 mm/s. Therefore, the results of isomer shift can be explained in $Fe_{80}B_{20}$ and Fe–P–B with a large boron content, whereas with increasing phosphorus content we have to assume a progressive decrease of 4s density; it can be estimated as 0.05 electrons for $Fe_{79}P_{16}B_5$ according to Fatseas' calibration [21].

Table 1

Fe–P–B and Fe–P–C system: Average isomer shift I_s related to pure iron, and average hyperfine field $\bar{H}$; the uncertainties are $\Delta I_s = 0.03$ mm/s and $\Delta \bar{H} = 2$ kOe

amorphous compound	I_s (mm/s)	$\bar{H}$ (kOe) ($T = 300$ K)	amorphous compound	I_s (mm/s)	$\bar{H}$ (kOe) ($T = 300$ K)
$Fe_{79}P_{16}B_5$	0.22	249	$Fe_{75}P_6B_{19}$	0.16	250
$Fe_{79}P_{14}B_7$	0.19	247	$Fe_{81}P_6B_{13}$	0.12	263
$Fe_{79}P_{13}B_8$	0.14	247	$Fe_{83}P_6B_{11}$	0.10	263
$Fe_{79}P_9B_{12}$	0.14	256	$Fe_{80}B_{20}$	0.07	260
$Fe_{79}P_7B_{14}$	0.12	258	$Fe_{75}P_{15}C_{10}$	0.22	232
$Fe_{79}P_5B_{16}$	0.10	255			

4. Quantitative Analysis: Hyperfine Field Distribution

The experimental spectra of $Fe_{80}B_{20}$ and Fe–P–B systems may likely be fitted by a distribution of hyperfine field where a given shape is assumed a priori, as previously shown for Fe–P–C [4], and for $Fe_{80}P_{17}C_3$ and $Fe_{80}P_{13}C_7$ [25]. Although for these alloys such methods give a good fit of the Mössbauer data, there are not enough physical reasons to assume a priori a given HFD in the amorphous state. In the present case it is more appropriate to use other procedures, such as the one proposed by Window [26] or by Hesse and Rübartsch [27].

In order to fit our experimental data we used two methods: The first one is that of Hesse and Rübartsch, where a continuous HFD is assumed, and in the second method a discrete HFD is assumed; the Mössbauer spectrum is then fitted by a minimum number of hyperfine field values. In both procedures the absorption intensity $I(v)$ is supposed to be the sum of various Lorentzian sextuplets $L_i(H_i, v)$ with a probability $P(H_i)$. The relative intensities of the Lorentzian lines were 3: R: 1, where R is an adjustable parameter. The linewidth Δ was first assumed to be 0.28 mm/s as in thin foil of pure iron. But the obtained fits showed noticeable discrepancies for the inner lines (3 and 4). Therefore, the linewidth was progressively increased to reduce such discrepancies, and a reasonable agreement was obtained for $\Delta = 0.38$ mm/s; such a value can mainly be explained, as pointed out in Section 3, by the quadrupolar splitting effect. In the Hesse and Rübartsch method, in order to avoid solutions with unphysical oscillations, a smoothening factor γ is introduced; this factor involves the existence of a second continuous derivative of $P(H)$, the value of γ determines the smoothness of the solutions. The best γ-value, i.e. the best HFD, is obtained when the deviations of the fit from the experimental data correspond to the statistical error. However, in the present case it was not possible to determine accurately the best γ due to the small dissymetries in $I(v)$. In order to account for these dissymmetries it would be necessary to introduce more complicated and likely non unique parameters. With the simple previous assumption, the dissymmetries are averaged and the obtained fits with $0.4 \leqq \gamma \leqq 4$ are reasonably good (cf. Fig. 1). The experimental spectra have been fitted for 256 $I(v)$ values and 50 $P(H_i)$ equidistant values.

In our experimental data analysis, each spectrum was first fitted by assuming a hyperfine field which lies between 0 and 400 kOe. The results show that no significant probabilitiy occurred near zero fields and above 340 kOe. Therefore, the field range was progressively reduced in order to get a better accuracy on $P(H)$, and the finally obtained distribution corresponds to the minimum range of hyperfine field. This range is typically 130 to 330 kOe for Fe–P–B and 100 to 330 kOe for Fe–P–C. The $P(H)$ curves are asymmetrical with a broad tail towards the low fields; this asymmetry decreases when the boron content is increased in the Fe–P–B system. Moreover, the HFD exhibits some bumps for large values of γ (≈ 4), and these bumps become

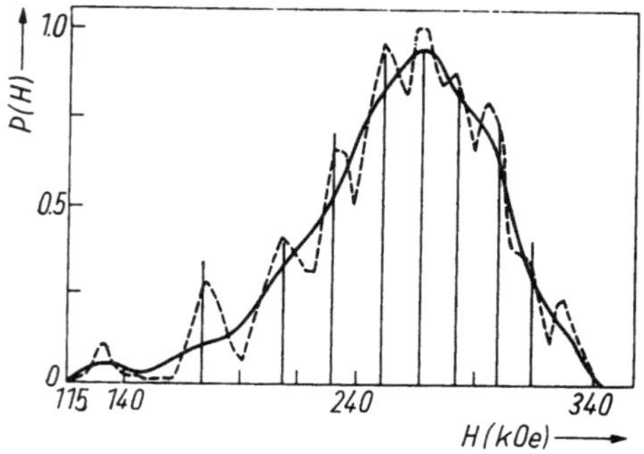

Fig. 2. $Fe_{83}P_6B_{11}$: various hyperfine field distributions. The full and dashed curves represent a continuous distribution (see fit in Fig. 1); the straight lines correspond to a discrete distribution, where the contribution at 330 kOe is attributed to iron precipitate ($I_s = 0$)

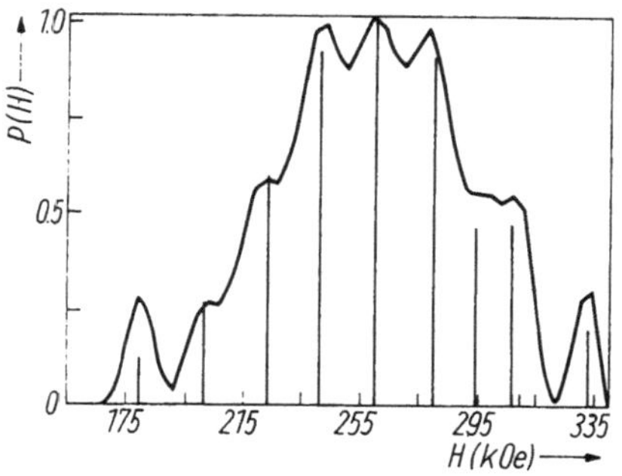

Fig. 3

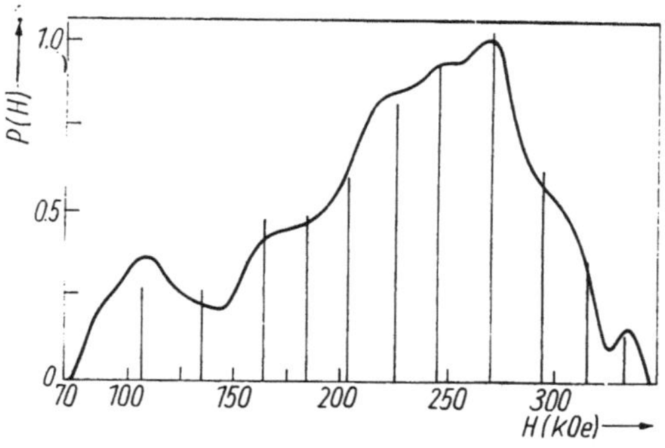

Fig. 4

Fig. 3. $Fe_{80}B_{20}$: continuous HFD obtained with $I_s = 0.07$ mm/s, $R = 2.0$, $\gamma = 0.7$, and discrete HFD obtained with the same R and I_s, except for $H = 330$ kOe where $I_s = 0.0$ (iron precipitate)

Fig. 4. $Fe_{75}P_{15}C_{10}$: continuous HFD obtained with $I_s = 0.20$ mm/s, $R = 2.1$, $\gamma = 3$, and discrete HFD obtained with the same R and I_s, except for $H = 330$ kOe where $I_s = 0.0$ (iron precipitate)

apparent peaks better and better defined as far as the γ-value is lowered, although the quality of the fit is not noticeably changed (Fig. 2). The origin of such shapes is discussed in the next section.

For each alloy we define the average hyperfine field $\bar{H}$, which corresponds to the centre of gravity of HFD: $\bar{H} = \sum_i H_i P(H_i) / \sum_i P(H_i)$. This value is found to be practically independent of the γ-values (less than 0.5%).

The second method used to analyse the experimental spectra is to assume a discrete HFD with a minimum number of field values, the characteristics of Lorentzian lines being the same as previously. The number and the values of the hyperfine field are adjusted in order to get a good fit. Such a procedure was mainly used in order to obtain a hyperfine field distribution with a different method and to evaluate the limit of the possible $P(H)$.

Although quantitatively different (Fig. 2, 3, and 4), the results obtained by such a method show that the hyperfine field range and the general shape of the discrete HFD are the same as those of the continuous HFD; the average field $\bar{H}$ is also the same within 0.5%. For all the investigated samples, the experimental spectrum can be fitted with 8 to 11 hyperfine field values.

Finally, we note in both analyses the existence of a small contribution at 330 kOe, which is attributed to iron precipitate since for a given sample foil this contribution is larger in the centre than in the edges where the cooling rate of the quenching is known to be faster.

5. Discussion

Although peaks, discrete values, or large bumps could be generated by the fitting procedures, an averaging of the obtained curves shows that the HFD is meaningful and presents some shoulders which are more or less pronounced. As a matter of fact, in the Hesse and Rübartsch method, when large values are taken (i.e. when a very smooth $P(H)$ is assumed), the calculated spectrum exhibits departures from the experimental data. Therefore, the obtained $P(H)$ curves suggest the existence of a structure in the hyperfine field distribution of the amorphous state. This result may seem surprising since the X-ray spectra present a first "flat" maximum without structure. This can be explained by the fact that the X-ray diffraction is mainly sensitive to the interatomic distance and the diffusion factor, whereas the absorption concerns mainly the magnetic state of atoms. As pointed out previously, many previous studies in the crystalline or amorphous state showed that the magnetic properties are related to the local arrangement of atoms, and the interatomic distances do not seem to be a major factor. In a number of amorphous alloys, diffraction studies showed that the average local arrangement of atoms is equivalent to the real or hypothetical crystalline counterpart [2, 28]. Although no detailed structure studies were yet performed on the alloy we investigated, it is reasonable to compare our Mössbauer results with those obtained in crystalline equivalents. Such a comparison is also suggested by the magnetic studies (summarized in Section 2).

5.1 Fe–P–B system

In the crystalline Fe(P–B) system, Mössbauer studies showed that hyperfine field values can be associated with magnetic sites of iron [23, 30] (Fig. 5). The correspondence between the magnetic iron sites and the crystallographic iron sites is ambiguous. However, the determined hyperfine field values and their variations with boron content can be considered as characteristic of the system. The difference between the extreme values decreases when the boron content is increased, and the average hyperfine field value increases significantly with increasing c_B (from 230 to 260 kOe) for $0 \leqq c_B \leqq 24$ at%.

A comparison of these results with those obtained in the amorphous $Fe_{80}B_{20}$ and in the Fe–P–B system shows that the field distribution lies in the same range. It is wider in the amorphous case mainly due to large tails of the distribution in the low- and high-field regions. The bumps or shoulders, peaks or discrete values with large probabilities occur at fields which are close to those obtained for the crystalline alloys (Fig. 5; see also [9]). Moreover, the average HF values are close to those found by Wäppling et al. [23, 30]. This suggests that the iron atoms in the amorphous state have on an average the same local environment as in the crystalline state with ε or ε_1-SRO. The tails at high- and low-field regions of $P(H)$ may be related to fluctuations of the local environments, i.e. departures from the average SRO. The low-field part, whose probability can be roughly estimated as 8%, is related to iron atoms with metalloid-rich environments, such as, for example, an Fe_5PB_2 SRO, where HF components are observed at 149 (68%), 182 (10%), and 203 kOe (23%); the percentages in parentheses indicate the relative intensities. A very rough estimation of the high-field part gives around 10%. If we assume that $P(H)$ varies with temperature in the same way as the magnetization [24], this high-field contribution would be around 330 kOe at 0 K and could be related to iron atoms with iron-rich environments.

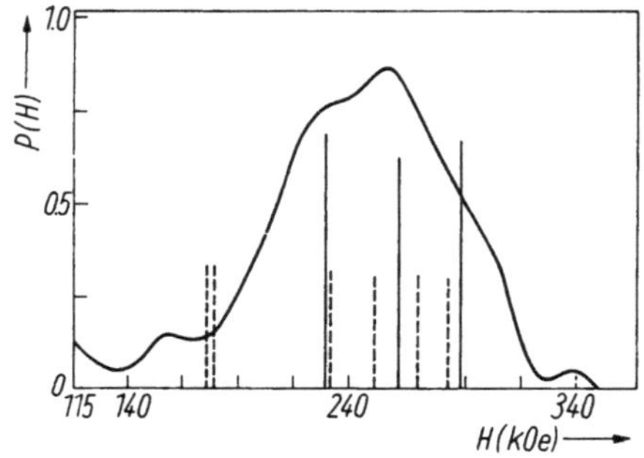

Fig. 5. $Fe_{75}P_6B_{19}$: HFD with $I_s = 0.16$ mm/s, $R = 2.3$, $\gamma = 3$ (full curve) compared to the values determined in crystalline Fe_3P after [23] (dashed lines) and in $Fe_{2.85}P_{0.05}B_{0.95}$ after [29] (full lines)

In the $Fe_{79}P_{21-c}B_c$ system, our results indicate a trend to the change of the SRO, as suggested by previous studies [8, 9]. However, the variation of the HFD and the average field $\bar{H}$ as a function of metalloid composition occurs in a smooth way. This suggests that the observed properties arise from the coexistence of the ε- and ε_1-SRO in the amorphous alloy, the former being predominant for low boron concentrations, whereas the second one becomes predominant for high c_B.

In the $Fe_{94-x}P_6B_x$ system, the width of the HFD is roughly constant. But when the iron content is increased, $P(H)$ is slightly shifted toward higher fields and the average field increases from 240 to 260 kOe. This variation is likely resulting from two combined effects: the decrease of c_B/c_P which tends to decrease $\bar{H}$, and the increase of iron concentration which tends to increase $\bar{H}$.

5.2 Fe–P–C system

For $Fe_{75}P_{15}C_{10}$, the hyperfine field distribution is more extended toward low fields than in the Fe–P–B system, being between 80 and 315 kOe (cf. Fig. 4), and the average HF is lower (230 kOe). Our results can be compared to those obtained for various crystalline compounds such as Fe_3P, Fe_3C, and likely Fe_5C_2, which has a similar structure; the results on Fe_3P have been previously discussed. In Fe_3C there is only one magnetic site corresponding to a hyperfine field of 208 kOe [31]. In Fe_5C_2 three magnetic sites are found: 222, 184, and 110 kOe. This comparison suggests that the short-range order in amorphous $Fe_{75}P_{15}C_{10}$ is predominantly of Fe_3P- and Fe_3C-type. The low part of the HFD, which is estimated to be around 10% of iron atoms, could be related to the existence of an Fe_5C_2 short-range order; the high part of the HFD, which is also estimated to be 5% to 10%, can be interpreted in the same way as in the Fe–P–B system, i.e. by the presence of iron-rich environments.

5.3 Magnetic moment and hyperfine field

We point out that for all the investigated alloys the ratio between the average hyperfine field and the magnetic moment, $\bar{H}/\bar{\mu}$, is nearly constant and equal to $\approx$ 135 kOe/μ_B at 300 K. This value is close to that found in equivalent crystalline alloys and a large number of crystalline compounds of transition metals with s–p elements. In Fe_3P such a ratio holds approximately for each iron site [29]. In transition metals and alloys, the magnetic moment is mainly due to unpaired d-electrons, whereas the hyperfine field is the sum of three contributions, the core polarization term, the contact term, and the orbital term. Magnetization measurements [24] suggest that the considered alloys are strongly ferromagnetic, therefore the orbital contribution is equal to zero.

Since a compensation between all the previous terms unlikely in many compounds, the constancy of the $\bar{H}/\bar{\mu}$ ratio suggests, as for crystalline compounds [31], that the moment distribution is constant around the iron nucleus. For the investigated alloys the HFD may reflect the distribution of magnetic moments.

Finally, in the Fe–P–B system the $\bar{H}$-values extrapolated to 0 K are in good agreement with a crude donor model; the $\bar{H}$-values at 0 K were obtained by assuming a same variation as $\bar{\mu}$, the average number of transferred electrons is taken as 1.4 for boron and 1.6 for phosphorous [24], and the field change is taken as 130 kOe per transferred electron, as for intermetallic compounds [13]. For $Fe_{75}P_{15}C_{10}$ this model is valid for $d \approx 1.6$ for phosphorus (as in the Fe–P–B system) and $d \approx 2.1$ for carbon, which is reasonably given by the magnetic moment results.

Acknowledgements

The author wishes to express his appreciation to Prof. P. Duwez for his advice and encouragement throughout this work. Thanks are also due to Dr. J. Durand for helpful discussions. Expert technical assistance was supplied by C. Geremia, S. Kotake, and J. Wisocky.

References

[1] M. H. Cohen and D. Turnbull, Nature (London) **203**, 964 (1964).
[2] J. F. Sadoc, Ph.D. Thesis, Paris 1976.
[3] A. Amamou, F. Gautier, and B. Loegel, J. Phys. F **5**, 1342 (1975).
[4] C. C. Tsuei and H. Lilienthal, Phys. Rev. B **13**, 4899 (1976).
[5] A. Amamou, IEEE Trans. Magnetics **12**, 948 (1976).
[6] R. Hasegawa and C. C. Tsuei, Phys. Rev. B **2**, 1631 (1970).
[7] R. Harris, M. Plischke, and M. J. Zuckerman, Phys. Rev. Letters **31**, 160 (1973).
[8] J. Durand, IEEE Trans. Magnetics **12**, 945 (1976).
[9] K. Raj, A. Amamou, J. Durand, J. I. Budnick, and R. Hasegawa, in: Amorphous Magnetism, Vol. II, Ed. R. A. Levy and R. Hasegawa, Plenum Press, New York/London 1977 (pp. 221 to 233).
[10] K. Raj, J. Durand, J. I. Budnick, and S. Skalski, J. appl. Phys. **49**, 1671 (1978).
[11] H. S. Chen, phys. stat. sol. (a) **17**, 561 (1973).
[12] K. Raj, J. Durand, J. I. Budnick, C. C. Tsuei, and S. Skalski, Solid State Commun. **24**, 189 (1978).
[13] L. Takacs, paper presented at Internat. Conf. Soft Magnetic Materials, Bratislava 1977, see also Solid State Commun. **21**, 611 (1977).
[14] I. Vincze, Solid State Commun. **25**, 93 (1978).
[15] U. Gonser, M. Ghafari, and H. G. Wagner, J. Magnetism magnetic Mater. **8**, 175 (1978).
[16] C. L. Chien, Phys. Rev. B **18**, 1003 (1978).
[17] J. M. Dubois and G. Le Caer, to be published.
[18] P. Duwez, Ann. Rev. Mat. Sci. **6**, 83 (1976).
[19] E. Kankeleit, Rev. sci. Instrum. **35**, 194 (1964).
[20] P. Mangin, Ph.D. Thesis, Nancy 1977.
[21] G. A. Fatseas, Phys. Rev. B **8**, 43 (1973).
[22] L. R. Walker, G. K. Wertheim, and V. Jaccarino, Phys. Rev. Letters **6**, 98 (1961).
[23] R. Wäppling, L. Häggström, S. Lundqvist, and E. Karlsson, J. Solid State Chem. **3**, 276 (1971).
[24] J. Durand and M. Yung, see [9] (pp. 275 to 288).
[25] F. E. Fujita, T. Masumoto, M. Ritaguchi, and M. Ura, Japan. J. appl. Phys. **16**, 1731 (1977).
[26] B. Window, J. Phys. E **4**, 401 (1971).
[27] J. Hesse and A. Rübartsch, J. Phys. E **7**, 526 (1974).
[28] R. Alben, J. I. Budnick, and J. S. Cargill III, in: Metallic Glasses, Amer. Soc. Metals, Metals Park (Ohio) 1976 (p. 304).
[29] E. J. Lisher, C. Wilkinson, T. Ericsson, L. Häggström, L. Lyndgren, and R. Wäppling, J. Phys. C **7**, 1344 (1974).
[30] R. Wäppling, T. Ericsson, L. Häggström, and Y. Anderson, J. Physique **37**, C-6, 591 (1976).
[31] H. Bernas, I. A. Campbell, and R. Fruchart, J. Phys. Chem. Solids **28**, 17 (1967).

(Received June 7, 1979)

phys. stat. sol. (a) 54, 573 (1979)

Subject classification: 1.2; 1.4; 21.6

Laboratoire de Cristallographie et Physique des Matériaux, Université de Paris XI, Orsay[1]) *(a), CECM, Vitry sur Seine*[2]) *(b) et Département de Technologie, SRMA, CEA, Saclay*[3]) *(c)*

Fluctuations prétransitionnelles dans AuCu

Par

M. GUYMONT (a), D. GRATIAS (b) et A. BISSON (c)

Un comportement structurel différent suivant les traitements thermiques appliqués est mis en évidence dans AuCu pour la même composition (50 at%) par diffraction électronique. Ce phénomène est attribué à l'existence de fluctuations de composition spinodales dans la zone désordonnée au voisinage immédiat du point de mise en ordre.

The structural behaviour of AuCu 50 at%, as investigated by electron diffraction, is different according to the thermal treatments performed. This phenomenon is ascribed to the existence of spinodal composition fluctuations in the disordered area at the immediate vicinity of the ordering transition point.

1. Introduction

AuCu II a été le premier composé à période irrationnelle étudié systématiquement. Jehanno et Pério [2, 3] ont déterminé pour la demi-longue période M (mesurée en unités de la maille de Bravais de l'alliage désordonné) la valeur 5,13 dans un domaine de 43 à 54 at% Au (Fig. 1). Dans une lettre précédente [1] nous avons signalé l'observation — pour la première fois — d'une longue période $M = 5$ exacte[3]) et émis l'hypothèse que cette différence de comportement pourrait provenir d'une différence de traitements thermiques. Jehanno, en effet, pour éviter un groississement des grains défavorable à l'exploitation des diagrammes de poudre en rayons X, a recuit ses échantillons à 450 °C préalablement à la mise en ordre, alors que les nôtres, recuits vers 550 à 600 °C, sont ensuite portés directement à 400 °C qui est la température de mise en ordre de la surstructure à antiphases périodiques (APP).

Nous avons vérifié expérimentalement cette hypothèse, et ce sont les résultats de ces expériences que nous présentons ici.

2. Résultats expérimentaux

Le schéma expérimental suivant a été appliqué: on prend deux rubans laminés d'environ 1×10 cm^2 provenant du même alliage AuCu 50 at%. (Il est en effet nécessaire de prendre deux échantillons à chaque expérience, car l'échantillon aminci électrolytiquement pour observation n'est plus récupérable pour un traitement ultérieur.) Après mise sous vide en ampoule de quartz, on leur fait subir la succession des traitements suivants:

1. Recuit à 550 °C environ 24 h pour éliminer l'écrouissage;
2. recuit à 450 °C pendant une vingtaine d'heures (zone de recuit de Jehanno);
3. recuit à 400 °C pendant une vingtaine d'heures (température de mise en ordre des APP).

[1]) Bât. 490, 91405 Orsay Cédex, France.

[2]) 15 rue Georges Urbain, 94400 Vitrysur Seine, France.

[3]) La valeur 5 mesurée pour M sur le diffractogramme n'a d'ailleurs pas de signification physique particulière: c'est une valeur *statistique* (elle ne correspond pas au modèle de Johansson et Linde qui décrit une structure *rationnelle*) [5].

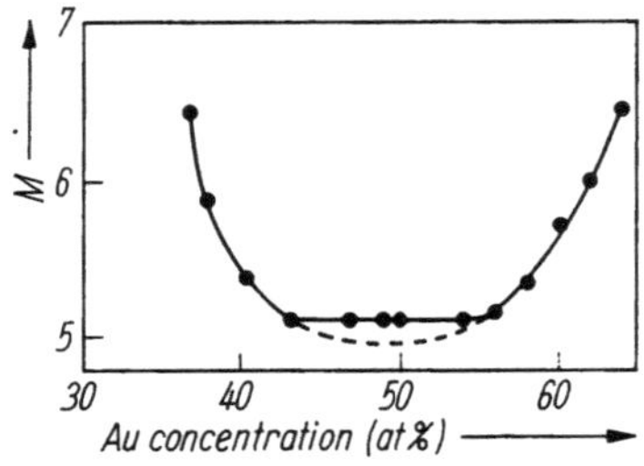

Fig. 1. Variation de M avec la composition dans AuCu II suivant Jehanno [3]. En pointillé: courbe prévue en l'absence de fluctuations

4. On recueille l'un des échantillons (α) pour observation. L'autre (β) est remis en ampoule de quartz sous vide, et soumis à nouveau aux traitements suivants:
1. Recuit à 550 °C quelques heures (pour le remettre en désordre);
2. recuit à 400 °C de un à plusieurs jours.
3. L'échantillon β est alors récupéré pour observation.

Les deux échantillons sont observés, après amincissement électrolytique, en diffraction électronique (microscopes utilisés: Philips EM 300 et Jéol JEM 200C).

Les résultats ont confirmé notre hypothèse. Les Fig. 2a et b montrent respectivement les diagrammes de diffraction électronique du plan (001)* d'un échantillon α et d'un échantillon β. On voit que l'échantillon β présente les mêmes caractéristiques structurales ($M = 5$ *exactement*) que nos échantillons stoechiométriques ordinaires (c'est à dire, ayant été mis en ordre APP à 400 °C *directement* après le recuit à 550 °C) [1]. Par contre l'échantillon α présente un diagramme de diffraction analogue à ceux qu'on obtient avec des échantillons AuCu II hors stoechiométrie: on n'a plus superposition exacte des taches de diffraction APP des séries $-1 + (2n + 1)/2M$, 1,0 et $1 - (2n + 1)/2M$, 1,0, mais chevauchement (voir l'agrandissement contrasté: Fig. 3). Ainsi, sur la Fig. 3, on mesure $M = 5{,}05$. D'autre part, comme nous l'avons constaté pour des échantillons de AuCu II 55 at% Au [1], les rangées réciproques de réflexions APP ne sont pas strictement perpendiculaires à $\boldsymbol{C}^*$ mais sont légèrement inclinées, ce qui confère à l'ensemble du diagramme un caractère "ondulé" dans cette direction.

Un autre fait à signaler est que, bien que l'aspect qualitatif du diffractogramme des échantillons α (à savoir, la légère inclinaison et $M \neq 5$) soit reproductible, on trouve des valeurs différentes pour M: par exemple, la valeur $M = 5{,}05$ mesurée sur la Fig. 3 est différente de la valeur 5,13 obtenue par Jehanno. Ceci est dû, à notre avis, à des raisons cinétiques d'une part ("trempe" à l'air plus ou moins rapide après le recuit à 450 °C, temps de recuit plus ou moins long), et, d'autre part, à une incertitude éventuelle sur la température de 450 °C: nous n'avons pas cherché à avoir une température précise de recuit, mais une température dans la zone précédent immédiatement la température de mise en ordre (415 °C), zone où Jehanno avait opéré. Il est fort possible (cf. discussion,) si on a affaire à une zone de fluctuations faciles dans le voisinage de la température de transition désordre → APP, que les valeurs obtenues pour M ensuite (c'est à dire *apres* mise en ordre à 400 °C) dépendent *et* de cette température "de fluctuation" *et* de conditions cinétiques.

3. Essai d'interprétation et discussion

L'interprétation que nous avançons pour expliquer ces phénomènes est qu'on a affaire à des fluctuations de concentration dans la zone de désordre précédant immédiatement la transition vers la structure APP [4]. Notre ligne de raisonnement est la suivante:

Nous partons du fait que M est fonction de la concentration x; cette fonction doit avoir l'allure grossièrement parabolique (en réalité l'allure est dissymétrique) suggérée par la Fig. 1 mais corrigée de la partie rectiligne et avec un minimum $M = 5$ à la

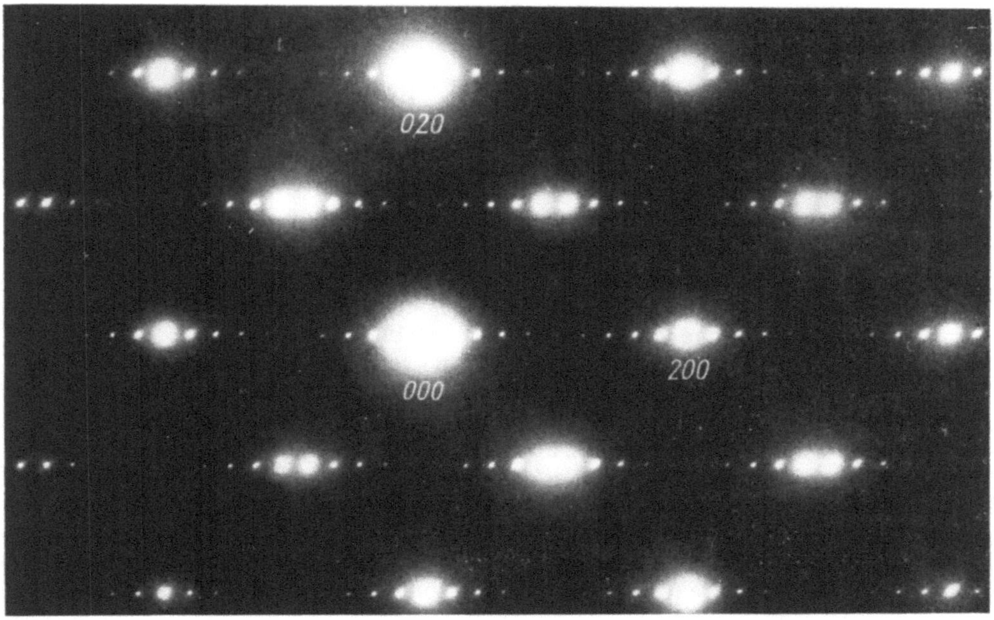

a

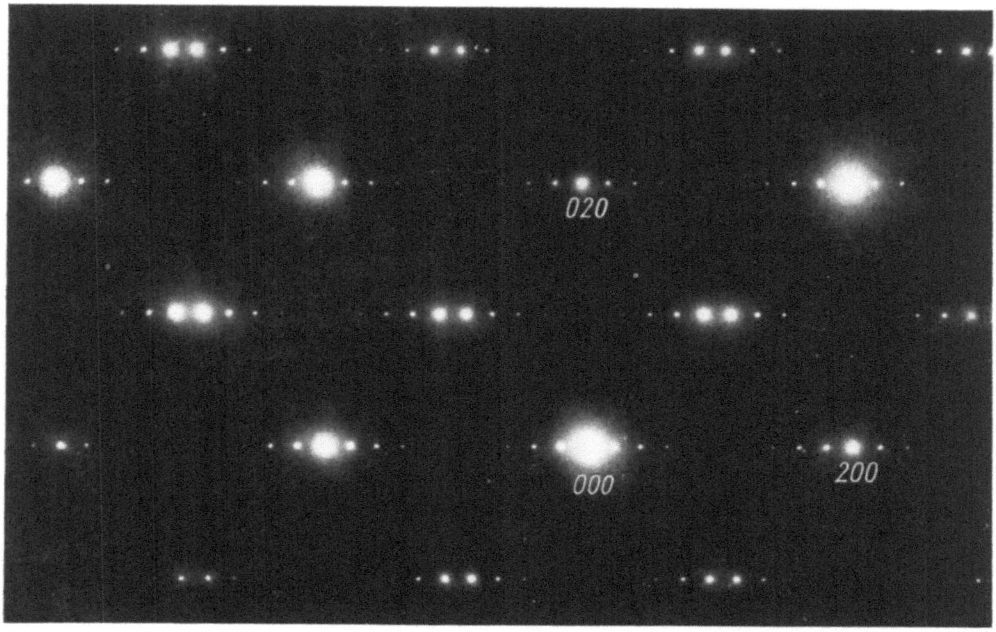

b

Fig. 2. Diffractogramme d'un échantillon a) α, b) β

Fig. 3. Agrandissement contrasté d'une rangée de taches APP de la Fig. 2a

stoechiométrie (ligne en pointillé). M constitue donc une véritable "sonde" interne renseignant sur la concentration. L'hypothèse heuristique que nous faisons est que M peut être considéré comme une fonction d'état, ne dépendant que de x. Dans les expériences de Jehanno, entre 43 et 54 at% Au, M ne varie pas et vaut constamment 5,13. Le système est alors invariant: l'analogie avec la règle des phases à température et pression constantes suggère que le système est *biphasé* dans la zone où M est constant: donc, on aurait dans toute cette zone, un mélange en proportion variable de deux structures (toutes les deux avec $M = 5{,}13$) de compositions 43 et 54 at% Au. En réalité l'équilibre n'est réalisé qu'avant la transition, car des fluctuations (dynamiques) de concentration sont compatibles avec le désordre: l'homogénéité reste assurée sur une période de temps suffisamment longue par la diffusion des lacunes et des atomes. Une fois l'alliage mis en ordre, on "ancre" les fluctuations de concentration dans la surstructure·APP: on n'est alors plus à l'équilibre. Ces fluctuations de composition, faciles avant la transformation en APP, sont maintenant immobilisées de façon métastable par la surstructure. On conçoit donc que la valeur trouvée pour M puisse varier suivant les conditions ayant permis ces fluctuations.

Pour que de telles fluctuations prétransitionnelles soient possibles, il faut admettre que l'enthalpie libre G de l'état "fluctué" est très voisine de l'enthalpie libre de l'état "monophasé" normal (correspondant à $M = 5$). Par conséquent G doit être la moyenne en proportion convenable des enthalpies libres G_1 et G_2 correspondant aux deux concentrations mélangées. Donc, dans cette zone de concentration, l'enthalpie libre doit être une fonction linéaire de la concentration. D'où

$$\frac{\partial G}{\partial x} = \text{const} \rightarrow \frac{\partial^2 G}{\partial x^2} = 0 .$$

Ces fluctuations de composition seraient donc de type spinodal.

Une remarque finale peut être faite: l'observation par diffraction électronique des echantillons est destructive à cause de l'amincissement. L'expérience non destructive idéale serait, bien entendu, l'observation de l'apparition et de la disparition par traitement thermique approprié d'une diffusion centrale associée aux fluctuations.

Remerciements

Nous remercions le Prof. P. Pério qui a eu l'idée de ce travail et Melle M. P. Van Kalck pour son aimable collaboration.

Bibliographie

[1] M. GUYMONT et D. GRATIAS, J. Physique, Lettres **39**, 437 (1978).
[2] G. JEHANNO et P. PÉRIO, J. Physique Radium **23**, 854 (1962).
[3] G. JEHANNO, Thèse, Univ. d'Orsay, Rapp. CEA R2812 (1965).
[4] P. PÉRIO, communication privée.
[5] M. GUYMONT, R. PORTIER et D. GRATIAS, soumis à Acta cryst. A.

(Received June 8, 1979)

phys. stat. sol. (a) 54, 577 (1979)

Subject classification: 13.4 and 20.3; 22.2.1

Institute of Electronics, Academy of Sciences of the Belorussian SSR, Minsk

Anomalous Temperature Dependence of Recombination Radiation of Gallium Arsenide

By

V. I. Osinskii and A. Ya. Peshko

The results of investigations on the anomalous temperature dependence of recombination radiation of GaAs structures produced by epitaxy from gaseous and liquid phase and by laser doping are presented. On the basis of the experimental measurements and computer calculations the luminescence mechanisms with positive and zero temperature coefficients of intensity are discussed.

Приведены результаты исследований аномальной температурной зависимости рекомбинационного излучения в арсенидгаллиевых структурах, полученных методами газовой, жидкостной эпитаксии, а также методом лазерного легирования. На основе экспериментальных и расчётных данных предложены и обсуждены механизмы люминесценции с положительным и нулевым температурными коэффициентами интенсивности.

1. Introduction

The external quantum yield of recombination radiation of gallium arsenide decreases by several hundred times as the temperature increases from helium to room temperature. It is considered that this is associated with the increasing proportion of non-radiative recombination and the large temperature shift of the main absorption edge of the semiconductor. From numerous experimental observations it has been empirically established that there is an exponential dependence between the intensity of radiative recombination and temperature:

$$\eta_e(T) = \frac{1}{1 + A \exp(-E/kT)}\,. \tag{1}$$

Analysis of the Shockly-Reed equation by Moll [1] and Winogradoff [2] has shown that the temperature dependence of the rate of radiative recombination has a minimum and a maximum, whose positions are determined by the ratio of recombination sections in two competitive channels. Instead of a monotonically decreasing dependence (1), an anomalous temperature dependence (ATD) of luminescence intensity is possible, in which the external quantum yield of recombination radiation increases or remains constant as the temperature rises within a certain range.

The ATD of recombination radiation of gallium arsenide diodes obtained by gaseous epitaxy was observed experimentally at low injection levels [3] in the temperature range from 293 to 383 K. As injection level increases up to 5 A/cm^2 the character of the temperature dependence of radiation changes to the normal one. Such temperature dependences of luminescence intensity are observed in diffused [4] and alloyed [5] GaAs diodes. Investigations of the photoluminescence of n- and p-type layers of the laser structure have shown that in n-type gallium arsenide a region of positive temperature dependence (PTD) of recombination radiation is observed over a wide excitation intensity range, while in p-type gallium arsenide the radiation intensity falls steadily over the whole temperature range as temperature increases [6] in accordance with the empirical dependence (1). ATD is apparently characteristic of p–n structures when recombination takes place in the n-region. In [1 to 6] ATD of intensities was observed at low injection levels.

The purpose of this work is a systematic investigation of structures and conditions under which the intensity of recombination radiation of gallium arsenide rises or remains constant, and, also, to clear up the mechanisms of electron transitions responsible for such an anomalous temperature dependence.

2. Experimental Techniques

The experiments were performed on GaAs layers epitaxially grown from gaseous and liquid phase. This allowed to determine the effect of impurity–vacancy complexes appearing under certain growth conditions on the ATD mechanism. Doping was done in the process of epitaxy, by diffusion, or by local laser heating of the gallium arsenide plate covered with a thin zinc film [7]. By the latter method p^+–n junctions with hole injection into the n-type region were obtained. Also, the temperature dependence of electroluminescence intensity in double heterostructures has been studied. Such structures allow one to realize an effective injection of current carriers from wide-gap $Ga_{1-x}Al_xAs$ emitters into the narrow-gap active region of n- or p-GaAs.

Samples were fixed to a copper holder in the cryostat whose temperature was maintained automatically at fixed temperatures in the range 77 to 500 K. The temperature was controlled by means of a copper–constantan thermocouple with an accuracy of ± 3 K.

Ohmic heating of p–n junctions did not exceed 5 to 10 K and was evaluated from the shift of the recombination radiation peak.

Luminescence intensity was measured with a ФЭУ-28 photomultiplier with appropriate correction of its spectral sensitivity. During integral measurements, part of the sample radiation was directed to a ИСП-51 spectrometer by means of a semitransparent plate. Spectra were registered automatically with preliminary synchronized detection of the signal.

Photoluminescence was excited by a helium–cadmium laser or a high-pressure mercury lamp using appropriate filters to select the excitation spectrum.

3. Experimental Results

Deep levels responsible for the low-energy band in the photoluminescence spectra of epitaxial GaAs films with energy corresponding to the radiation maximum at 1.2 eV are considered to be formed by donor impurity–Ga vacancy complexes appearing during epitaxy from the gaseous phase due to Ga atom deficiency. The donor impurity concentration determines the concentration of complexes, thus controlling the re-

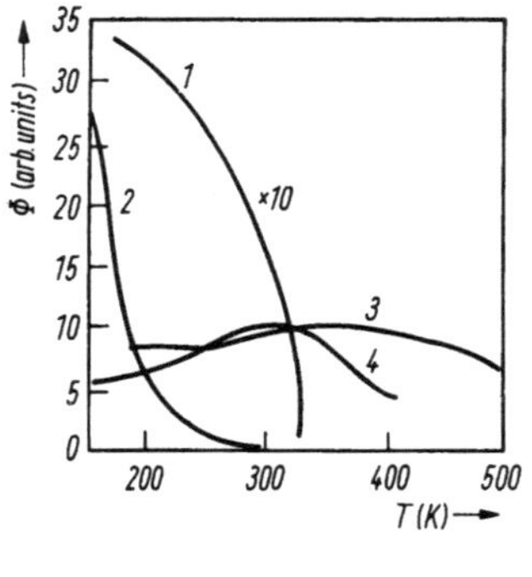

Fig. 1. Temperature dependences of the intensity of photoluminescence bands in gallium arsenide for two tellurium concentrations. (1), (2) 1.2×10^{15} and (3), (4) 5×10^{18} cm^{-3}

combination rate in the low-energy band. Fig. 1 shows the temperature dependence of the photoluminescence intensity for both radiation bands for two tellurium concentrations. An increase in Te concentration shifts the PTD portion of the edge radiation to higher temperatures, and the intensity of the low-energy band decreases in all portions where the edge radiation increases.

In gallium arsenide grown by liquid epitaxy from a solution in gallium melt, the low-energy band is absent, both in photo- [8] and electroluminescence [9]. In our experiments only one photoluminescence band was observed over the whole temperature range investigated. As temperature increases from 77 to 220 K, the radiation intensity decreases (Fig. 2, curve 1), passes through a minimum, after which intensity oscillations with a period of 30 to 50 K and an increase in average intensity are observed up to 350 K. It should be noted that at the temperatures of the starting PTD portion the energy of the radiation band maximum increases abruptly by 17 meV, and the temperature coefficient $\partial h\nu/\partial T$ (Fig. 2, curve 2) changes. Before the PTD portion starts the temperature coefficient of the radiation band maximum was equal to 0.36 meV/K and was determined mainly by the temperature coefficient of the band gap (Fig. 2, curve 3). After the abrupt energy change the temperature coefficient increases to 1.12 meV/K, and after the intensity has passed through a maximum it decreases again, approaching the value characteristic of the region of reduction of photoluminescence intensity with temperature rise.

The injection properties of structures with p–n junctions affect the number of recombining carriers and the spatial position of the recombination region. It is known that because of an essential difference in the mobility of electrons and holes the injection of electrons in gallium arsenide takes place primarily into the p-region of the p–n junction. The reverse injection can take place in a p^+–n structure, in which the hole concentration in the p-region is by two to three orders of magnitude higher than the concentration of electrons in the n-region. The number of holes passing through the p–n junction can then be considerably higher than the number of electrons travelling in the opposite direction. If the p^+–n junction is sufficiently abrupt, the recombination region is located in the n-layer, in which the ATD of radiation is observed experimentally. In diodes produced by the laser doping method the concentration of impurities in the n-region was equal to 1.3×10^{15} cm^{-3}, and in the p-region $-1 \times \times 10^{19}$ cm^{-3}. Voltage–capacitance measurements have shown that the width of the p–n junction does not exceed 0.1 μm. The same n-type material with carrier concentration of 1.3×10^{15} cm^{-3} was used for the production of diffused control diodes with smooth impurity distribution in the p–n junction the width of which was 2 μm. Fig. 3 shows temperature dependences of the integral radiation intensity for different currents passing through p–n junctions obtained by diffusion and laser doping. It can be seen that for all current densities shown in Fig. 3, abrupt p^+–n junctions have portions with PTD of radiation in one and the same temperature range (curves 1 to 3).

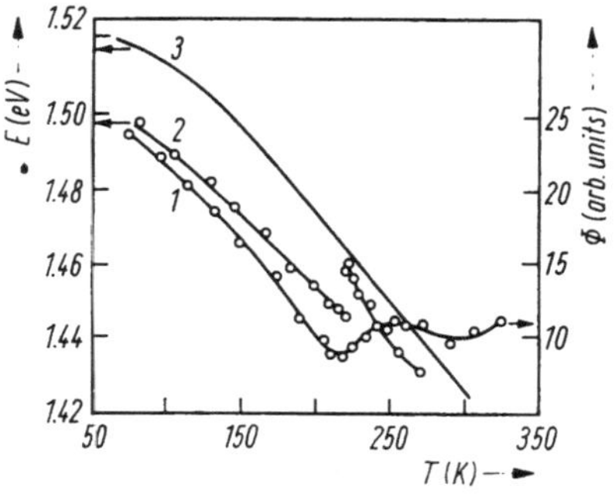

Fig. 2. Temperature dependences of (1) photoluminescence intensity, (2) energy in the maximum, (3) band gap of gallium arsenide

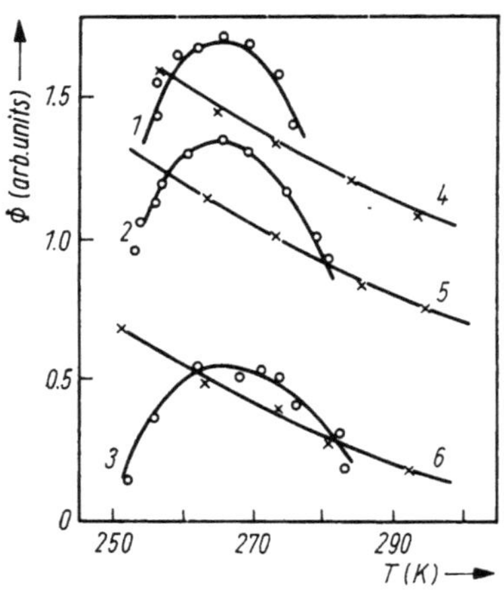

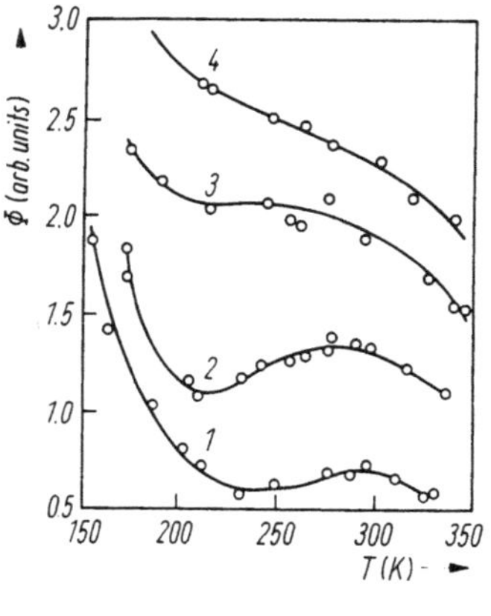

Fig. 3

Fig. 4

Fig. 3. Temperature dependences of radiation intensity of diodes obtained by laser doping (curves 1 to 3) and by diffusion (curves 4 to 6). Current density: (1), (4) 5, (2), (5) 3, and (3), (6) 1 A/cm^2

Fig. 4. Temperature dependences of electroluminescence intensity in GaAs diodes of $GaAs_{0.9}P_{0.1}$ type. Current density: (1) 5, (2) 10, (3) 15, and (4) 18 A/cm^2

With the same impurity concentration ratios in p- and n-layers and under identical injection conditions, control measurements of smooth diffused p–n junctions could not reveal a portion of PTD of radiation (curves 4 to 6).

With an increase in the injection level the PTD portion decreases, indicating the displacement of the recombination region into the p-layer of the structure.

As shown in Fig. 4 the character of the temperature dependence of the electroluminescence intensity in a radiative homostructure epitaxially grown from the gaseous phase changes as the current increases. For current densities up to 12 A/cm^2 a portion with $dI/dT = 0$ (curves 1 and 2, Fig. 4) is observed, which with increasing injection level transforms into an exponentially decreasing curve of the type given by (1) (curve 4, Fig. 4). Temperature stabilization observed for $j = 12$ to 16 A/cm^2 (curve 3, Fig. 4) is intermediate between these extremes.

The above experiments with homostructures show that the ATD of radiation can be obtained only in abrupt p–n junctions with hole injection into the n-layer. These conditions are fulfilled for small currents and also small absolute values of radiation intensity of the diodes.

Practically unlimited hole injection into the n-region and, hence, an ATD region at high absolute values of recombination radiation intensity are possible in heterostructures [10]. It is characteristic that the PTD portion is not observed experimentally at excitation levels from 1 to 150 A/cm^2 in the active layer of p-type, but with recombination in the n-type active layer it does not disappear at high injection levels as in homojunctions. With an increase in current density the PTD curve becomes slightly less steep. Thus, with electroluminescence in the active layer of an n-type heterostructure, as well as with photoluminescence, a PTD region is observed at high excitation levels and is not limited to low currents, at which in abrupt homojunctions primarily hole injection into the n-layer takes place.

4. Theory and Discussion

The recombination rate is proportional to the concentration of electrons and holes representing the majority and minority carriers in a certain type of semiconductor.

Therefore, the temperature dependence of internal quantum yield is greatly affected by an increase in carrier concentration with temperature. The concentration of equilibrium majority carriers (for n-type gallium arsenide with $n = 10^{17}$ cm^{-3} from 100 to 410 K) remains constant in a wide temperature range, while the concentration of minority carriers is small at low temperatures and increases abruptly above room temperature [11]. Therefore, the recombination rate is controlled by the minority carrier concentration in a wide temperature range.

Let us analyze the temperature dependence of the radiative recombination rate and the results of computer calculations for different parameters of gallium arsenide, and then compare the calculation results with experimental data.

Using the principle of detailed equilibrium for interband direct transitions, for which the selection rule for the wave vector $\boldsymbol{K}$ is valid, and taking into account the temperature dependence of E_g, the radiative recombination rate may be written as

$$r(E, T) = A \sqrt{\Delta E + \alpha T} \left\{ \frac{1}{1 + \exp\left[\frac{m_v}{m_v + m_c}\left(\frac{\Delta E + \alpha T}{kT}\right) - \left(\frac{F_c - E_c}{kT}\right)\right]} \right\} \times$$
$$\times \left\{ 1 - \frac{1}{1 - \exp\left[\frac{-m_c}{m_c - m_v}\left(\frac{\Delta E + \alpha T}{kT}\right) - \left(\frac{F_v - E_v}{kT}\right)\right]} \right\}, \qquad (2)$$

where ΔE is the difference between the energy of a quantum and the band gap, F_v and F_c are the Fermi quasi-levels for the valence and conduction bands, respectively. The coefficient A includes the semiconductor constants:

$$A = \frac{2\pi e^2 E (M_{cv})^2}{m_0^2 h^5 c^3 \pi^2} \left(\frac{2 m_c m_v}{m_c + m_v}\right)^{2/3}.$$

For gallium arsenide $A = 3 \times 10^{27}$ eV^{-1} s^1 cm^{-3}. Calculation of the radiative recombination rate from expression (2) for fixed temperatures in the range 70 to 450 K was performed on a "Nairi" computer for the following input values: $\alpha = 5 \times 10^{-4}$ eV/K, $k = 8.616 \times 10^{-5}$ eV/K, $m_v = 0.5 m_0$, and $m_c = 0.072 m_0$.

Temperature dependences of the radiative recombination rate were plotted from points calculated for constant values of $F_c - E_c$ and $F_v - E_v$. Plotted on a semilogarithmic scale they are shown in Fig. 5.

The families of curves shown characterize different values of parameter $b_2 = F_v - E_v$ at constant values of parameter $b_1 = E_c - F_c$. Negative values of b_1 correspond to the position of the recombination level within

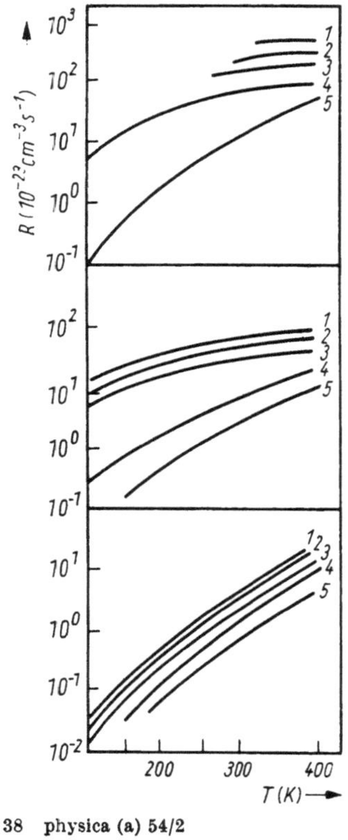

Fig. 5. Temperature dependences of integral radiative recombination rate. Curves 1 to 5 correspond to the values of parameter b_1 from -0.06 (curve 1) to 0.06 eV (curve 5) with 0.03 eV intervals

the valence band and positive values correspond to that within the band gap near the valence band. The signs for parameter b_2 for the conduction band are defined in an analogous way.

Analysis of the dependences obtained permits to arrive at two important conclusions. First, the radiative recombination rate through the low levels in the band gap always rises in the temperature range considered (100 to 450 K). Second, for large negative values of parameters b_1 and b_2 sections with decreasing temperature dependence are observed (Fig. 6), whose position on the temperature scale is different for different values of the parameters.

The fact that the luminescence intensity usually falls with temperature indicates a strong influence of nonradiative recombination processes. If the nonradiative recombination rate rises with temperature faster than the radiative recombination rate, a decreasing temperature dependence of the quantum yield of the luminescent structure is observed.

If the temperature dependences of radiative and nonradiative recombination are represented by curves 1 and 2 (Fig. 7), respectively, then the sum of their contributions to the total recombination rate will give the dependence (curve 3) corresponding to temperature quenching of luminescence. For two recombination channels with rates R_r and R_n, respectively, the quantum yield of luminescence can be determined from the following expression [13]:

$$\eta_e = \frac{\beta n_r p_r}{G} = \left[1 + \frac{R_n}{R_r} \exp\left(-\frac{E_r}{kT}\right)\right]^{-1}, \tag{3}$$

where n_r and p_r are the concentrations of free electrons and holes.

From (3) it follows that with increasing temperature the exponential term increases, which leads to a decrease in luminescence intensity. However, this dependence may take on another character, if with an increase in the exponential term the pre-exponential factor R_n/R_r decreases. A decrease in the value of this factor is possible if the local effect resulting from a change in temperature causes a change in the temperature dependences of the radiative and nonradiative recombination rates, as shown in Fig. 7 by the dashed curves 4 and 5. In such cases, there may appear a PTD portion of radiation intensity. This portion may be observed with an increase in radiative

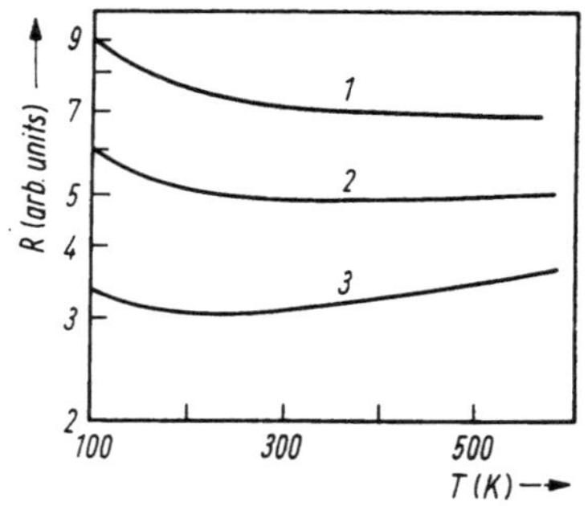

Fig. 6

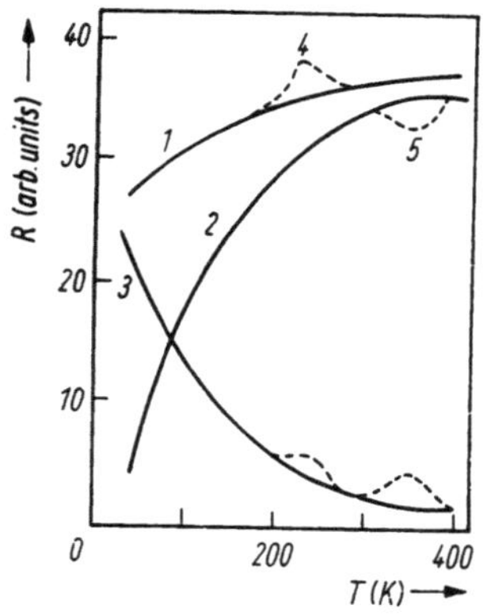

Fig. 7

Fig. 6. Temperature dependences of radiative recombination rate for large negative values of parameter b_1. (1) $b_1 = -0.12$, (2) -0.09, (3) -0.06 eV; $b_2 = -0.03$ eV

Fig. 7. Temperature dependences of (1) radiative, (2) nonradiative, and (3) total recombination rate

recombination rate due to an additional mechanism involved with rising temperature. The PTD portion may also be registered with a decreasing nonradiative recombination rate in a certain temperature range (curve 5, Fig. 7). However, the experimental investigation of this case is more difficult, because the study of the nonradiative recombination process is too complex.

The action of additional radiative channels in n-type GaAs may be associated with a rise in hole concentration caused by thermal capture of electrons from the valence band to the trap level (Fig. 8a) due to the impurity–vacancy complex. Such additional channels increase the integral radiation intensity until with rising temperature the trap level is filled. After this, the temperature dependence of radiation intensity is determined by a strong increase in the nonradiative recombination rate.

In materials with high impurity concentration, in which there are no trap levels, the PTD portion is associated with the participation of band tails caused by strong doping in recombination processes. In our case the doping degree of layers obtained by liquid-phase epitaxy was such that, in accordance with [14], the impurity band begins to overlap with the conduction band (Fig. 8b). The density-of-states function then becomes nonmonotonous [15, 16]. Its maximum for the impurity band has an energy position corresponding approximately to the ionization energy of the impurity. At sufficiently low temperatures radiative transitions take place from energy states corresponding to a density-of-states maximum near the conduction band to the levels of the valence band. An increase in temperature leads to a sharp change in the radiative recombination mechanism so that the latter becomes a band-to-band one, this change being characteristic only of samples obtained by liquid-phase epitaxy.

The energy jump in the dependence $E(T)$ can be explained by a radiative recombination mechanism, involved at a certain temperature, differing from that at lower temperatures. This is possible, for example, in the case of depletion of the impurity level near the conduction band with rising temperature due to impurity level-band radiative transitions designated by 1 in Fig. 8b. The energy levels in the conduction band are filled as the temperature increases. Since the density of states in the conduction band is considerably higher than in the impurity band [16], the probability of radiative transitions from the conduction band considerably increases. Transitions possible in this case are shown in Fig. 8b by 2. Therefore, a PTD portion appears in the temperature dependence of radiation intensity.

A discontinuity in the temperature dependence of radiation energy of an electron-beam-pumped semiconductor laser was also observed in [16]. The threshold density of such a laser decreased in the temperature range of the energy jump. These results may be compared with those obtained in this work. It is assumed that the laser threshold decrease and the increase in the intensity during photoexcitation of the active layer are consequences of analogous recombination mechanisms. Therefore, a study of the positive temperature dependence mechanism may be useful for lowering the laser threshold at higher temperatures or for reducing the effect of temperature on this important characteristic both in homo- and heterolasers. In both cases it is

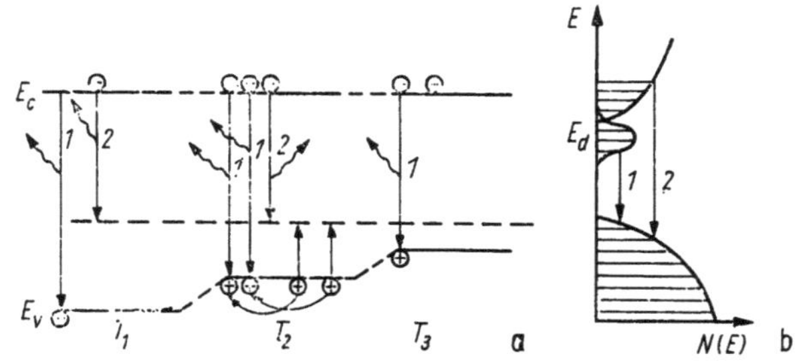

Fig. 8. Model of radiative transitions resulting in the appearance of the positive temperature dependence of radiation in samples obtained by a) gas and b) liquid-phase epitaxy

necessary to ensure conditions for directed injection of carriers into the active region having qualities necessary for the creation of a PTD region of luminescence.

5. Conclusions

1. Only two of the investigated mechanisms for the appearance of an ATD portion of recombination radiation are discussed above. Actually, other mechanisms are also possible, for which the radiative recombination rate rises with temperature so rapidly, in comparison with nonradiative recombination rate, that there appears an ATD portion of radiation.

2. Although theoretically ATD of radiation is not excluded in p-type gallium arsenide, this dependence has not yet been confirmed experimentally.

3. In conjunction with known methods, the experimental investigation of the recombination radiation intensity may provide additional information on the ratio of radiative to nonradiative energy transitions in the recombination and injection mechanisms in semiconductors.

References

[1] J. L. Moll, Physics of Semiconductors, McGraw-Hill Publ. Co., New York 1964.
[2] N. N. Winogradoff, Phys. Rev. **138**, A1502 (1965).
[3] N. N. Winogradoff and A. H. Neill, Jr., Internat. J. Electronics **28**, 401 (1970).
[4] V. I. Osinskii and N. N. Winogradoff, Phys. Rev. B **3**, 3341 (1971).
[5] A. A. Ptashchenko, V. A. Presnov, I. I. Kruglov, N. N. Golembevskii, A. P. Ovchinnikova, and L. F. Litovchenko, Fiz. Tekh. Poluprov. **6**, 215 (1972).
[6] Patentschrift BRD No. 1258980, class 21-89/03, applied for 24. 2. 1966, published 18. 1. 1968.
[7] V. A. Pilipovich, G. D. Ivlev, Y. F. Morgun, N. V. Nechaev, V. I. Osinskii, and A. Ya. Peshko, J. appl. Spectroscopy **22**, 431 (1975).
[8] M. B. Panich and H. J. Queisser, Solid State Electronics **9**, 311 (1966).
[9] M. Pilkuhn and H. Rupprecht, J. appl. Phys. **38**, 5 (1967).
[10] Zh. I. Alferov, V. Sbornik Poluprovodnikovye pribory i ikh primenenie, ed. 25, Ed. Ya. A. Fedotov, Sovetskoe radio, Moskva 1972.
[11] V. V. Novikov, Teoreticheskie osnovy mikroelektroniki, Vysshaya shkola, Moskva 1972.
[12] R. N. Hall, Solid State Electronics **6**, 405 (1963).
[13] S. M. Zi, Fizika poluprovodnikovykh priborov, Izd. Energiya, Moskva 1973.
[14] F. Stern and J. R. Dixon, J. appl. Phys. **30**, 286 (1959).
[15] V. P. Gribkovskii, Teoriya pogloshcheniya i ispuskaniya sveta v poluprovodnikakh, Izd. Nauka i Tekhnika, Minsk 1975.
[16] O. V. Bogdankevich, N. A. Borisov, I. V. Krukova, and B. M. Lavrushin, phys. stat. sol. **29**, 715 (1968).

(Received April 20, 1979)

phys. stat. sol. (a) 54, 585 (1979)

Subject calssification: 2 and 14.3.4; 22

A. F. Ioffe Physico-Technical Institute of Academy of Sciences of the USSR, Leningrad

Characteristic Features of the Thermal Switching Process in Chalcogenide Vitreous Semiconductors

By

B. T. Kolomiets, V. Ch. Shpunt, and E. A. Smorgonskaya

The current and temperature distributions during the thermal switching process in high electric fields exceeding threshold values are investigated using the experimental pulse technique and calculations based on the experimental data obtained for vitreous chalcogenide semiconductors (the vitreous material $Tl_2Se \cdot As_2Te_3$ is studied as an example). It is shown that in the course of the thermal switching process during the delay time a pronounced localization of the current and high temperature region occurs in a sufficiently narrow channel of the sample. The sharp increase of the total current flowing through the sample under switching itself (the transition to the low-resistance state) is associated with the high-conductive channel expansion till to the formation of a sufficiently wide channel in the on-state.

С помощью импульсной методики и расчета, базирующегося на экспериментальных данных на примере стеклообразного материала $Tl_2Se \cdot As_2Te_3$, исследовано распределение тока и температуры в образцах стеклообразных халькогенидных полупроводников при переключении в сильном электрическом поле, превышающем порог переключения. Показано, что в процессе развития теплового пробоя во время задержки имеет место ярковыраженная локализация тока и области повышенных температур в узком канале. Резкий рост тока при собственно включении образца (переходе в низкоомное состояние) связывается с расширением этого высокотемпературного канала и образованием достаточно широкого высокопроводящего канала включенного состояния.

1. Introduction

The high electric field behaviour, particularly the threshold switching effect (reversible breakdown), in chalcogenide semiconductors (ChVS) have been under extensive studies during the recent years in many laboratories (see, for example reviews [1, 2]). Nevertheless a universally accepted viewpoint is not available about the nature of the nonlinear current–voltage characteristics in high fields and the current instability resulting in switching. First of all this is connected with the lack of experimental data on details of the switching process in the course of time under applied voltage $U > U_{th}$ ($U/U_{th} \geqq 1$ is a parameter characterizing the excess of voltage above the threshold value U_{th}).

In this paper the results of a study of the current and temperature distributions in a ChVS sample during the direct switching process as well as the reverse transition from the low to high-resistance state are described. Some of the results have been included in our earlier paper [3]. Now as in [3] we restrict ourselves to the case of the thermal switching mechanism where the temperature dependence of conductivity $\sigma = \sigma_0 \exp(-E_g/2kT)$ plays the determining role.

2. Experiment

Samples of vitreous $Tl_2Se \cdot As_2Te_3$ were the object of our investigation. The thickness of the samples was $L = 110\ \mu m$.

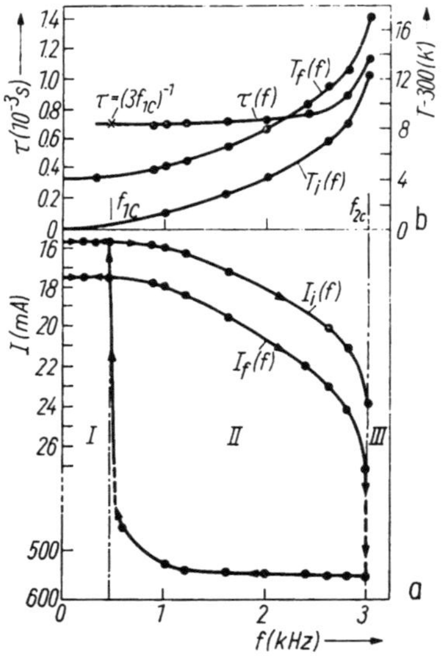

Fig. 1. a) Experimental frequency dependences of the initial (I_i) and final (I_f) current values during the pulse in the high-resistance state and of the current (I) in the low-resistance state. b) Frequency dependences of the initial (T_i) and final (T_f) temperatures of the sample during the pulse and of the cooling time constant (τ) for the high-resistance state. T_i, T_f, and τ have been calculated in the uniform model

According to the well-known dc dependence U_{th} on L [4], the thermal breakdown mechanism was reliably provided for this kind of samples. The voltage value satisfied the conditions: $U/U_{\text{th}} \geqq 1$ and $U/L < 10^4$ V/cm.

The measurements were performed using rectangular pulses following each other with different frequency (f). The duration of the pulses $t_p = 10^{-5}$ s was much less than the switching delay time (t_d): $t_p = (10^{-2}$ to $10^{-3})\, t_d$. The last condition excluded the thermal switching of the sample under the action of a single pulse, however, it was possible to observe the beginning of the same thermal process which should lead to switching in a sufficiently long pulse. This thermal process manifested itself as an increase of the current pulse from the initial value (I_i) to the final one (I_f). At low frequencies ($f < f_{1c}$, where $f_{1c} \approx 470$ Hz) the currents I_i and I_f did not change with f (Fig. 1a, region I), which showed that the cooling time constant of the sample (τ) is less than $(3f)^{-1}$. However, if $f > f_{1c}$ $(\tau > (3f)^{-1})$ the sample had no time to cool completely, so the currents I_i and I_f which were measured in the achieved regime of balance between heating and cooling of the sample at fixed frequency increased with increasing f (Fig. 1a, region II). At last if $f \geqq f_{2c}$ ($f_{2c} \approx 3000$ Hz) (Fig. 1a, region III) the sample switched into the low-resistance state, which was attended by the sharp increase of the currents I_i and I_f up to values restricted by the load resistance (R_l). For the low-resistance state (on-state) the current changed its value during the pulse, too, but these changes were relatively smaller than for the high-resistance state (off-state). Therefore, in Fig. 1a only one curve for the average value of the on-state current $I(f)$ is showed for the low-resistance branch. When the frequency decreased from the value f_{2c}, the current relatively weakly decreased but at frequencies close to f_{1c} sharply fell down to the values I_i and I_f observed at the corresponding frequencies with increasing frequency. The latter fact means that at $f \approx f_{1c}$ the reverse transition from the low-resistance state to the high-resistance state occurs. If the frequency decreased further the dependences $I_i(f)$ and $I_f(f)$ practically exactly copied those observed before switching.

Thus the measurements of the currents I_i, I_f, and I as functions of the frequency of the applied voltage pulses allowed us to trace both the thermal scwitching process till the switching itself and the reverse process of the transition from the on-state to the off-state.

3. The Analysis of the Experimental Results in the Uniform Model Approximation for Heating of the Sample

Using the magnitudes I_i and I_f we have readily calculated the conductivities Σ_i and Σ_f and the temperatures T_i and T_f which could occur in the sample during the switching process by the current being assumed to be distributed uniformly over the sample cross-section (S). The frequency behaviour of T_i and T_f is shown in Fig. 1b. According to [5] the gap has been taken as $E_g = 0.67$ eV.

From the magnitudes T_i and T_f the cooling time constant has been calculated under the assumption that the temperature drops exponentially in the absence of an applied voltage during the interval between the pulses: $T - T_0 = (T_f - T_0) \exp(-t/\tau)$, $T_0 = 300$ K (Fig. 1b). Let us note that for the frequency region, where τ does not depend on f, the magnitude of τ calculated in this manner is equal to $\approx 0.7 \times 10^{-3}$ s and closely coincides with the value of $(3f_{1c})^{-1}$. The further increase of τ with increaes of f seems to be logically injustified. Actually if one supposes the heat loss of the sample to take place through the electrodes, τ should not be frequency-dependent. The lateral heat loss (through the side surfaces of the sample) only should decrease τ with the increase of frequency.

Besides another fact should be considered as injustified. The energy (A) required for uniform heating of the whole sample from the temperature T_i to T_f, calculated above should be 3 to 4 times higher than the Joule-heating energy emitted in the corresponding pulse.

The reasons mentioned above essentially lead us to the idea of a non-uniform current density and temperature distribution over the sample cross-section under heating. Such non-uniform distribution may be associated, firstly, with the occurrence of the so-called forming region with the cross-section $S^* = \alpha S$ ($\alpha < 1$), and secondly, with non-uniform heating inside the forming region.

The totality of our data obtained with the pulse experimental arrangement allows to calculate the characteristic parameters of both these kinds of inhomogeneities of the current and temperature distributions. Corresponding calculations are presented in what follows. Of course, a more rigorous theoretical study of the kinetics of the thermal switching process is possible in principle by a treatment of the time-dependent heat conduction equation under specified assumptions about the character of the inhomogeneities. But it is extremely difficult to use such calculations for the comparison between theory and experimental data, particularly in the case of thick samples.

4. Calculation of the Current and Temperature Distributions under Non-Uniform Heating of the Sample (Non-Uniform Model)

4.1 Forming region

In determining the effective size of the forming region we assumed the conductivity parameters (σ_0^*, E_g^*) of this region in the material studied to behave in a way analogous to the widely studied composition $Si_{1.2}Ge_{1.0}As_{3.0}Te_{4.8}$ when these parameters are compared with the corresponding values of σ_0 and E_g of the initial material before forming. This means the assumption that $\sigma_0^* = \sigma_0$ and $E_g^* = 0.85E_g$ is valid for $Tl_2Se \cdot As_2Te_3$. Such relations provide the conductivity of the forming region at room temperature to be of an order of magnitude greater than the conductivity of the initial meterial, and this is in agreement with the experimental data [6]. The behaviour of the density (d) and specific heat capacity (c) under forming is unknown, so we supposed that $d^* = d = 6.7$ g/cm³ and $c^* = c = 0.21$ J/gK. The value of c was determined as an interpolation value with the use of the corresponding magnitudes measured for the compositions $TlAsSe_2$ and $TlAsTe_2$ in [7]. The parameter α and the temperatures inside the forming region (T_f) and outside it (T_f) (the outside region is called the ballast) in the end of the pulse were determined from the system of equations

$$\Sigma_{i,f} = \frac{\sigma_0 S}{L} \left[\alpha \exp\left(-E_g^*/2kT_{i,f}^*\right) + (1-\alpha) \exp\left(-E_g/2kT_{i,f}\right)\right], \tag{1}$$

$$A = cdLS[\alpha(T_f^* - T_i) + (1-\alpha)(T_f - T_i)]. \tag{2}$$

Here Σ_i, Σ_f, and $A = V(I_f + I_i/2)\, t_p$ were taken from the experiment at $f < f_{1c}$; $T_i^* = T_i = T_0 = 300$ K.

The numerical calculation of the system (1, 2) gave $\alpha = 0.14$, i.e. $D^* = 75\ \mu$m, $T_f^* = 305$ K, $T_f = 302$ K at $f < f_{1c}$.

4.2 Currents and temperatures inside the forming region in high-resistance state

As it was mentioned above the non-uniform heating inside the forming region should be taken into consideration if one wants to calculate the current and temperature distributions in the frequency range $f_{1c} < f < f_{2c}$. The simplest way to carry out such a consideration is to divide the forming region into two sections, which are strongly and less strongly heated, respectively, and have the cross-sections $S_1^* = \beta S^*$ and $S_2^* = (1-\beta)\, S^*$. The factor $\beta < 1$ was assumed to be unchanged during the pulse and therefore it characterizes the effective fraction of the forming region cross-section with the strongest heating during the pulse from T_{1i}^* to T_{1f}^*. The rest of the forming region is heated during t_p from T_{2i}^* to T_{2f}^*. The calculation is aimed at the determination of β, T_{1i}^*, T_{1f}^*, T_{2i}^*, T_{2f}^*. It was performed in the following approximations: 1. as the conductivity of the forming region is an order of magnitude higher than that of the ballast all the non-uniform heating is assumed to take place mainly inside the forming region, and the conditions for heating of the ballast are the same as at $f < f_{1c}$; 2. the conditions for cooling of the sample are equal across the cross-section and do not change with frequency, so $\tau \approx 0.7 \times 10^{-3}$ s in the whole frequency range up to f_{2c}. Then the starting system of equations takes the form

$$\Sigma_{i,f} = \frac{\sigma_0 S\alpha}{L} \left[\beta \exp\left(-E_g^*/2kT_{1i,f}^*\right) + (1-\beta) \exp\left(-E_g^*/2kT_{2i,f}^*\right)\right], \tag{3}$$

$$A^* = cdLS\alpha[\beta(T_{1f}^* - T_{1i}^*) + (1-\beta)(T_{2f}^* - T_{2i}^*)], \tag{4}$$

$$T_{1,2i}^* - T_0 = (T_{1,2f}^* - T_0) \exp(-1/f\tau). \tag{5}$$

Here $\Sigma_{i,f}^* = \Sigma_{i,f} - \Sigma_{i,f}'$, $A^* = A - A'$, where $\Sigma_{i,f}'$ and A' are the conductivity of the ballast and the energy expended in its heating. $\Sigma_{i,f}'$, A' have been found from the system (1–2). The indexes "1" and "2" relate to the section of the forming region with the strongest heating and the rest of the forming region, respectively.

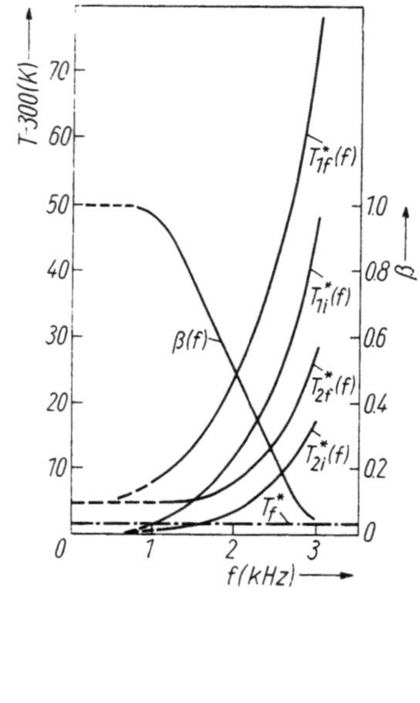

Fig. 2. Frequency behaviour of the initial (T_{1i}^*) and final (T_{1f}^*) temperatures during the pulse inside the hottest channel and outside it (T_{2i}^*, T_{2f}^*) in the forming region. T_f the final temperature for the ballast. The curve $\beta(f)$ demonstrates the decrease of the channel cross-section with frequency

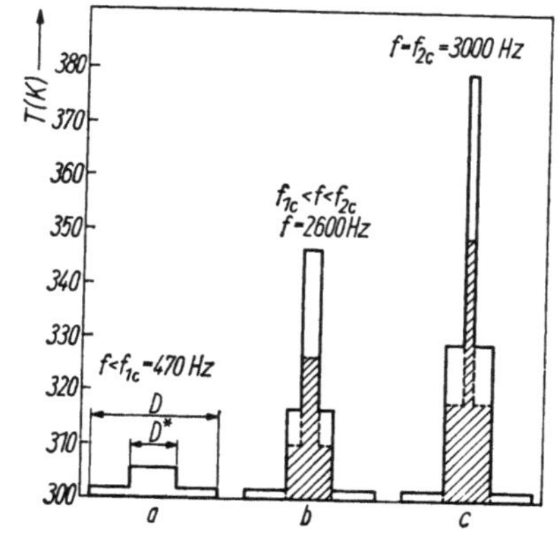

Fig. 3

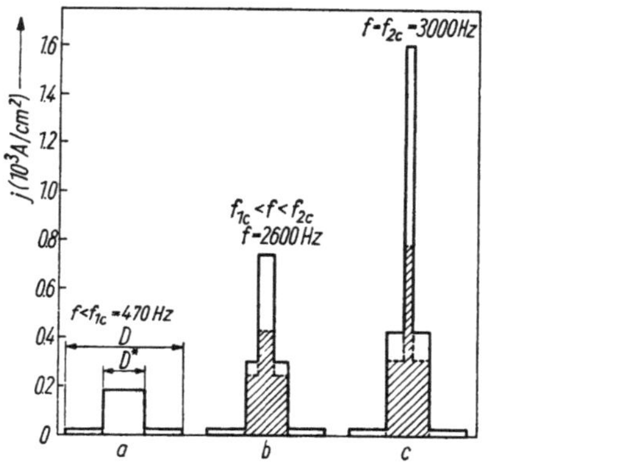

Fig. 4

Fig. 3. Diagram of the temperature distribution in the sample cross-section at the beginning and the end of the pulse (shadded and non-shaded regions, respectively) for three fixed frequency values

Fig. 4. The same as in Fig. 3 for the current density distribution

The results of the numerical solution of the system (4—8) are presented in Fig. 2 which shows the variations of the parameters β, T_{1i}^*, T_{1f}^*, T_{2i}^*, T_{2f}^* with frequency for the whole frequency range under study. It is evident from the curves of Fig. 2 that at $f \lessapprox 500$ Hz the non-uniform temperature distribution is associated only with the existence of the forming region, but at $f > 500$ Hz inside the forming region the section of higher temperatures T_{1i}^* and T_{1f}^* and current densities j_{1i}^* and j_{1f}^* stands out as a channel. When the frequency increases this channel essentially narrows (β drops) and temperatures and current densities inside it grow sharply. Of course, temperatures T_{2i}^* and T_{2f}^* and current densities j_{2i}^* and j_{2f}^* outside the channel increase with frequency, too, but not so sharply.

In Fig. 3 and 4 as an illustration the schematic representation of temperature and current density distributions over the sample cross-section at the beginning and the end of the pulse are shown for three different frequencies: a) $f < f_{1c}$, b) $f_{1c} < f < f_{2c}$, c) $f = f_{2c}$. It is evident from Fig. 3 and 4 that the degree of localization of current and high temperature region becomes stronger with frequency. At $f = f_{2c} = 3000$ Hz the channel diameter is $D_1^* = 13$ μm, the temperatures T_{1i}^* and T_{1f}^* inside the channel attain 350 and 380 K, respectively. The corresponding values of current densities j_{1i}^* and j_{1f}^* are 425 and 1600 A/cm^2.

It should be noted that roughly speaking, if one takes into account the current localization, i.e. the sharp decrease of the most strongly heated volume of the sample, account must be taken of the lateral heat loss (along with the heat loss through the electrodes) and hence the decrease of the real value of τ with frequency. For this case after solving the system of equations e.g. by an iteration method, one should get a still more pronounced localization. So generally saying the magnitudes of temperature and current density inside the channel reported above are underestimated.

4.3 The on-state (low-resistance state)

As mentioned above the on-state of the sample is held not only at frequencies equal to or higher than f_{2c}, but if the frequency decreases from f_{2c} to approximately f_{1c}. Most likely this fact is indicative of an increase of the cooling time constant τ on switching[1]) because of the expansion of the effective channel of the highest temperature and current density up to the size approximately corresponding to the forming region.

Estimations show that if such an expansion does not occur, the temperature should be as high as 1600 K in the channel of diameter $D_1^* = 13$ µm. It is unlikely that the sample of $Tl_2Se.As_2Te_3$ with the softing temperature $T_g \approx 380$ K can withstand so high temperatures without damage, while reversible switching can be many times repeated. On the other hand, if the channel does expand to a diameter $D^* = 75$ µm, the temperature inside it is estimated as ≈ 600 K. Probably it is just the temperature at which forming takes place and the forming region appears during the first switching cycles.

5. Discussion

Based on the data obtained one can describe the thermal switching process in the following way. Owing to structural inhomogeneities typical for vitreous materials and, besides, in the presence of the forming region and of fluctuations of the interelectrode distance, the current density is distributed non-uniformly over the cross-section of the sample in the electric field. When a voltage exceeding the threshold value is applied in the early stage of the switching process, such current density inhomogeneity may be only slightly pronounced (at least inside the forming region). But in the course of time, due to the positive feedback between conductivity and temperature, this current density inhomogeneity is amplified (or "self-multiplied"), so the process of localization of current and high temperature region occurs in a particular effective channel. The localization is of relative character, i.e. the relationship between the current density inside the channel and that outside it increases with time. At this time the total current flowing through the sample slightly increases and the voltage drop across it slightly decreases.

The temperature and the current density inside the channel increase much more rapidly than in the rest of the sample. Therefore, first of all fairly high temperatures are accomplished inside the channel. At last the moment comes when the further increase of the channel temperature slows down or generally stops. This may have two reasons: either the balance is established between Joule heating of the channel and the heat loss which is most probably superlinear with temperature at so high temperatures or the transition from semiconductor to metal occurs. At the same time in the rest of the sample the temperature increases nonlinearly up to the values, controlled by the conditions noted just above. These temperature values are achieved firstly in the regions nearest the channel and then more and more distant. Thus, the region of highest temperatures and current densities is expanded beginning from the moment of slowing down or stabilization of the values j_1^* and T_1^* inside the narrow channel. At this moment the current flowing through the sample sharply increases, the voltage sharply drops, and this leads to the cessation of the further expansion of the high-conduction channel — the sample is switched to the low-resistance state. Hence, the final values I and U and the width of the high-conduction channel are controlled by the load resistance R_l. Lowering the supply voltage below the maintaining voltage results in cooling of the previously developed wide channel and the reversible transition to the high-resistance state during the restoration time which corresponds to 3τ.[2])

---

[1]) We keep in mind the increase of τ above its real magnitude immediately before switching.

[2]) Let us note that the measurements of the restoration time for our samples with the use of conventional techniques give $t_r \approx (2 \text{ to } 3) \times 10^{-3}$ s.

Thus according to the ideas developed here the thermal switching delay time t_d for ChVS is governed by the time of the high-conductive narrow channel formation till the beginning of the channel expansion. From this time on switching itself occurs till the establishment of the wide stable channel during the interval t_s.

The picture of the thermal switching process considered differs essentially from the usual uniform model [8], which is known to assume a uniform distribution of current density during the delay time and the formation of a narrow current filament during the switching time, the narrow filament being conserved in the on-state.

The experimental data and the calculations presented above point to the fact that the non-uniform model is more preferable for disordered semiconductors, not only for thermal but also for electrothermal switching. In particular a number of specific properties which have no yet been explained by the uniform model of switching can be understood in electrothermal theory for thin samples using the non-uniform model [9].

In conclusion let us note that current inhomogeneities as a consequence of structural inhomogeneities in disordered semiconductors must be taken into account in investigations of the ChVS behaviour in high electric fields.

References

[1] V. B. Sandomirsky and A. A. Sukhanov, Zarubenaya radioelectronica 9, 68 (1976).
[2] H. Fritzsche, Amorphous and Liquid Semiconductors, London 1974 (p. 313).
[3] B. T. Kolomiets, E. A. Smorgonskaya, and V. Kh. Shpunt, Proc. Internat. Conf. Amorphous Semiconductors-78, Pardubice 1978 (p. 565).
[4] B. T. Kolomiets, E. A. Lebedev, and I. A. Taksami, Fiz. Tekh. Poluprov. **3**, 312 (1969).
[5] A. M. Andriesh and B. T. Kolomiets, Fiz. tverd. Tela **5**, 1461 (1963).
[6] R. F. Ormondroyd, M. J. Thompson, and J. Allison, J. non-crystall. Solids **18**, 375 (1975).
[7] V. V. Zigel, G. M. Orlova, and G. N. Diakova, Fizika Khimia Stekla **1**, 558 (1975).
[8] B. L. Gelmont and K. D. Tsendin, Fiz. Tekh. Poluprov. **10**, 1119 (1976).
[9] A. A. Sukhanov, Fiz. Tekh. Poluprov. **12**, 2338 (1978).

(Received June 6, 1979)

phys. stat. sol. (a) 54, 593 (1979)

Subject classification: 10 and 12; 22.2.2

Institute of Electronic Technology, Moscow

Investigation of Internal Friction in GaP

By

V. D. VERNER, D. K. NICHUGOVSKII, and V. YA. KOVYASIN

A calculation is made of the internal friction in GaP single crystals, produced by the reorientation of the divacancies of Ga atoms due to external stresses and the interaction between the vacancies. The ratio of the internal friction maxima in GaP, which correspond to the longitudinal oscillations in the [100] and [111] directions is found to be equal to 1.6, which is in good agreement with the results obtained experimentally for nondoped GaP single crystals with the help of the composite piezoelectrical oscillator method.

Es wird eine Berechnung der inneren Reibung in GaP-Einkristallen, die durch die Reorientierung der Doppelleerstellen der Ga-Atome infolge äußerer Spannungen hervorgerufen wird, und der Wechselwirkung zwischen den Leerstellen durchgeführt. Es wird gefunden, daß das Verhältnis der Maxima der inneren Reibung in GaP, die den longitudinalen Schwingungen in den [100]- und [111]-Richtungen entsprechen, 1,6 beträgt und gut mit den Ergebnissen übereinstimmt, die experimentell für undotierte GaP-Einkristalle mittels der Methode zusammengesetzter piezoelektrischer Schwingungen erhalten wurden.

Semiconductor compounds of the A^3B^5 type (including GaP) normally contain great numbers of vacancies due to the fact that the composition of real crystals does not strictly obey the laws of stoichiometry. The concentration of vacancies may be as high as 10^{19} cm^{-3} [1]. In case the vacancies are combined to complexes, their symmetry is lowered, therefore, external stresses applied to a crystal produce reorientation of the complexes, the process being accompanied by energy absorption. Hence, GaP crystals with a sufficiently high concentration of divacancies of Ga atoms may be expected to have a peak of internal friction resulting from the reorientation of the divacancies. In this connection it appears of interest to calculate the interal friction of this type and to make the necessary experimental investigations. This is the aim of the present paper. Calculations were based on the theory developed in [2] for divacancies in the f.c.c. lattice.

Similarly to the case of the f.c.c. lattice, divacancies of Ga atoms in GaP can have six different orientations corresponding to the [110], [1$\bar{1}$0], [101], [10$\bar{1}$] [011], and [01$\bar{1}$] directions. The external stresses σ_{ij} do not allow the distribution of the pairs of vacancies to be equally probable over the above orientations, the strain tensor components (according to [2]) being

$$u_{ij} = s_{ijkl}\sigma_{kl} + a_{ij}^{\nu} c_{\nu} \,, \tag{1}$$

where the coefficients s_{ijkl} are the usual components of the fourth-order tensor of elastic compliance, c_ν the excess concentration of divacancies of the ν orientation, while the tensor a_{ij}^{ν} describes the crystal strain by means of divacancies of the appropriate orientation.

For clarity we shall consider a divacancy of the [110] orientation. The index ν can be neglected. Then, in the Cartesian coordinate system with the axes of x_1, x_2, x_3 having [110], [100], and [001] directions, respectively, the tensor a_{ij} appears to be diagonal:

$$a_{ij} = \alpha\delta_{ij} + \beta\delta_{i1}\delta_{j1} + \gamma\delta_{i2}\delta_{j2} \,. \tag{2}$$

Here the constant α characterizes the symmetrical part of the strain and does not contribute to the internal friction. β and γ will be calculated as follows.

Assume that the interaction between the vacancies and the atoms of the GaP lattice is described by a pair potential and is characterized by the forces $\boldsymbol{F}(\boldsymbol{r}_p^l)$ exerted by the vacancies on the p-th atom in the l-th elementary cell determined by the vector $\boldsymbol{r}_p^l$. Then, the average value of the tensor stress components (P_{ij} being produced by the above forces when crystal surface stresses are absent) is defined by

$$P_{ij} = \frac{1}{V} \sum_{l,p} F_i(\boldsymbol{r}_p^l)\, r_{pj}^l\,, \tag{3}$$

where V is the crystal volume [2, 3].

We restrict ourselves to considering only the nearest neighbours and the next-nearest neighbours of the vacancies. Taking into account the symmetry requirements we find that the interaction of the vacancies with the atoms of the first coordination sphere is described by the force constant $F^{(1)}$, while that with the atoms of the second coordination sphere is described by $F^{(2)}$.

After the necessary summation of seven of the nearest neighbours and eighteen of the next-nearest neighbours in (3), we get the tensor components of the stresses produced by one divacancy of [110] orientation in the above Cartesian coordinate system

$$\left.\begin{aligned} P_{11} &= -\frac{2F^{(1)}a}{\sqrt{3}\,V} - \frac{7F^{(2)}a}{\sqrt{2}\,V}\,, \\ P_{22} = P_{33} &= -\frac{2F^{(1)}a}{\sqrt{3}\,V} - \frac{8F^{(2)}a}{\sqrt{2}\,V}\,, \qquad P_{12} = P_{13} = P_{23} = 0\,. \end{aligned}\right\} \tag{4}$$

Here a is the lattice constant.

The tensor components of the strain which are due to one divacancy are defined by

$$e_{ij} = \left\{ s_1 \delta_{ij} \delta_{kl} + s_2(\delta_{ik}\delta_{jl} + \delta_{il}\delta_{jk}) + s_3 \sum_p \alpha_{ip}\alpha_{jp}\alpha_{kp}\alpha_{lp} \right\} P_{kl}\,, \tag{5}$$

where the direction cosines in the above coordinate system are equal to

$$\alpha_{11} = \alpha_{22} = \alpha_{12} = -\alpha_{21} = \frac{1}{\sqrt{2}}\,, \qquad \alpha_{33} = 1\,,$$

$$\alpha_{13} = \alpha_{31} = \alpha_{23} = \alpha_{32} = 0\,,$$

while the relation between the coefficients s_i and the coefficients of elastic compliance is given by [2]

$$s_{11} = s_1 + 2s_2 + s_3\,, \qquad s_{12} = s_1\,, \qquad s_{44} = 4s_2\,.$$

If a crystal has an excess concentration of divacancies (c) of the given orientation, the stress due to one pair has to be multiplied by NVc, where N is the number of atoms in the crystal.

A comparison between the equations obtained and (1) and (2) yields

$$\beta - \gamma = \frac{NaF^{(2)}}{2\sqrt{2}}\, s_{44}\,, \qquad \beta + \gamma = \frac{NaF^{(2)}}{2\sqrt{2}}\,(s_{11} - s_{12})\,. \tag{6}$$

It should be noted that according to (6) the neasrest neighbours of the vacancy do not contribute to the internal friction of the given type. The constant $F^{(2)}$ was found with the help of the method described in [4]. The method was also used in [5] to determine the interaction energy of vacancies in semiconductor compounds of A^3B^5 type. Its value was found to be $F^{(2)} = 0.83 \times 10^{-4}$ dyn.

If a sinusoidal stress of frequency ω is applied to a crystal, the internal friction produced by the reorientation of the Ga divacancies can be expressed by the formula [2]

$$Q^{-1} = -\frac{\operatorname{Im} s_{ijkl}(\omega)}{\operatorname{Re} s_{ijkl}(\omega)}, \tag{7}$$

where $s_{ijkl}(\omega)$ is the fourth-order complex tensor of elastic compliance. The tensor is described by

$$s_{ijkl}(\omega) = s_1(\omega)\, \delta_{ij}\delta_{kl} = s_2(\omega)\, (\delta_{ik}\delta_{jl} + \delta_{il}\delta_{jk}) + s_3 \sum_p \delta_{ip}\delta_{jp}\delta_{kp}\delta_{lp}\,,$$

where

$$\left.\begin{aligned} s_i(\omega) &= s_i + \Delta s_i(\omega)\,,\\ \Delta s_1(\omega) &= -\frac{c_0(\beta+\gamma)^2}{36NkT(1+\omega\tau_1)}\,,\\ \Delta s_2(\omega) &= \frac{c_0(\beta-\gamma)^2}{12NkT(1+\omega\tau_2)}\,,\\ \Delta s_3(\omega) &= \frac{c_0}{12NkT}\left[\frac{(\beta+\gamma)^2}{1+i\omega\tau_1} - \frac{2(\beta-\gamma)^2}{1+i\omega\tau_2}\right]. \end{aligned}\right\} \tag{8}$$

Here c_0 is the concentration of divacancies, T the absolute temperature, τ_1 and τ_2 are the times of relaxation, k the Boltzmann constant.

If strain is applied in the [100] direction we have

$$Q^{-1} = -\frac{\operatorname{Im} s_{1111}(\omega)}{\operatorname{Re} s_{1111}(\omega)} \approx \frac{c_0(\beta+\gamma)^2}{18NkTs_{11}}\,\frac{\omega\tau_1^2}{1+\omega^2\tau_1^2}. \tag{9}$$

To determine the internal friction in case of the strain applied along the axis [111] we must find the complex tensor components of the coefficients of elastic compliance in the coordinate system with the axis x_1' parallel to the [111] direction and the axes x_2' and x_3' paallelr to the [01$\bar{1}$] and [$\bar{2}$11] directions, respectively. For this purpose we make the following transformation:

$$s'_{ijkl} = \alpha_{im}\alpha_{jn}\alpha_{ko}\alpha_{lp} = s_{mnop}\,, \tag{10}$$

where the direction cosines are

$$\alpha_{11} = \alpha_{12} = \alpha_{13} = \frac{1}{\sqrt{3}}, \quad \alpha_{22} = -\alpha_{23} = \frac{1}{\sqrt{2}},$$

$$\alpha_{32} = \alpha_{33} = \frac{1}{\sqrt{6}}, \quad \alpha_{31} = -\frac{2}{\sqrt{6}}, \quad \alpha_{21} = 0\,.$$

Having made the necessary calculations, we obtain

$$Q^{-1} \approx -\frac{c_0(\beta-\gamma)^2}{3NkT(4s_{11}+8s_{12}+s_{44})}\,\frac{\omega\tau_2}{1+\omega^2\tau_2^2}. \tag{11}$$

Thus, the internal friction in GaP crystals produced by the reorientation of Ga divacancies is characterized by two relaxation times and depends on the orientation of the stress applied. A divacancy of Ga in the GaP lattice as well as a divacancy in the f.c.c. lattice can change its position to any of the five other positions. If a divacancy of Ga is positioned along the [110] direction is should be rotated by 90° to have the [1$\bar{1}$0] orientation, while the transition to the four other orientations requires only

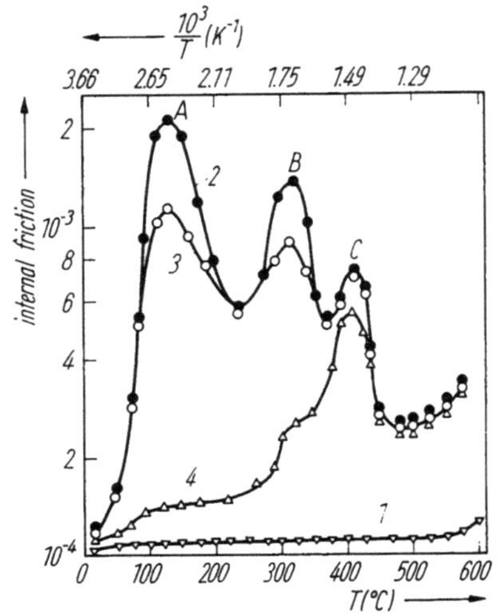

Fig. 1. The temperature dependence of internal friction. (1) Composite piezoelectrical oscillator; (2) GaP non-doped single crystal: [100] orientation and (3) [111] orientation; (4) specimen cooling

a 60° rotation. In this case according to [2] the ratio of the relaxation times for the 60° and 90° rotations will be less than 1.5, i.e. the associated relaxation peaks will not differ experimentally. For an arbitrary orientation of the crystallographic axes, however, the relaxation peak is expected to be widened, which can be observed experimentally.

Let us calculate the height of the internal friction peak in GaP crystals. In the present work we have already made an attempt to study the internal friction of a crystal by using the composite piezoelectrical oscillator method [6] for measurements in GaP non-doped single crystals. The results of the measurements are presented in Fig. 1. The internal friction peak was found to occur at a temperature of 130 °C and a frequency of 120 kHz. The peak height (without the background taken into account) was $Q^{-1}_{max} = 2.1 \times 10^{-3}$ for the [100] orientation and $Q^{-1}_{max} = 1.1 \times 10^{-3}$ for the [111] orientation.

Assuming the GaP values to be $s_{11} = 0.973 \times 10^{-12}$ cm² dyn⁻¹, $s_{12} = -0.299 \times \times 10^{-12}$ cm² dyn⁻¹, $s_{44} = 1.419 \times 10^{-12}$ cm² dyn⁻¹ [7]; $a = 5.4506$ Å [8] the values of the above maxima can be obtained from formulas (9) and (11) if $c_0 \approx 4 \times 10^{-3}$. Since crystals of A^3B^5 type compounds have an increased density of vacancies [1] the assumed concentration of divacancies appears to be possible.

The results of the above calculation imply that the ratio of the internal friction maxima in GaP corresponding to the longitudinal oscillations in the [100] and [111] directions is $Q^{-1}_{max}[100]/Q^{-1}_{max}[111] = 1.6$, which is in good agreement with the ratio experimentally determined to be 1.9. Thus, the experimental internal friction peak in GaP non-doped single crystals can be considered to be due to the reorientation of divacancies of the Ga atoms.

References

[1] V. B. OSVENSKII, L. P. KHOLODNYI, and M. G. MILVIDSKII, Izv. Akad. Nauk SSSR, Ser. neorg. Mater. 8, 802 (1972).
[2] S. K. TURKOV and T. D. SHERMERGOR, Soviet Phys. — Solid State 7, 2064 (1965).
[3] A. G. CHACHATURYAN, Soviet Phys. — Solid State 9, 2862 (1967).
[4] H. KANZAKI, J. Phys. Chem. Solids 2, 24 (1957).
[5] V. D. VERNER and D. K. NICHUGOVSKII, Soviet Phys. — Solid State 15, 3394 (1973).
[6] V. S. POSTNIKOV, Vnutrennee trenie v metallakh, Izd. Metallurgy, Moscow 1974.
[7] R. WELL and W. O. GROVES, J. appl. Phys. 39, 4049 (1968).
[8] O. MADELUNG, Physics of III-V Compounds, John Wiley & Sons, New York 1964.

(Received May 11, 1979)

phys. stat. sol. (a) 54, 597 (1979)

Subject classification: 10.2 and 13.5.3; 22.8.2

Jena, Carl-Zeiss-Straße 1

On the Polaron Character of the Coloration Centres in $LiNbO_3$ Crystals

By

HJ. BERNHARDT

The RT and low temperature colour centres which grow up in $LiNbO_3$ crystals after X-raying are described as bound polarons. It is shown that it is possible to describe a lot of centres in a generalized way. Following this model all coloration effects in $LiNbO_3$ are understood with the help of polarons the absorption maxima and thermal stabilities of which vary according to the variation of the bond strength. The properties of these colour centres in $LiNbO_3$ are analogous to those observed in YAG and $PbMoO_4$.

Die RT- und Tieftemperaturverfärbungszentren, die in $LiNbO_3$ Kristallen nach Röntgenbestrahlung entstehen, werden als gebundene Polaronen aufgefaßt. Auf diese Weise können sie mit Hilfe eines verallgemeinerten Modells dargestellt, beschrieben und verstanden werden. Entsprechend diesem Modell werden alle Röntgenverfärbungseffekte an $LiNbO_3$ durch Polaronen erklärt, deren Lage und Absorption infolge unterschiedlichen Bindungsgrades variiert. Die Grundeigenschaften dieser Zentren gleichen prinzipiell denen, die von uns an YAG- und $PbMoO_4$-Kristallen beobachtet wurden.

1. Introduction

Several coloration effects are observed in $LiNbO_3$ after X-raying at RT (room temperature) or at LNT. These effects have been described in some foregoing papers [1, 2]. The coloration centres which are created at RT show a great similarity to those observed in doped crystals after annealing in O_2 ($LiNbO_3$:Mn, Cr, Fe, Ni [3]). These show broad absorption bands at about 430 to 450 nm. The coloration after X-raying at LNT is not induced by impurities but influenced by their existence. Bollmann and Gernand [4] described another absorption in the IR region, which could be observed after annealing of $LiNbO_3$ in vacuum. They explained this absorption hypothetically by F_2 centres. This IR absorption will be discussed in this paper, we can describe it together with the other coloration centres although the IR centres cannot be created by X-raying. In the case of YAG crystals we succeeded in describing a similar phenomenon by the model of bound polarons. This model allows the qualitative description of detailed parameters as e.g. the phonon energy needed for self-trapping of holes. The algorithm used for YAG is capable to describe the lithium niobate centres in the same way. This observation supports the possibility of generalization of diverse coloration centres in several oxide crystals of different structure and composition. This has already been assumed [5, 6].

2. Results and Discussion

In [2] we described coloration effects which had been created by X-raying of $LiNbO_3$ crystals. The thermal bleaching curves showed a similar behaviour compared with those of other oxide crystals. The stepwise annealing of irradiated crystals and the following absorption measurements lead to the observation that several types of centres coexist, which have different absorption maxima and different thermal stabilities

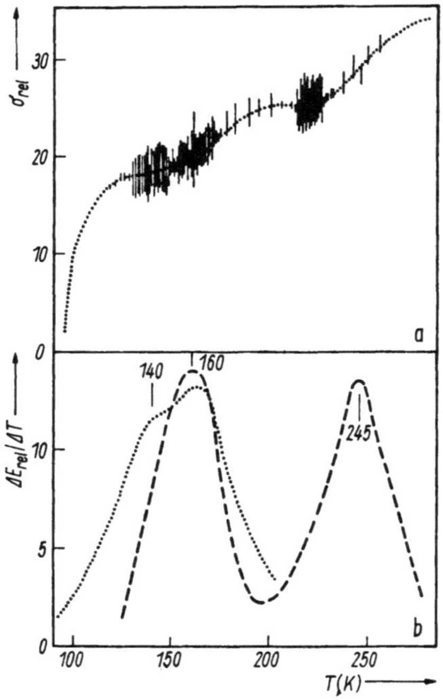

Fig. 1. Thermal bleaching curves (as the thermal bleaching curve we denote the bleaching velocity as a function of the bleaching of temperature. This bleaching curve corresponds to the glow curve thermoluminescence because of its proportionality to the decay rate of absorption centres) of $LiNbO_3$. a) Conductivity of $LiNbO_3$:Fe after reduction as a function of the temperature; b) $LiNbO_3$:10^{-3}% MgO ($\lambda_M = 800$ nm) dotted line; $LiNbO_3$:1% MgO ($\lambda_M = 450$ nm) dashed line

and are dichroic (see [2], Fig. 7). More precise investigations on this fact enabled us to generalize our models of these centres. Our experiments support the following observations which could be made on YAG and $PbMoO_4$ crystals, too [1, 5]:

1. The creation of short-wavelength absorption bands is supported by increasing impurity contamination; the centres are connected more or less with the impurities.

2. Increasing impurity contamination prevents the creation of long-wavelength centres (see [2], Fig. 7, curves 7 and 9) at about $\lambda_{max} = 850$ nm.

3. Short-wavelength centres, which have to be regarded in connection with impurities are more stable than the long-wavelength ones, which have been observed only in pure crystals.

The afore-mentioned impurities lead to a stabilization in an untypical way, that means the coloration can be produced by different impurities. A more precise investigation of the thermal bleaching showed that the glow peak described in [2], Fig. 8b has a structure. The comparison of the 160 to 170 K peaks of YAG, $YAlO_3$, and $LiNbO_3$ in [2] pointed out a strong broadening of the $LiNbO_3$ glow peak. A more detailed investigation is shown in Fig. 1b. The glow curves of two $LiNbO_3$ crystals are compared, one with strong and the other with weak Mg-doping (1 and 10^{-3}%, respectively). Mg ions cannot be detected by absorption measurements otherwise. They influence indirectly the creation of the coloration centres. The pure crystal has *two* glow maxima in the low-temperature region (at 140 and 160 K) which are analogous to the 138 and 160 to 166 K maxima of YAG crystals (see [2], Fig. 8c and [1]). Impurity contamination diminishes the lowest glow peak of $LiNbO_3$ as it has been observed for YAG and $PbMoO_4$, too. This observation has been confirmed by our investigations on YAG:Nd, YAG:Ce, YAG:Mn compared with pure YAG, on $PbMoO_4$:Mn and $PbMoO_4$:Fe compared with $PbMoO_4$, and on $LiNbO_3$:Mg, $LiNbO_3$:Ce, $LiNbO_3$:Cu, $LiNbO_3$:Fe, and $LiNbO_3$:Cr compared with pure $LiNbO_3$ (see also Fig. 1). In the case of $LiNbO_3$ the bleaching measurements are clearer than the thermoluminescence results.

The recombination process of centres created after X-raying of $LiNbO_3$ crystals takes place discontinuously (see Fig. 2). By heating the X-rayed crystals one observes a flash-like stochastic emission. (This observation has been made earlier by Schwotzer [7].) The knowledge of the thermal bleaching curves, however, enables us to show the enhanced probability for the emission in the vicinity of glow peaks. The flash-like emission is superimposed on a continuous one. The latter is vanishing every now and then,

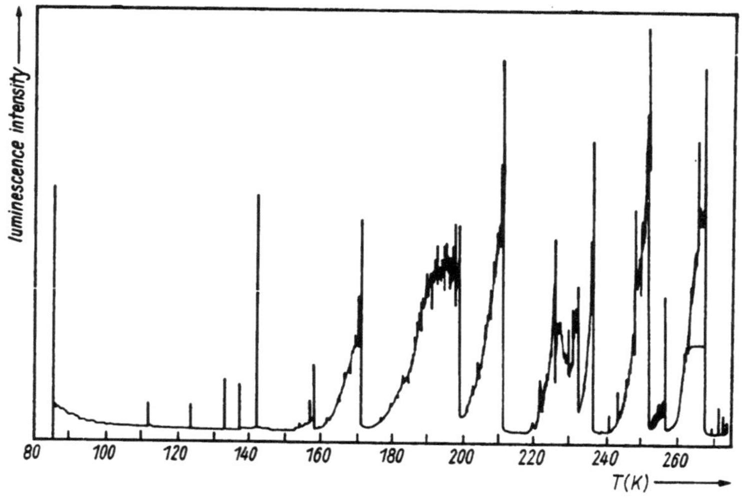

Fig. 2. Thermoluminescence of LNT X-rayed $LiNbO_3$

Fiedler [8] investigated the second harmonic generation of vacuum annealed crystals and observed the afore-mentioned flashes in the temperature region at about 140 K. He showed the flashes to be damped oscillations with maximum voltages of 180 mV, frequencies of about 5 kHz, and decay times of 8 to 16 ms. According to our results the temperature region in which the flashes diminish corresponds to the maximum of thermal bleaching of a room temperature stable coloration centre ($T_{glow} \approx 400$ to 420 K). This is the same region in which the thermal fixing of optical damage holograms takes place according to Phillips et al. [3]. The afore-mentioned discrepancy between the continuous bleaching of coloration and the discontinuous recombination process let us assume a "collective storage" of many charge carriers in an intermediate state; these have to be set free spontaneously. This should lead to a flash-like emission. Maybe, the domains of $LiNbO_3$ act as such intermediate traps, which set free a great quantity of charge carriers in umklapp processes.

The 140 K peak of pure crystals corresponds to the 850 nm absorption band which had been found before by investigating the dichroism of coloration centres [2]. This band could not be observed in $LiNbO_3$:1% MgO. The 160 K peak of $LiNbO_3$:1% MgO corresponds to the 650 nm coloration band. The related centre is connected with untypical impurities and is observed in crystals with other dopings, too. It is analogous to the 650 nm band in YAG crystals [1]. Annealing higher than 240 K destroys centres with absorption maxima at about 500 nm. Strongly coloured $LiNbO_3$:Fe (0.03%) shows a low-temperature absorption component at about 435 nm (Fig. 3). These com-

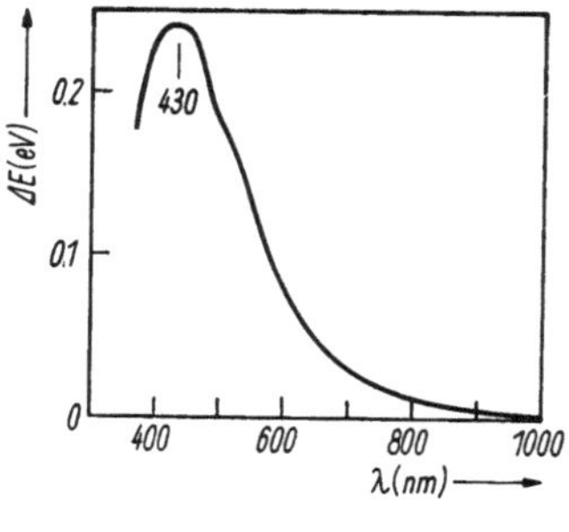

Fig. 3. LNT coloration of $LiNbO_3$:Fe (0.03% Fe) after X-raying at LNT

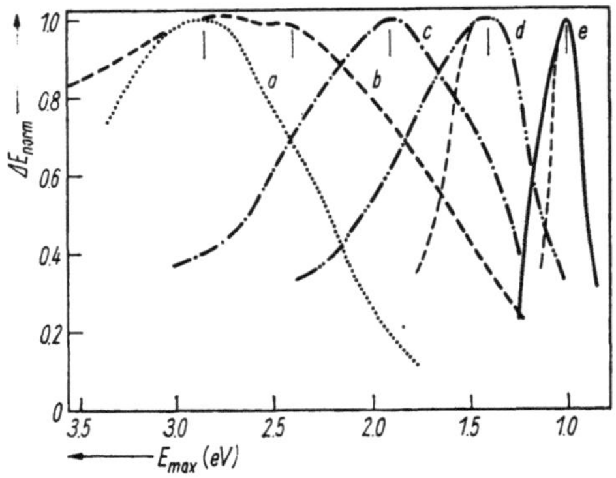

Fig. 4. Coloration components of $LiNbO_3$ crystals. (a) $LiNbO_3$: Fe(0.03%), $T_H = 200$ K; (b) $LiNbO_3$ pure, $T_H = 280$ K; (c) $LiNbO_3$: MgO(1%), $T_H = 190$ K; (d) $LiNbO_3$ pure, $T_H = 130$ K; (e) $LiNbO_3$:Fe, vacuum annealed, T_M (= LNT) − T_M (= RT). T_H temperature of annealing, T_M temperature of measurement

ponents are illustrated in Fig. 4. Inspite of the strong deformation of the bands by mutual superposition the following properties of the colour centres are observed:

1. Short-wavelength bands are broader.
2. Long-wavelength bands are less broad.
3. Centres which depend on impurities absorb at short wavelengths.
4. All bands are equidistant with energetic distances of 0.4 to 0.5 eV.

Curve d will be discussed later.

In [1] we succeeded in describing the coloration centres of YAG as differently strongly bound polarons by using a relation of Schirmer and Schnadt [11, 12]. Phenomenologically the coloration centres of $LiNbO_3$ (Fig. 4) correspond to those found in YAG (see [1] and Fig. 2). Using the method successfully applied in the case of YAG all $LiNbO_3$ absorption bands presented in Fig. 3 should be describable by the following relations:

$$I_{\text{abs}} = \hbar\omega \sqrt{\frac{w}{\pi}} \exp\left\{-w\left[\hbar\omega - \frac{8}{3} E_{\text{JT}}\right]^2\right\} \tag{1}$$

with

$$w^{-1} = \frac{16}{3} E_{\max}\hbar\omega_0 \coth\left(\frac{\hbar\omega_0}{2kT}\right) \tag{2}$$

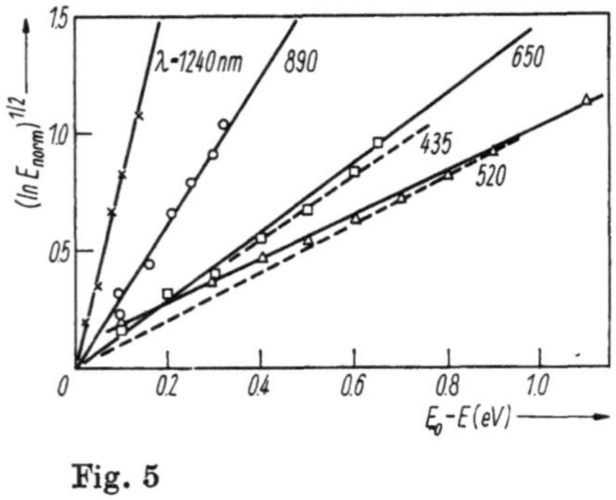

Fig. 5

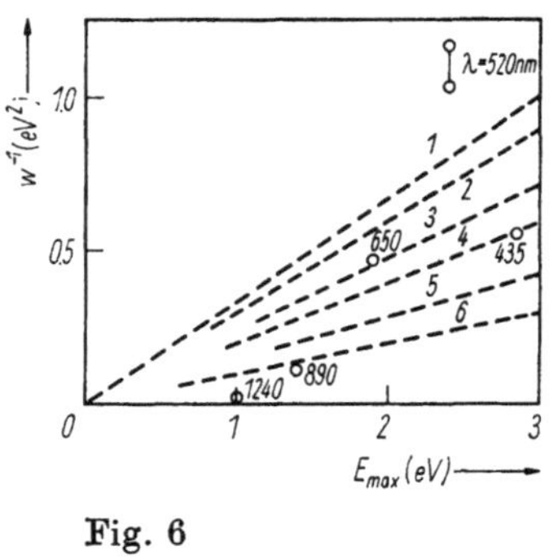

Fig. 6

Fig. 5. Checking of the Gaussian shape of the coloration components of $LiNbO_3$

Fig. 6. Checking of the relation (4) with help of the experimentally estimated parameters w^{-1} and $E_{\max}$. (1) $\hbar\omega = 0.17$, (2) 0.15, (3) 0.12, (4) 0.1, (5) 0.07, (6) 0.05 eV

and

$$\frac{E_H^2}{4} = \frac{16}{3} \hbar\omega_0 \ln 2 E_{max}, \quad (3)$$

$$w^{-1} = \frac{16}{3} E_{JT} \hbar\omega_0 = 2 E_{max} \hbar\omega_0, \quad (4)$$

where $\hbar\omega$ is the photon energy, w the "steepness" of the Gaussian peak, $\hbar\omega_0$ the energy of phonons which lead to self-trapping of holes or electrons, E_{JT} the "quasi-Jahn-Teller energy", E_H the half-width of the absorption band, $E_{max} = 8/3\, E_{JT}$ the energetic position of the absorption maximum (the factor 8/3 is right only for tetrahedral coordination).

The values E_H, E_{max}, and w^{-1} have to be estimated from the experiment. Neglecting the pre-exponential factor in (1) which is only weakly depending on ω the absorption band of the bound polarons has a Gaussian shape. This has been proved for all components of $LiNbO_3$ discussed in Fig. 5. From this figure we can easily get the values of the steepness w^{-1} for the different bands. Fig. 6 proves the relation (4) which is expected to correlate w^{-1} and E_{max}. The strong falsification of the steepness is due to the superposition and the incorrect knowledge of the real centre configuration ($E_{max} = 8/3\, E_{JT}$ is right only for tetrahedral coordination!). The dashed lines in the figure denote those phonon energies which lead to self-trapping according to relation (4). The check of (3) by the experiment leads to Fig. 7. The experimentally estimated correlation is linear within the accuracy of experiment. Using the steepness of this straight line we calculated the phonon energy $\hbar\omega_0 = 0.06$ eV in good agreement with the value estimated by Redfield and Burke [13] from UV edge measurements.

The properties of the IR band found on $LiNbO_3$:Fe after strong reduction [5, 9] shall be discussed now. Stöhr [9] observed the proportionality between the conductivity at RT and the absorption coefficient of differently strongly annealed crystals. Assuming a mobility enthalpy of $h_B = 0.13$ eV (estimated from conduction measurements and using a relation of Kushima [10]) he evaluated an electron effective mass of $M/m = 8$. The optical and electrical properties of these centres are analogous to those observed by Bogomolov et al. [14, 15] in doped rutile monocrystals (TiO_2:Nb or Li, P, respectively, and TiO_2 after strong reduction). The centres described in [14, 15] show absorption at 1.5 μm, their absorption coefficient is proportional to their conductivity. These IR centres both in $LiNbO_3$ and in rutile show a characteristic temperature dependence. The polaron character of the rutile centres has been proved in [14, 15]. If the IR absorption of reduced $LiNbO_3$:Fe (with absorption at about 1 to 1.1 μm) would be due to polarons weakly bound to Fe ions, they should get ionized by increasing the temperature. We observed indeed a very strong temperature dependence of the short-wavelength component at 1.0 eV (see Fig. 8). Fig. 1 shows the tempera-

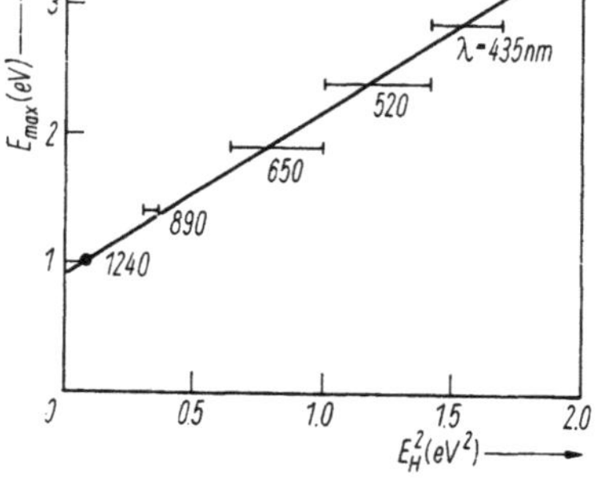

Fig. 7. Checking of the relation (3) correlating E_{max} and E_H^2 with help of the experimentally estimated data

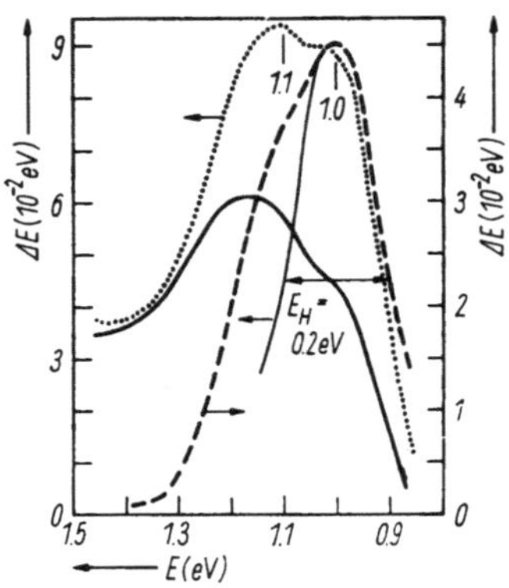

Fig. 8. Temperature dependence of the IR absorption of vacuum annealed $LiNbO_3$:Fe. ······ T_M = LNT, ——— T_M = = RT, ---- the difference between the afore-mentioned curves

ture dependence of the conductivity. The conductivity increases from LNT to RT. Owing to the thermal bleaching (Fig. 1b) one sees a clear connection between the "lightnings" (observed in conductivity), the points of inflection of the conductivity behaviour, and the bleaching maxima of X-rayed crystals. We assume that in the case of the vacuum annealed crystals the weakly bound polarons (IR absorption) are at first thermally ionized and then partially trapped again in deeper traps. With increasing temperature they get ionized again, and so on. It is very probable that in the case of very strongly reduced crystals (with absorption in the visual range of the spectrum) the deeper traps are occupied, too. This points to the fact that the lightnings observed by thermoluminescence measurements are *not* due to recombination processes of holes and electrons. They are due to the influence of the ferroelectric crystal lattice. We are sure that the lightnings observed by Fiedler are of the same nature.

Fig. 9 shows the temperature dependence of the 1240 nm band and Fig. 10 the correlation between conductivity and absorption. Following the afore-mentioned models we identify the 1240 nm absorption with electronic polarons (in contradiction to the model of a F_2 centre [5]). Consequently we discuss these centres together with the above-mentioned ones which have been created by X-raying. Following these conclusions the 1240 nm band is represented in Fig. 4 in the sequence of several supposed polaron bands. Fig. 5, 6, and 7 prove that this IR band fits sufficiently well into the ensemble of other bands and that it may be described by relation (1). The observation of the 1240 nm band in X-rayed but not reduced crystals should be expected, eventually, after X-raying at LHeT.

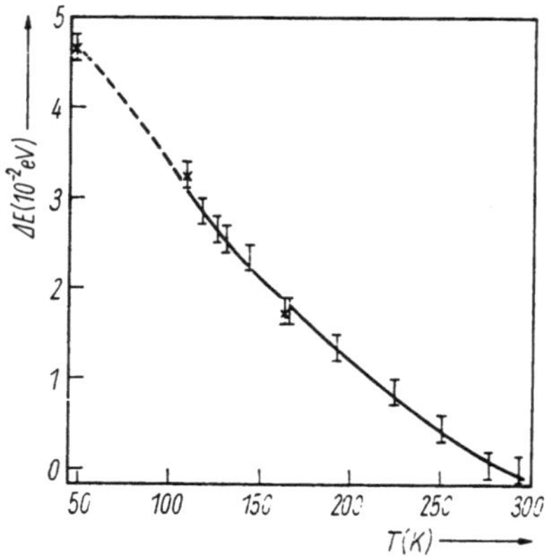

Fig. 9

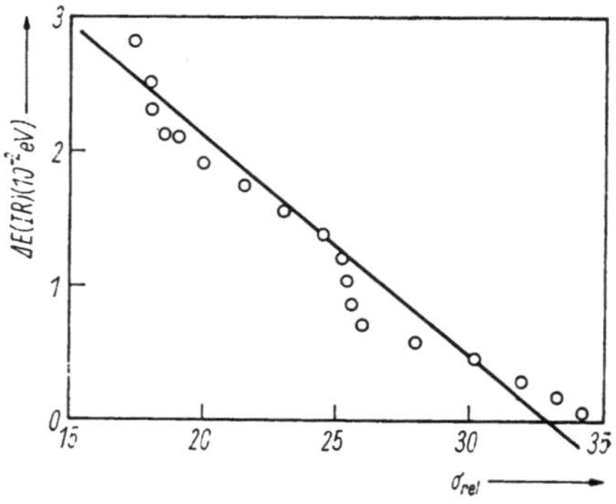

Fig. 10

Fig. 9. Temperature dependence of the 1240 nm absorption of reduced $LiNbO_3$:Fe

Fig. 10. Correlation between the 1240 nm absorption and the conductivity of reduced $LiNbO_3$:Fe in function of the temperature

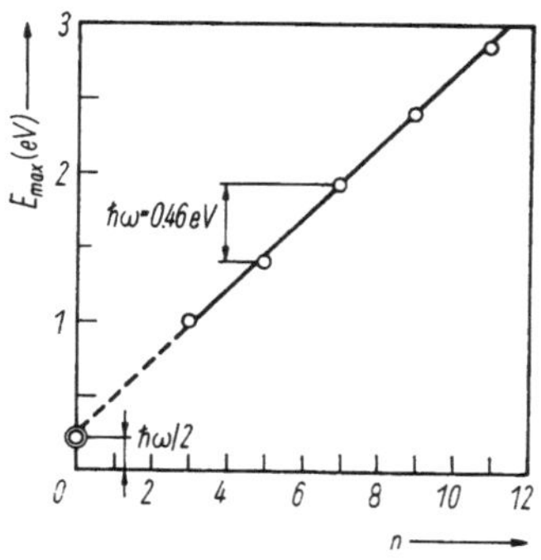

Fig. 11. Oscillator model of bound polarons in $LiNbO_3$ crystals

Following the 860 nm band created by X-raying of very pure crystals at LNT, the 1240 nm absorption represents the next longer-wavelength polaron component at equal energetic distance (Fig. 4). According to our recent investigations this component has no analogon in YAG crystals [1]. Using our polaron model we can draw a diagram with the polaron energy as ordinate and the supposed quantum number as abscissa (see Fig. 11 and [1]). This diagram proves the energetic equidistance of all components regarded (including the 1240 nm band). The energy quantum of the supposed harmonic oscillator is $\hbar\omega_0 = 0.46$ eV (compared with YAG 0.55 eV) in the case of $LiNbO_3$.

3. Conclusions

A lot of absorption components found in $LiNbO_3$ crystals prepared by different methods (X-raying at LNT, annealing at different temperatures between LNT and RT, reduction, etc.) are explained as absorption bands of differently strongly bound polaron centres. The shapes of these bands are described in a unified way with the help of a relation assuming bound polarons. The equal distance in the energetic scale may be explained by the model of a harmonic oscillator. The energy quantum as well as the energies of phonons which take part in the self-trapping have been estimated from experimental data.

The strong analogy of the present investigations to the properties of YAG crystals [1] supports the assumption of the high degree of generalizability of this type of centres for several oxide crystals.

References

[1] Hj. Bernhardt, phys. stat. sol. (b) 87, 213 (1978).
[2] Hj. Bernhardt, phys. stat. sol. (a) **33**, 211 (1976).
[3] W. Phillips, J. J. Amodei, and D. L. Staebler, RCA Rev. **33**, 94 (1972).
[4] W. Bollmann and M. Gernand, phys. stat. sol. (a) **9**, 301 (1972).
[5] Hj. Bernhardt, phys. stat. sol. (a) **52**, 223 (1979).
[6] Hj. Bernhardt, to be published.
[7] G. Schwotzer, private communication.
[8] F. Fiedler, unpublished.
[9] J. Stöhr, Diplomarbeit, Friedrich-Schiller-Universität Jena, 1976.
[10] I. Kushima, Japan. J. appl. Phys. **6**, 454 (1967).
[11] O. F. Schirmer and R. Schnadt, Solid State Commun. **18**, 1345 (1976).
[12] O. F. Schirmer, Solid State Commun. **18**, 1349 (1976).
[13] D. Redfield and W. J. Burke, J. appl. Phys. **45**, 4566 (1974).
[14] V. N. Bogomolov, E. K. Kudinov, D. N. Mirlin, and Yu. A. Firsov, Fiz. tverd. Tela **9**, 2077 (1967).
[15] V. N. Bogomolov and D. N. Mirlin, phys. stat. sol. **27**, 443 (1968).

(Received June 1, 1979)

phys. stat. sol. (a) 54, 605 (1979)

Subject classification: 10.2 and 20.3; 22.8.1

Laboratoire de Physique Expérimentale (a) and Laboratoire de Géologie Structurale (b), Université de Nice[1])

On the Thermoluminescent Properties of Hydroxyapatite $Ca_5(PO_4)_3OH$

By

D. LAPRAZ (a), A. BAUMER (b), and P. IACCONI (a)

The thermoluminescence (TL) properties of calcium hydroxyapatite obtained by hydrothermal synthesis are studied between 77 and 650 K after X-irradiation. Five out of the nine impurities added prove to be activators: Mn^{2+}, Ce^{3+}, Dy^{3+}, Pb^{2+}, and Sb^{3+}. The influence of synthesis conditions on the TL as well as that of the annealing procedure is shown. The 132 K peak is specific of calcium hydroxyapatite and its origin is discussed. The results are compared with those obtained for a natural sample (Holly Springs).

Les propriétés thermoluminescentes (TL) de l'hydroxyapatite calcique obtenue par synthèse hydrothermale sont étudiées entre 77 et 650 K après irradiation X. Cinq impuretés parmi les 9 introduites se sont révélées être des activateurs: Mn^{2+}, Ce^{3+}, Dy^{3+}, Pb^{2+} et Sb^{3+}. L'influence des conditions de synthèse sur la TL ainsi que celle du recuit thermique sont montrées. Le pic de TL à 132 K est une caractéristique de l'hydroxyapatite calcique et son origine est discutée. Des comparaisons sont faites avec un échantillon naturel (Holly Springs).

1. Introduction

Few papers have been published on the electronic properties of localized defects of calcium hydroxyapatite, $Ca_5(PO_4)_3OH$ (hereafter referred to as OHAP) [1, 2]. The luminescent properties of this crystal, in particular, are not well known. Jenkins et al. [3] who have studied the luminescent properties of $Ca_5(PO_4)_3(Cl, F)$: Mn, Sb, used in fluorescent lamps, briefly point out that Sb, Cu, and Tl are fluorescence activators (emissions from blue to violet-blue). Bachman and Ellis [4] mention the existence of two emission bands at 440 and 640 nm in the fluorescence of a bone excited by Hg 365 nm radiation and ascribe the 440 nm emission to Cu. Prener and Piper [1] have shown that, at 77 K, an intense UV absorption band associated with the 630 nm fluorescence band appears if the OHAP crystal is heated in vacuum at 1373 K; they link this with the presence of O^{2-} defects and OH^- vacancies formed during calcination, according to what was observed in fluoro- and chloroapatites containing OH^- impurities in halide sites [5].

After indicating that Mn-doped hydroxyapatite (OHAP:Mn) presents thermoluminescent properties after X-irradiation at 77 K [6], we extend this study to crystals obtained by hydrothermal synthesis and containing eight different impurities. Then we show that synthesis conditions (stoichiometry of reactive mixture, pressure, temperature, duration) and thermal treatments have a great influence on thermoluminescence (TL). Finally, we compare these results with the TL of a natural sample, originating from one of the well-known OHAP monocrystal deposits.[2])

[1]) Parc Valrose, 06034 Nice Cedex, France.

[2]) Holly Springs, U.S. National Museum No. 106217, Cherokee County, Georgia, USA.

2. Experimental Techniques

2.1 *Preparation of the crystals*

The samples are obtained by the hydrothermal method at 973 K, 750 bar during 72 h by means of a TEM-PRES HR IB type apparatus, according to the following reaction [7]:

$$Ca_{9.4}H_{0.6}(PO_4)_6(OH)_{1.4}\,,\quad 4\,H_2O + 0.6\,Ca(OH)_2 \rightarrow 2\,Ca_5(PO_4)_3OH + 4.6\,H_2O\,.$$

Each mixture (30 mg) is enclosed in a welded gold tube. The doping with Mg, Mn, Fe, Sn, Sb, Pb, Ce, Eu, or Dy (some of these elements are well-known phosphate activators [8]) is carried out by adding such an amount of MgO, MnO_2, Fe_2O_3, SnO_2, Sb_2O_3, PbO, CeO_2, Eu_2O_3, or Dy_2O_3 as impurity/calcium (mole) $= 2 \times 10^{-3}$ in the reactive mixture.

The crystals obtained are washed in distilled water, then dried at 325 K for several hours. Then they are controlled by X-diffractometry and IR spectrometry. We notice that the absorption bands at 3578 cm^{-1} (0.44 eV) and 630 cm^{-1} are to be attributed to the ν_s stretching and ν_l libration frequencies, respectively, of OH^-, as usually observed [9].

2.2 *Measurements*

2 mg of a sample are uniformly spread, after a slight crushing (1 to 5 μm) in a 10 mm diameter aluminium cupel, by sedimentation in acetone.

TL is measured in a cell allowing the variation of the sample temperature between 77 and 650 K at 10^{-5} Torr. The sample, previously submitted at 77 K to a weak X-ray dose (W target, 45 kV, 2 mA), ≈ 1 R, is heated in situ at 30 K/min. Simultaneously, the TL emission is detected by means of an S 13 or S 20R photomultiplier cooled to −20 °C. The spectral analysis is effected between 250 and 800 nm by 10 to 15 nm bandwidth interference filters; all the TL spectra presented are corrected according to the photomultiplier response and the filter bandwidths.

3. Results

3.1 *Influence of impurities*

3.1.1 *"Undoped" samples*

"Undoped" crystals are characterized by a weak TL peak at 132 K (Fig. 1(a)), with much weaker peaks at 100, 170, 275, 410, 445, 510, and 610 K. The spectral study of the 132 K peak (Fig. 2) shows an emission band centered at 3.5 eV, with a bandwidth at half-maximum from 1.3 to 1.4 eV; other maxima are to be attributed to extremely weak traces of impurities (in particular Dy^{3+}) originating in the tricalcium phosphate[3]) used for synthesis (see below).

3.1.2 *Doped samples*

Under the conditions of our experiments, only, Mn, Sb, Pb, Ce, and Dy among the nine impurities added have proved to be TL activators of OHAP; the most intense peak is in any case, as for "undoped" OHAP, that at 132 K.

a) Mn: In the Mn-doped crystal [6], we observe four main TL peaks at 132, 275, 410, and 445 K, as well as some much weaker peaks around 100, 170, 195, and 600 K (Fig. 1(b)). The emission of the main peaks (Fig. 2) is formed by a band whose maximum varies from 2.16 (132 K) to 2.18 eV (445 K) and its width from 0.21 (132 K) to 0.29 eV (445 K); in addition, the 132 K emission has a very weak band around 3.5 eV similar

[3]) PROLABO product, (Mn) < 5 ppm; (Pb) < 15 ppm; (Fe) < 50 ppm.

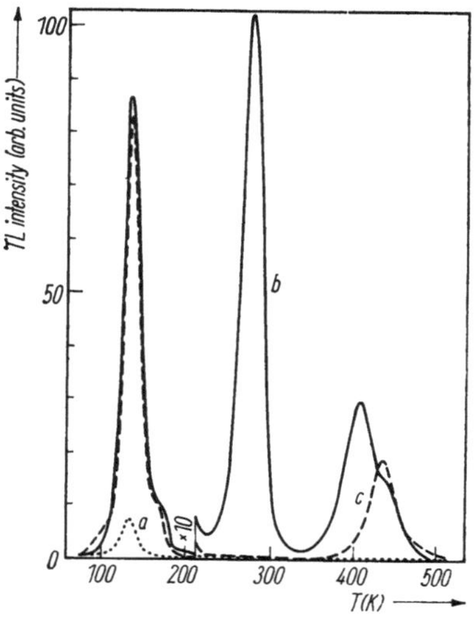

Fig. 1

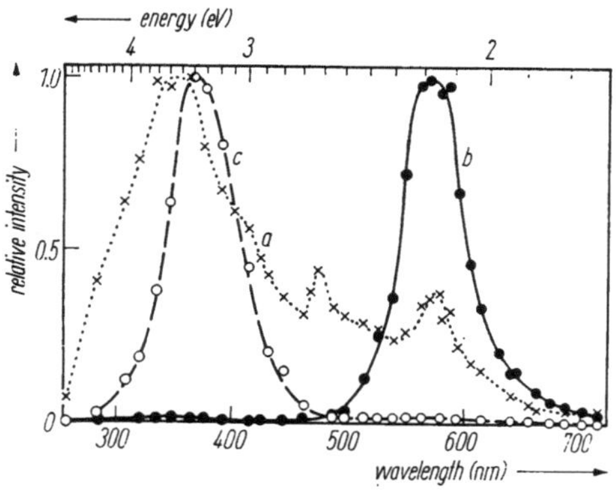

Fig. 2

Fig. 1. Glow curves of $Ca_5(PO_4)_3OH$ after X-irradiation at 77 K: (a) "undoped", (b) Mn-, and (c) Ce-doped samples

Fig. 2. Emission spectra of the 132 K TL peak: (a) "undoped", (b) Mn-, or (c) Ce-doped samples

to that of "undoped" OHAP but whose intensity is four times weaker. The parameters E (activation energy), s (frequency factor), and l (kinetic order) of the main TL peaks, determined by three methods, are displayed in Table 1, the first two being independent of l: "Initial rise" [10], Chen [11, 12].

We have noticed that the intensities of the different TL peaks increase linearly in relation to the irradiation dose (10^{-2} to 10^2 R) but that the parameters E, s, and l remain unchanged.

The TL efficiency proved maximum for $Mn/Ca = 2 \times 10^{-3}$ in the reactive mixture. Besides, we have been able to observe a decrease in the intensity of the wide band centered at 3.5 eV when the Mn concentration increases.

N.B.: The addition of Fe ($Fe/Ca = 10^{-3}$) in the course of OHAP:Mn synthesis causes a decrease by one half of the intensities of the peaks at 132 and 275 K, thereby confirming the inhibiting influence of Fe.

b) Ce: Two TL peaks are observed mainly at 132 and 430 K, with other much weaker ones around 100, 170, and 275 K (Fig. 1(c)). The emissions of the peaks at 132 and 430 K consist of a band at about 3.33 and 3.43 eV, respectively, 0.60 and 0.64 eV wide (Fig. 2). The parameters E, l, s relative to these two TL peaks are displayed in Table 1. As in the case of Mn, the TL efficiency is maximum for $Ce/Ca \approx 2 \times 10^{-3}$.

N.B.: If OHAP is doped both with Mn and Ce ($Mn/Ca = Ce/Ca = 10^{-3}$), the TL intensity ratios for two interference filters centered at 365 nm (Ce emission) and 579 nm (Mn emission) are 1, 0.05, and 0.15 for the peaks at 132, 275, and 430 to 445 K, respectively, thus showing the prominence of the Mn emission for the peaks that are not situated at 132 K.

c) Dy, Pb, or Sb: The TL of Dy-, Pb-, or Sb-doped samples is weak and similar to that of "undoped" OHAP. The peak at 132 K, however, is more intense, and for OHAP: Pb, an extra peak appears at 380 K.

Table 1
Average values of the parameters E, s, and l, calculated by various methods, for TL peaks in synthetic and natural hydroxyapatite samples

sample	method	average values of E (eV), s (Hz), and l for TL peaks														
		132 K			275 K			380 K			410 K			430−445 K		
		E	s	l	E	s	l	E	s	l	E	s	l	E	s	l
"undoped"	[11]	0.18	10^6	1.4												
	[12]	0.14−0.21		1−2												
Mn	[10]	0.28	10^9		0.72	10^{12}					1.17			1.34	10^{13}	
	[11]	0.21	10^7	>2	0.72	10^{12}	1.4				1.11	10^{12}	1.5	1.05	10^{10}	1.1
	[12]	0.14−0.21		1−2	0.57−0.83		1−2				0.85−1.22		1−2	0.93−1.34		1−2
Ce	[10]	0.25	10^8											1.41	10^{14}	
	[11]	0.26	10^8	>2										1.03	10^{10}	1.2
	[12]	0.17−0.25		1−2										0.90−1.30		1−2
Pb	[10]	0.23	10^8					0.96	10^{11}							
	[11]	0.32	10^{11}	1.9				0.97	10^{11}	1.5						
	[12]	0.23−0.33		1−2				0.75−1.08		1−2						
Dy	[11]	0.20	10^7	1.1												
	[12]	0.18−0.26		1−2												
Holly	[11]	0.17	10^5	>2												
Springs	[12]	0.08−0.13		1−2												

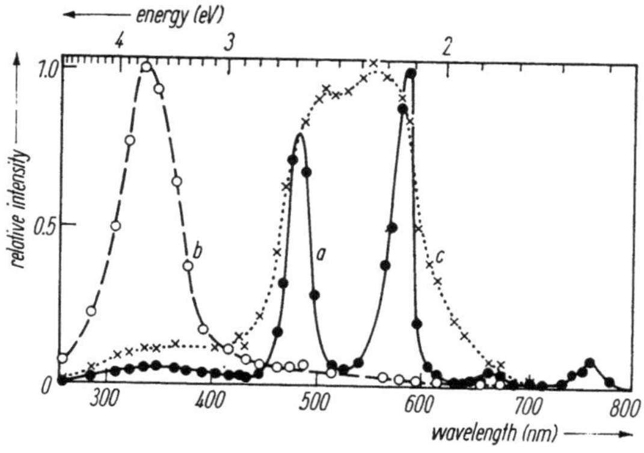

Fig. 3. Emission spectra of the 132 K TL peak of OHAP doped with (a) Dy, (b) Pb, or (c) Sb

The spectral study of the peak at 132 K (Fig. 3) shows:

(i) for Dy, narrow emission bands centered at 2.57, 2.13, 1.86, and 1.63 eV, with the wide UV band observed in the "undoped" crystal, whose intensity, however, has decreased by one half;

(ii) for Pb, only one band at 3.67 eV, with 0.70 eV width (for the peak at 380 K, the emission is shifted to 3.80 eV and it is 1.1 eV wide);

(iii) for Sb, one band with two ill-defined maxima around 2.4 and 2.2 eV, with $\approx$ 0.6 eV width; it is also associated with the wide UV band observed in the "undoped" samples but its intensity is four times weaker.

N.B.: All the crystals, whether doped or "undoped" which have not yet undergone any X-irradiation but which have been kept in ambient atmosphere, present a TL peak around 630 K characterized by an emission at 2.35 eV, with 0.8 eV width, which can be attributed to TL caused by oxygen adsorption in surface defects [13].

3.2 Influence of synthesis conditions on TL

A variation of Ca/P in the reactive mixture causes an important modification in the TL of OHAP:Mn; this is why, as soon as the ratio is no longer stoichiometric (Ca/P = $= \frac{5}{3} = 1.67$), other phases appear, such as $Ca_2P_2O_7$ or $Ca(OH)_2$, upsetting the TL of OHAP:Mn [7]. Fig. 4 shows the variation of the intensity of the peak at 132 K of OHAP:Mn in relation to Ca/P: the ratio of intensities for the 347 nm (UV band) and 579 nm (Mn) filters of the 132 K peak reaches a specific minimum when Ca/P is 1.67 (see the mark in Fig. 4). Further, we notice that the ratio of intensities of the peaks at 275 and 132 K reaches a maximum ($\approx$ 0.14) for Ca/P = 1.67. In the case of Ce doping, no crystallized phase other than OHAP has been detected for Ca/P of the reactive mixture of 1.65 ($< \frac{5}{3}$), which can be explained by a replacement of the type $3\,Ca^{2+} \rightarrow 2\,Ce^{3+} + \square$.

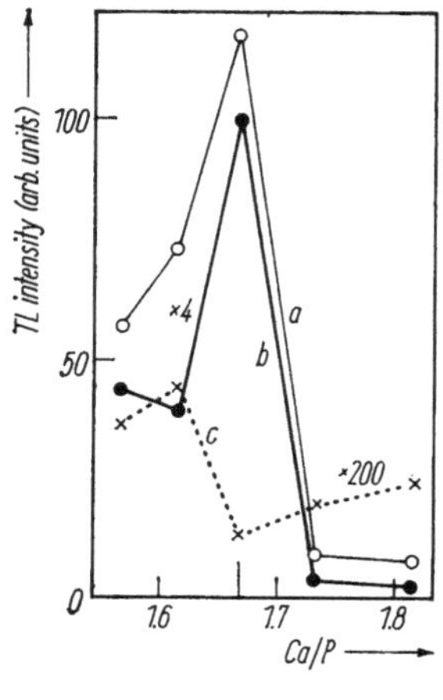

Fig. 4. Intensity of the 132 K TL peak of OHAP:Mn as a function of Ca/P ratio. (a) Without filter, (b) with 347 nm ("UV" emission), and (c) 579 nm (Mn) filters

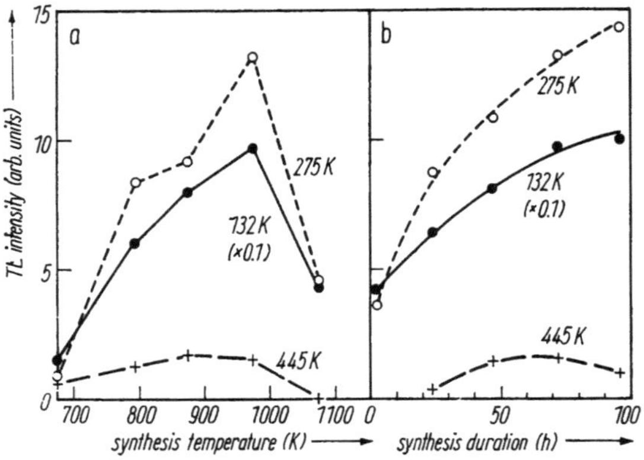

Fig. 5. Variation of OHAP:Mn TL intensities as a function of a) the synthesis temperature at 750 bar for 72 h and b) the synthesis duration at 750 bar and 973 K

Fig. 5a shows that, in relation to the synthesis temperature, the intensities of the main TL peaks of OHAP:Mn obtained at 750 bar during 72 h reach a maximum around 973 K and then decrease beyond. On Fig. 5b we indicate the variations of the intensities of the peaks at 132 and 275 K of OHAP:Mn obtained at 973 K and 750 bar for synthesis durations between 2 and 100 h.

Yet, the rapidity with which the cooling process is executed (i.e. the return to room temperature at the end of synthesis) seems to have no bearing either on the position or on the intensity of the TL peaks for cooling durations between 2 and 150 min.

3.3 Influence of thermal annealing

Annealing at atmospheric pressure has a great influence upon the TL of doped or "undoped" OHAP crystals but does not seem to vary with the type of atmosphere used (argon or air).

Fig. 6 shows how the intensities of the TL peaks at 132 and 275 K of OHAP:Mn decrease in relation to the annealing temperature (a) and duration (b). Thus, for annealing temperatures superior to 1073 K, the TL is noticeably weakened and OHAP:Mn crystals become slightly brownish.

If, after annealing at 1073 K, the Mn-doped sample is once again submitted to synthesis conditions (973 K, 750 bar, 72 h) the TL re-appears but with a weaker intensity than that shown by the sample before annealing; in particular, we observe a decrease in the emission band due to Mn.

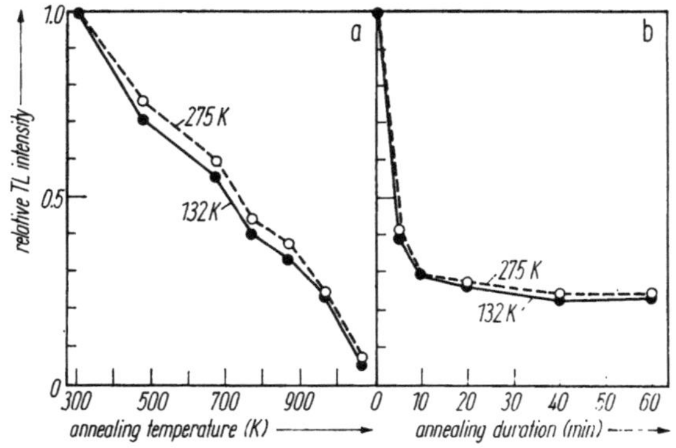

Fig. 6. Ratio of TL intensities at 132 and 275 K after and before annealing of OHAP:Mn as a function of a) the annealing temperature for a 60 min duration and b) the annealing duration at 973 K

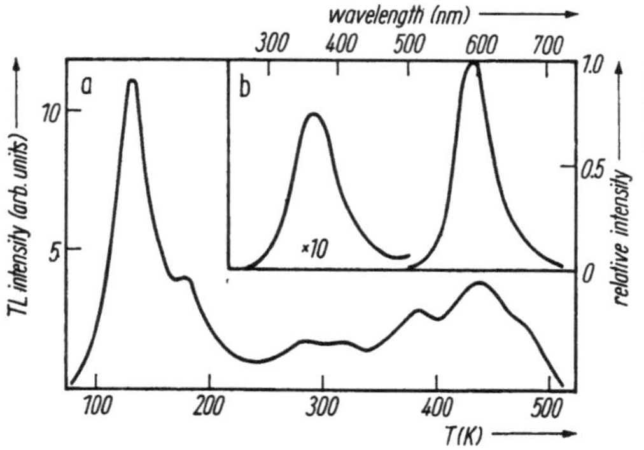

Fig. 7. Holly Springs hydroxyapatite. a) Glow curve after X-irradiation at 77 K, b) emission spectrum of the 132 K TL peak

Similar observations were made for Ce-doped samples but with no change in colour. A thermogravimetric analysis realized between 293 and 1173 K at 240 K/h for OHAP:Mn only shows a very slight continuous loss of mass (1.8%). In IR spectrometry, the intensity of the OH band (3578 cm^{-1}) is not significantly different after annealing at 1073 K.

3.4 *Thermoluminescence of a natural hydroxyapatite*

Holly Springs hydroxyapatite, containing Mn (Mn/Ca $\approx 10^{-3}$ [14]), displays the principal TL characteristics of OHAP:Mn, especially the peak at 132 K (Fig. 7a and Table 1). A few differences can be noticed, however: a peak at 180 K (instead of 170 K) which we attribute to the presence of F^- ions replacing OH^-, since F/OH $\approx$ ≈ 0.04 [14]; a peak at 320 K, sometimes detected in "undoped" OHAP, and a peak at 670 K, the only one to be found in natural thermoluminescence ($E = 0.18$ eV; $l = 1.4$, and $s \approx 10^6$, values determined by Chen's method [11]). Two emission types can be observed (Fig. 7b):

(i) a main band centered at 2.09 eV (2.15 eV), with 0.23 eV (0.29 eV) width for the TL peak at 132 K (445 K), which we can attribute to Mn^{2+} (see above);

(ii) a much weaker band centered around 3.4 eV, with 0.70 eV width at 132 K which could be attributed to Ce^{3+} (see above).

As in the case of Mn- or Ce-doped crystals, a thermal annealing at 1073 K under argon atmosphere strongly weakens the intensities of all the TL peaks. If the annealed sample undergoes the same synthesis as that of our doped samples, we observe once again the TL of the sample before annealing almost entirely.

4. Discussion and Conclusion

The nature of most traps responsible for the TL of hydroxyapatite formed by X-irradiation appears to be specific of the crystal lattice and independent of the chemical impurities present; this was tested with other crystals obtained by hydrothermal synthesis but starting from different reactive mixtures and with the Holly Springs sample. This is true, in particular, for the defect associated with the TL peak at 132 K $\left((0.24 \pm 0.08)\text{ eV}\right)$ which might be considered as characteristic of calcium hydroxyapatite. Its position in temperature is modified neither by the various impurities enclosed nor by synthesis conditions; but its intensity depends closely on the type of impurity.

The very wide UV emission band of the 132 K peak is clearly weakened in the presence of activating impurities, especially Mn, Dy, and Sb which do not emit in that region. This could indicate a transfer of energy from the defect responsible for that emission to the activators, since there is partial overlapping of the excitation spectra (FAP:Mn^{2+} [15]; ClAP:Sb^{3+} [16]; Dy^{3+} [17]) and of this UV emission band.

Several hypotheses can be suggested to explain the origin of this UV emission band observed in the TL peak at 132 K:

(i) either a recombination of the PO_4^{3-} group having captured a hole after X-irradiation ($PO_4^{3-} + h \rightarrow PO_4^{2-}$) with an electron of the lattice: $PO_4^{2-} + e^- \rightarrow PO_4^{3-} + h\nu$. In phosphate $Ba_3(PO_4)_2$, a similar TL emission at 135 K, with 1.35 eV width and with two ill-defined maxima ($\approx$ 3.8 and 4.2 eV) has been coordinated with the disappearance of an EPR signal (due to a hole) and interpreted according to that mechanism, supposing a diffusion of the hole towards a localized e^- [18]. But this type of mechanism does not account for the emission being more intense for "undoped" ClAP than for "undoped" OHAP [19];

(ii) or a recombination of H^0 and O^- species, formed by the irradiation of OH^- at 77 K, according to

$$H^0 + O^- \rightarrow (OH^-)^*, \rightarrow OH^- + h\nu .$$

A certain vibrational structure of the emission band (maxima equally distant by 0.44 eV, the ν_s vibration energy of OH^-) and its width (Fig. 2) could suggest an emission of OH^-, as it has been observed in the alkali halides containing OH^- [20]. The weak intensity of the UV emission observed in OHAP compared to that of "undoped" ClAP, which contains OH^- traces due to the hydrothermal method [19], can be explained by the process of concentration quenching; yet, we cannot explain why the first emission band centered around 4.0 eV and corresponding to the 0–0 transition of OH^-, and therefore the most likely, is less intense than the second one.

(iii) or again an electron–hole recombination caused by defects in Ca^{2+} and/or OH^- sites. This type of emission band is also observed in the TL of CaF_2:Lu^{3+} at 160 K [21], of SrF_2:Y^{3+} between 100 and 180 K [22] or of "undoped" CaF_2 at 100 and 155 K (see our own studies). Now, the TL below 150 K in CaF_2 crystals has been associated with the decline of V_K centres observed by means of EPR after X-irradiation, the recombination emission being centered at 4.28 eV [23, 24]. The existence of V_K-like centres seems plausible if one supposes that the widening of the emission band is due to strong perturbations in the lattice. But such centres disappearing at about 150 K have not been observed as yet in apatites; this is why Knotterus [25] concludes that the process of defect formation by means X-irradiation for apatites is probably different from that for alkali halides [26].

Among the nine impurities added, five proved to be activators, two of which in a significant way, Mn and Ce. Thus, the various emission bands observed can be attributed:

for Mn, to the transition $T_1(^4G) \rightarrow {}^6A_1(^6S)$ of Mn^{2+};

for Ce, to the transitions $5d(^2D_{3/2}) \rightarrow 4f(^2F_{5/2}, {}^2F_{7/2})$ of Ce^{3+};

for Dy, to the transitions $^4F_{9/2} \rightarrow {}^6H_{15/2}$ (2.57 eV), $^6H_{13/2}$ (2.13 eV), $^6H_{11/2}$ (1.86 eV), and $^6H_{9/2}$ (1.63 eV) of Dy^{3+};

for Pb, to the transition $^3P_1 \rightarrow {}^1S_0$ of Pb^{2+};

and for Sb, to the transitions $(^3P_1, {}^1P_1) \rightarrow {}^1S_0$ of Sb^{3+}.

The slight shift of the Mn emission band between synthetized and natural samples can be explained by a different localization of Mn on the two inequivalent sites for Ca^{2+}, if one excludes the possibility of interstitial sites. Such a shift is observed in the fluorescence emission bands at 1.8 K of Mn(I) (2.17 eV) and Mn(II) (2.20 eV) doped fluoroapatites [15].

The simultaneous decreasing of all the TL peaks according to the same kinetics, could originate in a diffusion of the activators outside the lattice; this would be particularly true for Mn^{2+} which has a small ionic radius (0.80 Å) compared to that of Ca^{2+} (0.98 Å), which is not the case for Ce^{3+} (1.07 Å) or Pb^{2+} (1.20 Å) [27]. Experiments carried out with an electronic microprobe on Mn-doped chloroapatite crystals did not evidence an Mn^{2+} shift after annealing. A partial decomposition of OHAP at 1073 to 1173 K is also possible, but this would imply [28] that OHAP before annealing is deficient in Ca^{2+} ions, a statement contradicted by our own studies in X-diffraction and TL [7]. Nevertheless, traces of CaO have been detected in certain cases by X-diffraction, which shows that a decomposition of the type $Ca_{10}(PO_4)_6(OH)_2 \rightarrow$ $\rightarrow Ca_{10-x}(PO_4)_6(OH)_{2-2x} + xCaO + xH_2O$ is possible. We can, therefore, admit that most of the trapping centres are likely to be located in the tunnels, where ions enjoy a relatively loose connection with the rest of the apatitic structure, and will hence, forth be easily perturbed by thermal annealing.

Acknowledgements

The authors wish to express their thanks to Prof. G. Turco, director of the Laboratoire de Pétrologie-Minéralogie of Nice, where the crystals have been synthetized, and Prof. Keller for his assistance in this work.

References

[1] J. S. Prener and W. W. Piper, Electronic Properties of Localized Defects in Apa tites General Electric R.D.C. Report No. 68-C 398, Schenectady (New York) 1968 (p. 14).
[2] M. Mengeot, R. H. Bartram, and O. R. Gilliam, Phys. Rev. B **11**, 4110 (1975).
[3] H. G. Jenkins, Mc A. H. Keag, and R. W. Ranby, Proc. Phys. Soc. **96**, 1 (1949).
[4] C. H. Bachman and E. H. Ellis, Nature (London) **206**, 1328 (1965).
[5] J. S. Prener, W. W. Piper, and R. M. Chrenko, J. Phys. Chem. Solids **30**, 1465 (1969).
[6] D. Lapraz, A. Baumer, P. Keller, and G. Turco, C.R. Acad. Sci. (France) **285B**, 291 (1977),
[7] A. Baumer and D. Lapraz, Bull. Soc. Franc. Minér. Crist. **101**, 53 (1978).
[8] W. L. Wanmaker and D. Radielovic, Coll. Internat. Phosphates minéraux solides, Toulouse 1967; Bull. Soc. Chim. Franc. 1657 (1968).
[9] F. Freund and R. M. Knobel, J. Chem. Soc. 1136 (1977).
[10] G. F. J. Garlick and A. F. Gibson, Proc. Phys. Soc. **A60**, 574 (1948).
[11] R. Chen, J. Electrochem. Soc. **116**, 1254 (1969).
[12] R. Chen, J. appl. Phys. **40**, 570 (1969).
[13] P. Iacconi, D. Lapraz, and P. Keller, C.R. Acad. Sci. (France) **273B**, 674 (1971).
[14] W. A. Deer, R. A. Howie, and J. Zussman, Rock Forming Minerals, Vol. 5, 4th edition, Longmans 1965.
[15] F. M. Ryan, R. C. Ohlmann, J. Murphy, R. Mazelsky, G. R. Wagner, and R. W. Warren, Phys. Rev. B **2**, 2341 (1970).
[16] T. S. Davis, E. R. Kreidler, J. A. Parodi, and T. F. Soules, J. Lum. **4**, 48 (1971).
[17] G. H. Dieke, Spectra Energy Levels of Rare-Earth Ions in Crystals, Interscience Publishers, New York 1968.
[18] V. Kronghauz, P. Kulis, and I. Tale, Internat. Conf. Luminescence, Paris 1978 (p. 32).
[19] D. Lapraz, A. Baumer, and R. Caruba, Communication, 7ème Réunion Annuelle des Sciences de la Terre, Lyon 1979 (p. 276).
[20] H. Pick, Proc. Internat. Conf. Luminescence, Budapest 1966, Akademiai Kiado, Budapest 1968 (p. 659).
[21] J. Kiessling and A. Scharmann, phys. stat. sol. (a) **32**, 459 (1975).
[22] P. Görlich, H. Karras, G. Kotitz, and P. Ullmann, see [20] (p. 762).
[23] W. Hayes and J. W. Twidell, Proc. Phys. Soc. **79**, 1295 (1962).
[24] J. L. Merz and P. S. Pershan, Phys. Rev. **162**, 217 (1967).
[25] D. I. M. Knotterus, Thesis, University of Groningen, The Netherlands, 1976.
[26] D. Pooley, Brit. J. appl. Phys. **17**, 855 (1966).
[27] R. C. Weast (Ed.), Handbook of Chemistry and Physics, 55th edition, C.R.S. Press, Cleveland (Ohio) 1974/1975.
[28] J. C. Heughebaert and G. Montel, Coll. Internat. C.N.R.S. no 230, Paris 1973 (p. 283).

(Received May 10, 1979)

phys. stat. sol. (a) 54, 615 (1979)

Subject classification: 14.1; 10.1; 21.1

Sektion Mathematik/Physik der Pädagogischen Hochschule Halle (Saale)[1])

On the Extra-Resistivity of Coherent Precipitates

By

C. RADOMSKY and H. LÖFFLER

By means of a formula recently developed by the authors which allows one to calculate the extra-resistivity of GP zones ($\Delta\varrho_{ex}$) by the help of theoretical XSAS intensities and XSAS intensity distribution functions of single spherical as well as ellipsoidal GP zones the course of $\Delta\varrho_{ex}$ is calculated considering certain decomposition processes. For example it is shown that the shape of the zones is remarkably affecting $\Delta\varrho_{ex}$. This conclusion is supported by experimental investigations performed on an Al–Zn(6 at%)Mg(0.1 at%) alloy. Under special conditions $\Delta\varrho_{ex} \sim f^{2/3}$ (f volume fraction of the zones) holds in accordance with experimental results of some other authors.

Mittels eines kürzlich von den Autoren abgeleiteten Zusammenhangs zwischen dem Zusatzwiderstand $\Delta\varrho_{ex}$, der durch die Bildung von GP-Zonen hervorgerufen wird, und den RKWS-Intensitäten, wird mit Hilfe theoretischer RKWS-Verteilungsfunktionen einzelner sphärischer als auch elliptischer GP-Zonen der Verlauf von $\Delta\varrho_{ex}$ für verschiedene Entmischungsprozesse ermittelt. Es wird z. B. gezeigt, daß die Gestalt der Zonen $\Delta\varrho_{ex}$ wesentlich beeinflußt. Diese Folgerung kann experimentell an einer Al–Zn(6 At%)Mg(0,1 At%)-Legierung bestätigt werden. Unter speziellen Bedingungen gilt $\Delta\varrho_{ex} \sim f^{2/3}$ (f Volumenanteil der Zonen), in Übereinstimmung mit experimentellen Ergebnissen anderer Autoren.

1. Introduction

During the last decade a relatively large number of papers (see [1 to 11] in [1]) have considered the abnormal behaviour of the resistivity during the early stage of isothermal aging in Al alloys in which Guinier-Preston zones (GPZ) are formed. These papers took into account mainly the origin of the maximum of resistivity. But no uniform opinion about the reasons of this maximum can be found out in these works.

Recently the authors developed a relation between X-ray small-angle scattering (XSAS) spectra and electrical resistivity [1]. We want to remark that lately Mimault et al. [13] calculated in a way similar to the one described in [1] the ratio $\varrho(t)/\varrho(0)$, where $\varrho(0)$ is the resistivity of the supersaturated as-quenched alloy and $\varrho(t)$ the resistivity in dependence on aging time without phonon contribution. Using experimental XSAS intensities the authors tried to show that the appearance of a resistivity maximum of the alloys investigated can only be explained by a spinodal decomposition process. But it is somewhat uncertain whether the XSAS curves presented by [13] are really resulting from spinodal decomposition or rather from a nucleation and growth process.

Assuming (i) isotropic scattering of the electrons on a Fermi sphere and (ii) the Boltzmann equation in the relaxation time approximation the extra-resistivity caused by the formation of GPZ ($\Delta\varrho_{ex}$) is given (see equation (14a) in [1]) by

$$\Delta\varrho_{ex} = \frac{\hbar \exp(\mu d)\, n_0^{-8/3}}{12.3^{2/3}\,\pi^{1/3} e^2 V_X} \int\limits_0^{2k_F} \frac{I(h)}{I_e(h)}\, h^3\, dh\,, \tag{1}$$

[1]) Kröllwitzer Str. 44, DDR-402 Halle (Saale), DDR.

where μ is the absorption coefficient of the sample, d its thickness, n_0 the mean density of the conductivity electrons, V_X the X-ray irradiated volume of the sample, $I(h)$ the slit-length-corrected and standardized XSAS intensities, $h = 2\pi/\lambda_X$ (λ_X wavelength of the X-rays), k_F the Fermi wave number, and $I_e(h)$ is given after Guinier and Fournet [2] by

$$I_e(h) = (e^2/mc^2)^2 \, I_0 \, [1 + \cos^2 (\varepsilon_X/2)] \, /(2P^2) \, , \tag{2}$$

where I_0 is the intensity of the primary beam, ε_X the scattering angle, m and e the mass and the charge of the electron, respectively, and P the distance between sample and detector.

As derived in [1] (see equation (7)) the X-ray spectra must be known (at least approximately) from $\varepsilon_X = 0$ to $\varepsilon_{X,\max} = \arcsin [\lambda_X k_F/(2\pi)]$. Thus $\varepsilon_{X,\max} = 51°$ holds, for example, using a copper tube ($\lambda_X^{Cu} = 0.154$ nm) and an Al matrix ($k_F = 1.75 \times 10^{10}$ m^{-1}).

Unfortunately by means of the usually applied XSAS cameras the intensities can be measured only up to about 10°. Therefore, the remainder of the XSAS spectrum must be approximated by Porod's rule and for that reason this contribution to $\Delta\varrho_{ex}$ is rather uncertain.

In order to avoid the difficulties mentioned above in this paper we try to obtain information on $\Delta\varrho_{ex}$ by means of calculated XSAS intensities provided by the literature.

Within this frame the purpose of this paper is to find out information on $\Delta\varrho_{ex}$ in dependence on certain decomposition and growth conditions of GPZ.

2. Description of the Model Used

Assuming a monodisperse dilute particle system the XSAS intensities are given after Guinier [2] by

$$I(h) = I_e(h) \exp(-\mu d) \, N_p V_p^2 (\Delta n)^2 \, i(h) \, , \tag{3}$$

where N_p is the number of precipitates in V_X, V_p the volume of a single precipitate, and Δn the difference of the electron density between precipitates and matrix. $i(h)$ means the intensity distribution function of a single precipitate. $i(h)$ depends on the shape and size of the precipitates and is given in [2] for spherical and ellipsoidal GPZ, namely by

$$i_{sph}(h) = \Phi^2(hR) = \left[3 \, \frac{\sin(hR) - hR\cos(hR)}{(hR)^3}\right]^2 \tag{4}$$

and

$$i_{ell}(h) = \int_0^{\pi/2} \Phi^2[hR(\cos^2\vartheta + v^2 \sin^2\vartheta)^{1/2}] \cos\vartheta \, d\vartheta \, , \tag{5}$$

respectively, where R is the radius of the precipitates and $v = a/c$ the ratio of the large (a) and the small (c) half-axis of the rotation symmetrical ellipsoid.

The density of the particles (often not constant during the aging process) can be expressed by

$$n_p = N_p/V_X = f(R)/V_p \, , \tag{6}$$

where f is the volume fraction of the precipitates in the sample considered.

Inserting (2) to (6) into (1) we obtain for GPZ_{sph}

$$\Delta\varrho_{ex}^{sph} = f(R) \, (\Delta n)^2 \, \frac{\pi^{2/3} h}{3^{2/3} e^2 n_0^{8/3}} \, \Omega(2k_F R)/R \tag{7}$$

and for ellipsoidal GPZ with the three half-axes $Rv^{-1/3}$, $Rv^{-1/3}$, and $Rv^{2/3}$ and the same volume $(4\pi R^3/3)$ as GPZ_{sph} $(1 < v < \infty)$

$$\Delta\varrho_{\text{ex}}^{\text{ell}} = f(R)\,(\Delta n)^2 \frac{\pi^{2/3}\hbar}{3^{2/3}e^2 n_0^{8/3}} \frac{v^{4/3}}{(v^2-1)^{1/2}} \int\limits_1^v \frac{\Omega(2k_{\text{F}}Rzv^{-1/3})}{z^3(z^2-1)^{1/2}}\,(\text{d}z/R)\,, \tag{8}$$

where

$$\Omega(\alpha) = \int\limits_0^\alpha \frac{(\sin z - z\cos z)^2\,\text{d}z}{z^3}\,. \tag{9}$$

In order to obtain reliable statements on the time dependence of $\Delta\varrho_{\text{ex}}$ during the decomposition process reasonable assumptions concerning the time dependence of both, the volume fraction f and the radius R of the precipitates, are needed.

During isothermal ageing of supersaturated alloys the growth kinetics of the length or size of the precipitates can often be decribed by a relation of the type (see for example [3, 4])

$$R = R_0 t^m\,. \tag{10}$$

The time dependence of the volume fraction of the precipitates is assumed to be given by a relation of the Johnson-Mehl-Avrami type [5]:

$$f/f_0 = 1 - \exp(-Kt^n)\,. \tag{11}$$

Supposing diffusion-controlled three-dimensional growth of the precipitates at least in the early stage of decomposition [6, 7] $n = 3/2$ holds.

f_0 means the saturation value of the volume fraction and K a parameter determining the decomposition rate. K can be taken from the time dependence of the XSAS data (integral intensity) or be used as a fitting parameter.

3. Results and Conclusions

In this section some types of decomposition processes are considered and the time dependence of $\Delta\varrho_{\text{ex}}$ is calculated in the frame of various assumptions.

Case 1: Particle density n_{p} and electron density difference Δn remain constant in the early stage of aging and (6), (10) (assuming $m = \frac{1}{2}$), and (11) (applying $n = \frac{3}{2}$) are valid and only spherical GPZ are present in the matrix.

In the frame of these assumptions $\Delta\varrho_{\text{ex}}^{\text{sph}} \sim R^2\Omega(2k_{\text{F}}R) \sim f^{2/3}\Omega$ holds according to (7) as $f = n_{\text{p}}(4\pi R^3/3)$. The course of Ω versus R is shown in Fig. 1. It can be seen that $\Omega \approx$ const for precipitates in a radius range larger than 0.8 nm (Fig. 1). In this range $\Delta\varrho_{\text{ex}}^{\text{sph}} \sim f^{2/3}$ holds fairly well. Such a relation was found by Lendvai et al. [8]

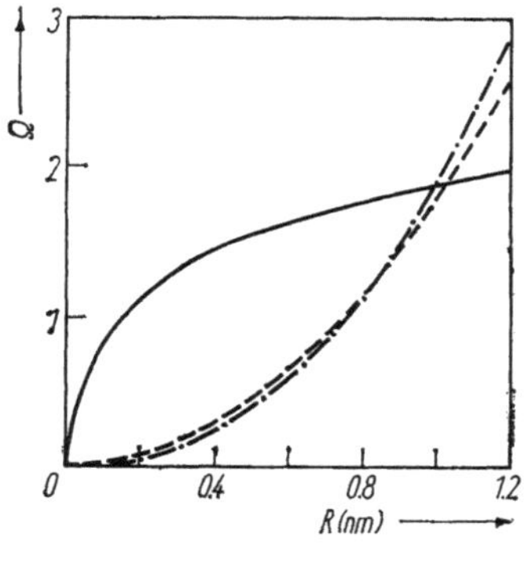

Fig. 1. $\Omega(2k_{\text{F}}R)$ as a function of the radius R of spherical precipitates (full line), relative resistivity change $(\Omega(2k_{\text{F}}R)\,R^2$, in units of 10^{-18} m^2) for a constant particle density (dashed-dotted line), approximation of the relative resistivity change $(\Omega(R = 0.8\text{ nm})R^2$, in units of 10^{-18} m^2) yielding $\Delta\varrho_{\text{ex}} \sim f^{2/3}$ (dashed line)

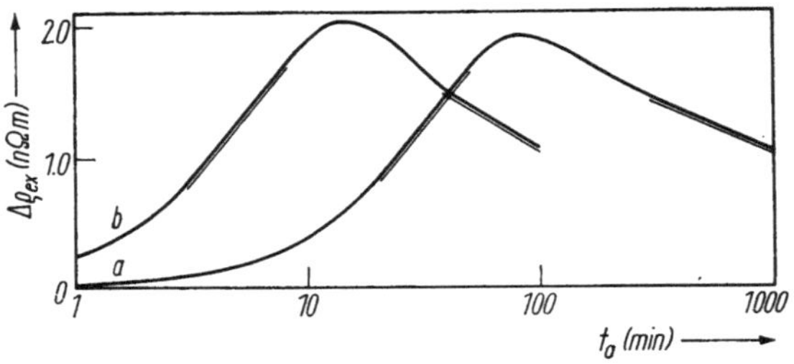

Fig. 2. Extra-resistivity $\Delta\varrho_{ex}$ of a spherical precipitate for the radius range 0.3 to 3.0 nm. (a) $m = 1/3$, (b) $m = 1/2$

investigating some Al–Zn(2.5 at%)–Mg(X) alloys in decomposition stages before the maximum of resistivity was obtained.

Case 2: Often n_p will not remain constant during the decomposition process, but is increasing during early aging stages (Osamura mechanism [9]) and later decreasing with R^{-3} (Ostwald ripening). This case can be described by (10) and (11) assuming $0 < m < 1/2$ (as often obtained from XSAS data (see for example [4]). Assuming spherical GPZ, $n = 3/2$, and several values of m we obtain straight lines plotting $\Delta\varrho_{ex}^{sph}$ versus $\ln t_a$ (t_a aging time) in a good approximation during the first increase of resistivity as well as during the decrease of resistivity after reaching the maximum. In Fig. 2 the results are shown valid for example for $m = 1/3$ and $1/2$. A straight line in the $\Delta\varrho_{ex}$–$\ln t_a$ plot was found experimentally by Bossac and Löffler [10] investigating Al–Zn–Mg alloys. In order to check whether the experimental curves after the appearance of the resistivity maximum during the decrease of $\Delta\varrho_{ex}$ can be described likewise by a straight line in the $\ln t_a$ plot, we investigated an Al–Zn(6 at%)–Mg(0.1 at%) alloy at $T_a = 5$ °C (after quenching from $T_q = 480$ °C). The results are shown in Fig. 3 (curve 1). Indeed in the $\Delta\varrho_{ex}$–$\ln t_a$ plot straight lines appear before as well as after the resistivity maximum, but the straight line after the maximum

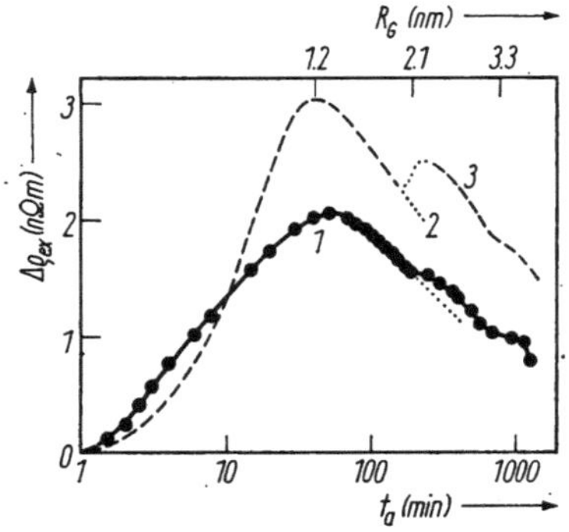

Fig. 3

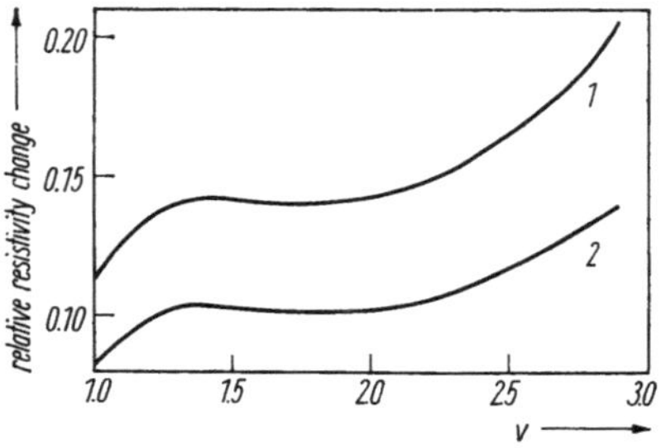

Fig. 4

Fig. 3. Experimental (1) and theoretical (2, 3) extra-resistivity $\Delta\varrho_{ex}$ of an Al–Zn(6 at%)–Mg(0.1 at%) alloy at aging temperature $T_a = 5$ °C. Curve 2 is valid for spherical GPZ, curve 3 corresponds to a shape transition $GPZ_{sph} \rightarrow GPZ_{ell}$ at a radius of 2 nm

Fig. 4. Relative resistivity change for an ellipsoid of revolution in dependence on its axis ratio v.

$$\left(\frac{v^{4/3}}{\sqrt{v^2 - 1}}\right) \int_1^v \frac{\Omega(2k_F Rz/v^{1/3})}{z^3 \sqrt{z^2 - 1}} \, dz/R$$

as a function of v for (1) $R = 2$ nm and (2) $R = 3$ nm, respectively

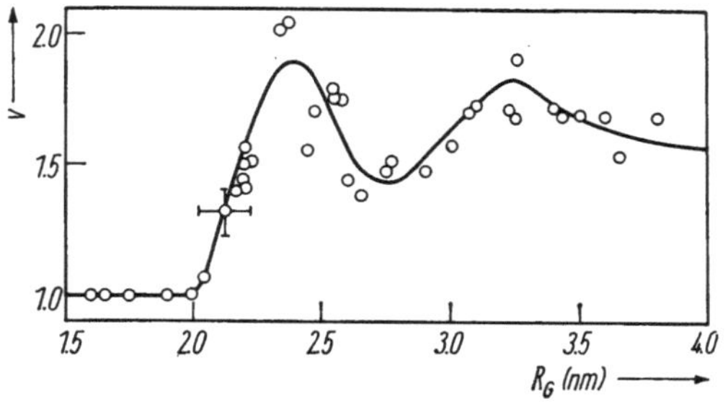

Fig. 5. Variation of axis ratio v experimentally obtained in [11]

is interrupted by a shoulder. This shoulder appears at a radius of about 2.2 nm. As investigated by Löffler et al. [11] in Al–Zn alloys at a critical radius of about 2 nm the transition from spherical to ellipsoidal GPZ takes place. The same authors showed [12] that in Al–Zn–Mg alloys this transition should also take place approximately at 1.8 nm. As shown in (8) this transition should influence $\Delta\varrho_{ex}$. Therefore, we are going to consider it.

Case 3: In order to obtain information on the influence of the shape of the zones on $\Delta\varrho_{ex}$ we calculated it using (8) for fixed f, Δn, and V_p varying only v. The results are shown in Fig. 4. It can be seen that $\Delta\varrho_{ex}^{ell}$ is increasing with rising deviation of the zones from the spherical shape.

Case 4: The course of the $\Delta\varrho_{ex}^{ell}$–ln t_a curve was calculated under the same conditions as mentioned in Case 2 ($\Delta\varrho_{ex}^{sph}$, curve 2 in Fig. 3), but taking into account additionally a possible change of the shape of the zones during the growth process. The possible variations of the shape (v-value) are taken from the data published in [11] (see Fig. 5). The results are shown in Fig. 3 by curve 3. The "second" maximum is caused by the maximum of the deviation from the spherical shape. By comparison of the calculated and experimental curves it can be seen that these two curves agree fairly well taking into account the large amount of assumptions needed to calculate $\Delta\varrho_{ex}$. Obviously in the experimental curve the increase of $\Delta\varrho_{ex}$ caused by the rising deviations from the spherical shape is not so distinctly marked as in the calculated one. Probably affected by the size distribution of the GPZ the transition from GPZ_{sph} to GPZ_{ell} does not go on simultaneously but within a certain time. Finally we would like to remark that the influence of the size distribution, the aging temperature, as well as the magnesium concentration will be considered in a following paper.

Acknowledgements

The authors would like to express their appreciation to the staff of the Computer Centre of the Pedagogical College in Halle for supporting numerical computations. We would also to thank Dr. Ing. W. Oettel for his careful resistivity measurements.

References

[1] C. Radomsky and H. Löffler, phys. stat. sol. (a) **50**, 123 (1978).
[2] A. Guinier and G. Fournet, in: Small Angle Scattering of X-Rays, John Wiley and Sons, New York 1955.
[3] Y. H. Chen and H. D. Doherty, Scripta metall. **11**, 725 (1977).
[4] H.-G. Fabian, R. Kroggel, H. Löffler, and O. Simmich, in preparation.
[5] C. Wert and C. Zener, J. appl. Phys. **21**, 5 (1950).
[6] F. S. Ham, J. Phys. Chem. Solids **6**, 635 (1958).
[7] O. Simmich and H. Löffler, Proc. 2th Internat. Conf. Age-Hardenable Al Alloys, Visegrad (Hungary) 1979 (in the press).

[8] J. Lendvai, T. Ungar, and I. Kovacs, Phil. Mag. **33**, 209 (1976).
[9] K. Osamura, Y. Hiraoka, and Y. Murakami, Phil. Mag. **28**, 809 (1973).
[10] H. Bossac and H. Löffler, phys. stat. sol. (a) **27**, K127 (1975).
[11] H. Löffler, O. Kabisch, and R. Kroggel, phys. stat. sol. (a) **49**, 299 (1978).
[12] H. Löffler and O. Kabisch, unpublished data.
[13] J. Mimault, J. Delafond, A. Jungua, A. Naudon, and J. Grilhé, Phil. Mag. **38**, 255 (1978).

(Received May 21, 1979)

phys. stat. sol. (a) 54, 621 (1979)

Subject classification: 10.1; 11; 21.1

Metallurgy Department, Risø National Laboratory, Roskilde[1])

Loop Growth and Point-Defect Profiles during HVEM Irradiation

By

T. LEFFERS and B. N. SINGH

The point-defect profile in a thin foil is included in the model for the growth of dislocation loops during HVEM irradiation suggested by Kiritani, Yoshida, Takata, and Maehara, and the possible effect of divacancies is discussed. It is found that there is a fairly wide transition range between the two extreme cases described by Kiritani et al. (the vacancy- and the surface-dominant case); this can directly (without the necessity of a divacancy effect) explain the observation of apparent activation energies for loop growth smaller than $\frac{1}{2} E_V^M$ (where E_V^M is the vacancy migration energy). Even after the inclusion of the point-defect profiles there are indications that the model cannot fully account for the loop growth behaviour in situations where surface losses and recombination losses are comparable.

Das Punktdefektprofil in einer dünnen Folie wird in das von Kiritani, Yoshida, Takata und Maehara vorgeschlagene Modell für das Wachstum der Versetzungsringe unter Bestrahlung im Hochspannungselektronenmikroskop eingeschlossen, und der mögliche Effekt von Doppelleerstellen wird diskutiert. Es wird gefunden, daß es zwischen den zwei von Kiritani u. a. beschriebenen extremen Fällen (dem Fall der Leerstellendominanz und dem Fall der Oberflächendominanz) ein ziemlich weites Übergangsgebiet gibt; dieses kann unmittelbar (ohne die Notwendigkeit eines Doppelleerstelleneffekts) die Beobachtung scheinbarer Aktivierungsenergien kleiner als $\frac{1}{2} E_V^M$ (wo E_V^M die Leerstellenwanderungsenergie ist) erklären. Auch nach dem Einschließen der Punktdefektprofile gibt es Anzeichen, daß das Modell das Wachstum der Versetzungsringe im Falle, wo Oberflächen- und Rekombinationsverluste vergleichbar sind, nicht völlig erklären kann.

1. Introduction

Kiritani et al. [1] have derived expressions for the growth rate of interstitial dislocation loops during irradiation in a high voltage electron microscope (HVEM). They described two extreme situations: "the vacancy-dominant case" (named so by Barlow and Leffers [2]) with a linear dependence of $\ln \dot{L}$ on $1/T$ ($\dot{L}$ and T being the growth rate and the irradiation temperature, respectively) and the "surface-dominant case" with a temperature-independent growth rate.

It is the aim of the present work to show that there is, between these extreme cases, a fairly wide transition range and that this range includes typical experimental situations. The experimental observations of loop growth in copper also made by Kiritani et al. [1] seem to fall in the transition range. The notion that the experiments of Kiritani et al. are in the transition range leads us to consider certain reinterpretations of their results.

2. Basic Equations for Loop Growth

The work of Kiritani et al. [1] is based on the following equation for the growth of an interstitial dislocation loop:

$$\dot{L} = 2a(Z_{IL} M_I C_I - Z_{VL} M_V C_V) , \tag{1}$$

[1]) Roskilde, Denmark.

where a is the change in loop radius by the absorption of one interstitial per atom site along the dislocation, Z_{IL} and Z_{VL} are the absorption cross-sections (per atom site) of the loop for interstitials and vacancies, M_I and M_V are the "mobilities" of interstitials and vacancies, and C_I and C_V are the concentrations of interstitials and vacancies (in defects per atom). In (1) and in all the equations to follow in Sections 2, 3, and 4 the effect of divacancies is neglected. The whole problem of the possible effect of divacancies will be discussed in Section 6 on the basis of experimental observations and theoretical considerations.

Kiritani et al. considered the number of point defects absorbed by dislocation loops to be insignificant for the overall point-defect population (i.e. the point-defect concentrations were determined by production rate, diffusion to the surfaces, and recombination). A combination of (1) with the equations for the formation and disappearance of interstitials and vacancies leads to the two extreme situations, viz. the vacancy-dominant case and the surface-dominant case for which $\dot{L}$ can be expressed as shown in (2) and (3), respectively (equation (3) was not written out explicitly by Kiritani et al.):

$$\dot{L} = 2a(Z_{IL} - Z_{VL})\,(PM_V/Z_{IV})^{1/2}\,, \tag{2}$$

$$\dot{L} = 2a(Z_{IL} - Z_{VL})\,P/C_S\,. \tag{3}$$

In these equations P is the production rate of Frenkel pairs, Z_{IV} is the capture-site number for recombination of interstitials and vacancies, and C_S is the surface sink efficiency (assumed constant in a given experiment).

The vacancy-dominant case refers to the situation away from the surface in a sufficiently thick foil. As seen from (2) loop growth is a thermally activated process with an apparent activation energy of $\frac{1}{2}E_V^M$, where E_V^M is the vacancy migration energy (M_V is proportional to $\exp(-E_V^M/kT)$, k being Boltzmann's constant).

The surface-dominant case refers to the situation close to the surface. Equation (3) shows that the loop growth rate is independent of temperature for these conditions.

Kiritani et al. determined the apparent activation energy for loop growth in copper to be (0.30 eV $\pm$ 0.02) eV, which is less than $\frac{1}{2}E_V^M$; recent measurements gave E_V^M values for Cu of about 0.7 eV (Bourassa and Lengeler [3], Wright and Evans [4], and Antesberger et al. [5] found 0.72, 0.71, and 0.70, respectively). Kiritani et al. ascribed this discrepancy to the presence of divacancies with lower migration energy than monovacancies. As it is to be shown, we can, on the basis of the present investigation, explain the results of Kiritani et al. without involving divacancies (for a theoretical discussion of the possible role of divacancies see Section 6).

3. Point-Defect Profiles and Their Effect on Loop Growth

Foreman [6] has calculated the point-defect concentrations in thin foils during HVEM irradiation as functions of foil thickness and position in the foil. The calculations are based on the assumption that the concentrations are determined by production rate, diffusion to the surface, and recombination, i.e. basically the same assumptions as those in the work of Kiritani et al. [1]. Thus, in order to include the transition range between the two extreme cases treated by Kiritani et al. one would just have to insert in (1) the point-defect concentrations calculated by Foreman.

The results of Foreman's calculations [6] were presented as the set of curves reproduced in Fig. 1. The curves give the factor by which the equilibrium point-defect concentrations in the interior of an infinitely thick foil should be multiplied to give the actual concentrations. This proportionality factor depends on the distance from the surface divided by the foil thickness (in the following designated d) and the

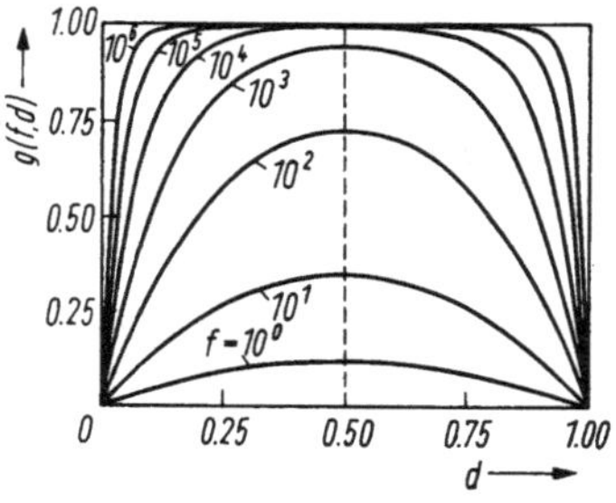

Fig. 1. The Foreman proportionality factor $g(f, d)$ as a function of f (equation (4)) and d (distance from foil surface divided by foil thickness). The point-defect concentration for a given combination of f and d is equal to the concentration in the central part of an infinitely thick foil multiplied by $g(f, d)$. From [6] with the courtesy of radiation effects

dimensionless parameter f defined by

$$f = Z_{IV}Pt^4/(\nu_V\gamma_I\gamma_V\lambda^4)\,, \tag{4}$$

where t is the foil thickness, ν_V is the vacancy jump frequency, and γ_I and γ_V are constants relating the interstitial and vacancy diffusion coefficients D_I and D_V and the jump distance λ to the corresponding jump frequencies by the equations

$$D_I = \gamma_I\lambda^2\nu_I\,, \tag{5}$$

$$D_V = \gamma_V\lambda^2\nu_V\,. \tag{6}$$

It should be noticed that for the parameters which are identical in the calculations of Kiritani et al. and Foreman we have adopted the symbols used by Kiritani et al. In the following we shall use the designation $g(f, d)$ for the Foreman proportionality factor given in Fig. 1.

Foreman's calculations (as those of Kiritani et al.) refer to steady-state conditions, i.e.

$$C_ID_I = C_VD_V\,. \tag{7}$$

This is justified for instance by the computer simulations presented by Kiritani et al. demonstrating that even at the lowest end of the temperature range in which HVEM loop-growth experiments are carried out in practice the time it takes to build up steady-state conditions is short compared with the duration of an experiment. In the experiments on loop growth in copper to be quoted in the present work the lowest irradiation temperature, for which the build-up time is less than 10 s, was 160 °C.

Foreman calculated the vacancy concentration in the interior of an infinitely thick foil to be

$$C_V^\infty = \left(P\gamma_I/(Z_{IV}\nu_V\gamma_V)\right)^{1/2}\,. \tag{8}$$

Inserting the Foreman factor $g(f, d)$ and equations (5), (6), (7), and (8) in equation (1) under the assumption that the mobilities M_I and M_V in (1) are identical to the jump frequencies ν_I and ν_V (the exact meaning of the term mobility was not specified by Kiritani et al.) produces our final equation for the loop-growth rate,

$$\dot{L} = 2a(Z_{IL}\gamma_V/\gamma_I - Z_{VL}) \cdot g(f, d) \cdot \left(P\gamma_I\nu_V/(Z_{IV}\gamma_V)\right)^{1/2}\,. \tag{9}$$

For the interior of an infinitely thick foil, i.e. for $g(f, d) = 1$, (9) is very close to (2); if $\gamma_I = \gamma_V = 1$ as assumed by Foreman, the two equations are identical. Thus, for the interior of an infinitely thick foil equation (9) gives the same loop growth rate as that quoted by Kiritani et al. [1] for the vacancy-dominant case.

For the other extreme situation (small values of $g(f, d)$ in (9) and the surface-dominant case of Kiritani et al. as described in (3)) a similar simple analytical comparison is not possible: Foreman did not give any analytical expression for $g(f, d)$, and Kiritani et al. did not define C_S in (3) in terms of the fundamental parameters. However, a comparison of the $\dot{L}$ values obtained from (9) for f equal to 1 and f equal to 10, the variation in f being produced by variation in ν_V or P, gave almost the same types of dependence on ν_V and P as those expressed in (3): no dependence on ν_V and proportionality with P.

4. The Relation between Apparent Activation Energy for Loop Growth and E_V^M

On the basis of Fig. 1 and equation (9) it is possible to construct a generalized chart for the relation between the apparent activation energy for loop growth (from now on designated E) and the vacancy migration energy E_V^M. In an experiment to determine E all parameters but the temperature will be constant. A sufficiently large change in irradiation temperature will, via the change in ν_V in (4), make f vary from the low values typical for the surface-dominant case to values of 10^5 or more typical for the vacancy-dominant case $(g(f, d) \approx 1)$.

Let us consider an experiment in which f takes the values 1, 10, 10^2, 10^3, 10^4, 10^5, and 10^6, i.e. the values for which Foreman has calculated $g(f, d)$ as shown in Fig. 1. We can rewrite equation (4) as

$$f(n) = k_1 \exp\left(E_V^M/kT(n)\right) = 10^n \tag{10}$$

with n taking the values 0, 1, 2, 3, 4, 5, and 6 and k_1 being a constant. The fulfilment of (10) requires that

$$E_V^M/kT(n) = n \cdot \ln 10 - \ln k_1 = n \cdot \ln 10 + E_V^M/kT(0)\,. \tag{11}$$

For our purpose it is irrelevant by which combination of the parameters k_1 and E_V^M these equations are fulfilled.

In order to determine E we plot $\ln \dot{L}$ versus $1/kT$. E is then the absolute value of the slope. By combining (9), (10), and (11) we get the following expression for $\ln \dot{L}(n)$ in our experiment (k_2 being a constant):

$$\ln \dot{L}(n) = k_2 + \ln\left(g(10^n, d)\right) - (n/2) \cdot \ln 10\,. \tag{12}$$

Thus, if we plot $\ln \dot{L}(n)$ from (12) versus n ,we are actually doing an activation-energy plot with one unit along the n-axis corresponding to $\ln 10/E_V^M$ units of $1/kT$ (according to (11)). Such plots are shown in Fig. 2 for positions 1/2, 1/4, and 1/10 foil thickness away from the surface (i.e. for d values of 0.5, 0.25, and 0.1). The plotted points are connected by straight lines; the absolute values of the slopes of these lines are given in Table 1. It should be underlined that Fig. 2 does not represent a basically new approach to loop growth. The basic model is still that of Kiritani et al. [1] — but now with

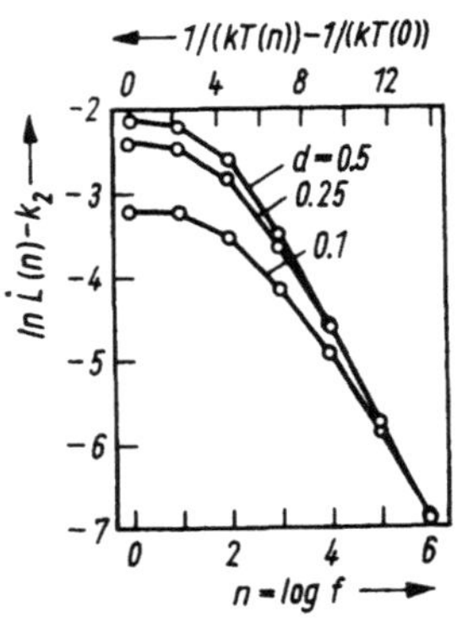

Fig. 2. Generalized theoretical activation-energy plot for three different positions in the foil. Absolute slopes for the individual straight sections are given in Table 1. $(1/(kT(n)) - 1/(kT(0))$ in units of $1/E_V^M)$

Table 1

Absolute values of the slopes of the straight sections in Fig. 2 (in units of E_V^M)

log f interval	$d = 0.1$	$d = 0.25$	$d = 0.5$
0–1	0.0001	0.0259	0.0421
1–2	0.1332	0.1615	0.1754
2–3	0.2747	0.3403	0.3886
3–4	0.3319	0.4349	0.4771
4–5	0.4031	0.4836	0.4944
5–6	0.4628	0.4967	0.5000

the theoretically indisputable improvement that the point-defect profile in the thin foil is included in a systematic way.

The significance of Fig. 2 is that the transition range between the surface-dominant case with $E = 0$ and the vacancy-dominant case with $E = \frac{1}{2} E_V^M$ covers a very wide range of f values — and that, as it is to be shown in Section 5, these values correspond to typical experimental situations. Furthermore, the deviation from linearity is so small in the upper end of the transition range that it would not be detected in the typical experiment, because the range covered is too narrow; the plot of ln $\dot{L}$ versus $1/kT$ would look like a straight line with a slope lower than $\frac{1}{2} E_V^M$. Already Urban and Wilkens [7] referred to the possibility of a transition range with E values below $\frac{1}{2} E_V^M$ (they quoted an E range of $\frac{1}{4} E_V^M$ to $\frac{1}{2} E_V^M$). However, they did not specify any relation between E and the other physical parameters.

It should be noticed that Fig. 2 is based on values of $g(f, d)$ taken from Fig. 1. This procedure will introduce some error particularly for low f values and the d values 0.25 and 0.1. The exact slopes of the lines in Fig. 2 as given in Table 1 may thus be questionable, but this is of no practical importance for our use of Fig. 2 and Table 1.

5. Comparison with Experimental Results and Discussion

As already mentioned Kiritani et al. [1] quoted an E value (apparent activation energy for loop growth) of 0.30 eV for copper. They also showed activation energy plots for various foil thicknesses without quoting the slopes. The same plots were shown by Kiritani [8], but this time with slopes quoted. These plots are reproduced in Fig. 3. There is a trend for E to increase with increasing foil thickness (or increasing f, cf.

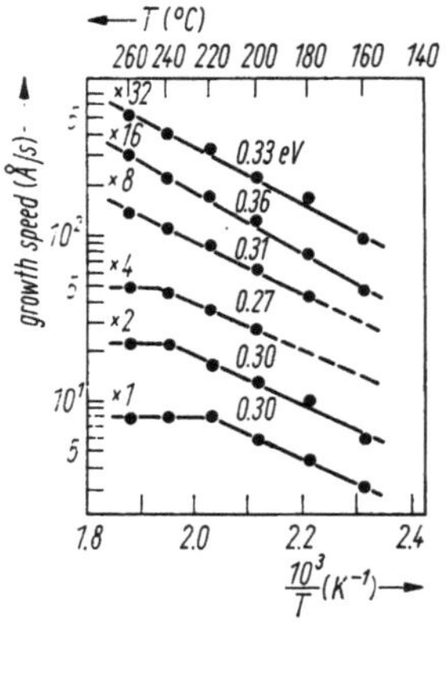

Fig. 3. Logarithmic loop growth rates in Cu foils of different thicknesses versus $1/T$. The slope (apparent activation energy) is given for each foil thickness. In order to separate the individual experiments different scaling factors (1, 2, 4, 8, 16, 32) have been used for each experiment. Specimen thickness (from top to bottom): 0.35, 0.25, 0.22, 0.18, 0.16, and 0.12 μm. From [8] with the courtesy of M. Kiritani

Table 2
E (apparent activation energy for loop growth) from various experiments on Cu divided by E_V^M (0.70 eV) and the corresponding experimental parameters and values of the Foreman parameter f

source	E/E_V^M	foil thickness (μm)	mean temperature for the sloping part of the plot (°C)	dose rate (dpa/s)	f
[8]	0.43 ± 0.03	0.12	190	1.6×10^{-3}	3.2×10^3
[8]	0.43 ± 0.03	0.16	200	1.6×10^{-3}	7.0×10^3
[8]	(0.39*)	0.18	220	1.6×10^{-3}	5.6×10^3
[8]	0.44 ± 0.03	0.22	220	1.6×10^{-3}	1.2×10^4
[8]	0.51 ± 0.03	0.25	210	1.6×10^{-3}	2.9×10^4
[8]	0.47 ± 0.03	0.35	210	1.6×10^{-3}	1.1×10^5
[9]	0.50 ± 0.03	≈ 0.5	325	1.3×10^{-2}	1.5×10^5

* Based on 2 to 3 points only.

Table 2), particularly when one omits the 0.18 μm foil for which the slope is based on very few points. Fig. 3 indicates that the E value of 0.30 eV quoted by Kiritani et al. refers to quite thin foils.

Recently Barlow et al. [9] found an E value of 0.35 eV in comparatively thick foils of Cu (and dilute Cu–Ni alloys). This result as well as the results in Fig. 3 can be ascribed to an activation energy for vacancy migration (E_V^M) of ≈ 0.7 eV with corresponding E values that vary with f according to Fig 2. To check this we have in Table 2 listed the E values for Cu from Fig. 3 and from [9] divided by E_V^M together with the experimental parameters and the derived values of Foreman's f parameter. The uncertainties quoted for the results from Fig. 3 are derived from the E uncertainty of ± 0.02 eV quoted by Kiritani et al.; in [9] the standard deviation on E was also found to be about 0.02 eV. In the calculation of f we have used $Z_{IV} = 10$ [6],

$$\nu_V = 10^{13} s^{-1} \cdot \exp(-E_V^M/kT) \text{ [1]}, \; E_V^M = 0.70 \text{ eV}, \; \gamma_I = \gamma_V = 1 \text{ [6]}, \; \lambda = 0.255 \text{ nm}.$$

In Table 2 E/E_V^M is significantly smaller than 0.5 for $f < 2 \times 10^4$. This is in perfect agreement with Fig. 2 and Table 1: if we consider loops in the centre of the foil ($d = 0.5$), E/E_V^M becomes significantly smaller than 0.5 in the f range 10^3 to 10^4; if we consider loops halfway between the foil centre and the surface ($d = 0.25$), the deviation from $E/E_V^M = 0.5$ becomes significant in the f range 10^4 to 10^5.

This comparison of Table 2 with Fig. 2 and Table 1 illustrates the use and the limitations of Fig. 2. Qualitatively Fig. 2 satisfactorily explains the observation of E/E_V^M values below 0.5. On the other hand, one cannot in practice get an exact value for E/E_V^M from Fig. 2. With reference to Fig. 3 the problem would arise already because we do not know the exact value of d. However, even if d is known and the curves in Fig. 2 are made smooth by the addition of points for non-integer values of n, there are other difficulties: (i) the exact value of Z_{IV} and of the pre-exponential factor in ν_V are not known; (ii) there is some ambiguity about the values of γ (Johnson and Lam [10] for instance used $\gamma_I = 2/3$ instead of 1); (iii) there is also some ambiguity about the calculation of the dose rate P from the beam current (for instance Kiritani et al. [1] refer to the problem of close-pair recombination). In practice Fig. 2 can therefore be used to check whether the experimental conditions are such that E/E_V^M is sufficiently close to 0.5, but it cannot be used to calculate E_V^M from E when E/E_V^M is significantly smaller than 0.5. However, these limitations in the use of Fig. 2 do not change the basic fact that Fig. 2 produces a convincing explanation of the observation of E/E_V^M values smaller than 0.5.

Table 3
Experimental parameters and f values for the transition point from temperature-dependent to temperature-independent loop growth (Fig. 3)

foil thickness (μm)	temperature (°C)	f
0.12	220	1.0×10^3
0.16	240	1.8×10^3
0.18	245	2.5×10^3

For certain experimental situations the loop growth rate becomes practically independent of temperature ($E = 0$) [1, 2, 8]; this is the pure surface-dominant case (cf. Section 2). According to Fig. 2 this situation should be reached for $f \approx 1$ to 10. In Table 3 we have listed the experimental parameters and the f values for the actual transition points for the Cu results from Fig. 3 (f is calculated with the same values of the constants as used in Table 2).

The experimental results in Table 3 and Fig. 3 clearly differ from the behaviour theoretically predicted in Fig. 2: the f value for which loop growth becomes temperature-independent is 10^2 to 10^3 times bigger than it should be according to Fig. 2, and there is a much sharper transition between the temperature-dependent and the temperature-independent regions than indicated in Fig. 2.

As already mentioned there are some uncertainties involved in the calculation of f. However, these uncertainties cannot lead to errors of 2 to 3 orders of magnitude in the calculated f values. Furthermore, the agreement between experimental and calculated values of E/E_V^M for $f \geqq 3 \times 10^3$ indicates that the calculated f values are reasonably correct. Thus the discrepancies between Table 3 and Fig. 3 and the theoretical predictions in Fig. 2 cannot be explained simply in terms of errors in the calculated f values.

Kiritani et al. [1] quoted an approximate expression for the "critical thickness", which is the thickness where the transition from temperature-dependent to temperature-independent loop growth should take place in a given material at given temperature and dose rate, and they found a satisfactory agreement with their Cu experiments. However, what they present is an approximate solution to the set of equations describing the physical situation common to their work and the present work, whereas in Fig. 2 we present the exact solution to these equations. Thus, when they find a satisfactory agreement with their experimental results and we do not, this simply illustrates the limitation in their approach. Their critical thickness refers to the situation where the loss of point-defects to the surface is equal to the recombination loss, i.e. it should correspond to the middle of the theoretical transition range in Fig. 2, which it actually also does.

Logically, these discrepancies can have their origin either in limitations in the correctness of Foreman's calculated profiles in Fig. 1 or in limitations in the correctness of equation (1). There is no obvious source of error in Foreman's calculations (as to the possibility that there is a significant effect of divacancies, see Section 6), whereas there are two basic shortcomings in equation (1): it does not consider the long-range interaction between the point defects and the dislocation, and it does not consider the local distortion of the point-defect population around the dislocation. Therefore, the most likely explanation of the discrepancies is that equation (1), even though it applies approximately for $g(f, d)$ values close to 1, does not give a correct description of loop growth when $g(f, d)$ is close to zero (small effective distance to the surface).

6. The Possible Effect of Divacancies

So far we have neglected the possible effect of divacancies. Kiritani et al. [1] suggested that values of E/E_V^M smaller than 0.5, as they observed in copper, are due to divacancies with higher mobility than monovacancies. As shown in the present work, these observations can be accounted for by the point-defect profile in the thin foil, which removes the necessity of involving divacancies. But, of course, this does not remove the possibility that divacancies can play a role in the loop-growth process.

In order to evaluate the effect of divacancies on loop growth one needs to consider two parameters: the concentration of divacancies and their mobility. It is obvious that the effect of divacancies would become significant only if a sufficiently large population of (fast moving) divacancies exists during the loop-growth experiments; if the divacancies dissociate to re-form monovacancies under the relevant experimental conditions, their influence can be neglected.

In the case of copper there is no clear and direct experimental evidence available to give reliable values of the binding and migration energies for divacancies. In a recent review Balluffi [11] for instance has abstained from quoting any divacancy binding and migration energies for Cu. From stage III recovery experiments Antesberger et al. [12] concluded that divacancies in Cu are not significantly more mobile than monovacancies. Recently Wienhold et al. [13] have reanalized basically the same experimental results as quoted by Antesberger et al. They claimed that the results can be consistently interpreted on the basis of either of the following two assumptions: (i) divacancies are much more mobile than single vacancies and are tightly bound (i.e. they do not under the given experimental conditions dissociate to any significant extent), or (ii) the migration energies of single- and divacancies are very similar.

Kiritani et al. [1] have included divacancies in their rate equations for loop growth, assuming that the conditions correspond to case (i) of Wienhold et al. Under these conditions Kiritani et al. found the apparent activation energy for loop growth E to be

$$E = (E_V^M + E_{2V}^M)/4\,, \tag{13}$$

E_{2V}^M being the divacancy migration energy. Equation (13) refers to conditions away from the surface in a sufficiently thick foil. If we take case (i) of Wienhold et al. to mean that E_{2V}^M for Cu is 0.50 eV, the resulting E value is ≈ 0.30 eV. This does not agree with the observations quoted in Table 2 of E values of ≈ 0.35 eV in Cu foils of sufficient thickness (E is E/E_V^M in Table 2 multiplied by 0.70 eV) — particularly not in view of the fact that the obvious experimental errors (the foils being thinner than estimated, the loops being too close to the foil surface) can only decrease the measured E values.

The absence of observable divacancy effects in the loop-growth experiments on Cu can have two possible explanations: either divacancies in Cu do not have significantly lower migration energy than monovacancies, or the concentration of divacancies is very low during the HVEM loop-growth experiments. The former is an obvious explanation that cannot be ruled out on the basis of existing evidence (cf. Wienhold et al.). In order to evaluate the latter we shall have to take a closer look at the rate equations that Kiritani et al. [1] derived for the case with a strong effect of divacancies. They found the following expression for the concentration of divacancies C_{2V}:

$$C_{2V} \approx 1/\sqrt{2}(Z_{VV}/Z_{I2V})^{1/2}\,(M_V/M_{2V})^{1/2}\,C_V \approx (M_V/M_{2V})^{1/2}\,C_V\,, \tag{14}$$

where the last approximation is made by the present authors. In this equation Z_{VV} and Z_{I2V} are the capture-site number for the reaction between two vacancies to form a divacancy and the reaction between an interstitial and a divacancy to form a monovacancy, respectively, and M_{2V} is the divacancy mobility. A combination of equations (8) and (11) of Kiritani et al. and our equation (14) with the assumption $M_I \gg M_{2V} \gg$

$\gg M_V$ produces an expression for the monovacancy concentration

$$C_V \approx (P/Z_{IV})^{1/2} (M_V M_{2V})^{-1/4} . \tag{15}$$

As already mentioned, Kiritani et al. claimed to consider a situation where the dissociation of divacancies to monovacancies is insignificant. Whether this is actually the case can be decided on the basis of (14) and (15) and Damask and Dienes's equation for the equilibrium between monovacancies and divacancies [14]

$$C_{2V}/C_V^2 = 6 \exp (E_{2V}^B/kT) ; \tag{16}$$

E_{2V}^B being the divacancy binding energy (and the vibrational entropy term being neglected). If the ratio C_{2V}/C_V^2 derived from (14) and (15) is clearly smaller than $6 \exp (E_{2V}^B/kT)$, i.e. if there is a surplus of monovacancies relative to divacancies, Kiritani et al. are justified in neglecting divacancy dissociation: such a situation would mean that the divacancies are removed by reaction with other defects before they dissociate to any significant extent, because the monovacancies cannot combine sufficiently fast to produce an equilibrium concentration of divacancies. If the ratio is not clearly smaller than $6 \exp (E_{2V}^B/kT)$ the equations of Kiritani et al. are not selfconsistent, exactly because they have neglected divacancy dissociation. Hence, the equations of Kiritani et al. would be consistent only if

$$(P/Z_{IV})^{-1/2} M_V^{3/4} M_{2V}^{-1/4} \ll 6 \exp (E_{2V}^B/kT) \tag{17}$$

or, expressed by the ratio R, if

$$R = (P/Z_{IV})^{-1/2} M_V^{3/4} M_{2V}^{-1/4} / (6 \exp (E_{2V}^B/kT)) \ll 1 . \tag{18}$$

In the case of Cu we have already used numerical values for most of the factors in conditions (17) or (18). In the following we shall use $Z_{IV} = 10$, $M_V = 10^{13}\ s^{-1}\ e^{-0.70 eV/kT}$, $M_{2V} = 10^{13}\ s^{-1}\ e^{-0.50 eV/kT}$, $E_{2V}^B = 0.20$ eV. The E_{2V}^B value of 0.20 eV is estimated on the basis of the E_{2V}^B values quoted for other f.c.c. metals in Balluffi's review (with $E_{2V}^B = 0.20$ eV, the value of E_{2V}^B/E_V^M for Cu becomes equal to the average of the E_{2V}^B/E_V^M values for the other f.c.c. metals for which E_{2V}^B is quoted). In the Cu loop-growth experiments of Kiritani et al. the mean temperature was 200 °C and the dose rate $1.6 \times 10^{-3}\ s^{-1}$. When these values are inserted, R in condition (18) gets a value of ≈ 17, i.e. the condition is very far from being fulfilled. This means that the divacancy concentration derived from the rate equations of Kiritani et al. is far too high compared with the monovacancy concentration and hence that the assumptions on which equation (13) is based are not valid.

The absence of any observable effect of divacancies in the Cu loop-growth experiments is therefore not surprising: even if E_{2V}^M is lower than E_V^M in Cu, one would not expect that (13) is obeyed, because the divacancies, contrary to the assumption of Kiritani et al., largely disappear by dissociation under the given experimental conditions.

Up to this point we have only considered the possible divacancy effect on the growth of loops away from the surface in sufficiently thick foils (in our terminology for $g(f, d) \approx 1$). However, when divacancies do not have any significant effect in Cu under such circumstances, there is no reason to believe that they should have a significant effect for $g(f, d) < 1$: with decreasing $g(f, d)$ the point-defect concentrations decrease, which will change the equilibrium monovacancies/divacancies even further towards monovacancies (cf. equation (16)).

In order to evaluate the applicability of our approach to loop growth in the HVEM, in which we neglect the effect of divacancies, to other metals (with $E_{2V}^M < E_V^M$) one has to apply conditions (17) or (18) to the metal in question. If they are fulfilled, it is not justified to neglect divacancies; if they are far from being fulfilled, i.e. if for in-

Table 4

Calculated values of R from condition (18) for the metals for which Balluffi [11] has quoted E_V^M, E_{2V}^B, and E_{2V}^M and for Cu

metal	$T/T_M = 0.35$	$T/T_M = 0.42$
Au ($E_{2V}^B = 0.25$ eV)	0.7	16
Au ($E_{2V}^B = 0.57$ eV)	3×10^{-4}	0.02
Al	0.08	3
Pt	6	95
Cu	17	223

stance R in condition (18) is much bigger than unity, our approach as summarized in Fig. 2 should apply. In Table 4 we have calculated R for the metals for which Balluffi [11] has quoted E_V^M, E_{2V}^M, and E_{2V}^B. In the cases where a range is quoted we have taken the midpoint of the range except for E_{2V}^B in gold where we have done the calculations for the two extreme values of the range quoted. The calculations have been done for $T/T_M = 0.35$ (T_M being the melting point) corresponding to 200 °C for Cu which is the average temperature in the experiments of Kiritani et al. [1] and for $T/T_M = 0.42$ corresponding to 300 °C for copper which is about the average temperature in the experiments of Barlow et al. [9]. We have taken P to be 1.6×10^{-3} s^{-1}, Z_{IV} to be 10, and the pre-exponential factor in the M's to be 10^{13} s^{-1}. The results for copper are also given in Table 4.

Table 4 demonstrates that there is no general answer to the question whether the divacances can be neglected in HVEM loop-growth experiments as we have done in our generalized chart in Fig. 2. If the parameters involved are known with sufficient accuracy, one can in each individual case get a theoretical answer from conditions (17) or (18) (with inverted signs of inequality when used as conditions for the applicability of Fig. 2). Experimentally, the situations where the divacancies can be neglected are characterized by an apparent activation energy for loop growth of $\frac{1}{2}E_V^M$ for loops away from the surface in sufficiently thick foils. Table 4 indicates that the situation where one gets the full effect of divacancies as expressed in (13), characterized by R values much smaller than unity, is the exception rather than the rule.

In the cases where a significant divacancy effect is to be expected ($M_{2V} \gg M_V$, $R \ll 1$), there is a high probability that the divacancies react with monovacancies to form trivacancies. If the trivacancies have migration and binding energies similar to those of divacancies, they will in turn react to form tetravacancies. This chain reaction will continue until the clusters reach a size where they become immobile and hence loose their importance for vacancy transport (which probably happens at the tetravacancy stage). Thus, equation (13) will probably overestimate the effect of divacancies even for $M_{2V} \gg M_V$ and $R \ll 1$, because the population of mobile vacancy clusters will be eroded by the above chain reaction.

7. Conclusions

It is shown theoretically that there is, within the basic framework of the work of Kiritani et al. [1], a relatively wide transition range between the vacancy-dominant and the surface-dominant cases.

It is demonstrated that the transition range corresponds to ordinary experimental situations and that this can account for the observation of apparent activation energies for loop growth smaller than half the vacancy migration energy.

In the "lower" end of the transition range, where point-defect loss to the surface becomes comparable to the recombination loss, the exact agreement between experiment and theory ceases: in the experiments there is a sudden transition to a behaviour typical for the surface-dominant case (temperature-independent loop growth) rather than the continued smooth transition predicted by theory.

The conclusions above are based on calculations in which the effect of divacancies has been neglected. All comparisons with experiments have been made for copper, and the theoretical and experimental evidence is that it is justified to neglect the divacancies in this metal. We suggest general theoretical and experimental criteria to decide whether or not it is justified to neglect the divacancies in a specific experiment with a specific material.

Acknowledgements

The authors wish to thank A. J. E. Foreman and H. v. Harrach for helpful discussions.

References

[1] M. Kiritani, N. Yoshida, H. Takata, and Y. Maehara, J. Phys. Soc. Japan **38**, 1677 (1975).
[2] P. Barlow and T. Leffers, Phil. Mag. **36**, 565 (1977).
[3] R. R. Bourassa and B. Lengeler, J. Phys. F **6**, 1405 (1976).
[4] P. Wright and J. H. Evans, Phil. Mag. **13**, 521 (1966).
[5] G. Antesberger, K. Sonnenberg, and P. Wienhold, J. nuclear Mater. **69/70**, 660 (1978).
[6] A. J. E. Foreman, Radiat. Eff. **14**, 175 (1972).
[7] K. Urban and M. Wilkens, phys. stat. sol. (a) **6**, 173 (1971).
[8] M. Kiritani, in: Fundamental Aspects of Radiation Damage in Metals, Ed. M. T. Robinson and F. W. Young, U.S. ERDA Conf. No. 751006, 1975 (p. 695).
[9] P. Barlow, T. Leffers, and B. N. Singh, Risø-M-2129 (1978).
[10] R. A. Johnson and N. Q. Lam, Phys. Rev. B **13**, 4364 (1976).
[11] R. W. Balluffi, J. nuclear Mater. **69/70**, 240 (1978).
[12] G. Antesberger, K. Sonnenberg, P. Wienhold, R. R. Coltman, C. E. Klabunde, J. M. Williams, and R. L. Chaplin, Fundamental Aspects of Radiation Damage in Metals, Ed. M. T. Robinson and F. W. Young, U.S. ERDA Conf. No. 751006, 1975 (p. 561).
[13] P. Wienhold, K. Sonnenberg, and G. Antesberger, Radiat. Eff. **36**, 235 (1978).
[14] A. C. Damask and G. J. Dienes, Point Defects in Metals, Gordon and Breach, New York 1963 (p. 5).

(Received December 13, 1978)

phys. stat. sol. (a) **54**, 633 (1979)

Subject classification: 17; 22.6

Uraler Polytechnisches Institut, Sverdlovsk (a) und Sektion Physik der Wilhelm-Pieck-Universität Rostock (b)

Untersuchung des Austritts von Exoelektronen aus geladenen Isolatorschichten mit Hilfe des Monte-Carlo-Verfahrens

Von

V. Kortov (a), V. Isakov (a), A. Gaprindoshvily (a), H.-J. Fitting (b), H. Glaefeke (b) und W. Wild (b)

Es wird der Prozeß der thermisch stimulierten Exoelektronenemission (TSEE) aus Isolatorschichten mit Hilfe der Monte-Carlo-Methode auf dem Computer simuliert und mit experimentellen Ergebnissen verglichen. Diskutiert wird der Einfluß innerer elektrischer Felder, der Elektron–Phonon-Wechselwirkung und des Einfangs von Elektronen durch Coulombzentren auf den Transport der thermisch befreiten Elektronen zur Oberfläche sowie der Einfluß der Elektronenaffinität des Festkörpers auf den Austritt der Elektronen ins Vakuum. Berechnete und gemessene Energieverteilungen der Exoelektronen zeigen ausreichende Übereinstimmung.

The process of thermally stimulated exoelectron emission (TSEE) from insulating layers is computer simulated with the aid of the Monte Carlo method and is compared with experimental results. The influence of inner electric fields, of the electron–phonon interaction, and a of the capture of electrons by Coulomb centers on the transport of thermally liberated electrons to the surface as well as the influence of the electron affinity of the solid on the ejection of electrons into vacuum are discussed. Calculated and measured energy distributions of exoelectrons show sufficient agreement.

1. Einleitung

Bei der Analyse von Isolatorschichten mit Hilfe von Elektronen- oder Photonensonden muß mit einer Aufladung des Isolators gerechnet werden, die die Energieverteilung quasifreier Elektronen innerhalb der Schicht und damit der austretenden Elektronen beeinflußt. Die an die Aufladung gekoppelten elektrischen Felder innerhalb und außerhalb der Isolatorschicht können einen Emissionsprozeß begünstigen oder hemmen. Bei Diskussion des Mechanismus der Elektronenemission aus geladenen Isolatorschichten muß der Verlauf der Feldstärke zumindest innerhalb der Isolatorschicht bekannt sein. Es ist mehrfach versucht worden, den Exoemissionsmechanismus unter diesem Aspekt zu deuten [1 bis 3]. Hier soll mit Hilfe des Monte-Carlo-Verfahrens der Emissionsvorgang für den feldbehafteten Fall computer-modelliert werden [4, 5]. Angenommen wird, daß der Effekt der thermische stimulierten Exoelektronen (TSEE) aus vier Teilprozessen besteht: Anregung der TSEE-aktiven Zentren in der emittierenden Schicht durch Beschuß mit mittelschnellen Elektronen (keV-Bereich), Stimulation der Freisetzung der Elektronen durch Zufuhr von Wärmeenergie, Diffusion oder Drift der quasifreien Elektronen unter Einbeziehung der Elektron–Phonon-Wechselwirkung, Elektronenaustritt über das Randpotential ins Vakuum.

Es werden reale Raumladungsverteilungen bzw. die damit verbundenen Feldstärkeprofile, wie sie in Isolatorschichten nach Elektronenbeschuß auftreten, berücksichtigt. Die Raumladungsstruktur ist experimentell mit Sekundärelektronenemission [6] und Kelvinmethode [7] und einer Computersimulation [8] untersucht worden. Die Monte-Carlo-Methode wurde zur Modellierung der Energie- und Winkelverteilung der quasifreien Elektronen in SiO_2- und Al_2O_3-Schichten benutzt und mit experimentellen Ergebnissen der TSEE [9 bis 11] verglichen. Der Einfluß der Größe der Elektronenaffinität auf Emissionsrate und Energieverteilung wird betrachtet.

2. Die Raumladung $\varrho(x)$

Wird ein Isolator mit Elektronen beschossen, so bildet sich in ihm eine Raumladungsstruktur aus, die neben Materialparametern im wesentlichen von der Primärenergie der Elektronen, der Schichtdicke des Isolators, der auf einem leitenden Material aufgebracht ist, und der Absaugspannung für die Sekundärelektronen abhängt. $\varrho(x)$ wurde durch eine Computersimulation der während des Elektronenbeschusses auftretenden, zeitabhängigen Ströme innerhalb und außerhalb des Targets nachgestaltet [8]. Ein Teil der aufgeschossenen Primärelektronen, die in das Target unter Energieverlust eindringen, wird durch Haftstellen eingefangen und bildet den Bereich der negativen Raumladung. Nahe der Oberfläche entsteht durch die äußere Sekundärelektronenemission ein Elektronenmangel, d.h. eine positive Raumladungszone. Zwischen den Bereichen mit entgegengesetzter Ladung existiert ein elektrisches Feld, das die Emission von Elektronen begünstigt [12]. In Fig. 1 ist dieser Prozeß schematisch wiedergegeben. Die Plus-Minus-Struktur der Raumladung kann im allgemeinsten Fall durch die analytische Näherung

$$\varrho(x) = \varrho_0 \left(1 - \frac{x}{x_1}\right) e^{-x/x_1} \tag{1}$$

beschrieben werden. Dabei ist x die Austrittstiefe des Elektrons und x_1 die Stelle des Übergangs der positiven in die negative Raumladungszone. Die entsprechende Feldstärkeverteilung lautet dann

$$F(x) = \frac{1}{\varepsilon_0 \varepsilon_r} \varrho_0 x \, e^{-x/x_1} . \tag{2}$$

x_1 ist eine Funktion der Primärenergie E_p der Elektronen. Diese Abhängigkeit läßt sich für Al_2O_3 und SiO_2 durch die Zahlenwertgleichung

$$x_1 = 38 E_p^{1,3} \tag{3}$$

darstellen. x_1 ergibt sich in nm, wenn E_p in keV eingesetzt wird. Mit $\varrho_0 \approx 1$ As cm^{-3} ergeben sich bei x_1 maximale Feldstärken F im Bereich von 10^5 bis 10^6 V cm^{-1}. Eine Elektronenemission begünstigende Feldverhältnisse existieren nur für den Fall $F > 0$. Unter der zusätzlichen Annahme, daß emittierte Elektronen nur aus der

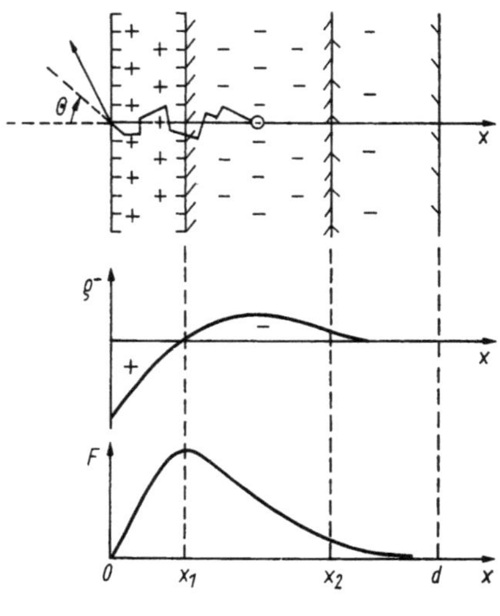

Fig. 1. Schematische Darstellung der Raumladung $\varrho(x)$ nach Elektronenbeschuß eines Isolators

negativen Raumladungszone stammen, ergeben sich die Emissionsbedingungen

$$F > 0\,,$$
$$\varrho(x) < 0\,.$$

Das entspricht in Fig. 1 einem Austrittstiefenbereich, der zwischen x_1 und x_2 liegt.

3. Elektroneneinfang durch Löcher

Getrappte Elektronen, die nach dem Elektronenbeschuß im Bereich $F > 0$ durch thermische Stimulation in Energiebereiche angehoben werden, in denen sie quasi frei beweglich sind, verursachen einen nichtstationären Strom in Richtung Oberfläche. Die Schwächung dieses Elektronenstroms im positiven Raumladungsgebiet kann mit einer einfachen exponentiellen Abhängigkeit beschrieben werden:

$$p_j(l) = \mathrm{e}^{-(\varrho(x)/e_0) s l_j(x)} \tag{4}$$

mit $\varrho(x) > 0$, e_0 als Elementarladung, s als Einfangsquerschnitt der Löcher für quasifreie Elektronen, l_j als mittlerer freier Weglänge zwischen zwei Elektron–Phonon-Wechselwirkungen [4] entsprechend

$$l_j(x) = \sqrt{\frac{2E_j}{m^*}}\, t_j\,. \tag{5}$$

E_j ist hierbei die kinetische Energie des sich bewegenden Elektrons nach j-ter Elektron–Phonon-Wechselwirkung, t_j die entsprechende Flugzeit. Der Einfangquerschnitt s ist T-abhängig und konnte aus [13] entnommen werden. s liegt im Bereich von 10^{-12} bis 10^{-13} cm^2 für SiO_2-Schichten bei Raumtemperaturen und Feldstärken von 5×10^5 bis 3×10^6 V cm^{-1}. Die Wahrscheinlichkeit für Elektronen, während der Bewegung in der positiven Raumladungszone nicht eingefangen zu werden, ist dann

$$p(l) = \mathrm{e}^{-(1/e_0) \sum\limits_j \varrho_j s_j l_j} \tag{6}$$

mit $\varrho_j > 0$.

4. Die Computermodellierung des Emissionsakts

Die Bedingungen für den Austritt eines Elektrons ins Vakuum lauten

$$\left.\begin{aligned} z &> x\,,\\ p(l) &> p_0 \approx \frac{1}{e}\,,\\ E_j &> \frac{\chi}{\cos^2 \Theta_j}\,, \end{aligned}\right\} \tag{7}$$

wobei z die Projektion der Elektronenbahn auf die Oberflächennormale, Θ_j der Winkel zwischen Elektronenbahn und Oberflächennormale bei j-ter Elektron–Phonon-Wechselwirkung und χ die Elektronenaffinität des Festkörpers sind.

Der Algorithmus der Berechnungen der Elektronenbahnen, d. h. der Energie- und Winkelverteilungen, nach dem Monte-Carlo-Verfahren wurde ausführlich in [4] beschrieben. Die neu eingebauten Veränderungen wurden in den Abschnitten 2 und 3 angegeben. Die Computermodellierung der Elektronenbewegung erfolgte mit Hilfe gleichmäßig im Intervall 0 bis 1 verteilter Zufallszahlen (Unterprogramm Randu). Die in Fortran programmierten Rechnungen wurden auf einem Computer vom Typ ES 1022 ausgeführt.

5. Ergebnisse der Computermodellierung

Fig. 2 zeigt das typische Histogramm der Verteilung der Energie emittierter Elektronen aus Al_2O_3 ohne Berücksichtigung des Elektroneneinfangs durch Löcher, aber mit realen Feldstärken in der Isolatorschicht, in der sich die Elektronen zur Oberfläche hin bewegen. Diese Verteilung wurde für ein TSEE Glowkurvenmaximum von Al_2O_3 bei 433 K, einer Energie der anregenden Pimärelektronen $E_p = 1$ keV und einem $\chi = 0{,}1$ eV berechnet. Die erhaltene Energieverteilung stimmt mit experimentellen Angaben [9, 10, 14] befriedigend überein.

Berechnungen der Energieverteilung emittierter Elektronen aus SiO_2 mit und ohne Berücksichtigung des Einfangs von Elektronen durch positive Löcher demonstriert Tabelle 1. Dabei wurde für die Rechnungen ein $\bar{\chi} = 0{,}5$ eV angenommen. Der Unterschied zum Literaturwert von $\chi_{SiO_2} = 0{,}9$ eV soll durch Struktur- und Adsorptionseffekte bedingt sein.

Tabelle 1

Parameter der Energieverteilung von Exoelektronen aus SiO_2

Parameter: $E_p = 1$ keV, $\bar{\chi} = 0{,}5$ eV, $T = 433$ K

	E_1 (eV)	$\bar{E}$ (eV)	E_2 (eV)	ΔE_h (eV)
$F = F(x), p = 1$	0,25	0,79	2,3	0,8
$F = F(x), p \leqq 1$	2,1	2,5	3,7	1,6
experimentell	0,2	0,6	1,8	0,7

E_1 wahrscheinlichste Energie, $\bar{E}$ mittlere Energie, E_2 maximale Energie, ΔE_h Halbwertsbreite der Energieverteilung.

Die Tabelle 1 zeigt, daß die berechneten Werte von E_1, $\bar{E}$, E_2 und ΔE_h bei Berücksichtigung der realen Feldstärke $F = F(x)$ die experimentellen Werte besser annähert als bei F-Verläufen, wie sie in [5] näherungsweise benutzt werden. Der Vergleich der beiden Fälle $p = 1$ (Wahrscheinlichkeit, daß die Elektronen *nicht* durch positive Löcher eingefangen werden) und $p \leqq 1$ mit dem Experiment läßt vermuten, daß Wiedereinfang bei TSEE-Prozessen aus Isolatoren vernachlässigt werden kann. Unter dieser Annahme stimmen berechnete und experimentelle Parameter in zunächst befriedigenden Grenzen überein.

Durch Variation von E_p läßt sich der Bereich $[x_2, x_1]$ verschieben. Dadurch ändern sich die Feldverhältnisse. Bei Erhöhung von E_p wird der Tiefenbereich $[x_2, x_1]$ der negativen Raumladungszone zu größeren Tiefen verschoben. Mit der Annahme, daß die emittierten Elektronen aus diesem Bereich stammen, folgt ein längerer Weg

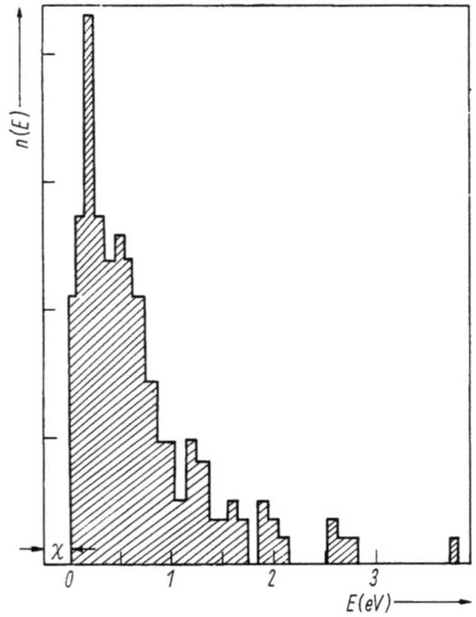

Fig. 2. Berechnete Energieverteilung für Exoelektronen aus Al_2O_3. $\chi = 0{,}1$ eV

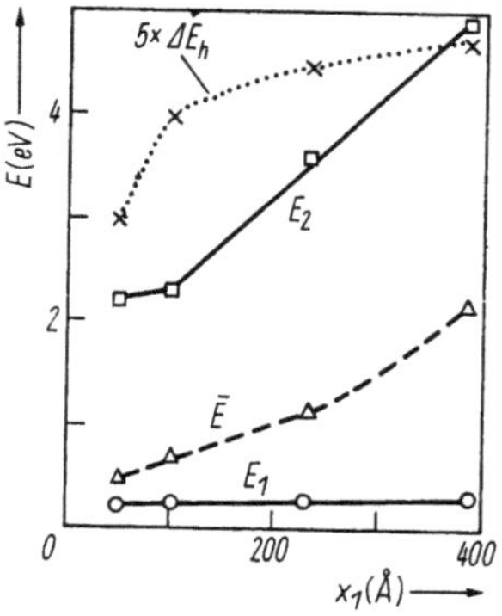

Fig. 3

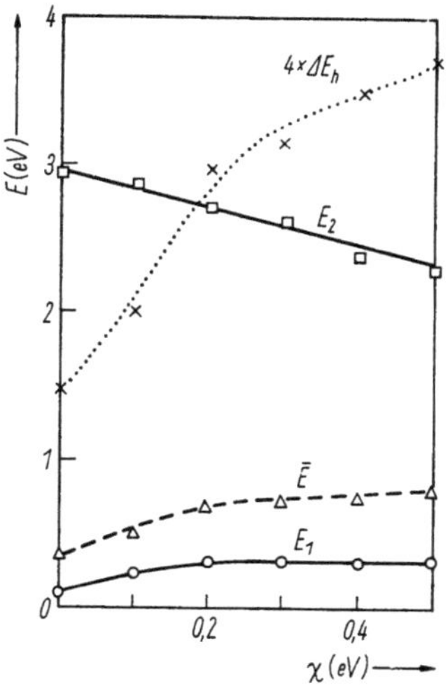

Fig. 4

Fig. 3. Berechnete Energieverteilungsparameter für Exoelektronen aus SiO_2 in Abhängigkeit von der Lage x_1 des Übergangs von der positiven zur negativen Raumladungszone

Fig. 4. Berechnete Energieverteilungsparameter für Exoelektronen aus SiO_2 in Abhängigkeit von der Elektronenaffinität χ

der Elektronen vom Entstehungsort zur Oberfläche und damit eine erhöhte Anzahl j von Elektron–Phonon-Wechselwirkungsprozessen. Das sollte zu einer Verbreiterung der Energieverteilung ΔE_h der emittierten Elektronen und wegen der erhöhten Potentialdifferenz zu einem wachsenden E_2 führen. Fig. 3 zeigt, daß diese Vorstellung mit den Computerergebnissen vereinbar ist.

Den Einfluß von χ auf die Energieverteilung der Exoelektronen aus Al_2O_3 und SiO_2 zeigen die Ergebnisse einer Computermodellierung in Tabelle 2 und Fig. 4.

Tabelle 2

Parameter der Energieverteilung von Exoelektronen aus Al_2O_3 bei variablem χ

χ (eV)	E_1 (eV)	$\bar{E}$ (eV)	E_2 (eV)	ΔE_h (eV)
0,1	0,2	0,63	3,8	0,6
0,2	0,3	0,70	3,7	0,7
0,3	0,3	0,80	3,6	0,75
0,4	0,3	0,85	3,6	0,8
0,5	0,3	0,92	3,5	0,85

Die Analyse dieser Ergebnisse ergibt ein Anwachsen von $\bar{E}$ und ΔE_h bzw. eine Abnahme von E_2 mit wachsendem χ. E_1 steigt nur sehr schwach an. Diese Ergebnisse lassen sich folgendermaßen deuten: Mit zunehmendem χ verarmt die Energieverteilung der emittierten Elektronen gegenüber denen innerhalb der Schicht an langsamen Elektronen. Daher steigt $\bar{E}$ an. Schneidet man von der Energieverteilung der Elektronen innerhalb der Schicht zunehmend niederenergetische Teile ab, so ergeben sich breitere

Relativverteilungen für die emittierten Elektronen. Mit zunehmendem χ weicht die Energieverteilung der emittierten Elektronen immer mehr von der annähernd Maxwellschen Verteilung im Volumen der Schicht ab [15]. Es sei bemerkt, daß der stärkste Einfluß von χ auf die Parameter der Energieverteilung im Bereich $\chi < 0{,}2$ eV beobachtet wird (Fig. 4). Für diesen Bereich ergibt sich eine sehr geringe Halbwertsbreite und ein relativ geringer Wert für $\bar{E}$. Dieses Resultat stimmt gut mit den Daten überein, die in [9] für NaF und BaO gewonnen wurden. Diese Emitter haben ein relativ geringes χ. Die oftmals beobachteten herabgesetzten χ-Werte anderer Emitter wurden auf Adsorption zurückgeführt, insbesondere bei lokalen Erniedrigungen von χ. Die Elektronenaffinität χ beeinflußt lediglich die an der Oberfläche bestehende innere Energie- und Winkelverteilung der Elektronen vor dem Austritt. Diese innere Energieverteilung ist das Produkt der Wechselwirkung der quasifreien Elektronen mit den Phononen und dem elektrischen Feld.

Die Ergebnisse zeigen, daß die Monte-Carlo-Methode gut geeignete ist, den Emissionsprozeß von Elektronen zu simulieren.

Literatur

[1] M. TANAKA, Proc. Phys.-Math. Soc. Japan **22**, 899 (1970).

[2] B. SUJAK, Acta phys. Polon. **20**, 969 (1961).

[3] J. DRENCKHAN, H. GROSS und H. GLAEFEKE, phys. stat. sol. (a) **2**, K51 (1970).

[4] V. S. KORTOV und P. P. ZOLNIKOV, phys. stat. sol. (a) **31**, 331 (1975).

[5] V. S. KORTOV und V. G. ISAKOV, Proc. V. Internat. Symp. Exoelectron Emission and Dosimetry, Zvikov (ČSSR) 1976 (S. 292).

[6] H.-J. FITTING, H. GLAEFEKE und W. WILD, Wiss. Z. WPU Rostock, Math.-Nat. Reihe **25**, 553 (1976).

[7] W. MEHNERT, IX. Arbeitstagung „Physik der Halbleiteroberfläche", Binz 1978 (S. 277).

[8] H.-J. FITTING, H. GLAEFEKE, W. WILD, M. FRANKE und W. MÜLLER, Exper. Tech. Phys. **27**, 13 (1979).

[9] U. BRUNSMANN und A. SCHARMANN, phys. stat. sol. (a) **26**, K123 (1974); **43**, 519 (1977).

[10] W. SCHLENK, phys. stat. sol. (a) **30**, K151 (1975).

[11] H.-J. FITTING, H. GLAEFEKE, W. WILD und J. LANGE, phys. stat. sol. (a) **42**, K75 (1977).

[12] H.-J. FITTING, H. GLAEFEKE und W. WILD, Exper. Tech. Phys. **26**, 253 (1978).

[13] T. H. NING, J. appl. Phys. **47**, 3203 (1976).

[14] L. I. SAMUELSSON, phys. stat. sol. (a) **32**, K155 (1975).

[15] H.-J. FITTING, H. GLAEFEKE und W. WILD, Surface Sci. **75**, 267 (1978).

(Received April 27, 1979)

phys. stat. sol. (a) 54, 639 (1979)

Subject classification: 14; 22.6

Materials Science Division, School of Engineering and Applied Sciences, University of Sussex, Falmer

Bulk and Grain Boundary Ionic Conductivity in Polycrystalline β″-Alumina

By

E. LILLEY and J. E. STRUTT[1])

Measurements of conductance and capacitance are made on polycrystalline β″-alumina over a range of frequencies from 10 to 10^7 Hz and temperatures from −130 to 400 °C. The results are analyzed using complex impedance plots which exhibited a bulk and grain boundary component. Above 220 °C the resistance of the grains exceeded that of the grain boundaries. The activation energy for bulk conductivity is found to be 0.18 eV. The low-temperature results indicated that the grain boundary impedance is being short-circuited either by easy paths through the grain boundaries and/or by some very low impedance boundaries leading to conduction, through preferred grains and grain boundaries.

On a éffectué des mesures de conductance et capacitance dans l'alumine β″ polycristalline pour une gamme de fréquences de 10 à 10^7 Hz et pour des températures comprises entre −130 et 400 °C. Les résultats ont été analysés en utilisant les courbes d'impédance dans le plan complexe distinguant ainsi la composante transgranulaire et la composante intergranulaire. Au dessus de 220 °C, la résistance des grains dépasse celle des joints de grain. L'énergie d'activation pour la conductivité transgranulaire est de 0,18 eV. Les résultats à basses températures indiquent que l'impédance de joint de grain est court circuitée soit par des raccourcis transgranulaires, et/ou par des joints de très faibles impédances, menant ainsi à une conduction à travers des grains et des joints de grains préférentiels.

1. Introduction

During the last decade considerable research has been carried out on β- and β″-alumina. These two materials are important fast ion conductors which are being used as solid electrolytes in the sodium sulphur battery. β″-alumina has a structure consisting of three spinel-like blocks separated by planes containing the highly conducting Na^+ ions [1, 2]. It is stabilised by additions of Li and Mg and has a composition approximately corresponding to $Na_20.5Al_2O_3$. In contrast β-alumina has a two-block structure with a lower Na^+ content [3].

Several ionic conductivity studies have been performed on both β- and β″-alumina. The stability of β-alumina, right up to its melting point, has allowed the growth of single crystals by Union Carbide [4], Bell Laboratories, and others. Large single crystal grains have been extracted from Monofrax bricks and used for various studies. β″ on the other hand is much less stable so that crystals cannot be simply solidified from the melt. Nevertheless, some small crystals have been grown at the Ford Motor Company and at the General Electric Company [5], and used for X-ray and conductivity studies.

Various approaches to the conductivity measurements have been made, including two- and four-probe techniques with reversible or blocking electrodes [6 to 15]. For example, Jones and Miles [8] studied polycrystalline β-alumina tubes, directly using

[1]) Now at the Cranfield Institute of Technology, Cranfield, Bedfordshire, Great Britain.

the four-probe technique. Two-electrode measurements have also been attempted using molten sodium or sodium salts as reversible electrodes, but these do restrict the temperature range that can be covered. Nevertheless, they are useful at 300 to 350 °C, the operating temperature of the sodium–sulphur battery. A rather novel solid state reversible electrode system was devised by Whittingham and Huggins [6]. They used solid tungsten bronze electrodes and showed the system to be reversible to ac current at 10 kHz. Similarly, Ormrod and Kirk [15, 16] used reversible electrodes to study polycrystalline samples of β''-alumina.

An alternative approach to using reversible electrodes is to use completely blocking electrodes, which under ideal circumstances behave as simple capacitors. The properties of the electrolyte are then obtained from the conductances and capacitances measured over a wide frequency range. This method was pioneered by Baurle [17] for solid electrolytes, although it has been widely used for liquid electrolytes. Under favourable circumstances it is possible, by plotting the data in the complex impedance or admittance plane, to determine an equivalent circuit corresponding to the system being measured. Armstrong and co-workers [7] have applied this approach to a number of polycrystalline β-alumina specimens.

The early work at Ford Motor Company (see [14]) indicated that polycrystalline β''-alumina had a resistivity considerably higher than that of single crystals, and that this was the result of a high grain boundary resistance. (The same effect was found for β-alumina.) The difficulty with dc measurements on polycrystalline material is that bulk and grain resistances are lumped together and cannot easily be separated. Moreover, having different activation energies, such measurements will show a change of slope from bulk conductivity at high temperature to grain boundary conductivity at low temperature. This effect is shown in the work of Gibson [18]. On the other hand, by ac methods it is possible to resolve both bulk and grain coundary components with the aid of a complex plane analysis [7]. Recently, Powers and Mitoff [12] measured the impedance of polycrystalline β-alumina as a function of frequency to determine bulk and grain boundary components. Their measurements, made over the temperature range −50 to +353 °C and from 10 Hz to 1 MHz, led to activation energies of 0.19 eV for the bulk and 0.29 eV for the grain boundary process. Important points to note about their work are, firstly, that at room temperature and below, the bulk conductivity is an order of magnitude higher than the grain boundary conductivity. Secondly, with increasing temperature the grain boundary and bulk conductivity converge; however, only at 1000 °C is the bulk conductivity lower than that of the grain boundary when extrapolated. From the Powers and Mitoff study it would appear, therefore, that grain boundary impedances can be highly significant at 350 °C, the temperature of operation of the sodium–sulphur battery. Consequently the microstructure and its relationship to the preparation processes will play a decisive role in determining the internal resistance of the battery.

In general, the ionic conductivity studies of β-alumina appear to be more reliable and in better agreement than those of β''-alumina. For example, the measurements of Whittingham and Huggins [6] on Monofrax single crystals of β-alumina were carried out between −150 and +800 °C giving a linear Arrhenius plot over six decades of conductivity. The activation energy was 0.165 eV, in good agreement with diffusion measurements [14]. A similar activation energy was determined by Fielder et al. [9] studying Union Carbide single crystals. Somewhat lower activation energies were found by Kim [10] and Hooper [11] also studying Union Carbide crystals. These latter studies were all performed over a more limited temperature range, above room temperature.

The discrepancies in the studies of β''-alumina are portrayed in a table in the paper of Ormrod and Kirk [5], in which the activation energies from different investigations range from 0.1 to 0.303 eV. In their own study activation energies of (0.226 ± 0.002) eV and (0.247 ± 0.002) eV were found for two specimens of different compositions. No attempt was made to separate the bulk and grain boundary conductivities. However, the authors concluded that their activation energies were representative of conductivity within the β″-grains. A subsequent dc study by Virkar et al. [13] of polycrystalline β″ showed that samples with smaller grain sizes exhibited lower conductivities and higher activation energies. In fact, the Arrhenius plots were slightly curved as might be expected for two conductivity mechanisms, and by subtraction the bulk and grain boundary contributions were determined.

In some of these earlier studies of poylcrystalline specimens it seems likely that some of the variations in results arise from a mixture of grain boundary and bulk impedances that have not been separated. There may also be variations due to different compositions, but at the present time these are not understood clearly. The following study seeks to determine both the bulk and grain boundary contributions to the ionic conductivity, making use of complex plane plots. By taking measurements over a wide range of temperatures we expect to obtain more accurate activation energies and to understand better the role of grain boundaries.

2. Experimental

The polycrystalline β″-alumina used in this study was provided by British Rail (Derby). We were informed that it contained 98% β″, with the remainder β. All of our specimens came from the same sample. The results for only one specimen will be reported in detail in this paper. The specimen had an $l/A = 6.61\ \text{cm}^{-1}$ (length 0.925 cm, area $0.350 \times 0.40\ \text{cm}^2$).

In order to prepare the β″-alumina specimens for the ionic conductivity measurements, the surfaces to which the electrodes would be attached had first to be polished. This involved grinding with carborundum grit, followed by polishing with 0.1 μm ceria rouge with water as the lubricant. The polished samples were ultrasonically cleaned in methyl alcohol, dried, and then heated to 700 °C in a vacuum of 10^{-5} Torr and held for 12 h. After slow cooling, the specimens were quickly mounted in an evaporation unit and either Al or Au deposited onto the two polished surfaces. The specimens were then transferred to the ionic conductivity cell or stored in a dessicator. The ionic conductivity cell, recently described by Strutt et al. [19], consisted of a pair of gold-plated electrodes which held the β″-alumina specimen under the pressure of a spring. The specimen could be held at temperatures between −150 and +400 °C under a vacuum of 10^{-6} Torr. The procedure we adopted was to hold the specimen between 300 and 400 °C for one or two days before commencing measurements. We have found that the presence of moisture can significantly affect the impedance measurements, and the second in situ heat treatment was necessary to remove the effect of any extra moisture picked up by our specimens after the first heat treatment. Accurate temperature control for the low-temperature measurements was achieved by a digital setting 090 Eurotherm controller. The electrical measurements were performed on a General Radio 1616 capacitance system for frequencies of 10 to 10^5 Hz. For conductivities which exceeded the range of this bridge, its generator and detecting system were used in conjunction with a Wayne Kerr B 221 Universal bridge. A Wayne Kerr 602 bridge was used for measurements in the frequency range 10^5 to 10^7 Hz.

In order to examine the microstructure, some of the polished specimens were etched by heating in air for 15 min at 1500 °C. These were examined and photographed in an optical microscope.

3. Equivalent Circuits

The interpretation of frequency-dependent conductances and capacitances can be greatly facilitated by the combined use of complex plane diagrams and equivalent circuits. For the case of polycrystalline β''-alumina bounded by blocking electrodes a physically reasonable equivalent circuit is that shown in Fig. 1 together with the corresponding complex impedance plot [7, 12, 17]. The impedance of this circuit is given by

$$Z^* = \frac{R_b + j\omega R_b C_b}{1 + \omega^2 R_b^2 C_b^2} + \frac{R_{gb} + j\omega R_{gb} C_{gb}}{1 + \omega^2 R_{gb}^2 C_{gb}^2} + \frac{j}{\omega C_{dl}}, \tag{1}$$

where R_b and C_b are bulk resistances and capcitances, R_{gb} and C_{gb} are grain boundary resistances and capacitances, C_{dl} is the double layer capacitance, ω is the angular frequency, and $j = \sqrt{-1}$. If the time constants (RC) for the first two terms are sufficiently different, then, in a plot of the real versus the imaginary part of the impedance, two clearly defined semicircles and a vertical straight line will result, otherwise the semicircles overlap.

In the experimental measurements an ac bridge is used to measure the real (R_s) and imaginary part (X_s) of the impedance

$$Z^* = R_s + \frac{j}{wC_s}, \tag{2}$$

where R_s is the series resistance and C_s the series capacitance. R_s is then plotted against $1/\omega C_s$.

In most experimental studies perfect simicircles and vertical straight lines are not found. Instead depressed semicircles with their centres below the abscissa and straight lines at angles less than 90° to the abscissa are frequently seen. Nevertheless, from the intercepts of the semicircles and the straight lines it is possible to determine the values of the resistances in the equivalent circuits, and from the maxima of the semicircles to obtain the capacitance values ($\omega_{max} RC = 1$). For the straight lines we can plot X_s versus $1/\omega$ whose slope gives $1/C_{dl}$.

An alternative method of analysis is to plot data in the complex admittance plane. This gives rise to the same kind of diagram as that shown in Fig. 1 for the same equivalent circuit. However, the impedance plot is preferable to the admittance plot for simplicity of interpretation. Both complex impedance and complex admittance plots will give good separation of the semicircles provided that the capacitances are very different even if the resistances are similar. But when the capacitances are similar and resistances are different, the complex modulus plot appears more appropriate [20].

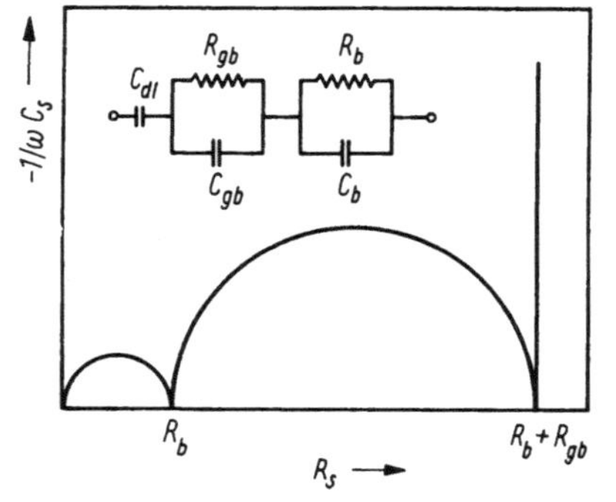

Fig. 1. Equivalent circuit representing the bulk and grain boundary resistances and capacitances, and the interface capacitance. The ideal complex impedance plot for this circuit is also shown $C_{dl} > C_{gb} > C_b$

4. Results

A typical microstructure of one of our specimens revealed that there is a wide distribution of grain sizes. The majority of the grains are about 1 μm in size, but there are many larger grains up to about 20 μm in length. These large grains appear to grow preferentially in planes normal to the C-axis.

Electrical measurements were performed over a wide temperature range, and in all cases the series resistance and capacitance were plotted as a complex impedance plot. Only a few of these plots will be shown. The impedance diagram for β″-alumina at −135 °C is shown in Fig. 2. Two depressed semicircles, one passing through the origin, and a straight line agree in principle with the model in Fig. 1. The rather shallow angle of the straight line is surprising, but it will be seen that this becomes vertical at higher temperatures. (It is interesting to compare this with measurements made with Au electrodes, which always gave angles greater than 70°). At the somewhat higher temperature of −21 °C in Fig. 3 the spur is at 72°, and good extrapolations can be made. As the temperature is raised, the frequency spectrum shifts to the right so that the first semicircle due to the bulk impedance is not visible, falling outside of the limit of the rf bridge. The first intercept with the abscissa gives the bulk resistance, and the length of the depressed semicircle intersecting the abscissa gives the grain boundary resistance. By extrapolation of the central semicircle it is possible to obtain reasonable values of both the bulk and grain boundary resistances.

At the highest temperatures of our study we see an interesting change taking place. Fig. 4 shows impedance data at temperatures ranging from 230 to 379 °C. As the temperature is increased, the grain boundary resistance becomes too small to measure. Consequently at the two highest temperatures shown here only the impedance due to the bulk resistance and the double layer capacitance is observed and the extrapolated lines give only the bulk resistance.

The various values of bulk and grain boundary resistance determined from the above impedance plots can be plotted as log (σT) versus $1/T$, as shown in Fig. 5. In this Arrhenius plot the triangles represent the bulk conductivity and the circles the grain boundary conductivity. The grain boundary plot has a slope corresponding to roughly 0.45 eV at the highest temperatures, but it may well be that this lines continues to curve upwards. Inevitably, it was not possible to resolve the grain boundary resistance at highest temperatures. At the lower temperatures both the bulk and grain boundary plots are parallel, and have the same activation energy of 0.18 eV. The data presented here were found to be reproducible on both heating and cooling of the polycrystalline specimens.

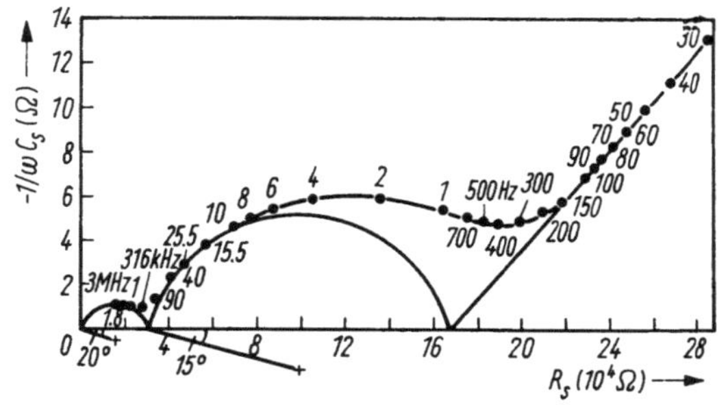

Fig. 2

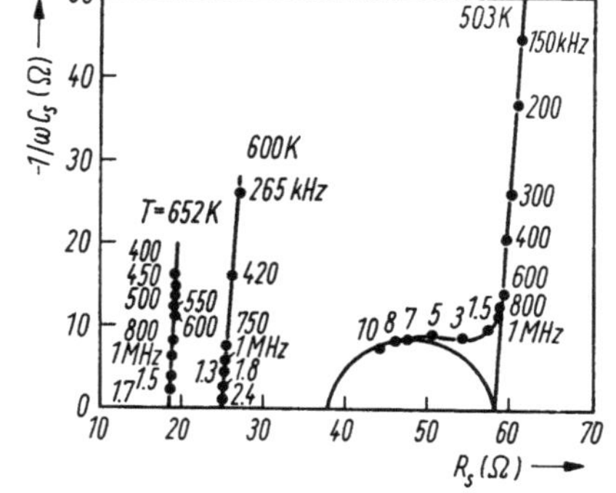

Fig. 3

Fig. 2. Complex impedance plot at −135 °C showing the features idealized in Fig. 1. $T = 138$ K
Fig. 3. Complex impedance plot at −21 °C

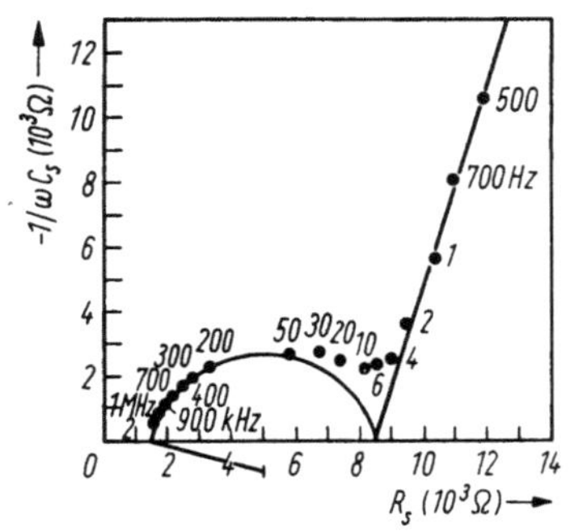

Fig. 4

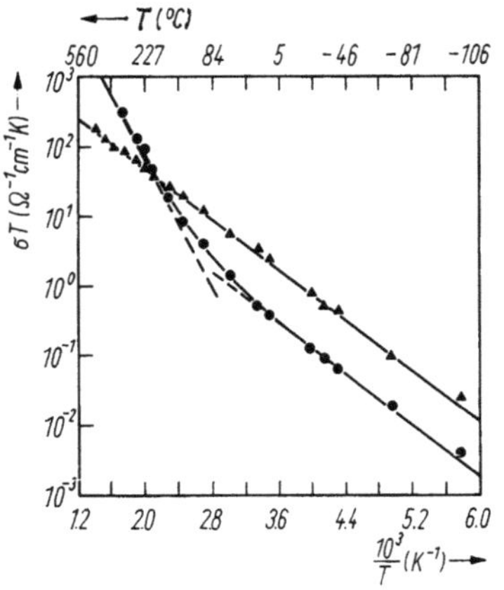

Fig. 5

Fig. 4. Complex impedance plots at 230, 327, and 379 °C

Fig. 5. Arrhenius plot of β″-bulk and the grain boundary associated conductivity. Bulk activation energy is 0.18 eV

In determining the bulk resistance by extrapolating the semicircle at temperatures below about 150 °C and the interface straight line at the highest temperatures, no serious arrors are expected to result. However, in the intermediate temperature range (e.g., 230 °C in Fig. 4) such an extrapolation of the semicircle would be liable to be very inaccurate. To get round this problem we have assumed that the bulk resistance could be taken from the linear Arrhenius plot in Fig. 5 at any particular temperature. Then, but subtracting this deduced bulk resistance value from the measured total resistance the grain boundary resistance could be determined. This procedure was necessary for the data points at 203, 230, 267, and 327 °C.

The bulk, grain boundary, and interface capacitances for the β″-material with Al electrodes are listed below in Table 1 at different temperatures.

Table 1

Values of the double layer C_{dl}, the grain boundary C_{gb}, and the bulk C_b capacitances measured at different temperatures

temp. (K)	C_{dl} (Fcm^{-2})	C_{gb} (Fcm^{-1})	C_b (Fcm^{-1})
138	1.5×10^{-7}	7×10^{-10}	4×10^{-12}
149	1.5×10^{-7}	8×10^{-10}	—
371	1×10^{-7}	8×10^{-10}	—
652	0.8×10^{-7}	—	—

The C_{dl} capacitances were difficult to determine with accuracy because plots of X_s versus $1/\omega$ did not always give good straight lines. The grain boundary capacitances show virtually no variation with temperature. A β″-specimen with Au electrodes had a grain boundary capacitance of 3×10^{-10} F cm^{-1}, and a 80% β″-specimen a value of 5×10^{-10} F cm^{-1}, varying little with temperature.

Earlier in our work another specimen was studied in a similar manner to that just described, but the material was not given a thorough bake-out treatment. The overall results were similar to those in Fig. 5, but there were some significant differences. The apparent bulk activation energy was 0.21 eV, and the low-temperature grain boundary conductivity was about a factor of ten, rather than six, less than the bulk conductivity.

One feature of Fig. 2 to 4 is that the angle of the low-frequency spur decreases away from the vertical as the temperature decreases. Similar measurements using evaporated gold electrodes did not show this trend. They have a fairly constant angle of 78°.

5. Discussion

Let us first consider the complex impedance plots shown in Fig. 2 to 4. Only at the lowest temperature were we able to obtain data for the first semicircle, which we have identified with the bulk impedance. Nevertheless, this and other data do show that there is a semicircle here and that this is followed by a second semicircle and a spur. The central grain boundary semicircle shown here is slightly depressed, a feature found widely in the literature. The reason for the depression is not yet understood, although explanations of 'non ideal' equivalent circuits have been made [7, 21]. There is, however another factor which may explain some of these depressions. In poorly baked-out specimens we have found that the central semicircle is broad, which is usually interpreted to be a large depression. However, following vacuum annealing in the cell, the breadth decreases and a better semicircle is observed, with a more clearly defined minimum between it and the adjacent spur. One way to interpret this effect is in terms of a small additional semicircle, associated with moisture pick-up. If this overlaps the true grain boundary semicircle, the two together would appear as one broadened and depressed semicircle. Large apparent depressions may therefore be a consequence of sample preparation.

It is frequently found experimentally that the angle of the spur to the abscissa deviates from 90°, particularly with polycrystalline specimens. For gold electrodes we found the angle to change only slightly with temperature and become a maximum at low temperatures. However, Al electrodes were used on the β''-sample in the detailed study described here, and this sample showed a reproducible change in the slope of the spur from 88° at 379 °C to 47° at −135 °C. This may be related to differential thermal expansion between the β''-alumina and the aluminium. During polishing some of the grains will be loosened, and if these grains adhere strongly to the aluminium coat, when the crystal temperature is varied, it is possible that some crystals will be dislodged. This will lead to an increase in surface roughness and hence alter the slope of the spur [25]. It is interesting to note in this respect that gold has very poor adhesion compared with aluminium.

The conductance data, extracted from the complex plane diagrams, are represented in an Arrhenius plot shown in Fig. 5. At 300 °C the bulk resistivity is 7 Ω cm. The bulk conductivity is linear over four decades of conductivity $\sigma(T)$ and has an activation energy of 0.18 eV. This value is similar to the activation energy for β-alumina of 0.165 eV [6]. The difference between our value and that of Ormrod and Kirk may arise from grain boundary effects influencing their measurements, or from differences in composition or preparation of the specimens. The linear plot in Fig. 5 is in marked contrast to the Ford β'' single crystal measurements, which showed a strong curvature (cf. [14]).

The grain boundary data in Fig. 5 show a linear region at low temperatures, again with a slope of 0.18 eV, and a curved region at high temperatures. The grain boundary line intersects the bulk line at about 200 °C. At this temperature the bulk and the grain boundaries offer the same resistance to the migration of Na^+ ions. However, at higher temperatures the grain boundary resistance becomes much less than the bulk resistance, due to the differences in activation energy. At the temperature at which the Na–S battery operates (e.g., 300 to 350 °C), the grain boundary resistance is about a third of the bulk resistance in this kind of sample.

6. Grain Boundary Easy Path Model

To explain the two regions of the grain boundary plot, let us first consider the model shown in Fig. 6. In this model we consider that there are basically two paths for a Na^+ ion to pass through a grain boundary. Firstly, they can cross the grain boundary at points which require a higher activation energy than motion in the bulk. Secondly, the Na^+ ions may cross at some other positions, which we will call easy paths, which have the same activation energy as the bulk. Given this model, we can fit our grain boundary data to the two straight lines in Fig. 5 with activation energies of 0.18 and 0.45 eV.

To take these ideas a little further and extend the model of Fig. 6 we should consider the nature of the boundary between two adjacent β″-alumina grains. The layer structure of β″-alumina suggests that most of the grain boundary will consist of impenetrable spinel block material. An idealized boundary is shown in Fig. 7, where the fast planes in a pair of adjacent grains are shown to intersect at points such as C, and there are many such positions for Na^+ ions to cross the boundary. However, in real boundaries local distortions can be expected and, in most cases, the Na^+ will be impeded in crossing the boundary even at these intersection points. The activation energy for motion across the boundary is therefore expected to be high. We imagine that when an electric field is applied to a β″-alumina sample, Na^+ ions will move within the grains and pile up at the grain boundaries. There will, however, be some points at which the ions can pass through the boundary. These points will have a resistance made up of two components: a constrictive resistance and a passage resistance. The constrictive resistance, R_c, arises from the narrowing of field lines and is geometric in origin. The activation energy associated with any constrictive resistance will be equal to that of the bulk material. The grain boundary passage resistance, R_p, on the other hand, is due to a limited cross-sectional area and depth at the point on the grain boundary that a Na^+ ion passes through in going from one grain to an adjacent one. This is illustrated in Fig. 8.

In general, there will be a number of paths, having a range of activation energies. The resistance due to n such passages in parallel is given by

$$R_{gb} = \left(\Sigma \frac{n_i}{R_i}\right)^{-1} = \left[\Sigma \left(\frac{n_i}{R_{ip} + R_{ic}}\right)\right]^{-1}. \tag{3}$$

n_i is the number of connecting paths of type i having passage resistances R_{ip} and constrictive resistances R_{ic}. The constrictive resistances will have the same activation

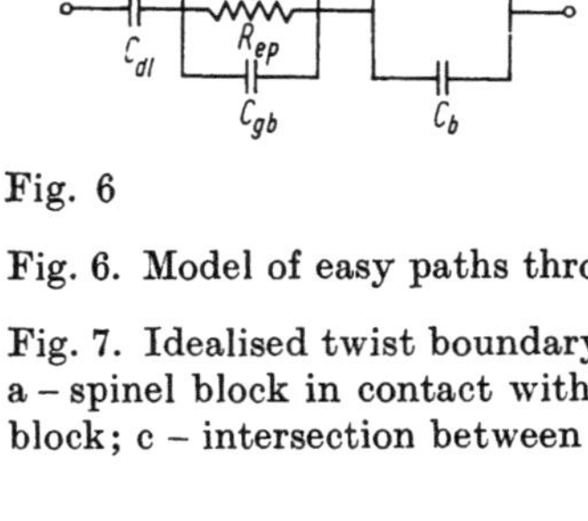

Fig. 6

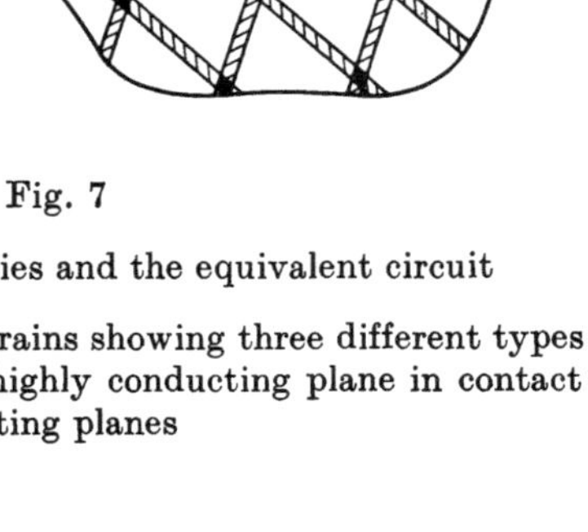

Fig. 7

Fig. 6. Model of easy paths through grain boundaries and the equivalent circuit

Fig. 7. Idealised twist boundary of two adjacent grains showing three different types of contact: a – spinel block in contact with spinel block; b – highly conducting plane in contact with spinel block; c – intersection between two highly conducting planes

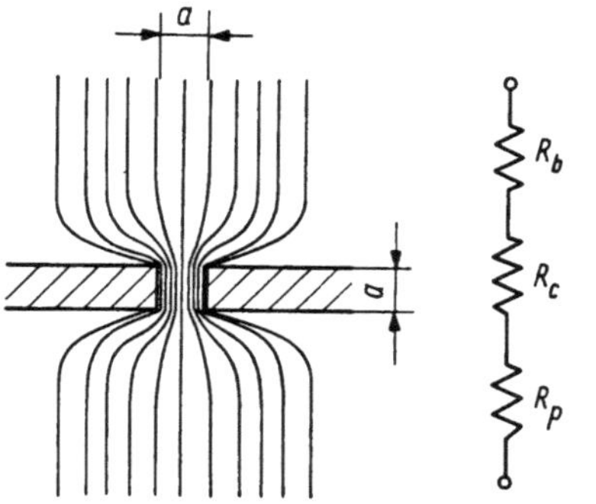

Fig. 8. Schematic representation of the flow lines through a passage across a grain boundary. The bulk resistance is R_b, the passage resistance R_p, and the constrictive resistance R_c

energy as the bulk. The passage resistances will be associated with activation energies such that $E_{1p}(= E_b) < E_{2p} < E_{3p}$. An equivalent circuit for the resistance of one plane within one grain and n_i connections to the adjacent grain is shown in Fig. 9.

In order to quantify the model further, we shall consider a single conducting plane with the geometry of a disk with a single entrance and exit point for the ions. It will also be convenient to define the resistivity of the plane and the connecting paths in terms of two-dimensional resistances. Consider first the constrictive resistance. The most useful analogue to us in electromagnetic theory is that of a conducting disk with two cylindrical connections jointed at right angles to the disk with their centres lying on the circumference [22, 23]. The fast conducting planes correspond to the disk, and the entry and exit points to the cylindrical connections. The constrictive resistance of the disk R_c is then related to the resistivity of the fast conducting plane ϱ_b (i.e. the bulk resistivity) by the expression

$$\frac{R_c}{\varrho_b} = \frac{2}{\pi} \ln \left(\frac{d}{e}\right), \tag{4}$$

where d is the distance between the connections of radius e as shown in Fig. 10. For our samples d can be taken approximately as the grain size (1 μm) and e as the radius of a Na^+ ion (1 Å). Substituting these numbers into (4) gives a value of 6 for R_c/ϱ_p.

In order to keep the terminology constant, it is also convenient to consider the grain boundary passage as a square for the calculation of the passage resistance (e.g.,

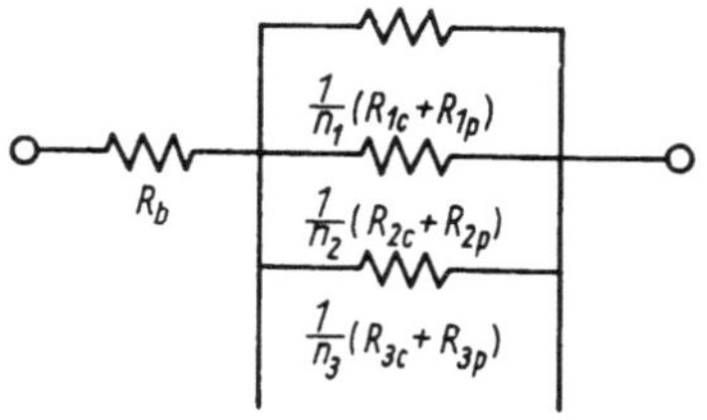

Fig. 9

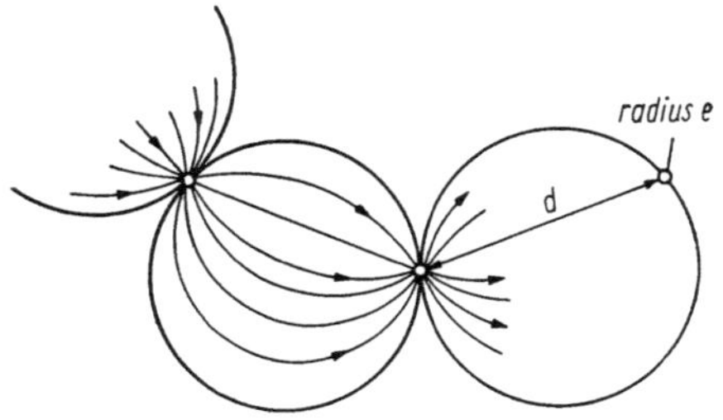

Fig. 10

Fig. 9. Circuit representing the parallel paths through a grain boundary in series with the bulk resistance (R_b). R_{ic} and R_{ip} represent the constrictive and passage resistances, respectively

Fig. 10. Electrical analogue for the restricted passage of Na^+ ions into and out of a single fast conducting plane in a grain of β-alumina. This is represented by a conducting disk, with two perpendicular cylindrical connections of radius e, with centres on the circumference at a distance d

$a \cdot a$ in Fig. 8). On this definition $\varrho_p = R_p(a/a)$ giving $R_p = \varrho_p$. In the special case of an easy path (R_{1p}), where there is no extra grain boundary activation energy above that for the bulk, $R_p = \varrho_b$. The grain boundary passage may possibly be longer than it is wide, which would consequently increase the value of R_p. However, for the case of an easy path in which the activation Energy $E_{1p} = E_b$, it is likely that no more than two jumps will take a Na^+ across the boundary, so the estimate of $R_p = \varrho_b$ will not be greatly in error.

When passage resistance and constrictive resistance are taken together, the total grain boundary resistance for easy paths is given by

$$R_{gb} = R_{1c} + R_{1p} \approx 7\varrho_b \,. \tag{5}$$

How does this compare with the measured values of R_{gb} and R_b? Experimentally it is found that $R_{gb}/R_b = R_{gb}/\varrho_b \approx 6$ at low temperatures. It is also found that the activation energy associated with the grain boundary at low temperatures is the same as that of the bulk activation energy. The agreement indicates that on average there is one exit and entrance easy path for each fast plane in a grain. In addition to the easy path resistance there will be other terms in equation (3) (other grain boundary passages) to consider. However, since these E_{ip} values are associated with higher activation energy passages, they should not make a significant contribution to the total conductivity at low temperatures.

It is interesting to consider the effect of increasing the number of easy path connection points across grain boundaries. In this case the total resistance can be taken as n constrictive and passage resistances ($R_p + R_c$) in parallel. Provided the constrictional flow lines do not overlap, the resistance is then given by

$$R_{gb} = \frac{7\varrho_b}{n}. \tag{6}$$

For a high density of easy paths the constrictional resistance becomes negligible and

$$R_{gb} = \frac{R_p}{n} = \frac{\varrho_b}{n}. \tag{7}$$

Clearly the ratio of the bulk and grain boundary resistance is very sensitive to the number of paths (n), but much less sensitive to the grain size (equal to d in (4)) and the sodium ion radius because they appear in the logarithmic term.

Let us now consider the effect of increasing the temperature on the mechanism of grain boundary conduction. The increased thermal energy will allow the migration of Na^+ ions across the grain boundary through passages possessing higher activation energies than the easy paths. If the number of these paths is significantly greater than those of the easy paths, then they will at some temperature come to transmit most of the Na^+ ions across the boundary. We would expect a distribution both in the number n_i and the activation energy E_{ip} of these passages. When the passage resistance is high, the associated constrictive resistance becomes less important. The increasing slope of the grain boundary conductivity with increasing temperature is thus accounted for.

In the discussion of grain boundary resistance at both low and high temperatures we have assumed that the treatment for a single fast conducting plane and a grain boundary can be generalised to the case of a polycrystalline sample. On this basis we can write an expression for the resistance of a unit length cube polycrystalline specimen:

$$R_s = \frac{\varrho_b}{2} + \frac{1}{G}\left[\sum \frac{n_i}{R_{ip} + R_{ic}}\right]^{-1} \tag{8}$$

where n_i is the number of the i-paths per unit length of fast conducting plane boundary, G the grain size, R_{ip} the resistance of the i-th passage, R_{ic} the associated constrictive resistance, and ϱ_b the bulk resistivity (which is divided by two to take into account tortuosity). The sample grain boundary resistance is expected to vary as the reciprocal of grain size at high temperatures where the constrictive resistances became negligible.

In the model discussed above we have mentioned that there will be variations in the number of entry and exit points for each fast conducting plane and variations in the grain boundary passage resistances. These will give rise to a distribution of grain boundary resistance from grain to grain implying a distribution of relaxation times for the grain boundary impedance. A distribution of relaxation times has been postulated to lead to a depression of the bulk impedance plane semicircle [7], and it is possible that the depressions we have noted in our grain boundary semicircles are caused by the variations in grain boundary resistance. It should be noted in the above discussion that simple macroscopic electrical models have been applied in the limit of atomic dimensions. The values calculated should therefore be regarded as semi-quantitative.

An alternative approach to the interpretation of grain boundary conductivity has been taken by De Jonghe [24]. On the basis of his electron microscopy studies he proposed that there are inhomogeneous conduction paths across the polycrystalline samples due to some grain boundaries being highly conducting with respect to others, so that a parallel circuit must be used to represent this situation. The simplest example is to consider a grain boundary short-circuited by grains with either no boundaries, or boundaries with zero impedance. This gives rise to the same kind of complex impedance plot as shown earlier in Fig. 1 and found experimentally. The intersection of the first semicircle with the abscissa still corresponds to the bulk resistance. However, the intersection of the second semicircle with the abscissa is now proportional to R_b and does not include R_{gb}. The maximum of the second semicircle is such that $1 = \omega_{max} R_b C_{gb}$, so that this does not change the interpretation of the grain boundary capacitance. With this model the intersection of the second semicircle will vary with temperature as a function of the bulk activation energy, independent of the resistance of blocking grain boundaries. At low temperatures this is in fact what we observe with the bulk and "grain boundary" terms having the same activation energy.

Just how much the Na^+ ions are flowing along microscopic or macroscopic easy paths is an interesting question. The two approaches can be combined so that we can think of inhomogeneous paths through the specimens arising from a distribution of easy paths across grain boundaries so that some boundaries are more conducting than others.

For the Na–S battery it is the grain boundary impedance at the operating temperature of β-alumina which is important. In the case of the British Rail material studied here, the grain boundary impedance is less than that of the bulk at 350 °C, but this is not true for the Powers and Mitoff material [12]. A finer grain size which strengthens the β-alumina could consequently raise the grain boundary impedance of a specimen. Not only can differences in grain boundary conductivities arise from grain size, but they can also be determined by the number and type of passages across the grain boundaries. An understanding of how impurity ions can block up passages and how this can be avoided is therefore of considerable technological importance.

Acknowledgements

We would like to thank Dr. B. Redfern of British Rail, Derby, for providing the specimens investigated in this study, Mr. J. Pal and Mr. R. Walters for taking some of the electrical measurements, and the Science Research Council for providing J.E.S. with a fellowship.

References

[1] J. H. KENNEDY, in: Topics in Applied Physics, Vol. XXI, Ed. S. GELLER, Springer-Verlag, 1978 (p. 81).
[2] C. R. PETERS, M. BETTMAN, J. W. MOORE, and M. D. GLICK, Acta cryst. **B27**, 1826 (1970).
[3] M. BETTMAN and C. R. PETERS, J. phys. Chem. **73**, 1774 (1969).
[4] L. R. ROTHROCK, J. Crystal Growth **39**, 180 (1977).
[5] G. C. FARRINGTON and J. L. BRIANT, Mater. Res. Bull. **13**, 763 (1978).
[6] M. S. WHITTINGHAM and R. A. HUGGINS, J. Chem. Phys. **54**, 414 (1971).
[7] R. D. ARMSTRONG, T. DICKINSON, and J. TURNER, J. electroanal. Chem. **118**, 1135 (1971).
[8] I. W. JONES and L. J. MILES, Proc. Brit. Ceram. Soc. **19**, 161 (1970).
[9] W. L. FIELDER, H. E. KAUTZ, J. S. FORDYCE, and J. SINGER, J. Electrochem. Soc. **122**, 528 (1975).
[10] K. K. KIM, private communication 1978.
[11] A. HOOPER, J. Phys. D **10**, 1487 (1977).
[12] R. W. POWERS and S. P. MITOFF, J. Electrochem. Soc. **122**, 226 (1975).
[13] A. V. VIRKAR, G. R. MILLER, and R. S. GORDON, J. Amer. Ceram. Soc. **61**, 250 (1978).
[14] J. T. KUMMER, Progr. Solid State Chem. **7**, 41 (1973).
[15] S. E. ORMROD and D. L. KIRK, J. Phys. D **10**, 1769 (1977).
[16] S. E. ORMROD and D. L. KIRK, J. Phys. D **10**, 1497 (1977).
[17] J. E. BAUERLE, J. Phys. Chem. Solids **30**, 2657 (1969).
[18] A. GIBSON, in: Power Sources, Ed. D. H. COLLINS, Vol. X, Oriel Press, Newcastle/Tyne 1977 (p. 10).
[19] J. E. STRUTT, M. W. WEIGHTMAN, and E. LILLEY, J. Phys. E **9**, 683 (1976).
[20] I. M. HODGE, M. D. INGRAM, and A. R. WEST, J. electroanal. Chem. **74**, 125 (1976).
[21] A. K. JONSCHER, J. Mater. Sci. **13**, 553 (1978).
[22] R. HOLM, Electrical Contacts, Springer-Verlag, 1967.
[23] J. H. JEANS, Mathematical Theory of Electricity and Magnetism, Cambridge University Press, 1933.
[24] L. C. DE JONGHE, J. Mater. Sci. **14**, 33 (1979).
[25] R. D. ARMSTRONG and R. A. BURNHAM, J. electroanal. Chem. **72**, 257 (1976).

(Received May 22, 1979)

phys. stat. sol. (a) 54, 651 (1979)

Subject classification: 3; 9; 21; 21.1

Institute of Physical Chemistry, Academy of Sciences of the USSR, Moscow

Fractionation of Stable Isotopes and Impurities during Zone Recrystallization of Metals and Preparation of Single Crystals Rich in Individual Isotopes

By

O. A. Troitskii, V. I. Spitsyn, M. M. Moiseenko, and A. D. Tsiganov

It is shown that a small number of zone melting (recrystallization) passes results in a periodical distribution of individual isotopes along the material. The effect is observed with the Zn^{64} and Sn^{119} isotopes in zinc and tin, respectively. An increase of the number of passes to about ten leads to some preferential final isotope distribution. In the case of Sn^{119}, the tail portion of the material becomes enriched (the portion to which the zone movement is directed), provided that the recrystallization rate is high enough (3 mm/min), or the head portion is enriched if the rate is slow (about 0.5 mm/min). In the presence of bismuth impurities, the isotope Sn^{119} is in all cases displaced towards the head portion of the material, which suggests a complex nature of the behaviour of individual isotopes during zone melting and an effect of impurities on this process. The Zn^{64} and Cd^{119} isotopes may be characterized, in accordance with their respective behavior in the course of zone melting, by some effective distribution coefficients which are less than unity ($K < 1$) in the former case, and greater than unity ($K > 1$) in the latter.

Es wird gezeigt, daß eine geringe Anzahl von Zonenschmelz(Rekristallisations)schritten eine periodische Verteilung individueller Isotope längs des Materials ergibt. Der Effekt wird mit den Zn^{64}- und Sn^{119}-Isotopen in Zink- bzw. Zinn beobachtet. Eine Erhöhung der Zahl der Schritte auf etwa zehn führt zur bevorzugten Endverteilung der Isotope. Im Fall von Sn^{119} wird der Schwanzbereich des Materials angereichert (das Ende, zu dem die Ziehbewegung gerichtet ist), vorausgesetzt, daß die Rekristallisationsrate hoch genug (3 mm/min) ist, oder der Kopfbereich ist angereichert, wenn die Rate klein ist (etwa 0,5 mm/min). Bei Anwesenheit von Wismutstörstellen wird das Isotop Sn^{119} in allen Fällen zum Kopfteil des Materials verlagert, was auf eine komplexe Natur des Verhaltens der individuellen Isotope während des Zonenschmelzens und auf einen Einfluß von Störstellen auf diesen Prozeß hinweist. Die Zn^{64}- und Cd^{119}-Isotope können in Übereinstimmung mit ihrem entsprechenden Verhalten während des Zonenschmelzens durch einen effektiven Verteilungskoeffiizienten, der kleiner als eins ($K < 1$) im ersteren Fall und größer als eins ($K > 1$) im letzteren Fall ist, charakterisiert werden.

1. Introduction

The isotopic effect due to electric current has first been discovered in mercury [1] and later it has been investigated in cadmium [2], lithium [3], indium [4], and other metals. The necessary time amounted to tens of hours.

In [5], zone metling (recrystallization) was for the first time used for isotopic enrichment of zinc and cadmium. The recrystallization method was also used in [6] for the separation of mixed normal and heavy water. The heavy isotope is predominantly concentrated in the solid phase, but the separation coefficient is small (about 1.021 for water and 1.0043 for zinc). The process rate is small (0.5 to 3.0 mm/min) and therefore the time necessary for the multi-pass zone melting is large.

For this reason, in [7 to 9] a combination of zone melting and electrotransport was employed which produced about the same magnitude of effect as in [1 to 4] but in much shorter times (of the order of a few hours).

Further intensification of the process was achieved in [10] by using an accelerated electron beam which has a twofold function: 1) creating a molten zone and 2) producing activated atoms in the molten metal. The electrotransport was effected as in earlier works [7 to 9].

The present paper continues the effort in studying the isotopic enrichment of metals during zone melting. The objects of the present investigation were tin (99.9995% Sn) and zinc (99.998% Zn). Since zone melting usually results in the formation of single crystals, the method in question is essentially a method of obtaining metal crystals with modified isotopic composition.

2. Experimental

The test material in the shape of 350 mm long and 1 mm diameter wire was subjected to zone melting in an annular furnace. The molten zone which had a thickness of 8 to 10 and 25 to 30 mm, was moved at a rate of 0.5 to 3.0 mm/min. The zone was subjected to 1 to 10 treatments. The crystals produced in this manner were then cut into 15 mm long sections which were subjected to neutron activation analysis. They were additinally analyzed by Mössbauer spectroscopy.

In case of zinc, the activation analysis was used in order to establish the content in the Zn^{65} isotope which was formed from Zn^{64} through the (n, γ)-reaction. The samples were irradiated in a nuclear reactor for a period of 1 h by a flux of $1.2 \times 10^{13} n/cm^2$ s. In natural zinc the Zn^{64} isotope is present in an amount of about 48.89%. The measurements were effected by means of a multichannel pulse-height analyzer (model LP 4840 of NOKIA) and repeated five times for each sample. The arithmetical mean value was assumed to be the real number of γ-quantum pulses from the radioactive samples. The measurement error was determined using the Student's distribution with confidence probability of 0.95. The measurement error thus determined in no case exceeded 1%.

In the case of tin, Mössbauer spectroscopy was used to determine the content of Sn^{119}. The natural content of this isotope in tin is 8.58%. The Mössbauer effect employed in this case for measuring the content of Sn^{119} allows a very high resolution owing to the very sharp resonance tuning (better than 10^{-9}). After the zone melting, the samples cut from long wires were processed into a 80 to 100 μm thick foil. This was done by a laboratory rolling mill. Disks of specific area and weight were cut from this foil. The γ-quantum spectral absorption line amplitudes were than compared between a standard sample and disks prepared from the treated and starting metallic material. The standard (reference) sample was SnO_2 dust placed into a disk-shaped cell.

If A is used as a symbol for spectral line amplitude of the reference sample having weight m, and a_x as that of the working sample (subjected or not to zone melting) having weight m_x, the concentration of the Sn^{119} isotope may be found by the formulas

$$C_0 = A/m \quad \text{in the reference sample ,}$$

$$c_x = a_x/m_x \quad \text{in the working samples .}$$

The c_x/c_0 ratio is characteristic of the relative variation of the Sn^{119} concentration. In the starting tin material it varied by about 3% as shown by the horizontal hatched area in Fig. 3.

The nuclear absorption in a thin layer with resonant nuclear density n is given by the following coefficient:

$$\mu(E) = \sigma(E)\, f'n = \frac{\sigma_0 f'n}{1 + \varepsilon^2}$$

which is resonance-dependent on energy. Here, $\sigma(E)$ is the phononless process cross-section, f_2 the recoilless absorption probability, $\varepsilon = (E - E_0)/\frac{1}{2}\,\Gamma$ the energy in $\Gamma/2$

units counted from the E_0 line centre. If the number of nuclei of the element in question per unit volume of the absorbant is denoted as n_Z, then $n = n_Z b$, where b is the resonant isotope content. Hence, the ratio of the resonant absorption coefficients per weight unit will be equal to the ratio of the resonant isotope concentrations in identical test matrices.

The experimental error was determined by triple measurements on a series of six samples and was found to be within $\pm 0.8\%$. The reproducibility of measurements was within $\pm 0.5\%$.

3. Results

Fig. 1 shows the distribution of Zn^{64} along the wires after five treatment cycles at $v_Z = 3.0$ mm/min and zone length $l = 25$ to 30 mm. It is seen that the isotope distribution is periodical. The isotope content begins to undergo wider fluctuations towards the end portion of the material.

The experimental data obtained for Sn^{119} are presented in Fig. 2. On the basis of these results the following conclusions can be made:

1. The Sn^{119} isotope distribution resulting from a one-pass zone melting may be referred to as periodical, as in the case of zinc. No such oscillations were observed in the initial samples where Sn^{119} content was on the natural level throughout the wire length.

2. Multiple zone melting levels down the fluctuations. For example, after seven passes the Sn^{119} isotope concentration increases in the second half of the wires but decreases in the first half.

The effect of bismuth impurity in zinc was also investigated. Alloys with different bismuth content were used (10^{-3}, 10^{-2}, 10^{-1}, 1, 3, and 10% Bi). Zone melting was carried out at $v_Z = 3.0$ mm/min and zone width $l = 8$ to 10 mm. Five passes were used each time. The head and tail portions of the wires were subsequently subjected to a chemical analysis.

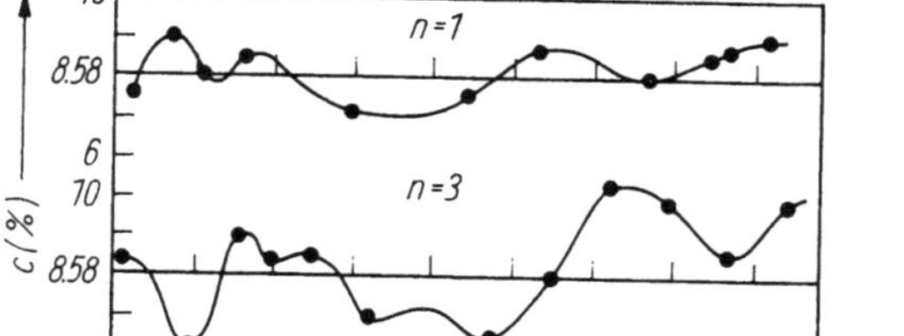

Fig. 1. Distribution of Zn^{46} along the zinc wire after five zone passes at $v_Z = 3.0$ mm/min and zone length $l = 25$ to 30 mm

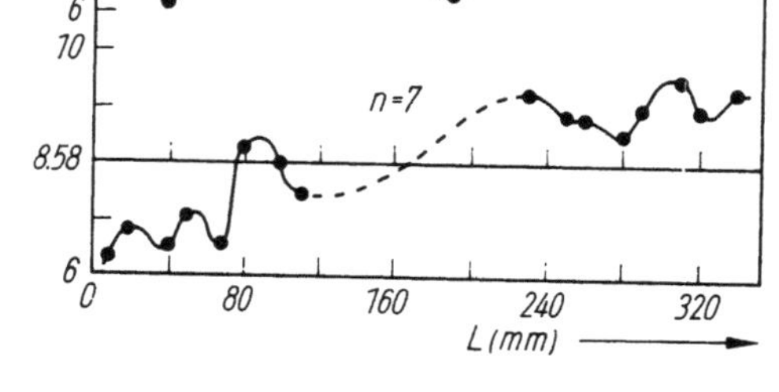

Fig. 2. Distribution of Sn^{119} along the tin wire after a different number of zone passes ($n = 1, 3, 7$) at $v_Z = 3.0$ mm/min

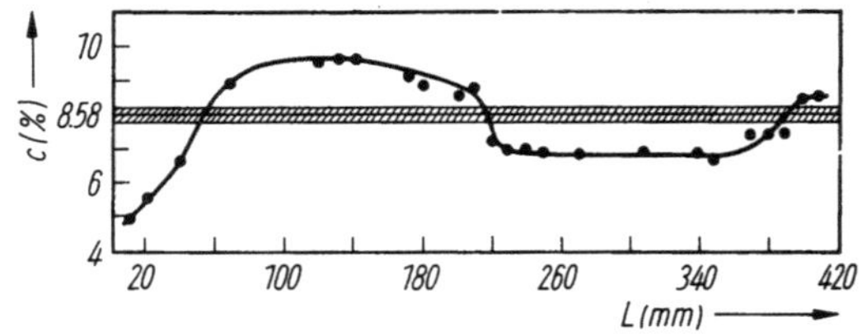

Fig. 3. Distribution of Sn^{119} along tin wire after ten passes at $v_Z = 0.5$ mm/min

It was found that under the effect of zone melting bismuth was preferentially "chasing" the zone. Thus, its distribution coefficient in tin $K < 1$.

The content of Sn^{119} was checked in these experiments only at the tail portions of the wires, i.e. the portions to which the movement of the zones was directed. It was found that the tail portions were poorer in Sn^{119}, i.e. the afore-mentioned effect was inversed in pure tin.

4. Discussion

Thus, the behaviour of isotopes in metal during a zone melting process is complex. If the number of zone melting runs (passes) is small, their distribution along the material is periodical. As the nuder of passes is increased, some final distribution results. For example, in pure tin the Sn^{119} isotope is displaced toward the tail portions of the wires to which the zone movement is directed, whereas its content in the head portions is decreased. In the presence of bismuth inpurity this effect is inversed which may be attributed to the displacement of Bi toward the wire ends. A similar result may be obtained in pure tin by reducing the zone movement rate (Fig. 3).

An equilibrium isotope concentration distribution may, to all appearance, be obtained under small zone movement rates. It may therefore be concluded that in the course of zone melting Sn^{119} behaves like an impurity with $K > 1$.

In all the experiments of this program, the crystallization front moved along the material with a rather high speed ($v_Z = 3.0$ mm/min), which was about one order of magnitude higher than the speeds normally used in zone melting. It is known at the same time [11] that under high zone movement rates the impurity-rich layer may be entrained by the crystallization front. This results in a diffusion layer preceding the crystallization front.

In view of these facts, it may be supposed that the fluctuating concentrations of individual isotopes along the wires observed in the present investigation may be due to their periodical entrainment from the diffusion layer, where they could be in excessive concentrations, by the crystallization front. We hope that future experiments will clarify the mechanism of this phenomenon.

References

[1] E. Haeffner, Nature (London) **172**, 775 (1953).
[2] A. Lunden, C. Reuterswärd, and A. Lodding, Z. Natruf. **11a**, 924 (1956).
[3] A. Lunden, A. Lodding, and W. Fischer, Z. Naturf. **12a**, 268 (1957).
[4] A. Lodding, A. Lunden, and H. van Ubisch, Z. Naturf. **11a**, 139 (1956).
[5] O. A. Troitskii, phys. stat. sol. (a) **35**, K151 (1976).
[6] B. Y. Hilton, A. Smith, and C. O. Thomas, J. phys. Chem. **63**, 445 (1959).
[7] O. A. Troitskii, V. I. Spitsyn, M. P. Glazunov, and R. M. Nifantieva, Dokl. Akad. Nauk SSSR **218**, 894 (1974).
[8] O. A. Troitskii, V. I. Spitsyn, and M. P. Glazunov, Zh. eksper. teor. Fiz. **68**, 1400 (1975).
[9] O. A. Troitskii, V. I. Spitsyn, and V. P. Shcheriodin, Dokl. Akad. Nauk SSSR, **233**, 1154 (1977).
[10] O. A. Troitskii, phys. stat. sol. (a) **48**, 229 (1978).
[11] V. Pfann, Zone Melting, Mir Publ., Moscow 1970 .

(Received May 10, 1979)

phys. stat. sol. (a) 54, 655 (1979)

Subject classification: 14.3.3; 13.4; 22.6

Departamento de Electricidad y Electrónica, Universidad Autónoma de Barcelona[1])

Zero Bias Anomaly Due to Elastic Tunneling in MIM Structures

By

X. Aymerich-Humet and F. Serra-Mestres

The elastic resonant tunneling process is analysed in a three-dimensional way, by means of a new stationary resonance model (SRM). Taking the Green's function of the potential barrier of an asymmetric MIM structure as a starting point, the resonant current density due to the localized states which are in the insulators is calculated. Peaks in the logarithmic conductance derivate (zero bias anomaly), as well as a great current sensitivity to temperature at low bias are proved to exist. The experimental results in the literature agree with the theoretical predictions.

Le processus tunnel résonant élastique est analysé tridimensionnellement par un noveau modèle de résonnance stationnaire (SRM). Au moyen de la fonction de Green du potentiel barrière d'une structure MIM asymétrique, on calcule la densité du courrant résonnant, provenant des états localisés dans l'isolant. On démontre l'existence de pics dans la derivée de la conductance logarithmique (anomalité à polarisation zero), ainsi que d'une grande sensibilité du courant par rapport a la temperature dans le cas de basse tension de polarisation. Les résultats des expériencies dans la littérature confirment nos prévisions théoriques.

1. Introduction

The study of the conductance in metal–insulator–metal structures has proved to be an effective method of establishing the conduction processes through the insulator. At present, the knowledge and analysis of such processes have been shown to be greatly interesting for the design of new electronic devices.

The study of the MIM structures, the main conduction mechanism of which is the tunneling effect, is being widely considered. Several authors [1 to 4] have experimentally proved that the relative increase in tunnel conductance against bias in such structures shows some peaks at low bias voltage, which are generally referred to as zero bias anomaly (ZBA).

As has been proved by Adler and Straus [1], at low temperatures (1 to 12 K) the ZBA can not be fully explained by inelastic tunneling processes or other types of processes such as quantization of the electronic states of the electrodes, non-zero relaxation time of tunneling states, etc. The above authors suggest that, in this Mg–MgO–Au, structures, the predominant process causing the ZBA is the resonant tunneling effect which is in turn caused by the existence of states localized in the insulator having energy around the Fermi level. Breitbarth has found similar results in metal–glow-polymer–metal structures [2, 3].

On the other hand, at higher temperatures, (80 to 300 K), Gundlach et al. [4], when studying the Al–Formvar–Pb structures, noticed the presence of a ZBA which can also be due to the same process.

In this paper we will apply our stationary resonance model (SRM) [5, 6] to the study of the resonant tunneling current in a MIM structure. Our work confirms that the experimental results concerning the ZBA can really be due to trap-assisted tunneling processes.

[1]) Bellaterra, Barcelona, Spain.

In Section 2 we establish a general expression for the resonant tunneling current created by a density of traps, which are located at parallel planes to the junction surface. As far as the SRM is concerned, we have only calculated the net current density scattered by a trap at every point of the device. We want now to extend our expressions to the instance where such a trap density exists.

In Section 3 we apply the above results to a MIM structure, which can be asymmetric. We work out the resonant tunneling current created by a volume density of traps.

In Section 4 we make an analysis of the j–V curve, showing the existence of an anomalous behaviour of the conductance at low bias voltage. In Section 5 we finally discuss the validity of our results.

2. Resonant Tunneling Current for a $N_{\|}$ Trap Density

The SRM obtains, by means of the three-dimensional resonant approximation, the stationary electronic states of the junction potential barrier in which there are short-range trap potentials. Through these states we calculate the density of the resonant tunneling current scattered by a trap and related to the current being "carried" by the three-dimensional Green's function of the junction potential barrier.

In most practical devices the volume density of N_T traps can be written as $N_T = N_{\|} N_{\perp}$ where $N_{\|}$ is the surface density of traps in a parallel plane to the junction and $N_{\perp}$ is a linear density in the perpendicular plane.

In order to introduce $N_{\|}$ it is suitable to proceed to a previous calculation of the electric current intensity carried by every electronic state. This will greatly simplify the Fourier transform of the three-dimensional Green's function.

For a trap located at $\boldsymbol{r}_0 = (x_0, y_0, z_0)$ the current density at the $\boldsymbol{r}$ point, carried by a state with energy E,

$$E = E_i + \frac{\hbar^2 k_{\|}^2}{2m}, \tag{1}$$

where m is the electronic effective mass and $\hbar \boldsymbol{k}_{\|}$ is the parallel momentum to the barrier plane, is [6]

$$\boldsymbol{j}_{i,\boldsymbol{k}_{\|}}(\boldsymbol{r}) = \frac{4\pi^2\hbar^4}{Sm^2} \frac{|\varphi_i(z_0)|^2 \boldsymbol{j}_G(E, \boldsymbol{r})}{[(E - E_0)\, Pg(E_0, z_0, z_0)]^2 + \left[\pi \sum\limits_0^E |\varphi_j(z_0)|^2\right]^2}. \tag{2}$$

S is the junction cross-section area, $\varphi_i(z)$ the one-dimensional wave function of the potential barrier, E_0 the resonant level energy and

$$\boldsymbol{j}_G(E, \boldsymbol{r}) = \langle \boldsymbol{r}_0 |\, G^+(E)\, \boldsymbol{J}(\boldsymbol{r})\, G(E)\, | \boldsymbol{r}_0 \rangle \tag{3}$$

being $\boldsymbol{J}$ the current density operator. $G(E)$ is the three-dimensional Green's operator and $g(E, z, z')$ the one-dimensional Green's function, both of them for the junction potential barrier. P means the Cauchy principal value.

The $\boldsymbol{j}_{i,\boldsymbol{k}_{\|}}$ flux ($I_{i,\boldsymbol{k}_{\|}}$) is obtained by integrating the current density "carried" by the $G(E)$ Green's function at a parallel plane to the junction, $z = \text{const}$:

$$\int\limits_{z=\text{const}} \boldsymbol{j}_G(E, \boldsymbol{r}) \cdot \mathrm{d}\boldsymbol{s} = \int\limits_{\boldsymbol{r}_{\|}} \langle \boldsymbol{r}_0 |\, G^+(E)\, J_z(\boldsymbol{r})\, G(E)\, | \boldsymbol{r}_0 \rangle \,\mathrm{d}^2 \boldsymbol{r}_{\|}\,. \tag{4}$$

Taking into account that the three-dimensional Green's function is the Fourier's transform of $g(E, z, z')$ at the $z = \text{const}$ plane, by straightforward operations we obtain

$$I_{i,\boldsymbol{k}_{\|}} = \frac{2\pi\hbar^2}{Sm} \frac{|\varphi_i(z_0)|^2\, j_g}{[(E - E_0)\, Pg(E_0, z_0, z_0)]^2 + \left[\pi \sum\limits_0^E |\varphi_j(z_0)|^2\right]^2}, \tag{5}$$

where

$$j_g = \int_{-\infty}^{E} \langle z_0| \, g^+(\varepsilon) \, J_z g(\varepsilon) \, |z_0\rangle \, d\varepsilon \, . \tag{6}$$

To calculate the current intensity through the junction, we distinguish at (5) between the $\varphi_i^+(z)$ states such as the terms $I_{i,\boldsymbol{k}_{||}} > 0$, $(I^+_{i,\boldsymbol{k}_{||}})$ and the $\varphi_i^-(z)$ states such as $I_{i,\boldsymbol{k}_{||}} < 0$, $(I^-_{i,\boldsymbol{k}_{||}})$.

The total current is given by

$$I = \sum_{i,\boldsymbol{k}_{||}} I^+_{i,\boldsymbol{k}_{||}} f^-(E) \, [1 - f^+(E)] + \sum_{i,\boldsymbol{k}_{||}} I^-_{i,\boldsymbol{k}_{||}} f^+(E) \, [1 - f^-(E)] \, , \tag{7}$$

where $f^{\pm}(E)$ are the distribution functions at both sides of the barrier. Considering that $I^+_{i,\boldsymbol{k}_{||}} = -I^-_{i,\boldsymbol{k}_{||}}$ then,

$$I = \sum_{i,\boldsymbol{k}_{||}} I^+_{i,\boldsymbol{k}_{||}} [f^-(E) - f^+(E)] = \sum_{i,\boldsymbol{k}_{||}} I^-_{i,\boldsymbol{k}_{||}} [f^+(E) - f^-(E)] \tag{8}$$

we make

$$j^+(E) = j_g(E, z > z_0) \, , \tag{9}$$

$$j^-(E) = -j_g(E, z < z_0) \, , \tag{10}$$

where, we have taken the retarded Green's function for j_g. Then, we can put the sum over $\boldsymbol{k}_{||}$, in the first equality of (8), into an energy integral. By approximating the resulting Lorentz's factor by means of a δ distribution, we can write

$$I = \int_0^\infty \frac{\left[\sum_0^\varepsilon |\varphi_i^+(z_0)|^2\right] j^+(\varepsilon) \, [f^-(\varepsilon) - f^+(\varepsilon)] \, d\varepsilon}{[(\varepsilon - E_0) \, Pg(E_0, z_0, z_0)]^2 + \left[\pi \sum_0^\varepsilon (|\varphi_i^+(z_0)|^2 + |\varphi_i^-(z_0)|^2)\right]^2} \approx$$

$$\approx \frac{\left[\sum_0^{E_0} |\varphi_i^+(z_0)|^2\right] j^+(E_0) \, [f^-(E_0) - f^+(E_0)]}{\left[\sum_0^{E_0} (|\varphi_i^+(z_0)|^2 + |\varphi_i^-(z_0)|^2)\right] |Pg(E_0, z_0, z_0)|} \, . \tag{11}$$

By a similar calculation using the second equality of (8) we can eliminate the terms $|\varphi_i^{\pm}(z_0)|^2$ in (11) thus finally obtaining, for the current intensity carried by a trap,

$$I_e = e \, \frac{j^+(E_0) \, j^-(E_0)}{j^+(E_0) + j^-(E_0)} \, \frac{f^-(E_0) - f^+(E_0)}{|Pg(E_0, z_0, z_0)|} \, , \tag{12}$$

e being the electron charge.

As this expression only depends on the z_0 trap coordinate, for a $N_{||}$ trap density, located at a $z_0 = \text{const}$ plane, the resonant current density through the junction will be

$$j = N \; Ie \, . \tag{13}$$

This expression is suitable for any flat junction device, within our effective mass approximation (1). However, we have assumed that the junction components gaps are large enough to avoid a two-band model [7].

We want to point out that the (12) expression has a similar symmetry to the one obtained by the non-stationary methods. For these methods it is necessary to consider a two-step model. The terms of interference between the probabilities of electrode trap transition [8] are neglected.

However, within the SRM we have always considered the trap-barrier system as a whole and each one of the terms j^+ and j^- in (12) are calculated by means of Green's function in which both electrodes are considered.

3. Calculation of the Resonant Tunneling Current in a MIM Structure

Let us apply the results of the above section to a metal–insulator–metal structure with a trapezoidal junction potential barrier. We will assume that there exist a $N_\perp(z_0)$ trap states density within it, which are located at the Fermi energy level, (at zero bias voltage) (cf. Fig. 1).

To calculate the Green's function concerning the potential barrier $U(z)$, we will perform the asymmetric rectangular barrier approximation.

By solving the differential equation

$$\left[-\frac{\hbar^2}{2m}\frac{\mathrm{d}^2}{\mathrm{d}z^2} + U(z) - E\right] g(E, z, z') = -\delta(z - z') \tag{14}$$

we obtain three expressions for the retarded Green's function at $z' = z_0$, $|z_0| < d$ (cf. Fig. 1). Making $g(E, z > d, z_0) \equiv g^+(E, z, z_0)$ and $g(E, z < -d, z_0) \equiv g^-(E, z, z_0)$ we have

$$g^\pm(E, z, z_0) = \frac{2m}{\hbar^2} \frac{[(\alpha + ik^\mp)\,\mathrm{e}^{-\alpha(d\pm z_0)} + (\alpha - ik^\mp)\,\mathrm{e}^{\alpha(d\pm z_0)}]\,\mathrm{e}^{-ik^\pm(d\mp z)}}{(\alpha + ik^+)(\alpha + ik^-)\,\mathrm{e}^{-2\alpha d} - (\alpha - ik^+)(\alpha - ik^-)\,\mathrm{e}^{2\alpha d}} \tag{15}$$

and for $|z| < d$

$$\begin{aligned} g(E, z, z_0) = {} & -\frac{m}{\hbar^2\alpha}\,\mathrm{e}^{-\alpha|z-z_0|} + \frac{m}{\hbar^2\alpha} \times \\ & \times [(\alpha + ik^+)(\alpha + ik^-)\,\mathrm{e}^{-2\alpha d} - (\alpha - ik^+)(\alpha - ik^-)\,\mathrm{e}^{2\alpha d}]^{-1} \times \\ & \times \{[(\alpha + ik^+)(\alpha - ik^-)\,\mathrm{e}^{\alpha(d+z_0)} + (\alpha + ik^+)(\alpha + ik^-)\,\mathrm{e}^{-\alpha(d+z_0)}]\,\mathrm{e}^{-\alpha(d-z)} + \\ & + [(\alpha - ik^+)(\alpha + ik^-)\,\mathrm{e}^{\alpha(d-z_0)} + (\alpha + ik^+)(\alpha + ik^-)\,\mathrm{e}^{-\alpha(d-z_0)}]\,\mathrm{e}^{-\alpha(d+z)}\}, \end{aligned} \tag{16}$$

where

$$k^\pm(E) = \left[\frac{2m}{\hbar^2}(E - U^\pm)\right]^{1/2}, \tag{17}$$

$$\alpha(E) = \left[\frac{2m}{\hbar^2}(U_0 - E)\right]^{1/2} \tag{18}$$

with $U^\pm$ and U_0 shown in Fig. 1.

Assuming that the barriers are high and wide enough, so that $\mathrm{e}^{-\alpha d} \ll 1$ we can write, using (15) for $j^\pm(E)$ defined under (9) and (10),

$$j^\pm = \frac{2}{\hbar(U_0 - U^\pm)}\,\mathrm{Re}\int_{-\infty}^{E_0} k^\pm\,\mathrm{e}^{[-2\alpha(d\mp z_0)]}\,\mathrm{d}E\,. \tag{19}$$

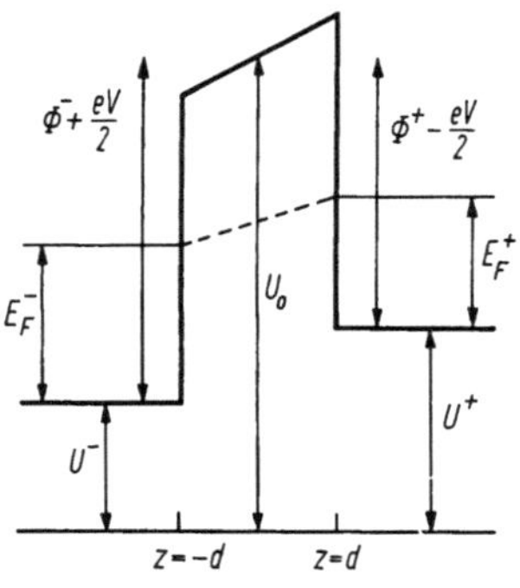

Fig. 1. Energy diagram of the metal–insulator–metal junction, with localized states (dashed line) at zero bias voltage at the Fermi level

Taking into account that these integrals take real values only in $[U^\pm, E_0]$ intervals, and performing the change of variable $\varepsilon = E_0 - E$ then,

$$j^\pm = \frac{2(2m)^{1/2}}{\hbar^2(U_0 - U^\pm)} \int\limits_0^{E_0 - U^\pm} (E_0 - U^\pm - \varepsilon)^{1/2}\, e^{-2\frac{d \mp z_0}{\hbar}[2m(U_0 - E_0 + \varepsilon)]^{1/2}}\, d\varepsilon \tag{20}$$

as only the small values of ε are important enough to be considered for the integrals, we can take

$$\alpha(E_0 - \varepsilon) \approx \alpha_0\left[1 + \frac{\varepsilon}{2(U_0 - E_0)}\right]; \quad \alpha_0 = \alpha(E_0) \tag{21}$$

and so

$$j^\pm = \frac{4k_0^\pm(E_0 - U^\pm)}{3\hbar(U_0 - U^\pm)}\, e^{-2\alpha_0(d \mp z_0)}\, M\left[1, 5/2, -\frac{(k_0^\pm)^2}{\alpha_0}(d \mp z_0)\right] \tag{22}$$

with $k_0^\pm = k^\pm(E_0)$ and $M(a, b, x)$ is the Kummer function [9].

We see that the traps which efficiently take part in the resonant tunneling process are far from both electrodes. We can take $(k_0^\pm)^2\, \alpha_0^{-1}(d \mp z_0) \gg 1$ and for $x > 0$

$$M(a, b, -x) = \frac{\Gamma(b)}{\Gamma(b - a)}\, x^{-a}[1 + O(x^{-1})] \tag{23}$$

obtaining

$$j^\pm = \frac{2(E_0 - U^\pm)}{\hbar(U_0 - U^\pm)}\, \frac{\alpha_0}{k_0^\pm(d \mp z_0)}\, e^{-2\alpha_0(d \mp z_0)}\,. \tag{24}$$

On the other hand, in the $e^{-\alpha d} \ll 1$ approximation of (16) we obtain

$$Pg(E_0, z_0, z_0) \approx -\frac{m}{\hbar^2\alpha_0}. \tag{25}$$

By means of these latter expressions we can establish the current density (13) through the junction, due to a $N_{||}$ trap density:

$$j = \frac{8e\hbar^{-1}N_{||}(U_0 - E_0)\, e^{-2\alpha_0 d}\,[f^-(E_0) - f^+(E_0)]}{\frac{U_0 - U^-}{E_0 - U^-}\, k_0^-(d + z_0)\, e^{2\alpha_0 z_0} + \frac{U_0 - U^+}{E_0 - U^+}\, k_0^+(d - z_0)\, e^{-2\alpha_0 z_0}}, \tag{26}$$

where we have introduced a factor 2 to take into account the spin variables.

This latter expression gives us the value of the trap-assisted resonant tunneling current. These traps are located at the $z = z_0$ plane. However, in the case we are interested in, we must consider the existence of a trap density in the $N_\perp(z_0)$ insulator. As the ZBA is due to the existence of traps around the Fermi level [10, 11], we will assume a $N_\perp(z_0)$ density of monoenergetic traps located at this level at zero bias, and

$$j_T = \int\limits_{-d}^{+d} N_\perp(z_0)\, j(z_0)\, dz_0\,. \tag{27}$$

For non-zero bias, the difference of the electrodes Fermi-Dirac distribution functions, at E_0 energy related to the traps located $z = z_0$ plane, is

$$f^-(E_0) - f^+(E_0) = \frac{-\sinh s}{\cosh s + \cosh su}, \tag{28}$$

where $s = eV/2kT$, $u = z_0/d$, k is the Boltzmann constant, T the absolute temperature, and V the bias voltage referred to the $z < -d$ electrode.

The $j(z_0)$ current shows a sharp peak for $z_0 = \hat{z}$ which is defined by

$$\frac{d + \hat{z}}{d - \hat{z}} e^{4\alpha_0 \hat{z}} = \frac{U_0 - U^+}{U_0 - U^-} \left(\frac{E_0 - U^-}{E_0 - U^+} \right)^{1/2} . \tag{29}$$

Neglecting the z_0 dependence on unimportant factors in (26), we can write

$$j(z_0) = j(\hat{z}) \left[\cosh 2\alpha_0 (z_0 - \hat{z})\right]^{-1} . \tag{30}$$

As the distribution of traps located in the insulator of the MIM structures varies much more slowly than $e^{2\alpha_0 z_0}$ the expression (27) can be written

$$j_T = \frac{\pi \sqrt{2}\, |e|\, N_T \hbar^2 \alpha_0\, e^{-2\alpha_0 d} \left[(\hat{E}_0 - U^-)(\hat{E}_0 - U^+)\right]^{1/4} \sinh s}{2d(1 - \hat{u}^2)^{1/2} m^{3/2} \left[(U_0 - U^-)(U_0 - U^+)\right]^{1/2} (\cosh s + \cosh s\hat{u})} =$$

$$= j_0 \frac{\sinh s}{\cosh s + \cosh s\hat{u}} \tag{31}$$

being $\hat{u} = \hat{z}/d$ and $\hat{E}_0$ the energy related to the traps located at $\hat{z}$.

Let us notice that j_T dependence on the barrier thickness ($2d$) is mainly given by $j_T \sim e^{-2\alpha_0 d}$. This dependence is typical of the resonant tunneling transmission coefficient when calculated by means of the WKB method. On the other hand, the normal tunneling current has a dependence on the insulator thickness $j_N \sim e^{-4\alpha_0 d}$. Therefore, we can say that the resonant tunneling current is, in principle, much more important than the normal tunneling current. However, the dependences of both tunneling currents on the bias voltage are different as we will see in the next section.

4. Study of the *j*-*V* Curve at Low Bias

An analysis of the dependence on the resonant tunneling current in function of the bias voltage shows that we can recognize two different factors.

The first one, j_0, slightly depends on V through the variations of the barrier shape, which are given by

$$\alpha_0^2(V) = \frac{2m}{\hbar^2} \left(\Phi^- - E_F^- - \frac{eV}{2} \hat{u} \right), \tag{32}$$

$$\hat{E}_0 - U^{\pm} = E_F^{\pm} \mp \frac{eV}{2} (1 \mp \hat{u}) , \tag{33}$$

$$U_0 - U^{\pm} = \Phi^{\pm} \mp \frac{eV}{2}, \tag{34}$$

where Φ^+ and Φ^- are the average barrier heights as seen from both electrodes at zero bias voltage.

As we prove in the Appendix, $\hat{u}$ can be considered constant to the bias voltage. We can even consider $\hat{u} = 0$ for low asymmetry barriers.

The second factor of j_T (31) has a quick variation with the bias voltage within a $\pm 10kT/e$ range. For higher values it is practically constant to ± 1.

We have now realized that the resonant tunneling current quickly increases for low bias voltages due to the starting of the trap-assisted tunneling mechanism. At higher voltages this process becomes saturated and the small variations of current are due to the variation of the barrier shape (cf. Fig. 2). Therefore, although at low bias voltage the resonant tunneling current is prevailing, for higher voltages the normal tunneling

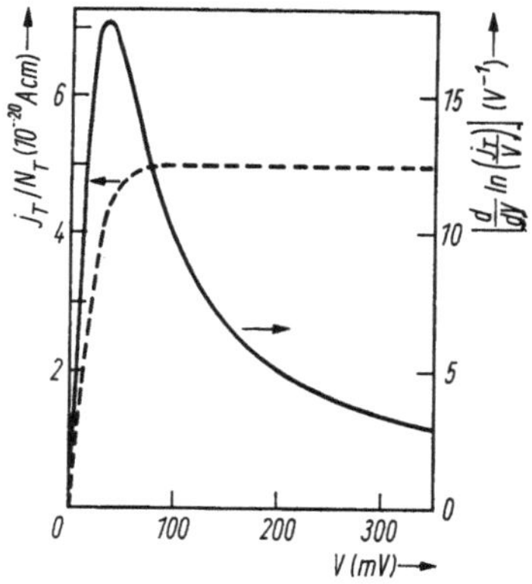

Fig. 2. Ratio between the resonant density current and the trap density versus bias (dashed line, left scale) for a typical MIM structure ($\Phi^+ = 11.8$ eV, $\Phi^- = 11.4$ eV, $E_F^+ = 8.8$ eV, $E_F^- = 8.4$ eV, $2d = 20$ Å; m is the free electron mass and $T = 77$ K). The saturation value, practically constant at low bias, is 4.97×10^{-20} Acm. The continuous line (right scale) is the absolute value of the logarithmic derivate of conductance against bias; it shown a peak at 36 mV (we only represent the positive bias because both curves are practically symmetric)

current will overcome the resonant one, due to the exponential dependence of the former [12] towards the saturation process of the latter.

The anomalous behaviour at low voltage (ZBA) has been shown up by working out $g(V) = \sigma^{-1}(\mathrm{d}\sigma/\mathrm{d}V)$, ($\sigma = j_T/V$) in function of V (cf. Fig. 2).

Considering (31), $g(V) = g_1(V) + g_2(V)$ where

$$g_1(V) = \frac{e}{2kT}\,\frac{s - \sinh s}{s \sinh s}\,, \tag{35}$$

$$g_2(V) = \frac{e}{8}\,(\Phi^+ - \Phi^- - eV)\;\frac{\Phi^+\Phi^- - 2E_F^+E_F^- + eV\left(\dfrac{eV}{4} + \dfrac{\Phi^+ - \Phi^-}{2}\right)}{\Phi^+\Phi^-E_F^+E_F^-}\,. \tag{36}$$

At usual temperatures, $kT \ll 1 eV$ then $g_1(1V) \sim 1V^{-1}$. Besides, $g_1(V) \sim V^{-1}$ for $|V| > 5kT/|e|$. The peak height (cf. Fig. 2) is $0.12\,|e|/kT$. At room temperature this height is similar to that of the peaks due to the normal tunneling conductance, which are located approximately at $\pm\Phi^\pm/e$ voltages, that is, far from the bias where the ZBA originates.

For a very low voltage, we see that $g(V)$ shows up a V_{off} tension offset, if the barrier is asymmetric, which is caused by the $g_2(V)$ term. This result is similar to the one found by the normal tunneling current [13, 14].

5. Discussion and Results

We have studied the metal–insulator–metal structure, assuming the existence of trap states in the insulator, which are located around the Fermi level. We have proved that these states make the j–V curve undergo a saturation process, due to this mechanism.

We have also seen that, although the trap-assisted tunneling current at low bias voltage is very important, when higher voltages are used, the normal tunneling current will prevail. This is because the normal tunneling current does not show the saturation process.

We have also studied the relative variation of conductance in function of the bias voltage, and we have proved that there exist two peaks around zero bias voltage. These peaks have been experimentally observed at low temperatures (1 to 4 K) in Mg–MgO–Au structures (see Fig. 6 in [1]) and in Sn–styrene-glow-polymer–Pb [3].

Both peak width and height are in accordance with our results.

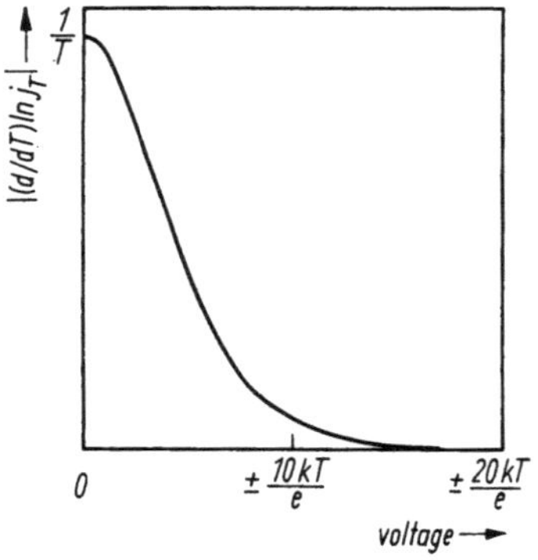

Fig. 3. Absolute value of relative increase in resonant tunneling current due to a temperature increase against bias, for an arbitrary MIM structure. The peak centered around zero volts is characteristic of the trap-assisted process

On the other hand, Gundlach et al. [4] have proved, at higher temperatures (80 to 300 K) the strong increase of $|(\mathrm{d}/\mathrm{d}V)\ln\sigma|$ centered around zero bias in Al–Formvar–Pb junctions. Their experimental results between $\pm 1\,V$ agree with our theoretical provisions, as it is at this bias interval where the trap assisted tunneling process is important.

In Fig. 3 we have shown $|(\mathrm{d}/\mathrm{d}T)\ln j_\mathrm{T}|$ and we notice that it shows a peak around zero bias, which height increase is inversely proportional to the temperature and its width is about $20kT/|e|$. This results shows that the current measurements around zero bias are very sensitive to temperature variations, as has been pointed out by Gundlach et al. [4].

This work also shows up the validity of the SRM method which has recently been suggested by the authors, as this satisfactorily explains the ZBA in MIM structures and its results agree with experimental data existing in the related literature. Besides, our method allows the calculation not only of the variations concerning the resonant tunneling current as function of the bias voltage and temperature, but also of its absolute value.

This way, we see that we can deal with the trap-assisted resonant tunneling problem in a stationary way. Therefore, we do not need to have recurse to pseudo-temporary interpretations in order to obtain satisfactory results.

Appendix

We deal with the dependence of $\hat{u}$ on bias, given in (29):

$$\left(\frac{1+\hat{u}}{1-\hat{u}}\right)^2 \exp\left[4\lambda_0\left(1-\frac{x\hat{u}}{q-1}\right)^{1/2}\hat{u}\right] = \left(\frac{\mu-x}{q+x}\right)^2 \frac{1+x\,(1+\hat{u})}{r-x\,(1-\hat{u})}, \tag{A1}$$

where

$$\mu = \frac{\Phi^+}{E_\mathrm{F}^-}, \qquad q = \frac{\Phi^-}{E_\mathrm{F}^-}, \qquad r = \frac{E_\mathrm{F}^+}{E_\mathrm{F}^-} \tag{A2}$$

and

$$x = \frac{eV}{2E_\mathrm{F}^-}, \qquad \hat{u} = \frac{\hat{z}}{d}, \qquad \lambda_0 = \frac{2d}{\hbar}\,[2m(\Phi^- - E_\mathrm{F}^-)]^{1/2}\,. \tag{A3}$$

Taking into account that $x \ll 1$, from the first-order expansion on x it follows

$$x = \left[\frac{rq^2}{\mu^2}\,\mathrm{e}^{4\lambda_0\hat{u}}\left(\frac{1+\hat{u}}{1-\hat{u}}\right)^2 - 1\right]\left[1+\hat{u}+\frac{1-\hat{u}}{r} - 2\left(\frac{1}{\mu}+\frac{1}{q}\right) + \frac{\hat{u}}{2(q-1)}\right]^{-1}. \tag{A4}$$

The variations of optimum position against bias are of the order $\Delta\hat{u} \sim \Delta x/4\lambda_0$. Since $x \ll 1$ and $\lambda_0 \gg 1$ then $\hat{u}$ remains constant, and we can take for $\hat{u}$ their value at zero bias voltage:

$$\hat{u} = \frac{1}{4(\lambda_0 + 1)} \ln \left[\frac{E_F^+}{E_F^-} \left(\frac{\Phi^+}{\Phi^-}\right)^2\right]. \tag{A5}$$

Hence, in the usual junction potential barriers $|\hat{u}| \ll 1$.

References

[1] J. G. ADLER and J. STRAUS, Phys. Rev. B **13**, 1377 (1976).
[2] F.-W. BREITBARTH, phys. stat. sol. (a) **25**, K117 (1974).
[3] F.-W. BREITBARTH and H.-J. TILLER, phys. stat. sol. (a) **41**, 151 (1977).
[4] K.H. GUNDLACH, G. FARACI, G. GIAQUINTA, and N. A. MANCINI, Phys. Letters A **43**, 27 (1973).
[5] X. AYMERICH-HUMET, F. SERRA-MESTRES, and J. MILLAN-GOMEZ, Proc. 75 Aniversario, Simposium 16, RSEFQ, Madrid 1978 (p. 61).
[6] X. AYMERICH-HUMET and F. SERRA-MESTRES, phys. stat. sol. (a) **51**, 583 (1979).
[7] K. H. GUNDLACH, J. appl. Phys. **44**, 5005 (1973).
[8] G. SARRABAYROUSE, Thèse d'Etat (Sciences), Université Paul Sabatier, n° ordre 854, Toulouse 1978.
[9] M. ABRAMOWITZ and I. A. STEGUN, Handbook of Mathematical Functions, Dover Publ., Inc., New York 1965 (p. 504).
[10] F.-W. BREITBARTH and H.-J. TILLER, Wiss. Z. Friedrich-Schiller-Univ. Jena, Math.-naturw. R. **25**, 871 (1976).
[11] R. COMBESCOT, J. Phys. C **4**, 2611 (1971).
[12] C. B. DUKE, Tunneling in Solids, Academic Press, New York 1969.
[13] P. LU and C. K. CHAN, Lettere Nuovo Cimento **22**, 410 (1978).
[14] J. RICHTER and P. SEIDEL, phys. stat. sol. (a) **46**, K25 (1978).

(Received May 25, 1979)

phys. stat. sol. (a) 54, 665 (1979)

Subject classification: 13.4 and 14.3.3; 22

Norges Tekniske Høgskole, Trondheim[1])

Theory of ISC Measurements for p-Type Schottky Diodes with Deep Acceptor Impurities

By

L. M. LAMBERT[2])

The behaviour of deep impurities in semiconductors has been studied for many years. In particular, the behavior of Schottky diodes containing deep impurity centers, especially gold, has been the subject of intense investigation. However, these studies are primarily concerned with the frequency behavior of the capacitance measurement. In this paper, we explore instead the static behavior of the ISC (inverse-square-capacitance) curve, that is, the ISC curve resulting from a condition of nonuniform equilibrium due to a slowly applied reverse bias voltage. We are especially interested in p-type Schottky diodes having ion implanted deep impurities, but a somewhat more general theory is developed which can be applied to other materials as well. Methods for interpreting experimental ISC curves will also be discussed and procedures for deducing doping profiles are presented.

Die Eigenschaften tiefer Verunreinigungen in Halbleitern werden seit mehreren Jahren untersucht. Insbesondere die Eigenschaften der Schottky-Dioden mit tiefen Verunreinigungszentren sind Gegenstand intensiver Untersuchungen gewesen. Diese Studien haben sich besonders auf die Frequenzabhängigkeit der elektrischen Kapazitätsmessungen konzentriert. Es werden statt dessen die statischen Eigenschaften der ISC (inverse quadratische-Kapazität)-Kurve untersucht, das heißt, die ISC-Kurve die das Resultat eines nichteinheitlichen Gleichgewichts infolge einer langsam angelegten Vorspannung ist. Es werden insbesondere p-Typ-Schottky-Dioden untersucht, die tiefe ioneninplantierte Verunreinigungen enthalten. Es wird jedoch auch eine etwas allgemeinere Theorie entwickelt, die auch auf andere Materialien angewendet werden kann. Methoden für die Interpretation der experimentellen ISC-Kurven werden diskutiert und Verfahren für die Herleitung der Dotierungsprofile angegeben.

1. Introduction

The subject of the capacitance behavior of deep impurities in semiconductors has been studied for many years. An excellent review of these works is given in [1]. In particular, the dynamic behavior of the capacitance measurement has been studied extensively [2 to 10]. Most of these studies have dealt with the behavior of Au in n-type diodes. In contrast to these works, we will develop a theory based on the static ISC characteristics of p-type Schottky diode structures with uniform and non-uniform impurity distributions. The method used is related to that used previously for MOS structures [11]. In practice, capacitance measurements are made with a small amplitude, high frequency signal superimposed on a static bias voltage. This approach will be used in developing the theory as it applies to the static ISC curves; however, it will become evident how the dynamic case could equally as well have been handled. For the sake of completeness, we will treat five specific cases of interest: (I) uniformly distributed shallow impurities only; (II) nonuniformly distributed shallow impurities only; (III) uniformly distributed deep and shallow impurities; (IV) uniform shallow impurities and nonuniform deep impurities; and (V) uniform deep impurities and nonuniform shallow impurities.

[1]) Trondheim, Norway.

[2]) Permanent address: Physics Department, University of Vermont, Burlington, Vt. 05401, USA.

It will be shown that although the familiar ISC curve is, in general, nonlinear, it can still be used to obtain information on impurity concentration and profiles. The specific case of zinc as the deep acceptor impurity in p-type silicon will be treated for purposes of illustration.

2. Theory

Fig. 1a shows the one-dimensional band structure and the associated parameters. We choose to define the Fermi level for p-type semiconductors as

$$\Phi_{\mathrm{Fp}} \equiv \frac{E_{\mathrm{g}}}{e} - \Phi_{\mathrm{Fn}}\,,$$

where the Fermi level for n-type material, Φ_{Fn}, is commonly used in other theories. In addition, we will define the barrier height for p-type material as

$$\Phi_{\mathrm{Bp}} \equiv \frac{E_{\mathrm{g}}}{e} - \Phi_{\mathrm{Bn}}\,,$$

where once again the barrier height for n-type material, $\Phi_{\mathrm{Bn'}}$, is more commonly used. The applied bias voltage is U and the deep acceptor impurity activation energy is $\varepsilon_{\mathrm{A}}^{\mathrm{d}}$. The other parameters have their usual meanings.

In Fig. 1b, the electrostatic potentials resulting from the various charge distributions are shown. Ψ_{S} is the usual surface potential which is given explicitly for a p-type Schottky diode in terms of the applied voltage, U, as

$$\Psi_{\mathrm{S}} = \Phi_{\mathrm{Bp}} - \Phi_{\mathrm{Fp}} + U\,. \tag{1}$$

The specific value of Ψ_{S} at which the deep impurities begin to ionize at the surface is designated as $\Psi_{\mathrm{S}}^{\mathrm{d}}$ and evaluated from Fig. 1a as

$$\Psi_{\mathrm{S}}^{\mathrm{d}} = \varepsilon_{\mathrm{A}}^{\mathrm{d}} - \Phi_{\mathrm{Fp}}\,. \tag{2}$$

One can now use (2) in (1) to find the value of applied voltage at which the deep impurities start to ionize as

$$U(\Psi_{\mathrm{S}}^{\mathrm{d}}) = \varepsilon_{\mathrm{A}}^{\mathrm{d}} - \Phi_{\mathrm{Bp}}\,. \tag{3}$$

In Fig. 1b, W is the actual width of the depletion region at a particular applied voltage, U. If only shallow impurities were present, the corresponding surface potential would be Ψ_{S}. If deep impurities are present, the surface potential is Ψ_{S} but W consists of two parts: W_{d}, the region in which only shallow impurities are ionized (with a corresponding surface potential, $\Psi_{\mathrm{S}}^{\mathrm{d}}$); and W_{Δ}, the region in which both deep and shallow impurities are ionized. If both the deep and shallow impurities are ionized throughout the volume of the semiconductor the width would be W_{λ}, as shown, for an applied voltage, U, giving rise to the same surface potential, Ψ_{S}.

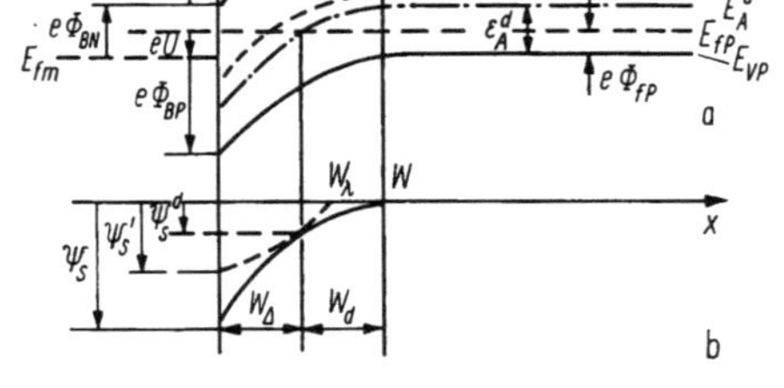

Fig. 1. a) One dimensional energy bands of a p-type Schottky diode; b) Nomenclature associated with the depletion regions and surface potentials

In order to find Ψ_S for a particular charge distribution, one must, of course, integrate Poisson's equation in the form

$$\nabla^2\Psi(x) = \frac{e}{\varepsilon_S}\,[N_A^S(x) + N_A^d(x)]\,, \tag{4}$$

where $N_A^S(x)$ and $N_A^d(x)$ are the distributions of the ionized shallow and deep acceptor impurities, respectively.

The integration of (4) twice and evaluated at $x = 0$ in order to give the surface potential, Ψ_S, becomes

$$\Psi_S = \frac{e}{\varepsilon_S}\,[\int \mathrm{d}x \int N_A^S(x)\,\mathrm{d}x + \int \mathrm{d}x \int N_A^d(x)\,\mathrm{d}x]_{x=0}\,, \tag{5}$$

where the boundary conditions $\nabla\Psi(x) = 0$ at $x = W$ and $\Psi(x) = 0$ at $x = W$ are to be used to evaluate the integration constants. Equation (5) cannot be integrated, in fact, unless the spatial distribution of impurities is explicitly known; however, we can examine specific cases of interest.

2.1 Case I: $N_A^S(x) \neq f(x)$; $N_A^d(x) = 0$

Equation (5) now easily becomes

$$\Psi_S = \Psi_S' = \frac{eN_A^S}{2\varepsilon_S}\,W^2\,. \tag{6}$$

Since we are interested in the ISC curve in particular, we will assume the capacitance is measured by means of a small ac voltage, $\tilde{\nu}$, superimposed on a bias voltage, U. By definition, the capacitance per unit area is given as

$$C_S = \frac{\partial Q_{SC}}{\partial\tilde{\nu}}\,,$$

where Q_{SC} is the semiconductor charge. In the depletion approximation, this is given as

$$Q_{SC} = eN_A^S W\,.$$

Substituting for Q_{SC} into the capacitance equation and taking the differential, one obtains

$$C_S(W) = eN_A^S\left(\frac{\mathrm{d}W}{\mathrm{d}\Psi_S'}\right)\left(\frac{\mathrm{d}\Psi_S'}{\mathrm{d}\Psi_S}\right)\left(\frac{\mathrm{d}\Psi_S}{\mathrm{d}\tilde{\nu}}\right). \tag{7}$$

One notes that W is a function of Ψ_S' and Ψ_S is related to the applied voltages through (1). Both $(\mathrm{d}\Psi_S'/\mathrm{d}\Psi_S)$ and $(\mathrm{d}\Psi_S/\mathrm{d}\tilde{\nu})$ are equal to unity for this case; therefore, (7) can be solved by substituting the differential of (6) with respect to Ψ_S' into it to give simply

$$C_S(W) = \frac{\varepsilon_S}{W}. \tag{8}$$

One now can invert (8), square both sides, substitute (6) for W^2, and then take the derivative with respect to the applied bias voltage, U, to obtain the desired ISC slope equation as

$$\left[\frac{\mathrm{d}}{\mathrm{d}U}\left(\frac{1}{C_S^2}\right)\right] = \frac{2}{e\varepsilon_S N_A^S}\left(\frac{\mathrm{d}\Psi_S'}{\mathrm{d}\Psi_S}\right)\left(\frac{\mathrm{d}\Psi_S}{\mathrm{d}U}\right). \tag{9}$$

Both differentials on the right are unity giving the well-known result

$$N_A^S = \frac{2}{e\varepsilon_S}\left[\frac{d}{dU}\left(\frac{1}{C_S^2}\right)\right]^{-1}. \tag{10}$$

Equation (10) shows that the slope of the ISC curve is inversely proportional to the shallow impurity concentration. This result has been known and used for many years.

2.2 Case II: $N_A^S(x) = f(x)$; $N_A^d(x) = 0$

Once again, we will use the depletion approximation to find the semiconductor charge. One can write, therefore,

$$Q_{SC} = e\int_0^W N_A^S(x)\, dx$$

and the capacitance C_S in (7) now becomes

$$C_S(W) = \left[e\frac{d}{dx}\int N_A^S(x)\, dx\right]_{x=W}\left(\frac{dW}{d\Psi_S'}\right)\left(\frac{d\Psi_S'}{d\Psi_S}\right)\left(\frac{d\Psi_S}{d\tilde{\nu}}\right),$$

giving for this case

$$C_S(W) = eN_A^S(W)\frac{dW}{d\Psi_S'}$$

since once again both $d\Psi_S'/d\Psi_S$ and $d\Psi_S'/d\tilde{\nu}$ are unity. Equation (6) can again be used to yield

$$N_A^S(W) = \frac{2}{e\varepsilon_S}\left[\frac{d}{dU}\left(\frac{1}{C_S^2}\right)\right]^{-1}. \tag{11}$$

One notes that now the instantaneous slope at voltage, U, is inversely proportional to the shallow impurity concentration at a distance W in from the surface. If $N_A^S(x)$ varies with depth, the slope of the ISC curve will vary with applied voltage. The value of W can be easily found for each value of C_S from (8); thus the experimental measurement of the shallow impurity profile can be obtained.

The theoretical ISC plot for a known impurity distribution can be obtained by using W as the independent variable. Equation (8) yields the ordinate; whereas, a double integration of (5) yields Ψ_S corresponding to this value of W. Equation (1) can now be used to obtain the abscissa for the plot in which Φ_{Fp} corresponds to the Fermi level at the surface concentration of impurities.

2.3 Case III: $N_A^S(x) \neq f(x)$; $N_A^d(x) \neq f(x)$

One can now integrate (5) directly giving

$$\Psi_S = \frac{e}{2\varepsilon_S}\left[N_A^S W^2 + N_A^d W_\Delta^2\right]. \tag{12}$$

We could have also integrated (5) from W to W_Δ only. Such an integration would yield

$$\Psi_S^d = \frac{eN_A^S}{2\varepsilon_S}(W - W_\Delta)^2$$

in which Ψ_S^d is the "surface" potential corresponding to the edge of the deep impurity ionization region. Since $W_d = W - W_\Delta$, one can evaluate W_d explicitly as

$$W_d = \left[\frac{2\varepsilon_S \Psi_S^d}{eN_A^S}\right]^{1/2}. \tag{13}$$

One recalls from (2) that Ψ_S^d is a constant for a given shallow impurity concentration and deep impurity type; therefore, W_d is also a constant. As Ψ_S increases, W_Δ and W increase with W_d maintaining a constant value.

For the sake of completeness, we will once again integrate (5), but this time assuming that both the deep and shallow impurities are completely ionized throughout the volume of the semiconductor. One obtains the parameter W_λ as

$$W_\lambda = \left[\frac{2\varepsilon_S \Psi_S}{e(N_A^S + N_A^d)}\right]^{1/2}.$$

Equation (6) can be written in a similar form as

$$W = \left[\frac{2\varepsilon_S \Psi_S'}{eN_A^S}\right]^{1/2}. \tag{14}$$

Substituting (14) into (12), one obtains the relationship between Ψ_S and Ψ_S' when deep impurities are present for a given applied bias voltage,

$$\Psi_S = \Psi_S' + \frac{eN_A^d}{2\varepsilon_S} W_\Delta^2 . \tag{15}$$

Taking the differential of (15) with respect to the applied bias voltage U, gives

$$\left(\frac{d\Psi_S'}{dU}\right) = \left[1 - \frac{eN_A^d}{\varepsilon_S} W_\Delta \frac{dW_\Delta}{dU}\right].$$

Substituting this into (9) gives

$$\left[\frac{d}{dU}\left(\frac{1}{C_S^2}\right)\right] = \frac{2}{e\varepsilon_S N_A^S}\left[1 - \frac{eN_A^d}{\varepsilon_S} W_\Delta \frac{dW_\Delta}{dU}\right]. \tag{16}$$

One notes in (16) that the slope of the ISC curve is reduced as the deep impurities ionize. Inherent in the derivation of (16) is the assumption that the ac measuring voltage, $\tilde{v}$, is so high in frequency that the deep impurities can not respond to this signal. In order to solve (16) explicitly, one needs to find dW_Δ/dU. Since $W = W_\Delta + W_d$ and W_d is a constant, $dW/dU = dW_\Delta/dU$. From (14) and (15), one obtains

$$\frac{dW}{dU} = \frac{W}{2\Psi_S'} \frac{d\Psi_S'}{dU}$$

and

$$\frac{dW_\Delta}{dU} = \frac{W_\Delta}{2(\Psi_S - \Psi_S')}\left(1 - \frac{d\Psi_S'}{dU}\right).$$

Equating dW/dU and dW_Δ/dU and using (13) and (14), one finds

$$\frac{d\Psi_S'}{dU} = \frac{\left[\sqrt{\dfrac{\Psi_S'}{\Psi_S^d}} - 1\right]}{\left[\sqrt{\dfrac{\Psi_S^2}{\Psi_S'\Psi_S^d}} - 1\right]}$$

giving finally

$$\frac{\mathrm{d}W_\Delta}{\mathrm{d}U} = \frac{W}{2\Psi_S'} \frac{\left[\sqrt{\frac{\Psi_S'}{\Psi_S^d}} - 1\right]}{\left[\sqrt{\frac{\Psi_S^2}{\Psi_S'\Psi_S^d}} - 1\right]}. \tag{17}$$

Equations (15) and (17) can now be used with (16) to find N_A^d if N_A^S is known for a particular ISC curve. One recalls that Ψ_S' is obtainable from (8) and (14) and Ψ_S is given by (1). If the bias voltage is applied as a pulse with $W_\Delta = 0$, then the initial values of $(1/C_S^2)$ versus U would correspond to the shallow impurity concentration only as given by (10) or (11) because the deep impurities require a finite time to respond to applied voltages. Thus N_A^S can be measured dynamically even with deep impurities present. This technique has been discussed in great detail in references [2, 6, 8]. Thus if N_A^S is known or measured dynamically and the static ISC curve obtained experimentally, one can find N_A^d. It is easier, however, to find N_A^d in an alternate but equivalent method as follows:

From the measured value of $(1/C_S^2)$ one can find Ψ_S' from (8) and (14) as

$$\Psi_S' = \frac{e\varepsilon_S N_A^S}{2} \left(\frac{1}{C_S^2}\right). \tag{18}$$

Also since $W_\Delta = W - W_d$, one can solve (15) to give

$$N_A^d = \frac{2\varepsilon_S}{eW_\Delta^2} (\Psi_S - \Psi_S'). \tag{19}$$

Fig. 2 shows a plot of the theoretical ISC curves for p-type silicon Schottky diodes having $N_A^S = 1 \times 10^{15}/\mathrm{cm}^3$ and zinc as the deep acceptor impurity of various constant concentrations. One notes that the slopes of the curves are not constant except for the case $N_A^d = 0$; thus, the slope is not as useful as it was in the case of MOS capacitor structures in determining the deep impurity concentrations [11]. One also notes in Fig.2 that all curves start from a point at $U = -0.18$ V, $(1/C_S^2) = 6.08 \times 10^{14}$ cm^4/f^2 corresponding to $U(\Psi_S^d)$ or the onset of deep impurity ionization at the surface. The shallow impurity curve is linear and continues through this point with an abscissa intercept of $U = -0.23$ V corresponding to $\Psi_S = 0$ in (1) or what is sometimes called flat-band. In computing theoretical curves it is easier if one uses Ψ_S' as the independent variable and equations (14), (18), (15), and (1) in that order.

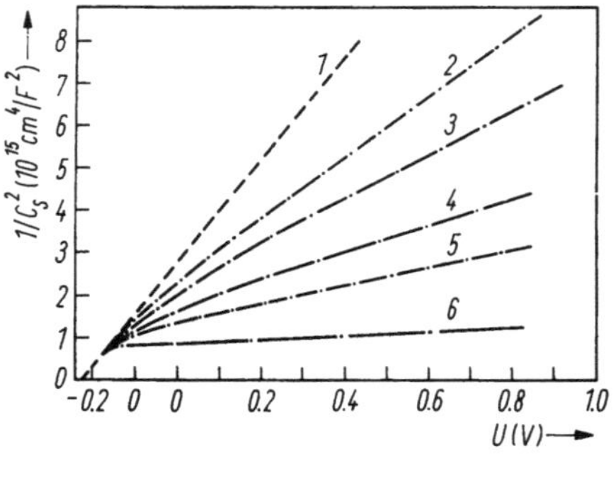

Fig. 2. Theoretical ISC curves for constant deep and shallow acceptor impurity concentrations (1) $N_A^d = 0$ ($N_A^S = 1 \times 10^{15}$ cm^{-3}), (2) 1×10^{15}, (3) 2×10^{15}, (4) 5×10^{15}, (5) 1×10^{16}, (6) 1×10^{17} cm^{-3}. $\Phi_{Bp} = 0.49$ eV, $\Phi_{Fp} = 0.26$ eV, $\varepsilon_A^d = 0.31$ eV, $\psi_S^d = 0.05$ eV

2.4 Case IV: $N_A^S(x) \neq f(x)$; $N_A^d(x) = f(x)$

Since, in general, $N_A^d(x)$ is an unknown function whose spatial dependency is sought, (5) can not be integrated completely as before in case III. We can, however, choose to define the results of such an integration of $N_A^d(x)$ in terms of some constant average value, $\langle N_A^d(W_\Delta)\rangle$, over the ionization distance W_Δ which is the limits of the integration. In particular, we can define it as

$$\langle N_A^d(W_\Delta)\rangle \equiv \frac{2}{W_\Delta^2} [\int dx \int N_A^d(x) \, dx]_{x=0} \,. \tag{20}$$

Now, however, W_Δ can reach a limiting value, $\hat{W}_\Delta$, as $N_A^d(x)$ approaches zero for increasing values of x. This behavior is shown in Fig. 3 along with the associated surface potentials. In this figure, W_d remains constant as previously defined and given by (13) but now $W = W_\nu + W_d$ when $W_\nu \geqq \hat{W}_\Delta$. Thus we will confine our interests to the region in which $W_\Delta < \hat{W}_\Delta$ since as W_Δ approaches $\hat{W}_\Delta$, dW_Δ/dU approaches zero in (16) and the ISC curve asymptotically approaches the slope of the shallow impurity curve of case I. This behavior is shown in Fig. 4 for an exponential distribution of deep impurities, where

$$N_A^d(x) = N_A^d(0) \, e^{-x/L}$$

and where $N_A^d(0) = 1 \times 10^{16}/cm^3$ and $L = 0.217 \times 10^{-4}$ cm.

The abscissa intercept of the shallow impurity slope asymptote can be found from (15) if $N_A^d(x)$ is a known function which can be evaluated at $W_\Delta \to \infty$ inside the integrals of (20) and $W_\Delta = \hat{W}_\Delta$ outside the integral. That is, since $\Psi_S' = 0$ on the abscissa, then

$$\Psi_S = \frac{e}{2\varepsilon_S} \langle N_A^d(\hat{W}_\Delta)\rangle \, \hat{W}_\Delta^2 \,.$$

Equation (1) can now be used to find the applied bias voltage, U.

For values of W_Δ less than $\hat{W}_\Delta$, (17) is still valid and (16) can still be used to calculate the slope if N_A^d is replaced by $\langle N_A^d(W_\Delta)\rangle$. The theoretical ISC curve can be calculated from (18) and (19) in the form

$$\left(\frac{1}{C_S^2}\right) = \frac{2}{e\varepsilon_S N_A^S} \left[\Psi_S - \frac{e}{2\varepsilon_S} \langle N_A^d(W_\Delta)\rangle \, W_\Delta^2\right] \tag{21}$$

using Ψ_S' as the independent variable once again.

The inverse process, namely, deducing the deep impurity profile from an experimental ISC curve, can be handled easily. First the asymptote corresponding to $\hat{W}_\Delta$ can be obtained graphically. For values of $W_\Delta < \hat{W}_\Delta$, one can find $\langle N_A^d(W_\Delta)\rangle \, W_\Delta^2$, the value of the integral for specific values of W_Δ. Since $W_\Delta = W - W_d$ and W is obtainable from (14) and (18) for a measured value of $(1/C_S^2)$, a scale of W_Δ values can be put on the measured ISC plots as shown at the right in Fig. 4. The intersection of a particular W_Δ value and the experimental ISC curve provides a U or Ψ_S value which can be used

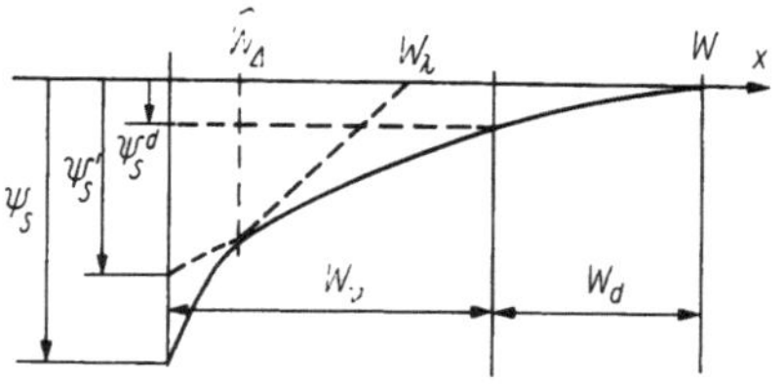

Fig. 3. The depletion regions for case IV in which the deep impurity concentration decreases with distance in from the surface

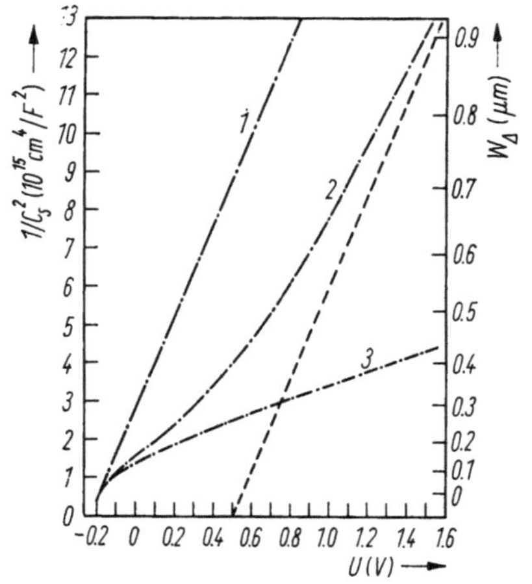

Fig. 4

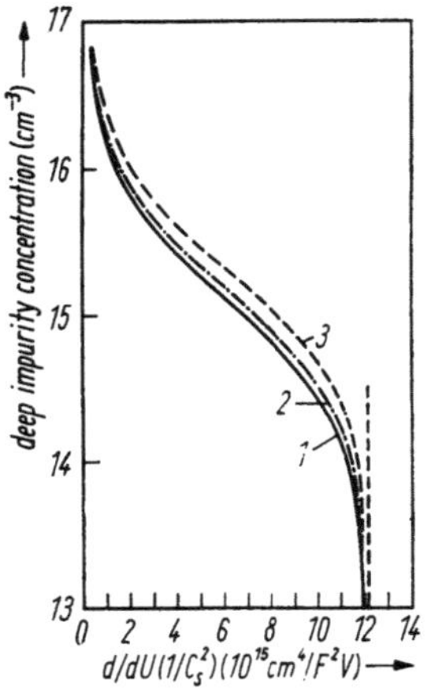

Fig. 5

Fig. 4. Theoretical ISC curves for a nonuniform deep impurity concentration and constant shallow impurity concentration. (1) $N_A^d = 0$ ($N_A^S = 1 \times 10^{15}$ cm^{-3}), (3) 1×10^{16} cm^{-3}. Curve 2 represents the calculated exponential profile. $\Phi_{Bp} = 0.49$ eV, $\Phi_{Fp} = 0.26$ eV, $\varepsilon_A^d = 0.31$ eV, $\psi_S^d = 0.05$ V

Fig. 5. The averaged deep impurity concentration versus the slope of the ISC curves. (1) $W_\Delta = 1.0$, (2) 0.5, (3) 0.25 μm. $N_A^S = 1 \times 10^{15}$ cm^{-3}, $\Phi_{Bp} = 0.49$ eV, $\Phi_{Fp} = 0.26$ eV, $\varepsilon_A^d = 0.31$ eV, $\psi_S^d =$ $= 0.05$ V

with (21) to give $\langle N_A^d(W_\Delta)\rangle$. It is easy to prove that the actual impurity concentration, $N_A^d(W_\Delta)$, is given approximately as

$$N_A^d(W_\Delta) \approx \frac{\langle N_A^d(W_{\Delta N})\rangle\, W_{\Delta N}^2 - \langle N_A^d(W_\Delta)_{N-1}\rangle\, W_{\Delta N-1}^2}{2 W_{\Delta N}\, \Delta W_{\Delta N}}, \qquad (22)$$

where N and $N - 1$ are two adjacent sections corresponding to the N-th slice which is ΔW_Δ thick. Clearly

$$W_{\Delta N} = \sum_{i=1}^{N} \Delta W_{\Delta i}$$

and corresponds to the coordinate x for plotting the impurity profile.

It may be more convenient to solve for $\langle N_A^d(W_\Delta)\rangle$ graphically if a large number of samples with the same shallow impurity concentration are to be examined. This can be done by use of (16) through (19) to make a curve of averaged impurity concentration, $\langle N_A^d(W_\Delta)\rangle$, versus slope as shown in Fig. 5 for various values of W_Δ. One notes in Fig. 5 that the width of the ionization region effects $\langle N_A^d(W_\Delta)\rangle$ only weakly; whereas, $\langle N_A^d(W_\Delta)\rangle$ is a strong function of the slope.

2.5 Case V: $N_A^S(x) = f(x)$; $N_A^d(x) \neq f(x)$

For this last case, we assume the deep impurity to be uniformly distributed throughout the volume of the semiconductor and the shallow impurity to have some unknown distribution or profile. Once again (5) cannot be integrated directly but one can now write the result of such an integration as

$$\Psi_S = \frac{e}{2\varepsilon_S} [\langle N_A^S(W)\rangle\, W^2 + N_A^d W_\Delta^2], \qquad (23)$$

where we have defined $\langle N_A^S(W)\rangle$ in a manner analogous to $\langle N_A^d(W_\Delta)\rangle$;

$$\langle N_A^S(W)\rangle \equiv \frac{2}{W^2} [\int \mathrm{d}x \int N_A^S(x)\,\mathrm{d}x]_{x=0}\,. \tag{24}$$

If $N_A^S(x)$ is known, (24) can be integrated to give $\langle N_A^S(W)\rangle$ for various values of W. Equations (14) and (23) can then be used to obtain Ψ_S' and Ψ_S, respectively. Finally, (21) can be used with $\langle N_A^d(W_\Delta)\rangle$ replaced by N_A^d and N_A^S replaced by $\langle N_A^S(W)\rangle$ to obtain the ordinate of the ISC plot.

To find the shallow impurity profile from an experimental ISC plot, (21) can be used in the form

$$\langle N_A^S(W)\rangle = \left[\frac{2\varepsilon_S\Psi_S}{eW^2} - N_A^d\left(\frac{W_\Delta^2}{W^2}\right)\right].$$

Ψ_S is obtained from the abscissa voltage for a particular value of $1/C_S^2$. Correspondingly, W^2 is obtained from this value of $1/C_S^2$ by use of (8) and W_Δ is obtained from $W_\Delta = W - W_d$ in which W_d corresponds to the Fermi level calculated from $N_A^S = \langle N_A^S(W)\rangle$. Thus an iteration may be required although Φ_{Fp} is not a strong function of N_A^S. Finally, $N_A^S(W)$ can be found by use of (22) as was done previously for the deep impurity profile.

3. Conclusion

The theory of the ISC measurement in p-type Schottky diodes with deep and shallow acceptor impurities has been developed in detail. The impetus for this study was a need to better understand the characteristics of ion implanted Schottky diode structures; however, a more general theory has been presented. In particular, it has been shown that the static ISC curve, in spite of its non-linearities, is very useful in deducing the impurity profiles of Schottky diode structures.

Acknowledgements

The author wishes to thank Dr. Kjell Ingebrigtsen of NTH for suggesting this problem; the Norges Tekniske-Naturvitenskapelige Forskningsråd for their support during my sabbatical leave stay at NTH; and finally a special thanks to NTH for the hospitality shown me during my stay.

References

[1] A. G. Milnes, Deep Impurities in Semiconductors, John Wiley & Sons, New York 1973 (p. 191).
[2] R. R. Senechal and J. Basinski, J. appl. Phys. **39**, 3723, 4581 (1968).
[3] J. A. Copeland, IEEE Trans. Electron Devices, **ED-16**, 445 (1969).
[4] Y. Zohta, Appl. Phys. Letters **17**, 284 (1970); J. appl. Phys. **43**, 1713 (1972); Solid State Electronics **16**, 1029 (1973).
[5] W. G. Oldham and S. S. Naik, Solid State Electronics **15**, 1085 (1972).
[6] K. Hesse and H. Strack, Solid State Electronics **15**, 767 (1972).
[7] C. R. Crowell and K. Nakano, Solid State Electronics **15**, 605 (1972).
[8] G. H. Glover, IEEE Trans. Electron Devices, **ED-19**, 138 (1972).
[9] M. Bleicher and E. Lange, Solid State Electronics **16**, 375 (1973).
[10] G. I. Roberts and C. R. Crowell, Solid State Electronics **16**, 29 (1973).
[11] L. M. Lambert and H. J. Kump, phys. stat. sol. (a) **21**, 235 (1974).

(Received May 3, 1979)

phys. stat. sol. (a) 54, 675 (1979)

Subject classification: 1.4 and 1.5; 21.6

Stanford/NASA–Ames Joint Institute for Surface and Microstructure Research, Stanford[1]) *(a) and Instituto de Fisica, Universidad Nacional Autónoma de México*[2]) *(b)*

A New Approach for Simultaneous Orientation and Three-Dimensional Shape Determination of Small Metal Particles

By

K. HEINEMANN (a), M. J. YACAMÁN (b), and H. POPPA (a)

The three-dimensional habit of small metal crystallites (> 5 nm in size) can be determined by a combination of selected-zone dark-field (SZDF) and weak-beam dark-field (WBDF) transmission electron miroscсopy. The information can be obtained from one electron micrograph which is taken in a defocused SZDF condition where all reflection images are clearly separated. The method is based on an evaluation of a) the general intensity of corresponding reflection images, which gives rise to the orientation of the particles with respect to the electron beam, and of b) the weak-beam thickness fringes which appear in those reflection images that were subject to a relatively large excitation error. The particular advantage over the WBDF method proposed earlier by Yacamán and Ocaña is that no tilting sequence is required.

Die dreidimensionale Form von kleinen Metall-Kristalliten (> 5 nm Größe) kann mit einer Kombination von Zonen-Dunkelfeld- (selected-zone dark-field, SZDF)- und Schwachstrahl-Dunkelfeld- (weak-beam dark-field, WBDF)-Elektronenmikroskopie bestimmt werden. Die gesamte Information ist in einer elektronenmikroskopischen SZDF-Aufnahme enthalten, die bei einer Defokussierung aufgenommen wird, bei welcher alle Dunkelfeldbilder räumlich getrennt abgebildet werden. Die Methode basiert auf einer Auswertung von a) der Intensität von zusammengehörigen Dunkelfeldbildern, welche über die Orientierung der Kristallite gegenüber dem abbildenden Elektronenstrahl Aufschluß gibt, und von b) Schwachstrahl-Dickenkonturen, welche in den Dunkelfeldbildern zu erkennen sind, die einem relativ großen Erregungsfehler unterworfen sind. Ein besonderer Vorteil gegenüber der WBDF-Methode, die Yacamán und Ocaña kürzlich vorschlugen, liegt darin, daß keine Kipp-Serien benötigt werden.

1. Introduction

The full characterization of small metal particles which are produced in the very early stages of thin film growth is highly desirable for a better understanding of the fundamental surface physical processes of thin film nucleation and growth. Habit and shape determinations of very small metal crystallites are also needed for new insights into chemical reaction mechanisms of supported metal catalysts. However, the determination of the habit, crystallographic structure, and orientation of metal particles in the 1 to 5 nm size range is quite complicated, and many of the conventional methods fail in this size range [1]. In this paper, a new experimental approach to the solution of this problem is described. It allows the simultaneous determination of the orientation and three-dimensional habit of individual small metal particles in the nm size range from one electron micrograph. The method is based on high-resolution transmission electron microscopy (TEM) employing simultaneously the selected-zone dark-field (SZDF) [2] and weak-beam dark-field (WBDF) [3] techniques.

[1]) Stanford, California 94305, USA.

[2]) México 20, D. F., México.

2. Approach

In SZDF microscopy, an annular objective aperture is used to select the diffraction zone that is allowed to contribute to the image formation. Usually, several diffracted beams of different azimuth are located within the selected zone. Each admitted beam gives rise to a separate dark-field ("reflection") image. The dark-field images are spatially displaced as a function of the focus setting of the TEM objective lens. The displacement direction is parallel to the respective **g**-directions as is described by Heinemann et al. [4]. For anyone zone, a focusing condition can be found for which the reflection images are superimposed. (The exact superposition of the reflection images is a pre-requisite for lattice fringe imaging in the SZDF mode [2]; it also establishes an extremely sensitive criterion for microscope alignment and astigmatism correction [5].) In the technique described here, the SZDF mode is employed in a defocused condition where all reflection images are clearly separated.

The TEM image contrast of a small metal particle is strongly dependent on its thickness profile (habit) and orientation with respect to the incident electron beam [6]. This dependence is most obvious under weak-beam dark-field conditions (large excitation error). The WBDF image intensity decreases with increasing excitation error, i.e. with increasing deviation from the Bragg condition. Superimposed on this effect is a concurrent rapid decrease of the extinction distance. As a consequence, equal thickness fringes can appear in the WBDF image and may correspond to thickness variations of as little as 1 nm. (The fringe spacing as function of excitation error actually goes through a minimum, which depends on the atomic number of the crystal material and the **g**-vector of the respective reflection and occurs at relatively large excitation errors [7].)

The detailed determination of the particle orientation, i.e. its corresponding Ewald sphere construction, can then be performed with the following steps from one defocused SZDF micrograph: First, the visually perceived intensity of the different dark-field images is used for a rough estimate of the position of the Laue circle in the diffraction pattern. (At this stage, the existence of thickness fringes is of no consequence, and only the average reflection image intensity is considered. The Laue circle will be near reflections that produce strong DF image intensities, and further away from reflections that are associated with weaker image intensities.) Second, the parameters for the excitation errors thus estimated are used for a first computer calculation in which the contrast of the observed SZDF reflection images is calculated, using the dynamical diffraction theory and a total of 49 beams that include all reflections near the estimated Laue circle position [7]. Third, the results are compared with the experimental SZDF reflection images, and a refined set of values for the excitation errors is generated as second-order approximation, which his then used for a refined computer calculation. The process is repeated until a satisfactory agreement is reached for the image contrast of all respective reflection images. Usually, already the first computer iteration gives quite satisfactory results. With some practice and experience in using this method one can usually determine the orientation of a particle with respect to the incident electron beam from the first visual inspection of the defocused SZDF images to well within 0.5°.

Once the excitation errors for the various reflection images have been determined, the observed thickness contour images can be used for the determination of the particle habit as described by Yacamán and Ocaña [3] and the present authors [7]. The many-beam dynamical calculations necessary for this determination require the knowledge of the particle orientation to within about $\pm 0.5°$ accuracy, which is well within the capability of the SZDF-WBDF method described here.

The strength of the particle characterization approach presented in this report is the fact that it can be performed from *one* electron micrograph and for as many individual particles as are contained in the imaged area, even if the orientations of individual particles, with respect to the incident electron beam, are quite different (see, for instance, Heinemann et al. [4] for gold particles nucleated in a vapor deposition process onto alkali halide surfaces). No tilting sequences are needed, as is required for the method reported in references [3] and [7], and the number of computer calculations per particle characterization is reduced considerably.

3. Experimental and Results

Gold islands in the nm size range were used as model particles. They were obtained by vapor deposition onto NaCl substrates in the 10^{-7} Pa residual-gas pressure range. Standard carbon coating and specimen preparation techniques were used to produce suitable TEM specimens. A Siemens Elmiskop 101 with a high-resolution objective polepiece arrangement and selected-zone dark-field objective apertures that admitted either the Au(111) + Au(200) zones or the Au(220) zone, was used.

Fig. 1a shows a square (100)-oriented particle in the superimposed ("focused") $\{(111) + (200)\}$ SZDF condition. A slight change in focus ($\delta z \approx 100$ nm) yields four spatially separated reflection images as shown in Fig. 1b. They correspond to $\boldsymbol{g}(0\bar{2}0)$, $\boldsymbol{g}(020)$, $\boldsymbol{g}(00\bar{2})$, and $\boldsymbol{g}(002)$. As can be clearly observed in Fig. 1b, the images are very different in contrast, which indicates that the [001] axis of the particle was not quite parallel to the imaging beam. Some of the reflection images also contain thickness fringes. (One of the four reflection images, D, is so weak, as the result of a particularly large excitation error, that it is just barely visible in the original photonegative.)

Fig. 1 c and d show computer simulations of the reflection images A and B in Fig. 1b. Agreement between the experimental images (Fig. 1b) and the computer-simulated images was obtained after the excitation errors had been determined for A and B in accordance with the method described above, and when the general particle shape of a truncated pyramid was assumed and entered into the calculations. The assumption of this particle shape was obvious from the defocussed SZDF images (Fig. 1b). They show the particle edges to be in [011] directions, and the particle faces were then likely to be (111) planes. The observed thickness fringes in the reflection image A of Fig. 1b suggested a truncation of the pyramid, and thus, the only uncertainty with respect to the shape was the degree of truncation. 50% ± 10% gave the best results in the actual computer image simulations (Fig. 1c). (Evidently, the accuracy with which the degree of truncation can be determined, is dependent on the number of spacings of the thickness fringes, and the accuracy increases with particle size and with optimization of the excitation error for minimum extinction thickness.)

A different example of the SZDF-WBDF thickness fringes method is illustrated in Fig. 2. The (111)-oriented gold particle (with $\{111\}$ planes parallel to the substrate surface) in the center of Fig. 2a was investigated. The bright-field image (Fig. 2a) shows the triangular projection of the particle. Fig. 2 b and c show (220) SZDF images in superimposed and spatially separated condition, respectively. Fig. 2d is a schematic representation of the reflection images shown in Fig. 2c and of the lattice planes responsible for the respective dark-field images (increasing excitation error is indicated by decreasing line thickness for the lattice planes). Fig. 2e shows the corresponding Ewald sphere construction. No thickness fringes are observed in the WBDF images, although thickness fringe calculations for the excitation errors corresponding to the different dark-field images (only two are shown in Fig. 2f, for $(02\bar{2})$ and $(\bar{2}02)$) (clearly indicated that thickness fringes should have been observed at least in the $(\bar{2}02)$ reflection image unless the particle thickness does *not* exceed a few nm. The only

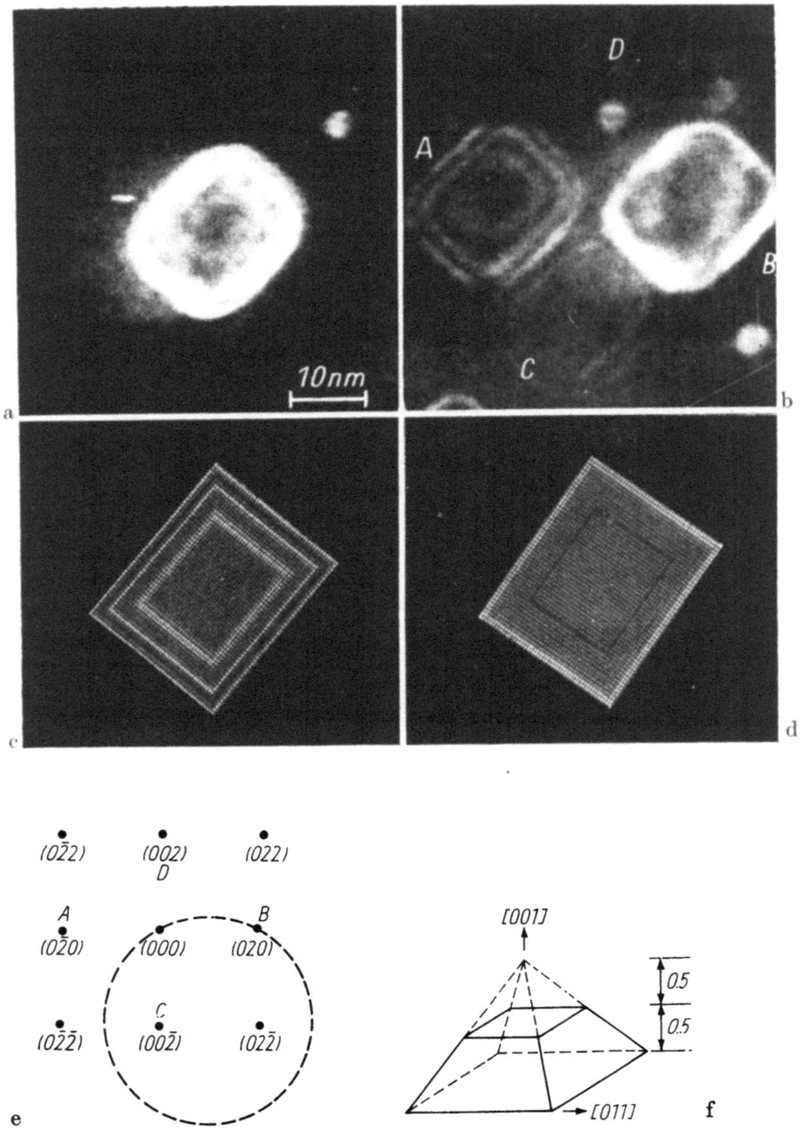

Fig. 1. Particle habit and orientation analysis by the SZDF-WBDF technique. a) and b) are focused and defocused SZDF micrographs of the gold particle. c) and d) show computer-simulated images for the reflection images A and B, respectively, e) illustrates the corresponding Laue circle and f) indicates the three-dimensional model of the particle resulting from the analysis (50% truncated pyramid)

explanation is that the particle is very flat and that its three sides are steep, as would be the case for a strongly truncated tetrahedron with {111} faces, as is shown in Fig. 2g.

4. Discussion and Conclusions

The comparison of experimentally observed image intensities with computer evaluations of the intensity distribution in a set of SZDF reflection images represents a means of particle orientation characterization which is more accurate than any other known electron optical method that can be applied to particles in the nm size range. In the case of very flat particles (Fig. 2) the accuracy of the determination of the particle height and orientation is reduced because it is not possible to match experimental extinction distances with theoretical calculations.

It is of major significance that the SZDF-WBDF method allows the determination of the habit of a crystalline particle without the need of a tilting sequence as in refer-

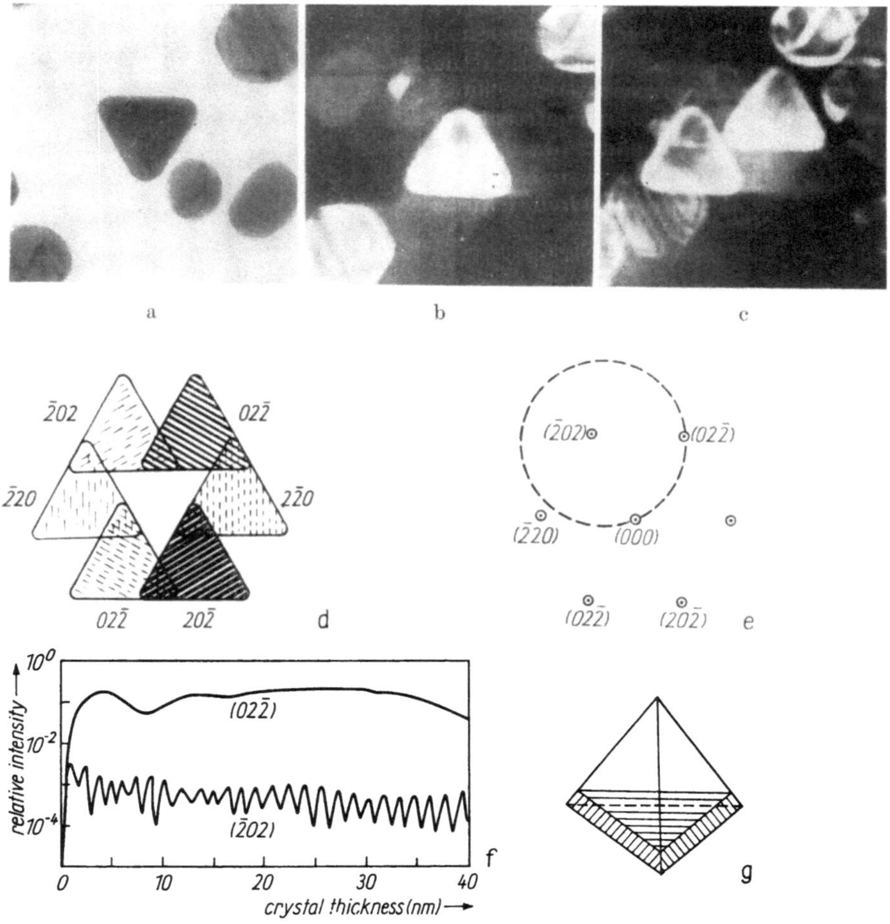

Fig. 2. Triangular gold particle in a) bright field and b) focused and c) defocused selected-zone dark field. d) shows schematically the SZDF images with the respective (220) lattice planes; e) indicates the corresponding Ewald sphere construction; f) is the result of computer calculations (see text); g) shows a three-dimensional model of the analyzed particle (strongly truncated tetrahedron)

ences [3] and [7]. Therefore, this technique can also be useful for determining the initial orientation of a particle at the start of any tilting sequence.

Obviously, the full analytical potential of this technique can only be realized if a high-resolution TEM equipped with annular objective apertures, and computer facilities for many-beam dynamical electron diffraction calculations are available. However, even without the computer calculations the method described here is still useful for particle orientation determinations to better than 1° with respect to the incident electron beam, and for good *qualitative* three-dimensional particle shape determinations.

The major restriction of this technique is that it can be applied to more or less isolated particles only. Closely spaced particles become difficult to analyze because their (defocused) reflection images will then overlap.

Acknowledgements

The authors are indebted to Prof. J. Cowley for fruitful discussions. One of the authors (M.J.Y.) was holding a National Research Council Senior Research Associationship while on leave of absence from the Universidad Nacional Autónoma de México. Support through NASA grant NSG 2249 is also acknowledged.

References

[1] J. R. ANDERSON, The Structure of Metallic Catalysts, Academic Press, 1975.
[2] K. HEINEMANN and H. POPPA, Appl. Phys. Letters **20**, 122 (1972).
[3] M. J. YACAMÁN and T. OCAÑA, phys. stat. sol. (a) **42**, 571 (1977).
[4] K. HEINEMANN, M. J. YACAMÁN, C. Y. YANG, and H. POPPA, J. Crystal Growth, in the press.
[5] K. HEINEMANN, Optik **34**, 113 (1971).
[6] J. M. COWLEY, Diffraction Physics, Elsevier Publ. Co., New York 1975.
[7] M. J. YACAMÁN, K. HEINEMANN, and H. POPPA, to be published.

(Received March 8, 1979)

phys. stat. sol. (a) 54, 681 (1979)

Subject classification: 10.2; 14.4; 20.1; 22.5.2

Physikalisches Institut der Universität Münster[1])

Photo-Induced Polarization and Photo-Induced Depolarization of S^{2-}–Anion Vacancy Dipoles in Potassium Iodide

By

J. PRAKASH[2])

The effect of irradiation in the presence/absence of an electric field is studied in S^{2-}-doped KI single crystals using ITC technique. It is observed that the irradiation in the 368 nm absorption band in the presence of an electric field creates a situation in which more dipoles are aligned along the electric field direction, whereas such an irradiation in the absence of an electric field destroys the preferred orientation of the frozen-in dipoles. These phenomena are named photo-induced polarization (PIP) and photo-induced depolarization (PID), respectively. PIP is studied as a function of the electric field strength. It is observed that PIP measurements below a limiting value of the electric field result in a decrease of the ITC peak whereas above they yield an increase of the ITC peak. PID is studied as a function of the irradiation temperature and irradiation time. It is observed that PID does not take place below 130 K. The results are qualitatively explained following a bistable model.

Der Einfluß von Bestrahlung in Anwesenheit/Abwesenheit eines elektrischen Feldes wird in Si^{2-}-dotierten KJ-Einkristallen mit der ITC-Technik untersucht. Es wird beobachtet, daß eine Einstrahlung in die 368 nm-Absorptionsbande bei Anwesenheit eines elektrischen Feldes eine Situation hervorruft, bei der mehr Dipole längs des elektrischen Feldes ausgerichtet werden, während eine derartige Bestrahlung bei Abwesenheit eines elektrischen Feldes die Vorzugsorientierung der eingefrorenen Dipole zerstört. Diese Erscheinungen werden photoinduzierte Polarisation (PIP) bzw. photoinduzierte Depolarisation (PID) genannt. PIP wird in Abhängigkeit von der elektrischen Feldstärke untersucht. Es wird beobachtet, daß PIP-Messungen unterhalb eines Grenzwertes des elektrischen Feldes eine Abnahme des ITC-Maximums ergeben, während sie oberhalb dieses Feldes zu einem Anstieg des ITC-Maximums führen. PID wird in Abhängigkeit von der Bestrahlungstemperatur und der Bestrahlungszeit untersucht. Es wird beobachtet, daß PID unterhalb 130 K nicht stattfindet. Die Ergebnisse werden qualitativ mit einem bistabilen Modell erklärt.

1. Introduction

The dielectric relaxation observed in alkali halide crystals doped with divalent impurity ions is due to impurity–vacancy (IV) dipoles. With respect to the divalent impurity ion, the IV dipole lies along one of the twelve equivalent orientations ⟨110⟩. The orientation of these dipoles is influenced by the temperature and the frequency of the applied electric field giving rise to dielectric dispersion and absorption. The method of ionic thermocurrent (ITC) [1] is widely used for determining the dielectric behaviour of IV dipoles. It has recently been reported by us [2 to 4] that the irradiation in the 368 nm absorption band in the presence of an electric field in the KI : S^{2-} system creates a situation in which more dipoles are aligned along the electric field direction, whereas such an irradiation in the absence of an electric field destroys the preferred orientation of the frozen-in dipoles. These phenomena are named as photo-induced polarization (PIP) and photo-induced depolarization (PID), respectively. PIP and PID have re-

[1]) Schloßplatz 7, D-4400 Münster, BRD.

[2]) DAAD post doctoral research fellow. Present address: Department of Physics, University of Gorakhpur, Gorakhpur-273001, India.

cently been reported also for the Suzuki phase by Benci et al. [5] in Pb^{2+}-doped KCl. The present paper reports the measurements on S^{2-}-doped potassium iodide single crystals which form a part of our systematic study of the phenomena of PIP and PID. PID has been studied as a function of the irradiation temperature and irradiation time. It has been observed that PID does not take place below 130 K. PIP has been studied as a function of the electric field strength. A limiting value of the electric field, E_0, is obtained below which the peak of the ITC spectrum decreases instead of increasing even during the PIP procedure.

2. Experimental Techniques

Single crystals of KI doped with S^{2-} are prepared by an almost similar procedure as described in an earlier paper [6]. For ionic thermocurrent measurements (for details see [3]) the crystal is polarized by applying a polarizing voltage U_p at the polarization temperature T_p for about 4 min, and with the electric field still on it is rapidly cooled down to LNT when the electric field is switched off. This experimental procedure for ITC measurement is mentioned in the present paper as: The crystal is polarized at T_p by applying a polarizing voltage U_p in the polarization temperature range T_p to LNT. The results reported here have been found to be repeatable for various single crystals of varying concentration.

3. Results and Discussions

The ITC spectrum in sulphur-doped potassium iodide single crystals due to S^{2-}–anion vacancy dipoles is shown by curve 1 of Fig. 1, where the depolarization current (I) has been plotted as a function of temperature (T). For recording the ITC spectrum the crystal is polarized at 210 K by applying a polarizing voltage of 1500 V in the polarization temperature range 210 K to LNT. It is apparent from the figure that the maximum current of the ITC spectrum lies at 204 K (T_{max}) for a heating rate of 0.05 K/s. The other relevant dielectric relaxation parameters are reported elsewhere [3].

The S^{2-}–anion vacancy dipole in KI has three characteristic absorption bands at 232, 278, and 368 nm [7]. Irradiation in the 232 and 278 nm absorption bands at 240 K results into a photochemical reaction [8]. No photochemical reaction is observed when the crystal is irradiated at LNT in the 232 and 278 nm absorption bands. Also, irradiation in the 368 nm absorption band from room temperature down to LNT has no effect at all.

However, irradiation in the 368 nm absorption band in the presence of an electric field ($U_{irr} = U_p$) creates a situation in which more dipoles are aligned along the electric field direction [3]. The effect of irradiation in the presence of an electric field (U_{irr}) is shown by curves 2, 3, and 4 of Fig. 1 [3]. It is apparent from Fig. 1 that the peak of the ITC spectrum increases. The difference in the area of ITC curves 2, 3, and 4 with respect to 1 gives the number of IV dipoles which have undergone further orientation. This phenomenon of increase in the ITC peak is called photo-induced polarization

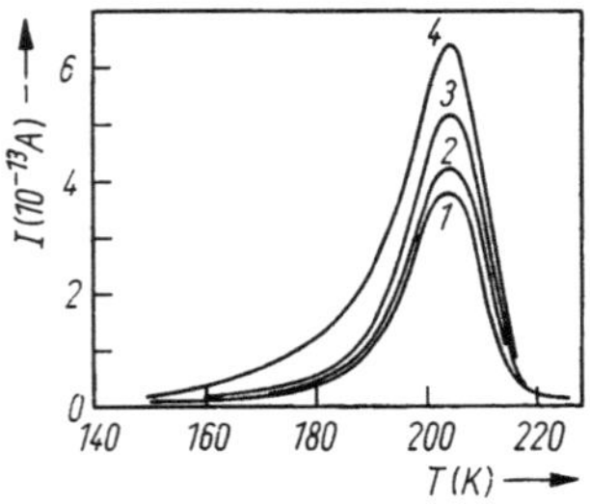

Fig. 1. Photo-induced polarization of the S^{2-}–anion vacancy dipole in sulphur-doped potassium iodide single crystals. $T_p = 210$ K, $U_p = 1500$ V, polarization temperature range 210 K to LNT, irradiation wavelength 368 nm, $U_{irr} = U_p = 1500$ V, and irradiation time 210 s. (1) Normal ITC spectrum, (2) irradiated at 107, (3) 185, (4) 179 K. For the normal ITC spectrum the crystal is not irradiated

(PIP). PIP has been studied as a function of irradiation temperature, irradiation time, and electric field strength and the results are reported in our earlier paper [3]. It is speculated that the further polarization during PIP procedure takes place through the electronic excited state where the activation energy for the orientation of IV dipoles (E_a') is sufficiently small in comparison to that required in the electronic ground state (E_a). It has been shown by Fischer [9] that $E_a' < E_a$ is not a sufficient condition for orientation in the electronic excited state. The phenomenon of PIP does take place only when E_a' is sufficiently less than E_a. If this is a mechanism responsible for the phenomenon of PIP, it should also happen that the polarized dipoles disorient when being irradiated in the absence of an electric field in the 368 nm absorption band. The effect of irradiation in the absence of an electric field ($U_{irr} = 0$ V) is shown by curve 2 of Fig. 2 [3]. It is apparent that the peak of the ITC spectrum decreases in comparison to curve 1 (normal ITC spectrum) of Fig. 2. The difference in the area of ITC curve 1 and 2 of Fig. 2 gives the concentration of IV dipoles which have undergone depolarization. This phenomenon of decrease in the ITC peak is named photo-induced depolarization (PID). PIP and PID have also been recently observed for the Suzuki phase (heterogeneous phase occlusions of IV dipoles) by Benci et al. [5] in Pb^{2+}-doped KCl. It should be mentioned, however, that according to them the maximum current of the ITC spectrum corresponding to an isolated IV dipole occurs at 222 K, whereas that corresponding to the Suzuki phase at 278 K. It is the ITC spectrum at 278 K corresponding to the Suzuki phase which shows the phenomena of PIP and PID. They have argued that PIP and PI Din the Suzuki phase are taking place via an electrically excited state. In another report on Pb^{2+}-doped KCl, Capelletti et al. [10] have emphasized that the disorientation of IV dipoles via the electronic excited state is the mechanism responsible for PID. In the present paper we report some more measurements on S^{2-}-doped potassium iodide single crystals which will help in building up a suitable model for understanding the mechanism responsible for PIP and PID.

In the case of Fig. 2, the crystal is polarized at 210 K by applying a polarizing voltage of 1500 V in the polarization temperature range 210 to 179 K (for details see [3]). For curve 2 the crystal after polarization is soon irradiated for 5 min at 179 K in the absence of an electric field ($U_{irr} = 0$ V) in the 368 nm absorption band. The decrease in the peak of the ITC spectrum comes out to be 23%. Just to insure that this depolarization is solely due to PID and does not have any contribution from the self-depolarization, the crystal after polarization is left as such at 179 K for 5 min. The depolarization current recorded showed a decrease in the ITC peak by 11% when compared with curve 1 of Fig. 2. Thus, the effective contribution due to PID at 179 K is only 12% whereas polarized dipoles depolarize due to thermal agitation at 179 K by 11%. For the time being let us not stick to the magnitudes just mentioned, but infer simply that there are temperatures at which the phenomena of PID and self-depolarization take place simultaneously. Although, it is not advisable to study the time dependence at 179 K because of the presence of simultaneous phenomena of self-depolarization and PID. Even then, just for the sake of information the data of Fig. 3 are

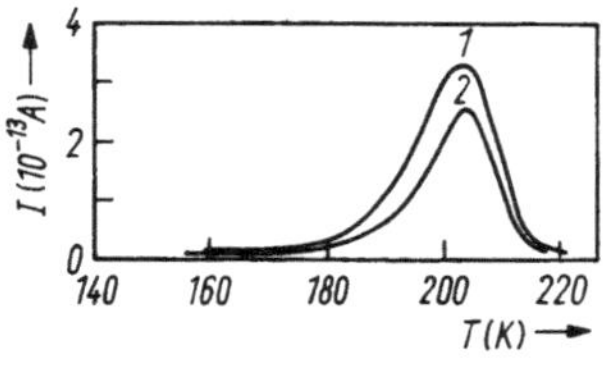

Fig. 2. Photo-induced depolarization of the S^{2-}–anion vacancy dipole in sulphur-doped potassium iodide single crystals. $T_p = 210$ K, $U_p = 1500$ V, $U_{irr} = 0$ V, irradiation temperature 179 K, irradiation wavelength 368 nm, and irradiation time 5 min. (1) Normal ITC spectrum when polarized in the temperature range 210 to 179 K; (2) ITC spectrum when polarized in the temperature range 210 to 179 K and soon irradiated at 179 K for 5 min in the absence of an electric field in the 368 nm absorption band

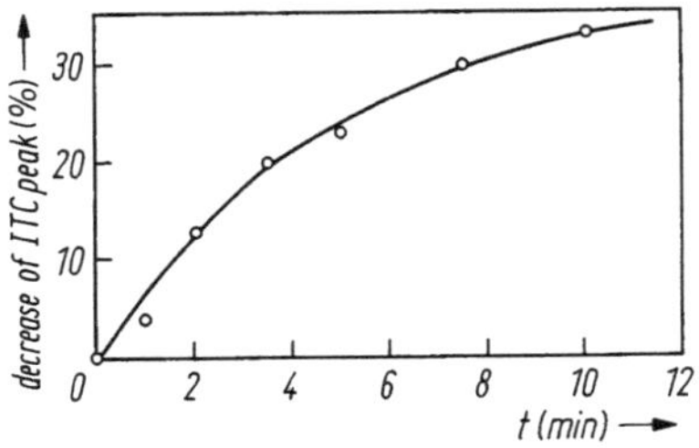

Fig. 3. The dependence of photo-induced depolarization of the S^{2-}–anion vacancy dipole on irradiation time (t) in the KI:S^{2-} system recorded at the irradiation temperature of 179 K where the phenomena of PID and self-depolarization take place simultaneously. $T_p = 210$ K, $U_p = 1500$ V, polarization temperature range 210 to 179 K, irradiation temperature 179 K, $U_{irr} = 0$ V, and irradiation wavelength 368 nm

collected. The crystal is polarized at 210 K by applying a polarizing voltage of 1500 V in the polarization temperature range 210 to 179 K. Soon after switching off the electric field at 179 K, the crystal is irradiated at 179 K in the absence of an electric field ($U_{irr} = 0$ V) in the 368 nm absorption band for varying irradiation time (t). It is obvious from Fig. 3 that the peak of the ITC spectrum initially decreases linearly as expected and then tends towards a constant value exponentially. The exponential decrease of the ITC peak is due to the simultaneous presence of self-depolarization along with PID. Since self-depolarization and PID are simultaneously present, it is planned to study the irradiation temperature dependence at first and then to study the time dependence of the phenomenon of PID. It will be worth mentioning that the self-depolarization due to thermal disturbances will take place through the electronic ground state, whereas PID takes place through the electronic excited state.

In order to study the dependence of PID on the irradiation temperature, the crystal is polarized at 210 K by applying a polarizing voltage of 1500 V in the polarization temperature range 210 K to the irradiation temperature. At the irradiation temperature (T_{irr}) the electric field is switched off and the crystal is soon irradiated in the absence of an electric field ($U_{irr} = 0$ V) in the 368 nm absorption band for 210 s. Just after the irradiation the crystal is rapidly cooled down to LNT. The results of such experiments corresponding to different temperatures of irradiation are shown in Fig. 4. It is apparent from the figure that from LNT up to 130 K the phenomenon of PID does not take place. The ITC peak starts decreasing slowly with increasing irradiation temperature above 130 K and then rapidly above 170 K. The rapid decrease of the ITC peak above 170 K is partly due to the self-depolarization of the polarized dipoles. Self-depolarization below 170 K is experimentally found to be absent. It is obvious that below 130 K the IV dipoles are unable to cross the potential energy barrier in the electronic excited state through which the phenomenon of PID takes place.

Let us try to explain the irradiation temperature dependence of PID following the bistable model of IV dipoles where there are two orientations in the electronic ground state [11] and similarly only two orientations in the electronic excited state. The corresponding potential energy barrier heights in the electronic ground state and in the electronic excited state are E_a and E'_a, respectively. During the PID procedure, we irradiate in a state where polarized dipoles are frozen in and thereby statistically the

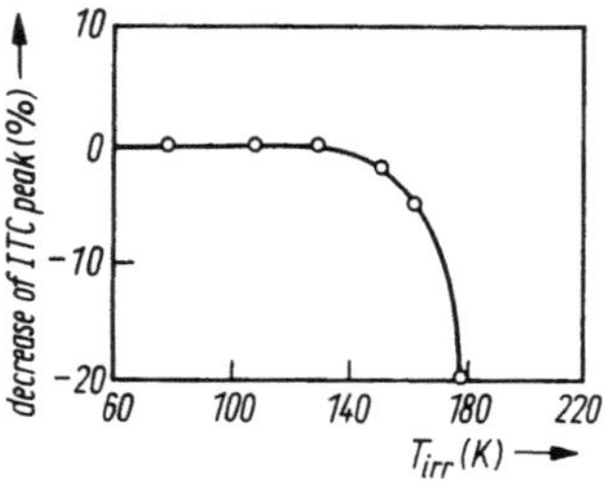

Fig. 4. The dependence of photo-induced depolarization of the S^{2-}–anion vacancy dipole on irradiation temperature (T_{irr}) in the KI:S^{2-} system. $T_p = 210$ K, $U_p = 1500$ V, polarization temperature range 210 K to T_{irr}, $U_{irr} = 0$ V, irradiation wavelength 368 nm, and irradiation time 210 s

number of dipoles in the frozen-in condition in one of the potential wells corresponding to the preferred orientation (say, well No. 1) will be larger than the number of dipoles in the other potential well which corresponds to the unpreferred orientation (say, well No. 2). This difference in the number of dipoles in the two potential wells is responsible for the frozen-in polarization. Also, in the electronic excited state the number of dipoles in potential well No. 1 will be larger than that present in the potential well No. 2 since dipoles from potential well No. 1 as well as from potential well No. 2 of the electronic ground state are excited simultaneously. The potential energy barrier height will not be modified during PID (since the electric field is absent) and hence the jump probabilites per second in going from one of the potential wells to the other, i.e. from No. 1 to 2 (P_{12}) or No. 2 to 1 (P_{21}), will be the same. So, statistically more dipoles will jump from potential well No. 1 to 2 through the electronic excited state resulting into the decrease of the polarization during PID. Since $E'_a < E_a$, there must still be a sufficiently high probability for depolarization via electronic excited state below 170 K. The relaxation time (τ') of the dipole centres in the electronic excited state depends on temperature in accordance with $\tau' = \tau'_0 \exp(E'_a/kT)$ where τ'_0 is the pre-exponential factor, whereas the lifetime τ_e ($\approx 10^{-8}$ s) of the dipole centres in the electronic excited state is temperature independent. In the cases when $\tau' < \tau_e$ (i.e. above 130 K), the dipole centres in the electronic excited state are able to attain the thermal equilibrium corresponding to the crystal temperature. As soon as $\tau' \geqq \tau_e$ (i.e. below 130 K), the thermal equilibrium in the electronic excited state cannot be obtained any more. The dipole centres then return to the electronic ground state befoie they could cross the potential energy barrier in the electronic excited state. This leads to the disappearance of the phenomenon of PID below 130 K.

We have seen that during PID above 170 K, self-depolarization also contributes and hence the effective temperature range in which further properties of PID can be investigated is 130 to 170 K (see Fig. 4). We choose a temperature of 162 K for studying the depencence of PID on irradiation time. In this case, the crystal is polarized at 210 K by applying a polarizing voltage of 1500 V in the polarization temperature range 210 to 162 K. At 162 K, the electric field is switched off and the crystal is soon irradiated at 162 K for varying irradiation time (t) in the absence of an electric field ($U_{irr} = 0$ V) in the 368 nm absorption band. It is apparent from Fig. 5 that the peak of the ITC spectrum initially decreases linearly with irradiation time and finally tends towards a constant value. We notice from the figure that the polarized dipoles depolarize to an extent of 8%. If there is a single mechanism responsible for the phenomenon of PID, all the dipole centres should behave alike. One should expect, then, a decrease of 100% after long enough irradiation following an exponential law. When only 8% of the polarization can be bleached one has to suspect that at least two different groups of dipole centres are involved. This value of 8% obtained at 162 K could depend on the temperature of irradiation. This aspect has not yet been checked. It will be justified to mention here, however, that in the present investigations the whole area of the crystal specimen is exposed to irradiation.

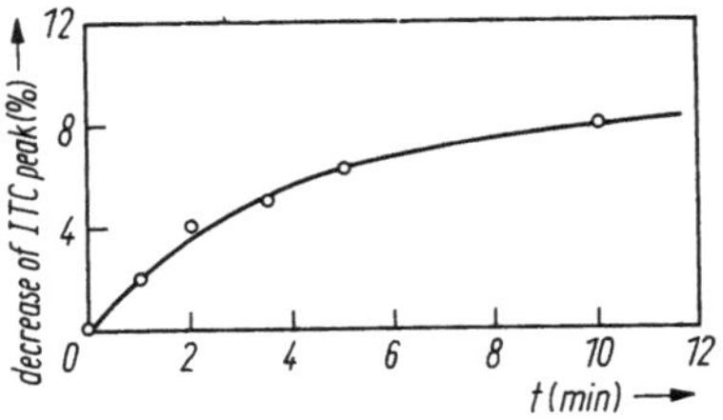

Fig. 5. The dependence of photo-induced depolarization of the S^{2-}–anion vacancy dipole on irradiation time (t) in the KI:S^{2-} system. $T_p = 210$ K, $U_p = 1500$ V, polarization temperature range 210 to 162 K, irradiation temperature 162 K, $U_{irr} = 0$ V, and irradiation wavelength 368 nm

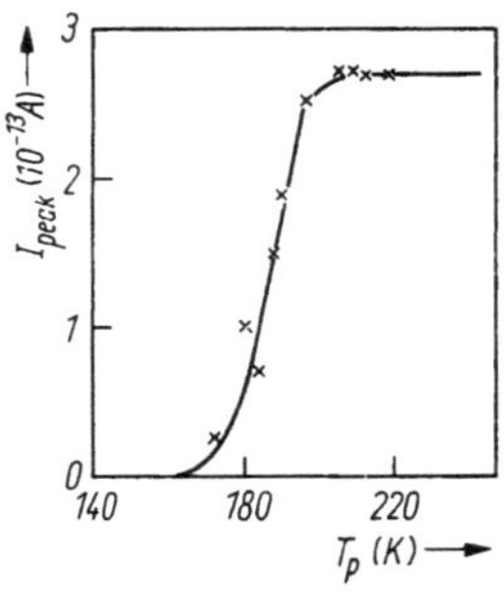

Fig. 6. Polarization of the S^{2-}–anion vacancy dipole in sulphur-doped potassium iodide single crystals at different polarizing temperatures (T_p). Polarized at T_p, $U_p = 1500$ V, and polarization temperature range T_p to LNT

To visualize the mechanism of orientation in the electronic ground state, the polarization of IV dipoles at varying polarization temperatures has been recorded. The results of such experiments are presented in Fig. 6. In this case the crystal is polarized at varying polarization temperature T_p by applying a polarizing voltage of 1500 V in the polarization temperature range T_p to LNT. It is apparent from the figure that when the crystal is polarized in the vicinity of T_{max} (204 K) such that $T_p \geqq T_{max}$, the peak of the ITC spectrum (I_{peak}) is unaffected and the crystal is polarized to the same extent. The polarization starts decreasing rapidly below 204 K. At about 170 K and at other lower temperatures the crystal ceases to be polarized by the application of the electric field. We have already observed that above 170 K self-depolarization also contributes to PID, whereas below 170 K self-depolarization is absent which obviously takes place through the electronic ground state. Thus, the inference that the crystal cannot be polarized at ≈ 170 K and at other lower temperatures by the application of the electric field, is in accordance with the results of PID measurements.

To record the dependence of PIP on the electric field strength a routine experiment was carried out which is reported in our earlier paper [3]. It has been observed that the peak of the ITC spectrum (I_{peak}) recorded when the crystal was not irradiated varies linearly with the applied polarizing voltage (U_p). Also, when the crystal was irradiated at 179 K in the presence of an electric field ($U_{irr} = U_p$) in the 368 nm absorption band, I_{peak} varies linearly with U_p. Although, there is a linear dependence of I_{peak} on U_p in the latter case, we had some doubt that at lower polarizing voltage the linear dependence does not hold. This deviation could not be emphasized with certainty as at lower voltages the accuracy was too poor to draw a definite conclusion. To decide whether the straight line dependence deviates at lower polarizing voltage during PIP, the following experiments are performed.

Based on the findings of PID (see Fig. 4 and 5), a temperature of 162 K was chosen for the purpose of irradiation. The crystal is polarized at 210 K by applying a polarizing voltage of 1500 V for ≈ 4 min, and with the electric field still on the crystal is cooled down to the irradiation temperature of 162 K. At 162 K the electric field is decreased to different irradiation voltages (say 500 V) and the crystal is irradiated in the presence of this irradiation voltage (500 V) in the 368 nm absorption band for 210 s. Soon after the irradiation the crystal is rapidly cooled down to LNT, where the electric field is switched off and the depolarization current is recorded. The results of such experiments are presented in Fig. 7. It should be worth mentioning that at 162 K the contribution of the self-depolarization is practically zero and hence just after the irradiation, increasing the irradiation voltage up to the polarizing voltage of 1500 V or leaving it at its values has no effect at all on the ITC spectrum. I_{peak} in Fig. 7 corresponds to the peak of the ITC spectrum when the crystal is polarized at 210 K and is irradiated at 162 K, whereas $I_{0\,peak}$ in Fig. 7 corresponds to the peak of the ITC spectrum when the crystal is polarized at 210 K and is not irradiated. In Fig. 7 the ratio $(I/I_0)_{peak}$ has

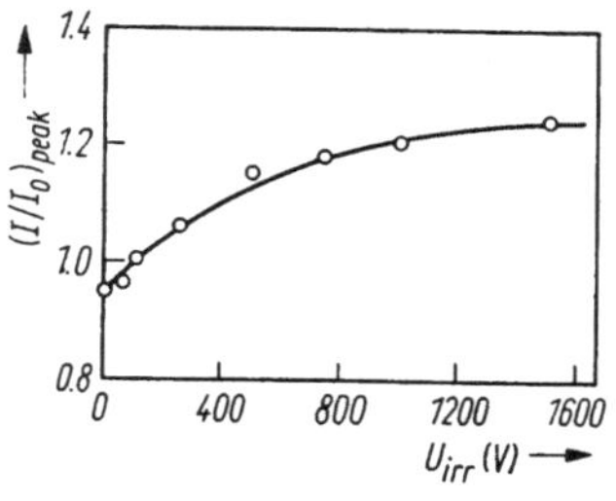

Fig. 7. Photo-induced polarization of the S^{2-}–anion vacancy dipole in sulphur-doped potassium iodide single crystals as a function of irradiation voltage (U_{irr}) in the presence of which the crystal is irradiated at 162 K in the 368 nm absorption band for 210 s. $T_p = 210$ K, $U_p = 1500$ V, irradiation temperature 162 K, voltage at the irradiation temperature U_{irr}, and polarization temperature range 210 K to LNT. I_{peak} corresponds to the peak of the ITC spectrum when the crystal is irradiated. $I_{0\,peak}$ corresponds to the peak of the ITC spectrum when the crystal is not irradiated

been plotted against the irradiation voltage (U_{irr}) kept at the irradiation temperature of 162 K, in the presence of which the crystal has been irradiated at 162 K in the 368 nm absorption band for 210 s. During PIP the peak of the ITC spectrum increases and hence the ratio of $(I/I_0)_{peak}$ should be larger than unity. The positive deviation from unity gives the magnitude of extra polarization which occurs during the PIP procedure. It is apparent from the figure that the value of $(I/I_0)_{peak}$ is larger than unity when the voltage during irradiation (U_{irr}) is kept above 100 V. At $U_{irr} = 100$ V, the ratio of $(I/I_0)_{peak}$ happens to be unity, which effectively speaks that there is no change in the ITC spectrum. At $U_{irr} = 100$ V, the phenomenon of PIP is absent. Further, when U_{irr} is kept equal to 50 V, we notice a decrease in the peak of the ITC spectrum resulting into a value below unity for $(I/I_0)_{peak}$. Obviously, when the crystal is irradiated in the presence of an electric field with $U_{irr} = 50$ V at 162 K, the peak of the ITC spectrum decreases instead of increasing. And hence, the phenomenon of PIP in the case $U_{irr} = 50$ V seems to be a PID phenomenon. Thus, there exists a limiting value of the electric field strength (100 V/d, where d is the thickness of the crystal specimen). If U_{irr}/d is larger than the limiting value of the electric field strength (E_0), the phenomenon of PIP will take place. In the cases $U_{irr}/d < E_0$ the phenomenon of PIP will behave as a PID phenomenon. Further, when $U_{irr}/d = E_0$ there will be no change in the ITC spectrum.

The behaviour of PIP as PID in the cases when $U_{irr}/d < E_0$ can be conceived following the bistable model of IV dipoles. When an electric field is applied, the potential wells are modified and the jump probabilities P_{12} and P_{21} now differ, resulting into the net polarization according to the relation

$$\frac{P_{12}}{P_{21}} \approx \exp\left(-\frac{\Delta E_a}{kT}\right),$$

where $\Delta E_a = Er$ is the change in the activation energy for the orientation of IV dipole (or in the potential energy barrier height E_a) in the presence of the electric field E and r the cation– anion separation. For irradiation voltages higher than $E_0 d$, the jump probabilities will differ to such an extent to give an extra polarization via an electric excited state. However, for irradiation voltages lower than $E_0 d$ the potential wells No. 1 and 2 will not be modified considerably leading to a very small difference in the jump probabilities P_{12} and P_{21} or practically the jump probabilities remain unaffected. Obviously enough, since the jump probabilities P_{12} and P_{21} are almost the same, the experimental conditions during the phenomenon of PIP will look like a condition prevailing during PID. In the case when $U_{irr}/d = E_0$, the jump probabilities are such that they overcome the contributions of each other leading to no change in the ITC spectrum or $(I/I_0)_{peak}$ equal to unity. Thus the results of PIP and PID in the KI : S^{2-} system presented here can be qualitatively explained following the bistable model of IV dipoles. It is plausible to presume that the limiting value of the electric field (E_0) depends partly on the local internal field. Further measurements on this system and various other systems, e.g. KI:Se^{2-}, KBr:S^{2-}, KBr:Se^{2-} etc. are carried out such that the proper understanding of the mechanisms responsible for the phenomena of PIP and PID can be conceived.

Acknowledgements

The author is grateful to Prof. Dr. F. Fischer, University of Münster (FRG) for valuable suggestions and for providing the necessary facilities for measurements in his laboratory. He is also thankful to Dr. J. Höner zu Siderdissen for his cooperation in setting the cryostat. Thanks are also due to Mr. R. Arends for his help during crystal growing.

References

[1] C. BUCCI, R. FIESCHI, and G. GUIDI, Phys. Rev. **148**, 816 (1966).
[2] J. PRAKASH and F. FISCHER, J. Physique **37**, C7-167 (1976).
[3] J. PRAKASH and F. FISCHER, phys. stat. sol. (a) **39**, 499 (1977).
[4] J. PRAKASH, Indian J. pure appl. Phys. **16**, 995 (1978).
[5] S. BENCI, R. CAPELLETTI, R. FERMI, and M. MANFREDI, J. Physique **37**, C7-138 (1976).
[6] F. FISCHER and H. GRÜNDIG, Z. Phys. **184**, 299 (1965).
[7] P. HENNL, phys. stat. sol. (a) **46**, 147 (1978).
[8] J. PRAKASH, Japan. J. appl. Phys. **17**, 1067 (1978).
[9] F. FISCHER, private communication.
[10] R. CAPELLETTI, R. FIESCHI, and E. OKUNO, Internat. Conf. Color Centers in Ionic Crystals, Sendai (Japan) 1974 (p. G138).
[11] A. S. NOWICK, Point Defects in Solids, Chap. 3, Ed. J. H. CRAWFORD, JR. and L. M. SLIFKIN, Plenum Press, New York 1972 (p. 166).

(Received April 5, 1979)

phys. stat. sol. (a) 54, 689 (1979)

Subject classification: 13.3 and 14.3.2; 22.2.1

Department of Electrical Engineering, Faculty of Engineering, Hokkaido University, Sapporo[1])

Small-Signal Admittance Study of GaAs Anodic MOS System

By

T. SAWADA and H. HASEGAWA

The small signal admittance of the GaAs anodic MOS system is analyzed using the Cole-Cole plotting technique. Among the various models for the Si MOS system including the single time constant model, the continuum model, the statistical model, and the tunneling model, excellent fit of the measured data can only be obtained with the tunneling model with an exponentially decaying trap distribution. The energy and spatial distribution of traps, and the capture rates are discussed. An anomalously large electron capture cross section is obtained, and it is explained by the interface-state-band (ISB) model recently proposed by the authors.

Die Admittanz des anodischen GaAs-MOS (Metal–Oxyd–Halbleiter)-Systems wird mit den Cole-Cole-Kurven analysiert. Von den verschiedenen Modellen für das Si-MOS System, das Modell mit einer Zeitkonstante, das Kontinuum-Modell, das statistische Modell und das Tunnelmodell, können die Meßwerte der MOS-Admittanz nur mit dem Tunnelmodell mit einer exponentiellen Haftstellenverteilung angepaßt werden. Die Energie- und räumliche Verteilung der Haftstellen und ihre Einfangkonstanten werden diskutiert. Es ergibt sich ein anomaler Querschnitt für den Elektroneneinfang, der mit dem Grenzflächenzustands-Band-Modell der Autoren erklärt wird.

1. Introduction

Formation of high quality insulators on GaAs is currently of great technological importance of its potential applications to surface passivation of devices, to construction of MISFET'S, and to realization of GaAs integrated circuits. Anodic oxidation either in liquid electrolytes [1] or in gaseous plasmas [2, 3] is known to be capable of producing uniform and dense native oxide films with excellent dielectric properties on GaAs. Successful operation of MOSFET's using these films have already been demonstrated in the GHz range at various laboratories [3 to 6]. However, the behavior of the GaAs anodic MOS system is anomalous [7, 8], and there exists no generally accepted understanding of the basic electronic processes taking place at the interface region.

In the present paper, a detailed study of small-signal admittance of the GaAs anodic MOS system is made in order to gain informations concerning the electronic processes at the interface. A detailed analysis of the capacitance/voltage properties was already presented elsewhere [7]. In the present paper, we are solely concerned with the frequency dependences of the MOS admittance. The experimental data is compared with the previously existing theories for silicon MOS systems, including (i) the single-time constant model, (ii) the state-continuum model, (iii) the statistical model [9], and (iv) the tunneling model [10]. The technique of Cole-Cole plotting [11] is used. It is shown that the data can only be fitted to the tunneling model with an exponentially decaying trap distribution. The energy distribution of trap density and the surface potential dependence of capture rate are discussed. It is shown that the effective electron capture cross-section is anomalously large, being of the order of 10^{-8} cm^2, and this anomaly is explained by the interface-state-band (ISB) model which was recently proposed by the present authors [7, 12, 13] to explain all the electrical anomalies in the GaAs MOS system in a consistent way.

[1]) Sapporo 060, Japan.

2. Theoretical Models for Interface-State Admittance

A computer program was developed to compute theoretical curves of interface-state admittance predicted by the various models for the silicon MOS system. We shall use the equivalent circuit representation of the MOS admittance shown in Fig. 1, and consider the behavior of the interface-state admittance Y_p defined by

$$Y_p = G_p(\omega) + j\omega C_p(\omega) . \tag{1}$$

As a convenient way to compare the experimental and theoretical variations with frequency of both real and imaginary parts of Y_p simultaneously, a Cole-Cole plotting technique [11] is employed. In this technique, frequency loci of Y_p/ω are drawn on the complex $C_p - G_p/\omega$ plane.

Examples of the computed loci are shown for the single-time constant model, the state-continuum model, and the statistical model in Fig. 2, and for the tunneling model in Fig. 3, respectively. Note that the real and imaginary axes are both normalized by the peak value of G_p/ω for the sake of comparison of the models. In Fig. 2, σ_s denotes the standard deviation of the surface potential fluctuation as expressed in the units of kT/q ($\sigma_s = 0$ reduces to the state-continuum model). For the tunneling model, two different spatial distributions of interfacial traps are assumed. A rectangular distribution with a constant volume density N_{T0} to a depth d_m into the oxide, is assumed for the computation of the loci in Fig. 3a. On the other hand, the loci in Fig. 3b correspond to an exponentially decaying distribution into the oxide with a functional form of

$$N_T(x) = N_{T0} \exp(-\alpha x) , \tag{2}$$

where N_T is the volume density of traps, x the distance from the interface into the oxide, and α a decay constant. In both cases, it is assumed that the time constant $\tau(x)$ of a trap located at distance x is given, according to [14], by

$$\tau(x) = \tau_0 \exp(2\varkappa_0 x) , \tag{3}$$

where τ_0 is the time constant of a trap at the interface, and $\varkappa_0$ a quantum-mechanical

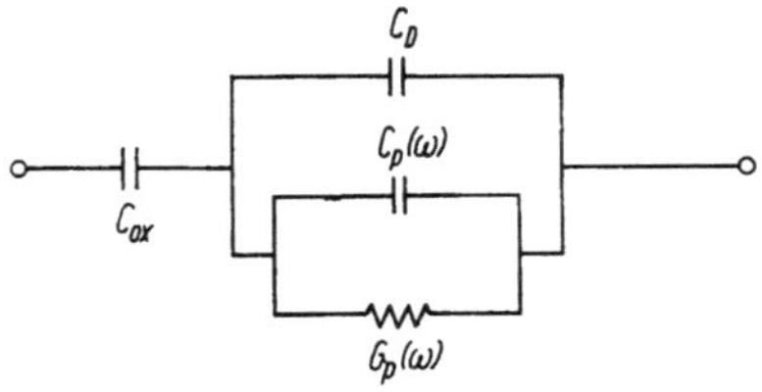

Fig. 1. Equivalent circuit of MOS system. C_{OX} oxide capacitance, C_D semiconductor depletionlayer capacitance, $C_p(\omega)$ interface-state capacitance (ω angular frequency), $G_p(\omega)$ interface-state conductance

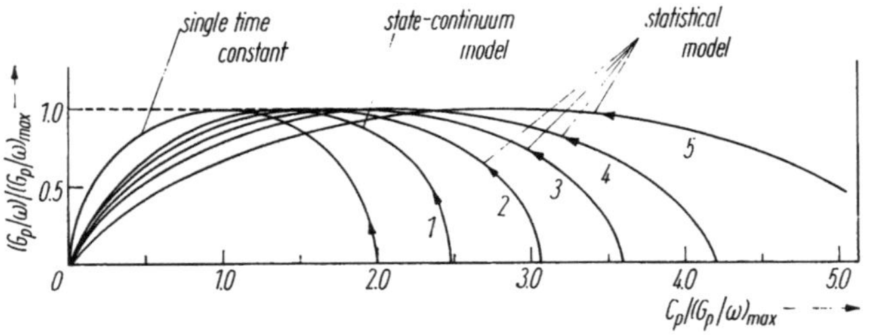

Fig. 2. Cole-Cole plotting of Y_p/ω for single-time constant model, state-continuum model, and statistical model. Both axes are normalized by the peak value of $G_p(\omega)/\omega$. (1) $\sigma_s = 0$, (2) 1, (3) 1.5, (4) 2, (5) 3

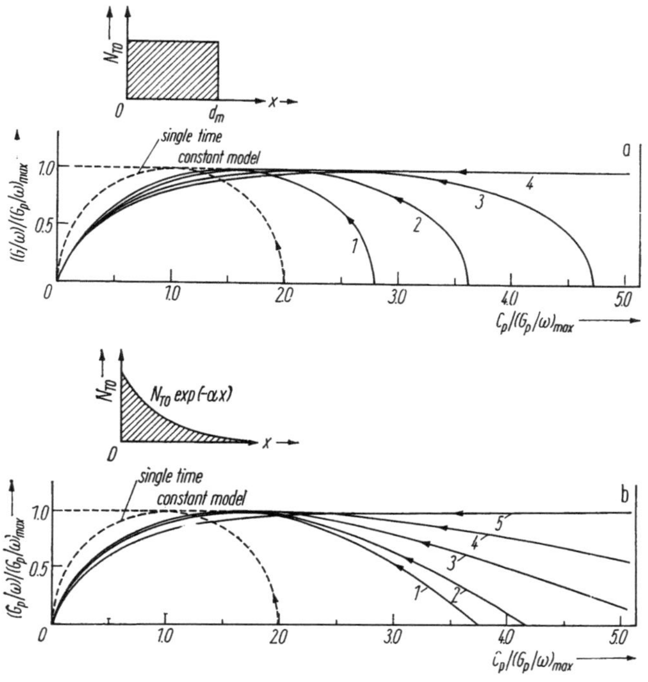

Fig. 3. Cole-Cole plotting of Y_p/ω for tunneling model with two different spatial distributions of interfacial traps. a) Rectangular distribution, (1) $d_m = (1/2\varkappa_0)\ln 10$, (2) $(1/2\varkappa_0)\ln 10^2$, (3) $(1/2\varkappa_0) \times \ln 10^3$, (4) ∞; b) exponential distribution, (1) $\alpha/2\varkappa_0 = 0.42$, (2) 0.34, (3) 0.2, (4) 0.1, (5) 0

decay constant of the electron wave function. τ_0 is related to the surface Fermi level position E_{Fs} through

$$\tau_0 = (\sigma v_{th} N_{eff})^{-1} \exp\left(|E_M - E_{Fs}|/kT\right) , \tag{4}$$

where σ is the capture cross-section, v_{th} the mean thermal velocity, N_{eff} the effective density of states, and E_M the majority carrier band edge (E_c or E_v).

To obtain Fig. 2 and 3a, mathematical expressions of [9] and [10] were used. To obtain the loci in Fig. 3b, the following expressions were derived :

$$C_p(\omega) = \frac{q^2 N_{T0}}{2\varkappa_0} (\omega\tau_0)^{\alpha/2\varkappa_0} \int\limits_0^{1/\omega\tau_0} z^{\alpha/2\varkappa_0} \tan^{-1}(z^{-1})\, dz , \tag{5a}$$

$$\frac{G_p(\omega)}{\omega} = \frac{q^2 N_{T0}}{4\varkappa_0} (\omega\tau_0)^{\alpha/2\varkappa_0} \int\limits_0^{1/\omega\tau_0} z^{\alpha/2\varkappa_0} \ln(1 + z^{-2})\, dz . \tag{5b}$$

In deriving the above expressions, the condition of $t_{ox} \gg \alpha^{-1}$ is assumed.

3. Experimental

The AGW process [1] was used for anodic oxidation of p- and n-type (100)-oriented horizontal Bridgman GaAs materials with the carrier concentration of 3 to 4 $\times$ $\times$ 10^{16} cm^{-3}. Anodization was done first in the constant current mode at 0.5 mA/cm^2, and then changed to the constant voltage mode at 70 V. The oxide thickness was

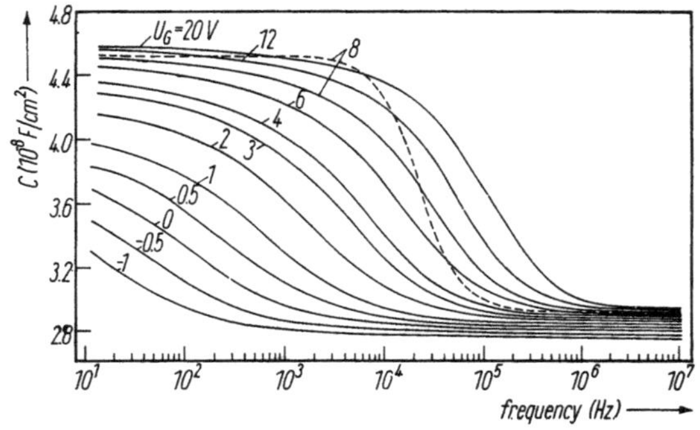

Fig. 4. Measured capacitance vs. frequency for an n-type (GaAs (100) sample. C MOS capacitance, U_G gate voltage. A theoretical curve for the single-time constant model is shown by the dashed curve (fitted to 8 V data). $N_D = 3.5 \times 10^{16}$ cm^{-3}, $C_{ox} = 4.84 \times \times 10^{-8}$ F/cm^2

about 1500 Å. MOS capacitors were formed by vacuum deposition of gold through a metal maks. The gate area was typically 1.2×10^{-3} cm^2. The samples were annealed at 300 °C for 3 h in a high-purity H_2 flow prior to the measurement.

The MOS admittance was measured at room temperature by an automatic C–U plotting system with phase sensitive detection for the frequency range of 1 Hz to 10 MHz. The ac signal amplitude applied across MOS samples, was kept less than 15 mV$_{p\text{-}p}$. Since the MOS samples show the so-called injection-type hysteresis effect, the samples were biased at each fixed bias point for 1 h prior to the measurement, and the admittance data for all the measurement frequencies were taken in a period of a few minutes. During the measurement, the bias was changed only in one direction from the accumulation side towards the inversion side.

For the extraction of the interface-state admittance Y_p from the measured MOS admittance, knowledge of the oxide capacitance C_{ox} and the depletion-layer capacitance C_D is required. C_{ox} was determined by biasing the samples at the highest possible bias to the accumulation side and measuring the capacitance at the lowest frequency. As shown in [7], such measurement gives a consistent result for the permittivity of the oxide bulk for both of n-type and p-type samples. On the other hand, C_D was determined from the high-frequency limit of the MOS capacitance on the assumption that it is a series connection of C_{ox} and C_D. Only a part of the measured data reached the high-frequency limits within the measured frequency range up to 10 MHz for both of n-type and p-type samples, and detailed analysis was made only on such data. An example of the measured MOS capacitance is shown versus frequency in Fig. 4 for an n-type sample.

For the determination of the surface potential, it is further necessary to know the ionized impurity concentration under the gate. This was measured by analyzing the deep-depletion MOS C–U curves [7] and Schottky C–U curves. The latter curves were taken by removing the oxide layer after the MOS measurement. Some of the samples showed appreciable frequency dispersion of Schottky capacitance, indicating the presence of a fairly large number of bulk traps. These samples also showed additional conductance peaks in the MOS admittance curves. Such sample were avoided from the present study to make sure that the observed frequency dispersion of MOS admittance originates from the oxide–semiconductor interface.

4. Results and Discussion

As is already clear from Fig. 4, the real and imaginary parts of Y_p/ω showed frequency dispersion which was much broader than that of the single-time constant model (shown by the dashed line). A careful analysis showed that, among the various models considered, excellent fits of the experimentally determined interface-state admittance

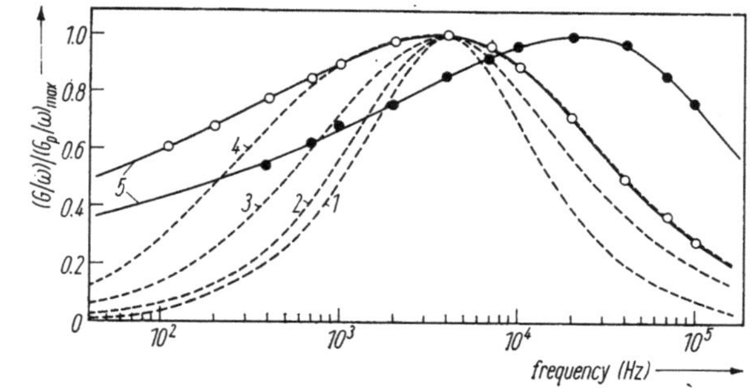

Fig. 5. Measured and theoretical curves of $G_p(\omega)/\omega$ vs. frequency. Theoretical curves (—— and - - - -): (1) single-time constant, (2) continuum, (3) statistical, (4) tunneling/rectangular, (5) tunneling/exponential. Experiment: ○ n-GaAs(100), $N_D = 3.5 \times 10^{16}$ cm^{-3}, $U_G = 8$ V; ● p-GaAs(100), $N_A = 4.0 \times 10^{16}$ cm^{-3}, $U_G = -2$ V.

data can only be obtained with the tunneling model with the exponentially decaying trap distribution. This is mainly due to the fact that the experimentally observed behavior of G_p/ω versus frequency is highly asymmetric with respect to the conductance peak frequency. In Fig. 5, the experimental conductance curves are compared with the various theoretical curves, and here we see highly asymmetric conductance curves.

The result of Cole-Cole plotting is shown for various gate biases in Fig. 6. The solid curves are the theoretical curves based on the tunneling model with exponential trap distribution. In both of n- and p-type samples, best fit was obtained at $\alpha/2\varkappa_0 = 0.2$ for all the bias voltages. For comparison, curves calculated for the rectangular trap distribution are also shown by the dashed curves. We see that large deviations of the data from the latter curves occur on the low-frequency side.

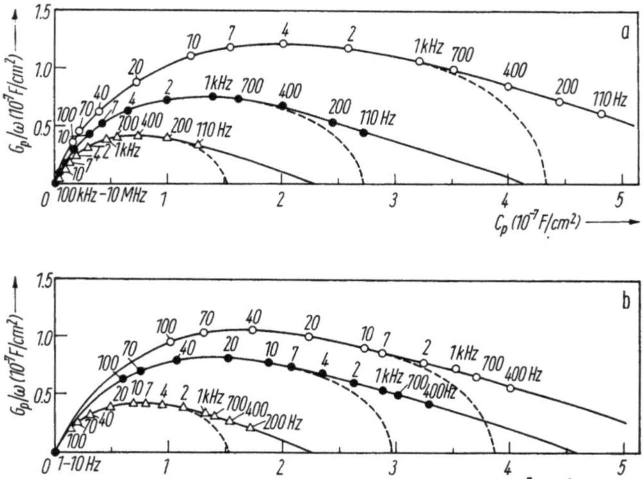

Fig. 6. Cole-Cole plotting of Y_p/ω fitted into the tunneling model. a) n-type sample, $N_D = 3.5 \times \times 10^{16}$ cm^{-3}, $C_{ox} = 4.84 \times 10^{-8}$ F/cm^2; ○ $U_G = 8$, ● 4, △ 2 V. b) p-type sample, $N_A = 4.0 \times \times 10^{16}$ cm^{-3}, $C_{ox} = 4.82 \times 10^{-8}$ F/cm^2; ○ $U_G = -4$, ● -2, △ 0 V. —— tunneling model/exponential, - - - - tunneling model/rectangular

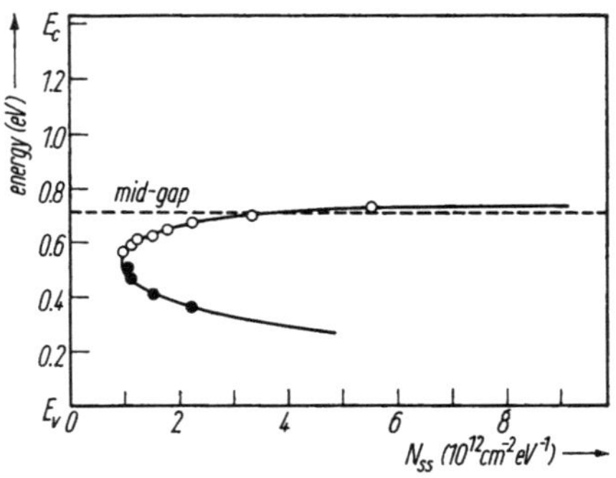

Fig. 7. Experimentally determined trap density N_{ss} per unit gate area per unit energy vs. energy in the band gap. ○ n-GaAs(100), $N_D = 3.5 \times 10^{16}$ cm^{-3}, $C_{ox} = 4.84 \times 10^{-8}$ F/cm^2; ● p-GaAs(100), $N_A =$ $= 4{,}0 \times 10^{16}$ cm^{-3}, $C_{ox} = 4.82 \times 10^{-8}$ F/cm^2

By curve fitting, the trap density per unit gate area per unit energy, N_{ss}, defined as

$$N_{ss} = \int_0^\infty N_T(x)\,dx = \frac{N_{T0}}{\alpha} \tag{6}$$

can be determined, and is plotted versus energy in Fig. 7. We see a U-shaped distribution in the lower half of the energy gap, and this is quite similar to what was obtained by Terman's method [3, 7, 12].

Since only the ratio $\alpha/2\varkappa_0$ can be determined by the curve fitting, the absolute values of the decay constant α of the trap distribution, and the volume density of the traps (N_{T0}) at the interface, cannot be determined. If we assume a barrier height for tunneling of 1 eV, the tunneling decay constant $\varkappa_0^{-1}$ will take a value of a few Angstroms [10, 14]. In this case, α^{-1} is estimated to be 10 to 20 Å, and N_{T0} is of the order of 10^{19} to 10^{20} cm^{-3} eV^{-1}.

On the other hand, one can determine the time constant τ_0 of the traps at the interface by curve fitting. The results is summarized in Fig. 8 for n- and p-type sample. As seen in Fig. 8, the measured time constants show in both cases of p- and n-type samples, an exponential dependence on E_{Fs} as is expected from (4). However, if we

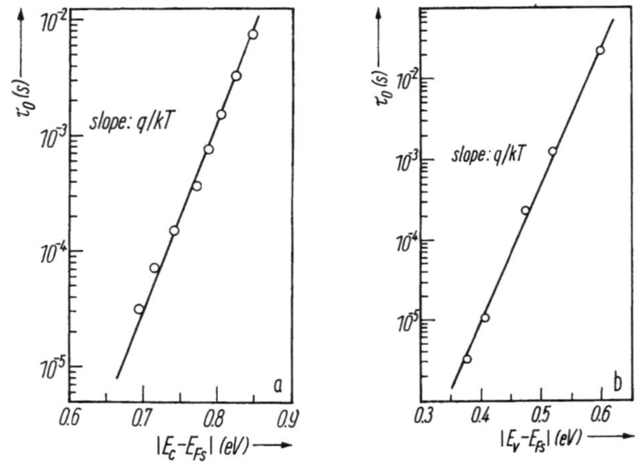

Fig. 8. Measured time constant τ_0 of traps at the interface vs. surface potential. a) n-type sample, $N_A = 3.5 \times 10^{16}$ cm^{-3}, $C_{ox} = 4.84 \times 10^{-8}$ F/cm^2; b) p-type sample, $N_D = 4.0 \times 10^{16}$ cm^{-3}, $C_{ox} =$ $= 4.82 \times 10^{-8}$ F/cm^2

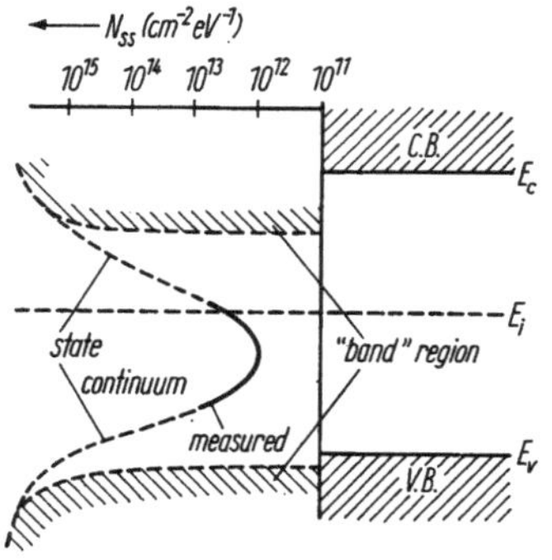

Fig. 9. A schematic representation of ISB model showing the presence of band regions and state continuum region (top of the lower-band region may be above E_V)

estimate the values of capture cross-section from this result with the use of (4), we obtain the value of hole capture cross-section of 8.8×10^{-15} cm^2 for the p-type sample, and the value of electron capture cross-section of 3.8×10^{-8} cm^2 for the n-type sample. Obviously, the value of the electron capture cross-section is too large to be physically acceptable, whereas the hole capture cross-section appears much more normal.

An obvious explanation for the above anomaly of gigantic cross-section would be an incorrect determination of the surface potential. We have to admit that the result presented here is quite sensitive to the choice of parameter values such as the oxide capacitance and the doping concentration. However, a careful check of the various errors possibly involved in the measurement led to the conclusion that the anomaly is too large to be accounted for by such errors. The maximum possible error in E_{Fs} determination is estimated to be 0.1 eV, and this would reduce the cross-section value to the range of 10^{-10} cm^2, which is, however, still too large. On the other hand, a recent measurement of surface potential by Meiners [15] seems to be in good agreement with ours.

Our interpretation of the above anomaly is that it is a strong indication that the interfacial traps do not actually make direct exchange of electrons with the GaAs conduction band. Recently, the present authors proposed the interface-state-band (ISB) model to explain various electrical anomalies in the GaAs MOS system [7, 12, 13] In this model, partial band formation by interface states is assumed as schematically shown in Fig. 9. If electrons are exchanged by way of such a band region, the rate of exchange will be enhanced drastically, and if such a process is represented by a cross-section for a direct jump, an anomalously large value will result. In other words, the presence of such a band region will partially "short-circuit" the energy difference $E_c - E_{Fs}$ by a certain amount ΔE owing to highly frequent state-to-state transitions within the band region. We have quite recently succeeded in detecting the presence on ΔE by a capacitance transient and DLTS analysis [16], and this provides a further support for the above interpretation. However, more work is needed to clarify the origin of such a band region. It may be due to intrinsic surface-state bands brought back into the band-gap region by oxidation [17], or due to formation of a thin amorphous semiconductor layer by, for instance, pile-up of non-oxidized As atoms [18].

Acknowledgements

We would like to thank Prof. H. Tagashira for his useful discussion. The present work is financially supported by a grant-in-aid from Ministry of Education, and by a Kurata research grant.

References

[1] H. HASEGAWA and H. L. HARTNAGEL, J. Electrochem. Soc. **123**, 713 (1975).
[2] R. P. H. CHANG and A. K. SINHA, Appl. Phys. Letters **29**, 56 (1976).
[3] F. KOSHIGA and T. SUGANO, Thin Solid Films **56**, 39 (1979).
T. SUGANO, F. KOSHIGA, and K. YAMASAKI, Tech. Digest XXIV. Internat. Electron Devices Meeting, Washington 1978, paper 6.7.
[4] D. L. LILE, A. R. CLAWSON, and D. A. COLLINS, Appl. Phys. Letters **29**, 207 (1976).
[5] H. TOKUDA, A. ADACHI, and T. IKOMA, Electronics Lett. **13**, 761 (1977).
[6] E. KOHN, A. COLQUHOUN, and H. L. HARTNAGEL, Solid State Electronics **21**, 877 (1978).
[7] T. SAWADA and H. HASEGAWA, Thin Solid Films **56**, 183 (1979).
[8] E. KOHN and H. L. HARTNAGEL, Solid State Electronics **21**, 409 (1978).
[9] E. H. NICOLLIAN and A. GOETZBERGER, Bell Syst. tech. J. **46**, 1055 (1967).
[10] H. PREIER, Appl. Phys. Letters **10**, 361 (1967).
[11] P. CHR. MALMIN, phys. stat. sol. (a) **8**, 597 (1971).
[12] H. HASEGAWA and T. SAWADA, Proc. VII. Internat. Vacuum Congr. and III. Internat. Conf. Solid Surface, Vol. 1, Vienna 1977 (p. 549).
[13] H. HASEGAWA, T. SAWADA, and T. SAKAI, Proc. Internat. Conf. Solid Films and Surfaces, Tokyo 1978, in: Surface Sci., to be published.
[14] F. P. HEIMAN and G. WARFIELD, IEEE Trans. Electron Devices **12**, 701 (1965).
[15] L. G. MEINERS, J. Vacuum Sci. Technol. **15**, 1402 (1978).
[16] H. HASEGAWA and T. SAWADA, Proc. VI. Conf. Physics of Compound Semiconductor Interfaces, Asilomar 1979, in: J. Vacuum Sci. Technol., to be published.
[17] W. SPICER, P. PIANETTA, I. LINDAU, and P. W. CHYE, J. Vacuum Sci. Technol. **14**, 917 (1977).
[18] R. P. H. CHANG, T. T. SHENG, C. C. CHANG, and J. J. COLEMEN, Appl. Phys. Letters **33**, 341 (1978).

(Received May 14, 1979)

phys. stat. sol. (a) 54, 697 (1979)

Subject classification: 10.2 and 13.4; 22.1.1

School of Engineering, Okayama University[1])

Degree of Copper Contamination in Germanium Heat-Treated at High Temperatures

By

Y. KAMIURA and F. HASHIMOTO

The degree of copper contamination in Ge heat-treated above 700 °C is studied by measuring the density of the copper levels in order to provide an experimental standard which serves to treat Ge at high temperatures. The degree of contamination is in the range 10^{13} to 10^{16} cm^{-3}, depending upon the heat-treatment methods and the purity of water used to rinse specimens before the heat treatments. Joule heating in a vacuum attains the lowest degree of contamination, 2×10^{13} cm^{-3}, above 900 °C. The results warn us to take into account the presence of an appreciable number of Cu atoms whenever thermally induced defects, impurity diffusion or precipitation, and dislocations are studied.

Kupferverunreinigung in oberhalb von 700 °C wärmebehandeltem Ge wird durch Messung der Dichte der Kupferniveaus untersucht, um einen experimentellen Standard für die Behandlung von Ge bei hohen Temperaturen zu erhalten. Die Verunreinigungskonzentration, die von der Wärmebehandlung und von der Reinheit des zur Probenwaschung gebrauchten Wassers abhängt, liegt im Bereich von 10^{13} bis 10^{16} cm^{-3}. Stromheizung der Probe im Vakuum ergibt die geringste Verunreinigung von 2×10^{13} cm^{-3} oberhalb von 900 °C. Die Ergebnisse zeigen, daß man bei der Untersuchung von durch Abschrecken gebildeten Fehlstellen, Diffusion oder Ausscheidung von Verunreinigungen und von Versetzungen auch immer eine Verunreinigung durch Kupfer zu berücksichtigen hat.

1. Introduction

It has long been known that fast diffusants like Cu, Ni, Ag, etc. contaminate germanium heat-treated at high temperatures, and greatly affect its electrical properties. We must therefore keep in mind the identity of contaminants and the degree of contamination, whenever we study the electrical characteristics of thermally induced defects, impurity diffusion or precipitation, and dislocations. However, such data have not been reported yet. The purpose of this paper is to provide the data on impurity contamination of Ge, which serve as an experimental standard for the high-temperature treatments of Ge.

2. Experimental Procedure

Germanium crystals doped with 4.7×10^{14} As atoms/cm^3 or 4.3×10^{14} Sb atoms/cm^3, and an undoped crystal containing 3×10^{12} cm^{-3} of an unknown acceptor were used. Their dislocation densities are about 10^3 cm^{-2}. The crystals were cut into rod specimens of $0.5 \times 2 \times 15$ mm^3.

Before heat treatments, all specimens were etched in an acid solution (HNO_3: HF = = 10:1) for a few minutes and then rinsed in twice distilled and de-ionized water ($\approx 10^7$ Ωcm). Following this, they were immersed in a strong aqueous KCN solution (10%) for an hour and rinsed again in twice distilled and de-ionized water. Three kinds of heat-treatment methods were carried out as follows:

[1]) 700 Okayama, Japan.

1. Each specimen was heated on a silica platform in a high-purity nitrogen atmosphere (99.999% up) for 3 to 4 h and then quenched into ethylene glycol kept at about 5 °C. This method is similar to those used by Logan [1], Samuelsson [2], and the authors [3].

2. Tantalum wires which had been pre-heated above 1000 °C in vacuum were spot-welded to the specimens. They were heated by sending a direct current through the wires in a vacuum of 1×10^{-5} Torr for 2 h and then cooled to room temperature only by radiation. This is similar to those of Mayburg [4], Hopkins and Clarke [5], Zhidkov [6, 7], and Ishino et al. [8].

3. Each specimen was heated on a silica platform in a wet-oxygen atmosphere (99.99%) for an hour and cooled into air. Some specimens were cracked when quenched in glycol. Above 850 °C specimens were melted in this atmosphere. This method is the same as those adopted by Hiraki [9, 10] and Konorova [11].

Hereafter, these three methods are called the first, second, and third method, respectively. In these methods, silica platforms and tantalum wires were also subjected to the same chemical treatments as used to clean the specimens.

Heat-treated specimens were then cleaned by grinding lightly and etching. Indium-soldered contacts were attached to p-type specimens, and indium containing 3% antimony was used for n-type ones. Hall coefficient measurements were made as a function of temperature over the range 20 to 350 K. A field of 5000 G was applied.

3. Results and Discussion

3.1 Principal contaminant

It has been known that quenched-in acceptors produce three energy levels near $E_v + 0.04$ eV, $E_v + 0.2$ eV, and $E_c - 0.2$ eV [1, 7, 10, 11]. We [12] recently determined the accurate positions of the two upper levels as $E_v + 0.33$ eV and $E_c - 0.26$ eV. Moreover, we have confirmed that the lowest level lies just at $E_v + 0.040$ eV (Fig. 1). These facts have established that the quenching effect observed in most previous investigations is due to copper contamination, indicating that the quenching of any

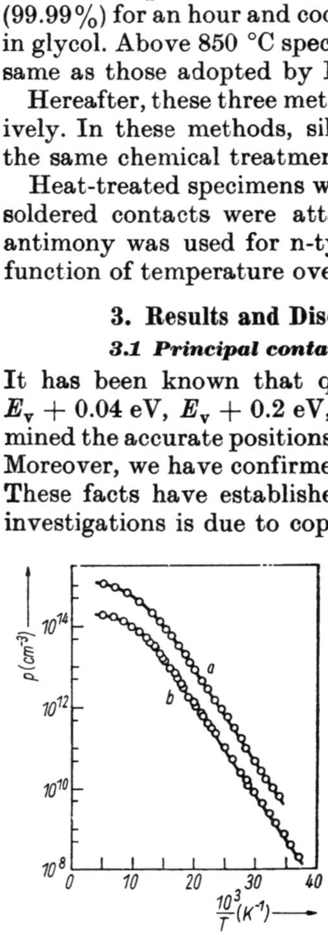

Fig. 1

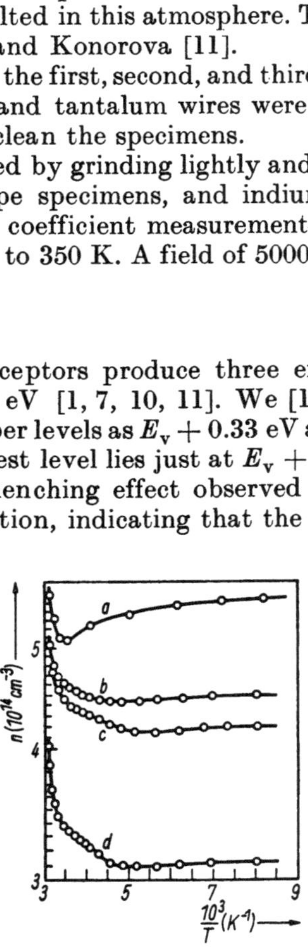

Fig. 2

Fig. 1. Hole density, p, versus temperature curves for As-doped specimens converted to p-type by quenching. (a) $T_Q = 800$ °C, slope 0.040 eV; (b) $T_Q = 730$ °C, slope 0.039 eV

Fig. 2. Electron density, n, versus temperature curves for Sb-doped specimens heat-treated by various methods. (a) Not quenched; (b) for the specimen heat-treated by the first method (see Section 2 of the text): $T_Q = 550$ °C; (c) for the specimen heat-treated by the second method: $T_Q = 900$ °C; (d) for the specimen heat-treated by the third method: $T_Q = 800$ °C. It is seen that the $E_v + 0.26$ eV level is ionized around 250 K

intrinsic thermal defects is always accompanied by the unavoidable introduction of copper.

Penning [13] found $E_v + 0.04$ eV levels in a specimen saturated with nickel at 800 °C and annealed at 430 °C although no copper was added deliberately. The density (1×10^{15} cm^{-3}) of the $E_v + 0.04$ eV level was comparable with that of nickel present in the specimen. Tweet [14] also observed $E_v + 0.04$ eV levels in plastically deformed Ge. These warn us that we should take the presence of copper into account when we investigate the diffusion or precipitation of impurities and the plastic deformation.

3.2 Degree of copper contaminantion

Fig. 2 shows that about 3×10^{13} cm^{-3} of the $E_c - 0.26$ eV level, the third level of substitutional copper, are introduced into Sb-doped specimens regardless of the difference in the method of heat treatment. It is noted, however, that the temperatures at which the specimens were heated are significantly different. This suggests that the safest method for preventing copper contamination is joule-heating in a vacuum. Further evidence is given in Fig. 3, where the density of $E_v + 0.04$ eV levels is used as a measure of the copper density.

Hall [15] used liquid metals as getters to avoid copper contamination in his quenching experiments, where specimens were quenched from 400 °C. We tried to extend his method to higher quenching temperatures. A specimen was heated in liquid Pb (99.9999%) at 700 °C, but fairly rapid dissolution of Ge into the liquid metal occurred, although we used as small an amount of Pb as possible. This may suggest that it is difficult to extend the Hall method to high temperatures above 700 °C, where the solubility of Ge in liquid metals rapidly increases with temperature.

Fig. 4 shows the effect of the purity of water which we used to rinse specimens before heat treatments. The specimens were heat-treated at 700 °C in a nitrogen atmosphere (the first method). The use of unpurified water ($\approx 10^5$ Ωcm) increases the density of the $E_v + 0.04$ eV level by a factor of 10, confirming that this level arises from in-diffusing of small traces of copper contaminating the specimen surface.

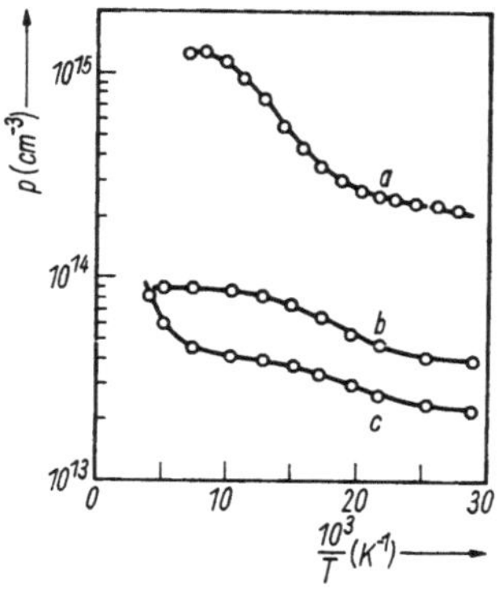

Fig. 3

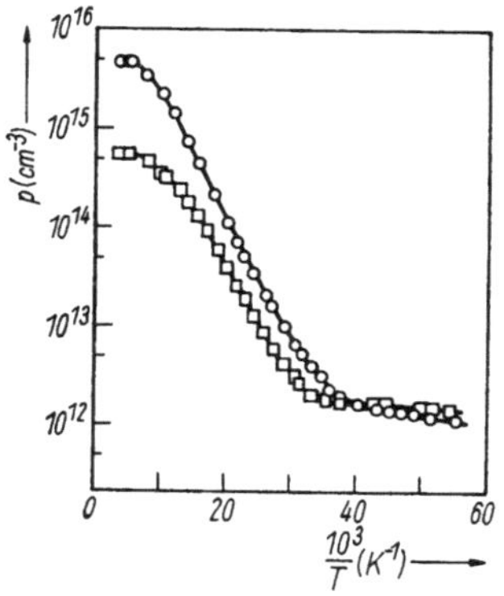

Fig. 4

Fig. 3. Hole density, p, versus temperature curves for undoped specimens heat-treated by various methods. $E_v + 0.04$ eV levels, whose density is denoted by $N_{0.04}$, are ionized between 40 and 100 K. (a) For the specimen heat-treated by the first method: $T_Q = 840$ °C, $N_{0.04} = 1 \times 10^{15}$ cm^{-3}; (b) for the specimen heat-treated by the third method: $T_Q = 845$ °C, $N_{0.04} = 4.9 \times 10^{13}$ cm^{-3}; (c) for the specimen heat-treated by the second method: $T_Q = 925$ °C, $N_{0.04} = 1.7 \times 10^{13}$ cm^{-3}

Fig. 4. Dependence of $E_v + 0.04$ eV level density, $N_{0.04}$, on the purity of water used to rinse undoped specimens before being heat-treated at 700 °C by the first method. ○ Unpurified water: $N_{0.04} = 4.2 \times 10^{15}$ cm^{-3}; □ purified water: $N_{0.04} = 5.6 \times 10^{14}$ cm^{-3}

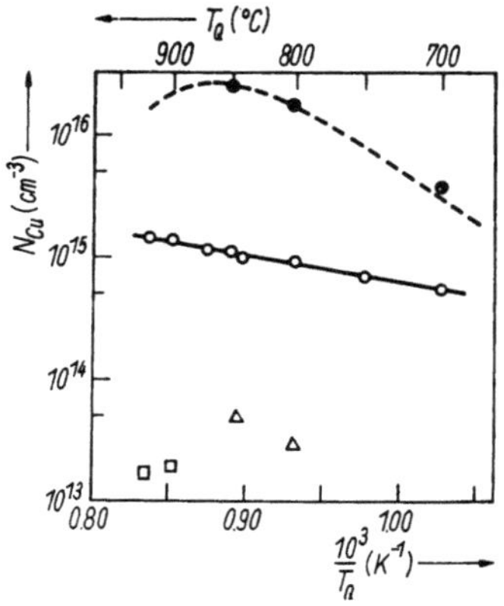

Fig. 5. Degree of copper contamination for various heat-treatment methods. N_{Cu} represents the density of the first level of substitutional copper. ○ The first method (purified water); ● the first method (unpurified water); □ the second method; △ the third method. The dashed line is the solubility curve of copper after Woodbury and Tyler [16]

The dependence of copper density on the quenching temperature is shown in Fig. 5, which indicates that the first method, even by the use of purified water, introduces contaminating Cu atoms as many as 1×10^{15} cm^{-3} over the range 700 to 900 °C. The results by the two other heat-treatment methods are also summarized in Fig. 5. Joule-heating in a vacuum attains the lowest degree of copper contamination, which is only 2×10^{13} cm^{-3} even above 900 °C. It may be possible to reduce the copper contamination to less than 10^{13} cm^{-3} by the use of more purified water and by heating specimens in a higher vacuum.

4. Conclusions

It is unavoidable that an appreciable number of copper atoms are introduced into germanium by the heat treatments at high temperatures above 700 °C. The degree of Cu contamination is in the range 10^{13} to 10^{16} cm^{-3}, depending upon the heat-treatment methods and the purity of water used to rinse specimens before the heat treatments. Joule-heating in a vacuum attains the lowest degree of contamination: 2×10^{13} cm^{-3} above 900 °C. We must take the presence of copper into account whenever we study thermally induced defects, impurity diffusion or precipitation, and dislocations.

References

[1] R. A. LOGAN, Phys. Rev. **101**, 1455 (1956).
[2] B. SAMUELSSON, Ark. Fys. (Sweden) **35**, 321 (1967).
[3] F. HASHIMOTO and Y. KAMIURA, Japan. J. appl. Phys. **13**, 762 (1974).
[4] S. MAYBURG, Phys. Rev. **95**, 38 (1954).
[5] R. L. HOPKINS and E. N. CLARKE, Phys. Rev. **100**, 1786 (1955).
[6] V. A. ZHIDKOV, Soviet Phys. — Solid State **3**, 335 (1961).
[7] V. A. ZHIDKOV, Soviet Phys. — Solid State **3**, 339 (1961).
[8] S. ISHINO, F. NAKAZAWA, and R. R. HASIGUTI, J. Phys. Soc. Japan **20**, 817 (1965).
[9] A. HIRAKI, J. Phys. Soc. Japan **21**, 34 (1966).
[10] A. HIRAKI and Y. INUISHI, Proc. Internat. Symp. Lattice Defects in Semiconductors, Tokyo 1966, Ed. R. R. HASIGUTI, University of Tokyo Press, Tokyo 1968 (p. 138).
[11] L. F. KONOROVA, Soviet Phys. — Solid State **10**, 2233 (1969).
[12] Y. KAMIURA and F. HASHIMOTO, Japan. J. appl. Phys. **17**, 433 (1978).
[13] P. PENNING, Philips Res. Rep. **13**, 17 (1958).
[14] A. G. TWEET, Phys. Rev. **99**, 1245 (1955).
[15] R. N. HALL, Proc. Internat. Conf. Lattice Defects in Semiconductors, Freiburg 1974, Inst. Phys., London 1975 (p. 190).
[16] H. H. WOODBURY and W. W. TYLER, Phys. Rev. **105**, 84 (1957).

(Received May 14, 1979)

phys. stat. sol. (a) 54, 701 (1979)

Subject classification: 1.1 and 10.2; 22.1.2

Department of Applied Physics, Faculty of Engineering, The University of Tokyo[1])

Applications of X-Ray Triple Crystal Diffractometry to Studies on the Diffusion-Induced Defects in Silicon Crystals

By

A. IIDA

Silicon single crystals containing minute lattice defects caused by the diffusion of impurities, Cu, Li, and Au, are investigated by using X-ray triple crystal diffractometry (TCD), in which the kinematical and dynamical components are measured separately. From the TCD intensity curves the relative defect volume is estimated, resulting in good agreement with the value obtained by topography. The TCD method is particularly useful for the case where the individual defect image cannot be separated by topography.

Siliziumeinkristalle, welche nach Diffusion mit Cu, Li und Au Mikrodefekte enthielten, werden mit einem Röntgendreikristalldiffraktometer (TCD) untersucht, das den kinematischen und dynamischen Anteil der Beugung getrennt zu messen gestattet. Aus den TCD-Kurven wird der relative Kristallbaufehlergehalt abgeschätzt, der befriedigend mit dem aus der Topographie abgeleiteten übereinstimmt. Die TCD-Methode ist besonders dann nützlich, wenn die Kristallbaufehler in Topographien nicht mehr aufgelöst werden.

1. Introduction

In a previous paper [1] we showed that X-ray triple crystal diffractometry of $(+n -n, +n)$ setting (TCD) is useful to measure the angular dependence of the diffracted intensity from perfect and surface-damaged silicon crystals, separating the kinematical and dynamical diffraction very near the reciprocal lattice point to study the thermal diffuse scattering and imperfect crystal structures. TCD method is expected to be suitable for the study of nearly perfect crystals containing various kinds of minute lattice defects.

In this paper, we show some applications of the TCD method to the studies on lattice defects in silicon crystals caused by the diffusions of copper, lithium, and gold. The relative defect volume is estimated from the analysis of the TCD intensity curves. TCD method is shown to be complement to topography.

TCD measurements were done in the same way as described in the previous paper, but the simple explanation is given here for the convenience of the following description. A tailless beam of about 6 s of arc in angular spread which was obtained by successive (five fold) diffractions from a grooved (first) crystal was used as the incident beam on the specimen (second) crystal. The angular distribution of diffracted X-rays from the specimen at various angular position (α) around the exact Bragg angle (θ_B) was measured by rotating the analyzer (third) crystal (θ). α and θ are the specimen and analyzer deviation angles from Bragg condition. 111 Bragg case diffraction of $CuK\alpha$ was studied using a rotating anode tube with apparent focal size 1×0.5 mm^2 operating at 45 kV and 180 mA. The specimen surfaces were perpendicular to the growth direction of $\langle 111 \rangle$.

[1]) Hongo, Bunkyo-ku, Tokyo, Japan.

2. Copper Diffused Silicon

A silicon single crystal containing copper precipitates was investigated. Dislocation free FZ silicon covered with $Cu(NO_3)_2$ was annealed at 1000 °C for 60 min in N_2, and then quenched to room temperature. The specimen was sliced, lapped, and then etched to remove all traces of surface damages. A Lang topograph of the specimen is shown in Fig. 1, where three different contrast regions, marked A, B, and C, are visible. In C-region the precipitate images are seen to form a concentric pattern. In A- and B-regions no isolated image is observed while the diffraction intensities are slightly different between them.

The diffraction intensity curves from three regions were measured by using TCD. The area of the specimen $1 \times 0.5\ mm^2$ was measured at about 20 angular positions of the specimen in the range of $|\alpha|$ from 0″ to 100″ for each region. Fig. 2 shows the typical diffraction intensity curves measured at a specimen angular position of $\alpha = 45''$. Three peaks are always observed in each curve. From each peak angular position three peaks are assigned as follows in the same manner as in the previous paper [1]: the peak appearing at $\theta = 2\alpha$ is the dynamical diffraction component, the peak at $\theta = \alpha$ is the pseudo-peak due to the slowly decreasing tail of the incident beam on the specimen, and the broad peak at $\theta = 0''$ is the kinematical diffraction component since the kinematical component has its peak at $\theta_{kin} = 2\alpha\,(1 - \cos^2 \theta_B)$. It is to be noted that the kinematical and dynamical components appear simultaneously.

In order to analyze the TCD curves it is assumed that the kinematical diffraction is dominant and the dynamical diffraction is negligible near the precipitate core so

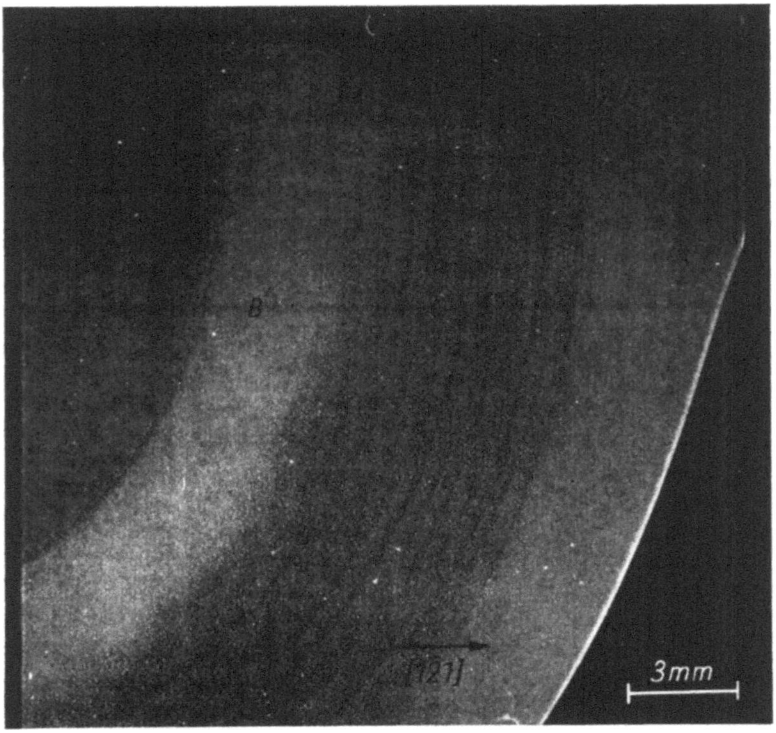

Fig. 1. Lang topograph of copper diffused silicon crystal. 111 reflection of $MoK\alpha_1$, specimen thickness ≈ 600 μm. Different contrast regions are marked A, B, and C

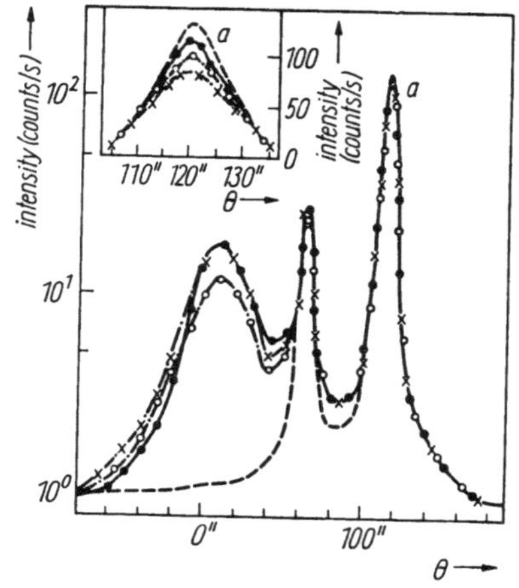

Fig. 2

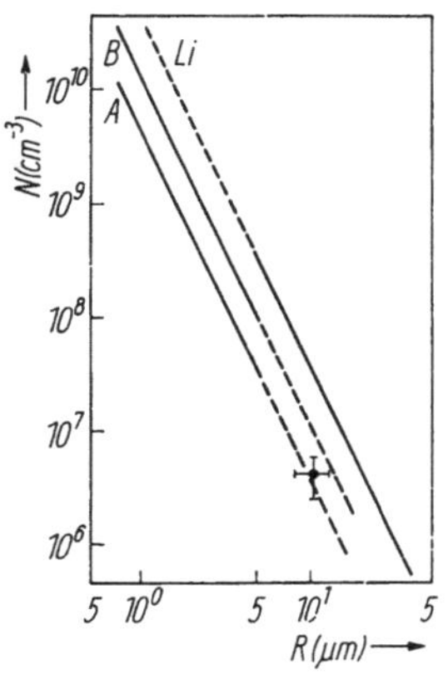

Fig. 3

Fig. 2. TCD diffraction intensity curves from copper diffused silicon crystal at $\alpha = 60''$; Bragg case 111 reflection of CuKα. •, ×, and ○ correspond to regions A, B, and C in Fig. 1, respectively. Dashed line corresponds to the perfect silicon. α and θ are the specimen and analyzer deviation angles from the Bragg condition. Details of the dynamical components at $\theta = 2\alpha$ are shown in the insert

Fig. 3. The relation between the average defect density N and the defect radius R for copper and lithium diffused silicon. Lines A, B, and Li correspond to A- and B-regions in Fig. 1 and lithium diffused silicon, respectively. Solid circle with error bars shows the result of C-region obtained from Fig. 1

that the decrease in the dynamical diffraction intensity directly correlates to the relative defect volume of the imperfect crystal. First it was verified that the ratio of the dynamical component of the copper diffused silicon to that of the perfect one, γ, was nearly constant independent of the α-value, where the integrated intensity of the dynamical diffraction component is approximately given by the product of the peak intensity and the half-width. This supports the present assumption used for the analysis. With this assumption, the defect volume ratios per unit volume, which are given by $1 - \gamma$, are obtained as 0.07 ± 0.01, 0.17 ± 0.01, and 0.08 ± 0.01 for regions A, B, and C, respectively. To make this analysis sure the relative defect volume in C-region is estimated from the topograph of Fig. 1, and is obtained as 0.07 ± 0.02, which agrees well with the result of TCD.

Furthermore, if the strain field of the precipitate is assumed to be spherical, the relation between the average radius R and the density N is obtained for each region as shown in Fig. 3. In this case R is assumed to be smaller than the spatial resolution of the topograph.

It is known that the FZ dislocation-free silicon contains the swirl defects [2], which are classified as A- and B-swirls from their size and density: the former has lower density and larger size and the latter has higher density and smaller size. It is speculated that the precipitate in region C corresponds to A-swirls and that in regions A and B to B-swirls from the defect densities in each region.

3. Lithium Diffused Silicon

FZ silicon diffused with lithium is also investigated. Lithium was diffused into a silicon wafer at 450 °C for 4 h and then annealed at about 200 °C for 150 h. By Lang topography no isolated image was observed but the integrated intensity was several

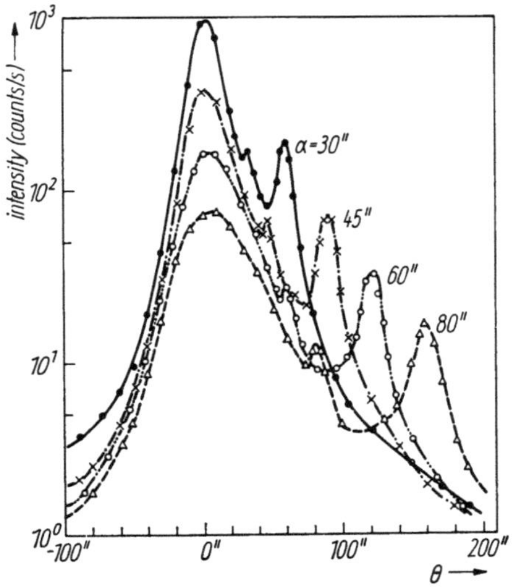

Fig. 4. TCD diffraction intensity curves from lithium diffused silicon; Bragg case 111 reflection of CuKα

times stronger than the perfect one. Fig. 4 shows the intensity curves obtained by TCD at $\alpha = 30''$, $45''$, $60'$, and $80''$. The kinematical and dynamical diffractions coexist at any value of α, except at $\alpha = 0''$, but the former intensity is much stronger than the latter in contrast to the copper diffused silicon.

Following the same procedure as applied to the case of copper diffusion, the defect volume ratio per unit volume was obtained as 0.44 ± 0.01. The relation between R and N is obtained as shown in Fig. 3, if the spherical precipitate is assumed. In this case it is considered that R is larger than the spatial resolution of the topograph but the isolated image cannot be observed because of the high density of the precipitates.

As lithium is known to precipitate preferentially at A-swirls [2], it is supposed that this specimen has much higher density of A-swirls than the copper diffused silicon described above.

4. Gold Diffused Silicon

Gold was diffused into silicon at high temperature. FZ and CZ silicon specimens (about $10 \times 20 \times 1$ mm^3) were evaporated with gold on the front surface and annealed at 1100 and 1000 °C for 5 h, 75 min, and 20 min.

Fig. 5a shows a section topograph of FZ silicon annealed at 1100 °C for 5 h. The left side of the topograph shows the strong strain due to the Au–Si alloy layer [3], which is easily removed by slight etching. Small defect images are seen in the specimen. The higher temperature and longer time annealing produced larger size defects. The average diameter of the defect is about 20 μm in the case of Fig. 5a. From the extinction rule, the type of the defect was determined as Frank sessile dislocation loop, which agrees with the previous report [4]. It is noted that the density of the Frank loop is highest at the middle part along the depth, and nearly zero defect image is seen near the surface. This defect distribution is not due to the diffraction phenomenon for thick specimens, as is confirmed by thinning the specimen. In the case of CZ crystals the defect density is much higher than in the FZ crystals by two orders of magnitude, while the dependence of the defect density on the depth shows the same tendency as in the FZ specimens.

This defect distribution seems to be due to the outward diffusion of the defect nuclei because of the high temperature and long time heat treatment [5]. It is to be noted that the lateral defect distribution in the specimen was not homogeneous and sometimes showed a concentric pattern.

The defect distribution along the depth was investigated by the TCD method combined with the step etching technique. Fig. 5b shows the diffraction intensity

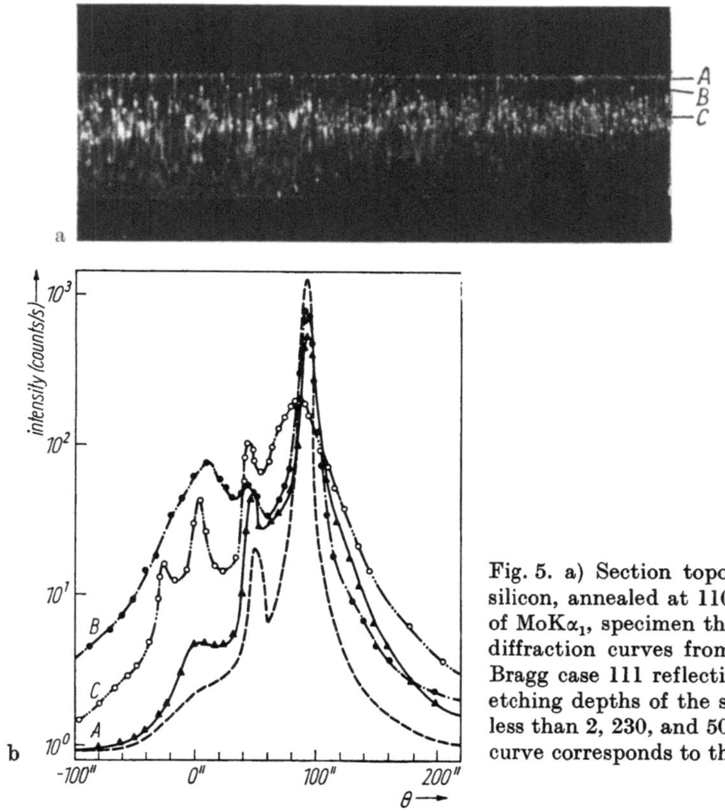

Fig. 5. a) Section topograph of gold diffused FZ silicon, annealed at 1100 °C for 5 h. $40\bar{4}$ reflection of $MoK\alpha_1$, specimen thickness $\approx$ 1.3 mm. b) TCD diffraction curves from gold diffused FZ silicon; Bragg case 111 reflection of $CuK\alpha$, $\alpha = 45''$. The etching depths of the specimens A, B, and C are less than 2, 230, and 500 μm, respectively. Dashed curve corresponds to the perfect silicon

curves measured at a specimen angular position of $\alpha = 45''$ for each etching step. After the removal of the surface Au–Si alloy layer, the TCD intensity profile (A) shows the slight broadening of the dynamical diffraction peak and also the slight increase of the kinematical diffraction component. After 230 μm etching (B), where no defect image was seen in the topograph, the strong and broad kinematical diffraction appears, while the dynamical diffraction is almost the same as the perfect one. The appearance of the kinematical component indicates the existence of lattice strain despite the lack of the topographic image. Lastly in the middle of the high defect density layer (C), about 500 μm etching, two extra peaks appear at $\theta \doteqdot 0''$ and $-\alpha/2$ while the total kinematical components is rather weak. These extra peaks seem to be due to the faulted loops, but a detailed analysis has not yet been done. The peak broadening of the dynamical component is due to the surface uneveness, from the same consideration as the previous paper [1].

From the series of TCD intensity curves it is concluded that even in the layer where no defect image is observed in the topograph, a minute lattice strain still exists and causes the kinematical diffraction. This means that the size as well as the density of the defects varies along the depth and that the defect distribution is broader than that expected from the topograph.

5. Comparison with the Double Crystal Method

The rocking curves obtained from the lithium and gold diffused specimens were measured by double crystal diffractometry of $(+n, -n)$ setting. Fig. 6 shows the logarithmic plot of the diffraction curves. In the case of the lithium diffused silicon

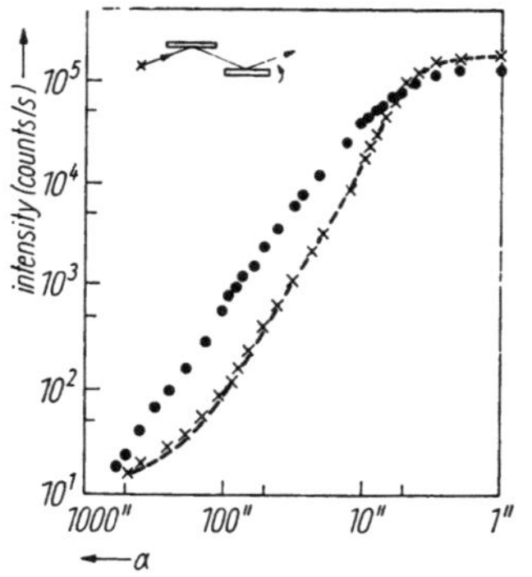

Fig. 6. Diffraction curves from lithium (●) and gold (×) diffused and perfect (dashed line) silicon crystals measured by a double crystal diffractometer of (+n, −n) setting. Bragg-case 111 reflection of CuKα

specimen the diffraction peak shows remarkable broadening due to the strong kinematical diffraction of Fig. 4. But two curves from perfect and gold diffused specimens are almost the same, even though the TCD curve corresponding to the same etching step shows a strong kinematical component (Fig. 5b). This phenomenon is explained as follows: the dynamically scattered component becomes weaker and the kinematically scattered one stronger so that the sum of both does not change so much in the case of slightly distorted crystals.

The present results show that in the TCD method the diffraction intensity could be treated more quantitatively compared with the previous experiments performed by the double crystal method [6], by separating the contributions of the dynamical and kinematical diffractions.

Acknowledgements

The author would like to thank Prof. K. Kohra for his interest, encouragement, and helpful discussions during this work. The author is also indebted to T. Abe, Y. Takano, and M. Azuma for providing specimens and for helpful discussions.

References

[1] A. IIDA and K. KOHRA, phys. stat. sol. (a) **51**, 533 (1979).
[2] A. J. R. DE KOCK, Philips Res. Rep., Suppl. **1**, 1 (1973).
[3] O. BRÜMMER and M. HOFMANN, phys. stat. sol. (a) **5**, 199 (1971).
[4] T. IIZUKA, Japan. J. appl. Phys. **4**, 1018 (1966).
[5] J. T. YUE and H. J. RUIZ, Semiconductor Silicon, Ed. H. R. HUFF and E. SIRTL, Electrochemical Soiety, Princeton, N.J. 1977 (p. 596).
[6] S. KIKUTA and K. KOHRA, J. Phys. Soc. Japan **21**, 1449 (1966).

(Received May 30, 1979)

phys. stat. sol. (a) **54**, 707 (1979)

Subject classification: 3; 14.3.1; 22.2.3

Department of Electrical Engineering, School of Sciences and Engineering, Waseda University, Tokyo[1])

Heteroepitaxial InSb Films Grown by Molecular Beam Epitaxy

By

M. YANO, T. TAKASE, and M. KIMATA

Films of InSb are heteroepitaxially grown on GaAs substrates by MBE. In spite of a large lattice mismatch, single-crystalline films with a fairly flat surface are obtained. Experimental data on electrical properties of these films suggest that dislocation scattering represents a significant mobility limiting process above 230 K. From in-depth profiles of the electron concentration and mobility, it is found that the layer within about 1.5 μm from the interface is inferior in crystallinity.

Les films de SbIn ont été crûs à la manière hétéroepitaxique sur le substratum de AsGa au moyen de MBE. Malgré la présence d'une grande discontinuité de réseau, les films monocristallins à la surface assez plate ont été obtenus. Les données expérimentales sur les propriétés électriques de ces films suggèrent que la diffusion par dislocation représente le procédé limitant une mobilité significative, supérieur à 230 K. D'après le profils à la profondeur de la concentration de electron et de la mobilité, il se trouve que la couche dans 1,5 μm environ à partir de l'interface est inférieure à la cristallinité.

1. Introduction

Indium-antimonide has the highest electron mobility in the group of III–V compound semiconductors, and is a useful material for galvanomagnetic devices. It is also a promising material for optical devices of 6 ≈ 7 μm infrared light. To fabricate such devices, it is important to prepare thin films of good quality. Many works have been made on InSb films prepared by means of vacuum deposition, sputtering, and so on. Especially the films grown by a microzone-recrystallization have a superior quality in electrical properties [1 to 3].

Thin film growth technique called molecular beam epitaxy (MBE) has been developed because of its utility in fabricating uniform and well-controlled thin films [4]. Reports on the MBE growth of InSb, however, are few. Recently we have presented a brief report about the MBE growth of InSb in literature of the growth of InGaSb [5].

In this paper, the MBE growth of undoped InSb films on GaAs substrates is described and the electrical properties of the films are discussed. It is revealed that the lattice mismatch between the GaAs substrate and InSb epilayer causes considerable influence on the grown layer.

2. Growth and Measurement Procedure

The MBE apparatus used in this study was reported previously [6]. The background pressure was less than 4×10^{-7} Pa prior to deposition and 2 to 1×10^{-6} Pa during the growth. The (100) surface of Cr-doped semi-insulating GaAs was used for the substrates. The source materials for molecular beams were In (6-N) and Sb (6-N).

[1]) Shinjuku, Tokyo, Japan.

The beam-intensities were derived from the equilibrium vapor pressure data at each cell temperature assuming the cosine law. Also the magnitude of the signal of a nude ion gauge was monitored during the deposition. The cell temperatures of In and Sb were 945 °C and 640 to 560 °C, respectively. The growth rate of films was about 1 μm/h. The crystallographic quality was examined by a refractive high energy electron diffraction (RHEED). The in-depth profile of the electrical properties of epilayers were examined from Van der Pauw method combined with a step etching. A technique of anodic oxidation followed by disolution of the oxide in a HCl solution was applied to the step etching. The thickness of the each step-etched layer was controlled with accuracy of $\pm 10\%$.

3. Experimental Results

The crystallinity of the grown films depends on the arrival rates of In(n_{In}) and Sb_4 (n_{Sb4}) molecules as well as on the substrate temperatures (T_s). A RHEED pattern of the epilayer is shown in Fig. 1. In spite of the large lattice mismatch, we have obtained epilayers with considerably smooth surfaces, whose crystallographic orientation is parallel to that of the substrates. Experimental results are summarized in Table 1. The optimum T_s has a tendency to decrease with the arrival ratio of each molecule, n_{Sb4}/n_{In}. Many droplets (probably indium) have been appeared at the surface when the ratio is decreased to 0.6. Epitaxial growth at low temperatures would be desirable because of the closer control of doping. In order to examine the epitaxial growth at lower deposition temperatures, T_s was reduced after the formation of single-crystal InSb films (about a few thousands of angstrom) at T_s larger than 330 °C. This method gives epitaxially grown films at $T_s = 300$ °C; however, films become polycrystalline at $T_s = 280$ °C. In the case where the deposition rate is reduced to 0.5 μm/h, smooth epilayers have been obtained even at $T_s = 180$ °C. Fig. 2

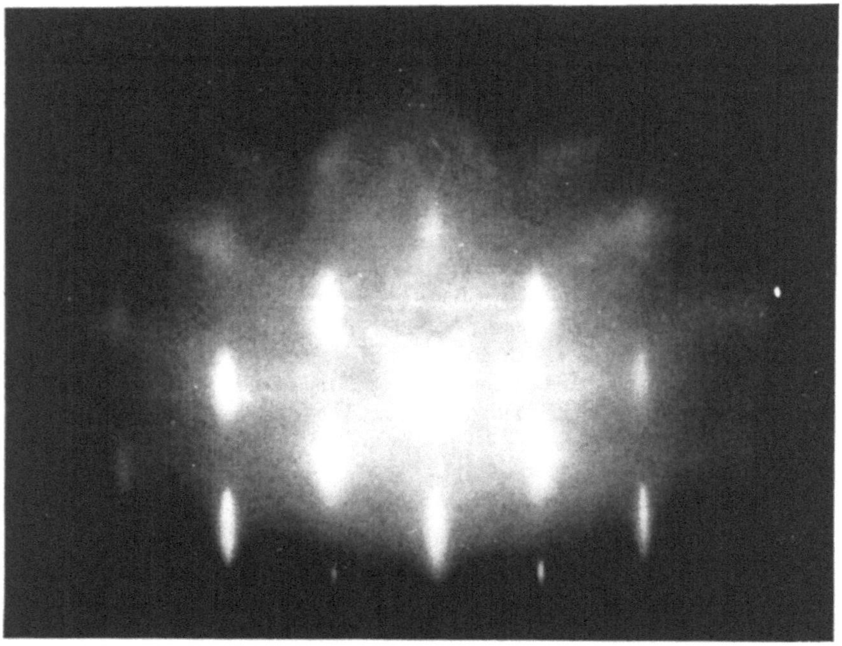

Fig. 1. RHEED pattern (50 keV, ⟨011⟩ azimuth), of InSb epilayer grown on GaAs (100) substrate. $T_s = 380$ °C and $n_{Sb4}/n_{In} = 1.0$

shows the surface morphology of the epilayer grown at various substrate temperatures, for the fixed ratio n_{Sb4}/n_{In} of unity. At 330 °C $\leqq T_s \leqq$ 400 °C, the grown layers are analyzed as single-crystals from RHEED observation. Smoth epilayers are obtained at T_s around 380 °C, and many stacking faults are appeared at $T_s = 330$ °C. Stoichiometric single-crystal films have no longer been obtained at $T_s = 450$ °C. The trend of the surface morphology upon the ratio n_{Sb4}/n_{In}, at fixed T_s, is shown in Fig. 3. When Sb_4 molecules are supplied excessively, step growth of the (100) orientation becomes dominant. The electrical properties at 300 K shown in Table 1 are representative results of films whose thickness is about 3 μm. The electrical conduction is n-type and the carrier mobility appears not to be sensitive to the growth conditions. The carrier concentrations at 300 K are a little higher than the intrinsic value ($1.6 \times \times 10^{16}$ cm^{-3}), and have a tendency to become higher when the films are deposited under low T_s or Sb-rich condition. Fig. 4 shows the temperature dependence of the electrical properties of the epilayer. The mobility at 300 K is somewhat low compared to that of single-crystal bulk InSb with the same impurity concentration. The mobility has the maximum at about room temperature. The carrier concentration slightly increases below 150 K. An annealing in vacuum or Sb atmosphere for 10 h at 200 °C has no noticeable influence on their properties.

Above-mentioned electrical properties are derived under the assumption of the uniformity of the grown layer. However, there will be inhomogeneity in the in-depth profiles of the electrical properties. The carrier concentration $n(x)$ and the mobility $\mu(x)$ (x denotes the distance from the surface) are estimated from the differential sheet conductivities and effective Hall coefficients measured at each time of step etching. The surface scattering of carriers would not be ignored because the mobility of InSb is considerably large. In order to extract the "bulk" electrical properties, we correct our experimental data assuming a simple model [7], where the surface

Table 1

Representative characteristics of deposited layers

n_{Sb4}/n_{In}	T_s (°C) 300	330	380	400	420	450
1.2		△	○ 4.0×10^4 3.5×10^{16}			
1.0	×	○ 4.5×10^4 4.0×10^{16}	⊙ 5×10^4 2×10^{16}	○ 4×10^4 2×10^{16}	○ 4×10^4 3×10^{16}	×
0.8	△	⊙ 4×10^4 4×10^{16}				
0.6	△	△				notation (crystallinity) (mobility) (concentration)

Films are about 3 μm in thickness, and their crystallinity is: ⊙ — single-crystal with a flat surface, ○ — single-crystal, △ — some are single-crystals and some are not, × — polycrystalline. Units of carrier concentrations and mobilities are cm^{-3} and cm^2/Vs, respectively.

Fig. 2 Fig. 3

Fig. 2. Surface morphology of deposited layers: the substrate temperatures are a) 450, b) 380, and c) 300 °C, and n_{Sb4}/n_{In} is unity. Thickness of films is about 3 μm

Fig. 3. Surface morphology of deposited layers: the ratios n_{Sb4}/n_{In} are a) 1.2, b) 0.8, c) 0.5, and T_s is 330 °C. Thickness of films is about 3 μm

(and the interface) scatters carriers randomly, and the observed average collision time τ_a is expected as

$$\frac{1}{\tau_a} = \frac{1}{\tau_b} + \frac{1}{\tau_i}, \tag{1}$$

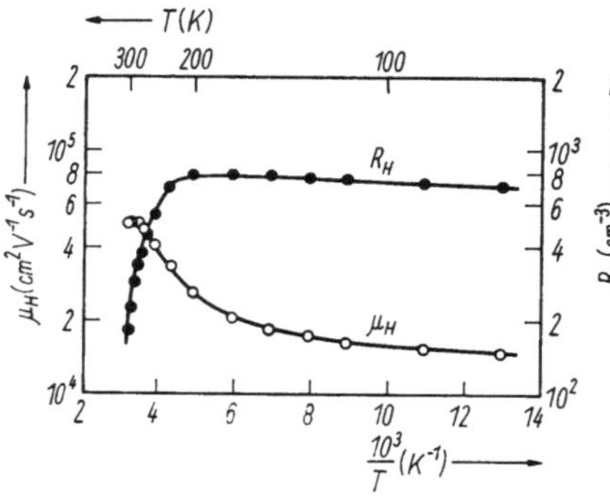

Fig. 4

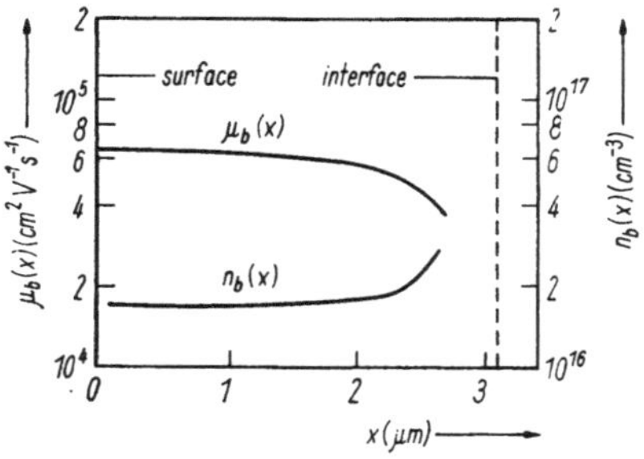

Fig. 5

Fig. 4. Temperature dependence of the Hall coefficient R_H and mobility μ, of a 3 μm thick, single-crystal InSb film

Fig. 5. Electron mobility and concentration distributions at 300 K in a grown layer as a function of the distance from the surface. The dashed line represents a metallurgical interface between the grown layer and the substrate

where τ_b is collision time in the bulk, and τ_i the time interval between collisions with the surface. If we assume that the average carrier velocity is the thermal velocity, τ_i is given by

$$\tau_i = \frac{d}{(2\,kT/\pi m^*)^{1/2}}, \tag{2}$$

where d is the residual film thickness, k the Boltzmann constant, T the absolute temperature, and m^* the effective mass of carriers. From τ_b, the "bulk" sheet conductivity $\sigma_b^s(x)$ and effective Hall coefficient $R_b^s(x)$ can easily be derived. Then $n_b(x)$ and $\sigma_b(x)$ become [8]

$$n_b(x) = -\,\Delta\{\sigma_b^s(x)\}/e\mu_b^s(x)\Delta x\,, \tag{3}$$

$$\mu_b(x) = \Delta[R_b^s(x)\,\{\sigma_b^s(x)\}^2]/\Delta\{\sigma_b^s(x)\}\,. \tag{4}$$

The correction effect of the surface scattering has been small because the magnitude of the mobility is small near the interface. The curves of $n_b(x)$ and $\mu_b(x)$ are shown in Fig. 5. We can see that there exist a low mobility and high carrier concentration region with thickness of about 1.5 μm from the interface.

4. Conclusion and Discussion

Smooth surface InSb epilayers have been grown on the GaAs substrates by MBE. The electrical properties of these films, however, are different from those of bulk InSb. The characteristics are

a) the conduction type is n,

b) the mobility approaches to the value of the bulk at high temperature and rapidly decreases with decreasing temperature,

c) the Hall carrier concentration becomes higher near the interface.

Such a rapid decrease in mobility below room temperature has generally been observed in InSb films prepared by the three-temperature technique [9], by the source-temperature-programmed vacuum deposition [10], or by the microzone-recrystallization [1 to 3], and cannot be explained simply in terms of impurity scattering. In order to obtain a qualitative description of this mechanism, we assume that the measured mobility μ is related to μ_b (characterized by the lattice scattering which dominates the mobility above 200 K in bulk InSb [11]) and μ^* (by an additional scattering mechanism) through

$$\frac{1}{\mu} = \frac{1}{\mu_b} + \frac{1}{\mu^*}. \tag{5}$$

We adopt $\mu_b = 7 \times 10^8 T^{-1.6}$ cm²/Vs for the electron mobility in bulk InSb [12]. The measured mobilities of three specimens have been used in conjunction with μ_b to determine μ^* from (5) as a function of temperature. The results shown in Fig. 6 indicate that μ^* increases linearly with temperature above 230 K. In our heteroepitaxial films, an adoptable mechanism to describe such a linear dependence on temperature is scattering by strain fields associated with dislocations. An expression of the mobility derived by Dexter and Seitz [13] for carriers with Maxwell-Boltzmann statistics is

$$\mu_d = \frac{32}{3\pi} \left(\frac{1 - \nu}{1 - 2\nu}\right)^2 \frac{kT\hbar e}{\varepsilon^2 \lambda^2 m^* N_d}, \tag{6}$$

where ν is Poisson's ratio, $\hbar$ Dirac's constant, ε the deformation potential, λ the lattice spacing, and N_d the density of dislocation lines.

The electrical properties described above also can be accounted for provided that our films contain not only shallow donors but also many dislocations which act as sources of dense acceptors. The Fermi level E_F is near the mid-band and the number of the conduction carriers is large at room temperature. Then the collision is dominated by the deformation-potential scattering associated with dislocations, because the impurity scattering is not so large. On the other hand, as the temperature is lowered, E_F approaches to the conduction-band edge except in the vicinity of the dislocations, which raise the potential around the dislocations. At low temperature, this high-potential fields around the randomly distributed dislocations and the degeneracy of the electrons make the mobility independent of temperature. Regions of high hole

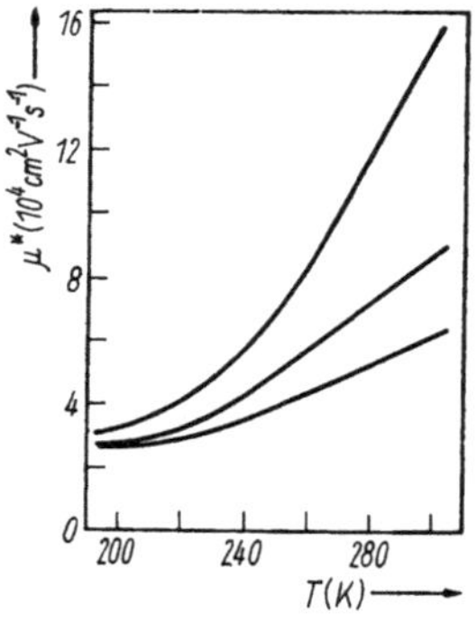

Fig. 6

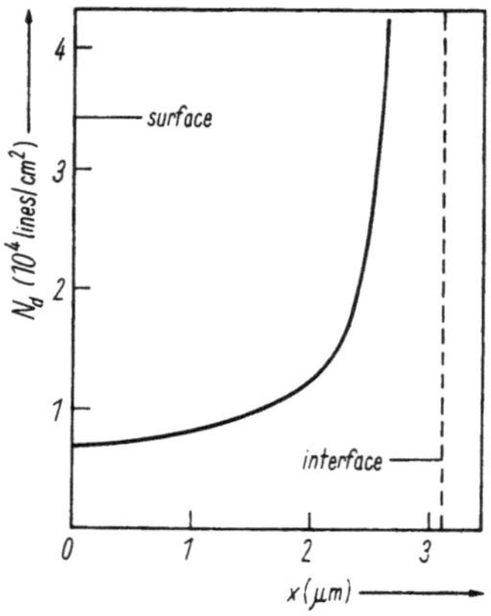

Fig. 7

Fig. 6. Temperature dependence of μ^*, the mobility associated with dislocations in InSb. Three curves represent different specimens

Fig. 7. Distribution of the dislocation density in a grown layer as a function of the distance from the surface. The dashed line represents the metallurgical interface between the grown layer and the substrate

density caused by dislocation acceptors would be near the interface as a result of the lattice mismatch. This reduces the magnitude of the Hall coefficient $|R_H|$, and then the carrier concentration calculated as $n = (eR_H)^{-1}$ becomes higher than the real value near the interface.

The above-mentioned model can give a qualitative interpretation of the electrical properties of our films. Other explanations will of course be possible. We can estimate the dislocation density from (5) and (6) in conjunction with the carrier mobility at 300 K. The result is shown in Fig. 7. There are dense dislocations near the interface, and dislocations in the order of 10^8 cm^{-2} are retained beyond 1.5 μm in thickness from the interface. This dislocation density evaluated are compatible with the results of Oi et al. [3]. They counted 10^7 cm^{-2} order etch pits due to dislocations on their microzone-recrystallized InSb which has a little higher mobility than our hetero-epitaxial films. We see from Fig. 7 that a rapid decrease in the dislocation density cannot be expected as the film-thickness exceeds a few microns.

Acknowledgements

The authors wish to express their thanks to H. Futami, T. Ishihara, and T. Suga for their experimental assistance. Thanks are also to A. Chan for continuing encouragements. This work has been done as part of co-operative research with three colleges (Keio Univ., Tokyo Institute of Tech., and Waseda Univ.).

References

[1] H. H. Wieder, Solid State Electronics **9**, 373 (1966).
[2] A. R. Clawson, Thin Solid Films **12**, 291 (1972).
[3] T. Oi, N. Kotera, J. Shigeta, N. Yamamoto, and M. Nakanishi, Japan J. appl. Phys. **17**, 407 (1978).
[4] A. Y. Cho and J. R. Arthur, Progr. Solid State Chem. **10**, 157 (1975).
[5] M. Yano, T. Takase, and M. Kimata, Japan J. appl. Phys. **18**, 387 (1979).
[6] M. Yano, M. Nogami, Y. Matsushima, and M. Kimata, Japan J. appl. Phys. **16**, 2131 (1977).
[7] F. Berz, Surface physics of phosphors and semiconductors, Ed. C. G. Scott and C. E. Reed, Academic Press, New York 1975 (p. 180).
[8] T. Arizumi, T. Nishinaga, and M. Kakehi, Japan J. appl. Phys. **7**, 468 (1968).
[9] H. H. Wieder, Intermetallic Semiconducting Films, Pergamon Press, New York 1970 (p. 122).
[10] Y. Amemiya, H. Terao, and Y. Sakaki, J. appl. Phys. **44**, 1625 (1973).
[11] K. I. Vinogradova, V. V. Galavonov, and D. N. Nasledov, Soviet Phys. — Solid State **4**, 1230 (1962).
[12] C. Hilsum and A. C. Rose-Innes, Semiconducting III–V Compounds, Pergamon Press, New York 1961 (p. 126).
[13] D. L. Dexter and F. Seitz, Phys. Rev. **86**, 964 (1952).

(Received May 29, 1979)

phys. stat. sol. (a) 54, 715 (1979)

Subject classification: 1.4 and 10

Engineering Materials Group and Department of Mechanical Engineering, University of Maryland, College Park[1])

Dislocations, Disclinations, and Grain Boundaries in a Finite Solid[2])

By

K. JAGANNADHAM and M. J. MARCINKOWSKI

An earlier surface dislocation analysis applied to infinite grain boundaries is extended to grain boundaries and disclinations in a finite solid. The coalescence of the grain boundary surfaces is described by grain boundary surface dislocation arrays. The conservation of Burgers vectors is satisfied by the screening dislocation array. Equilibrium positions of the screening array are determined by minimizing the total energy of the configuration. A disclination is described as a grain boundary ending inside a crystal. The differences between the distortion around a disclination and a grain boundary in a finite body are elucidated. The surface dislocation analysis of an elastically bent crystal is given for the first time. A plastically bent crystal is described in terms of regions containing disclinations or grain boundaries surrounded by stress-free surfaces. The stress-free surfaces form a network and cannot end inside the crystal.

Eine frühere Oberflächenversetzungsanalyse für unendliche Korngrenzen wird auf Korngrenzen und Versetzungen in einem endlichen Festkörper ausgedehnt. Die Koaleszenz der Korngrenzenoberflächen wird durch Korngrenzenoberflächen-Versetzungsanordnungen beschrieben. Die Erhaltung des Burgersvektors wird durch die abschirmende Versetzungsanordnung befriedigt. Gleichgewichtslagen der abschirmenden Anordnung werden durch Minimalisierung der Gesamtenergie der Anordnung bestimmt. Eine Disklination wird als Korngrenze, die im Inneren des Kristalls endet, beschrieben. Die Unterschiede zwischen einer Verzerrung um eine Disklination und einer Korngrenze in einem endlichen Körper werden aufgedeckt. Zum erstenmal wird eine Oberflächenversetzungsanalyse eines elastisch gebogenen Kristalls angegeben. Ein plastisch gebogener Kristall wird mit Bereichen beschrieben, die Disklinationen oder Korngrenzen enthalten, die von spannungsfreien Oberflächen umgeben sind. Die spannungsfreien Oberflächen bildene in Netzwerk und können nicht im Inneren des Kristalls enden.

1. Introduction

It is now well established that a disclination or a grain boundary can be described in terms of a more fundamental configuration namely that of a dislocation [1]. A grain boundary consists of a uniform array of dislocations termed grain boundary dislocations. The spacing between the dislocations in the array, the magnitude of the Burgers vectors of the dislocations together with their orientation with respect to the plane of the grain boundary completely determine the geometry of the configuration [2 to 5]. Similarly a disclination is an array of dislocations ending inside a crystal [1]. Alternately, a disclination can also be described as a grain boundary terminating inside a crystal [1]. Fig. 1a and b show a grain boundary and a disclination, respectively, in the completely torn configuration. These figures also illustrate the relationship between a disclination and a grain boundary.

[1]) College Park, Maryland 20742, USA.

[2]) The present research effort was supported by the United States Department of Energy under contract No. At-(40-1)-3935.

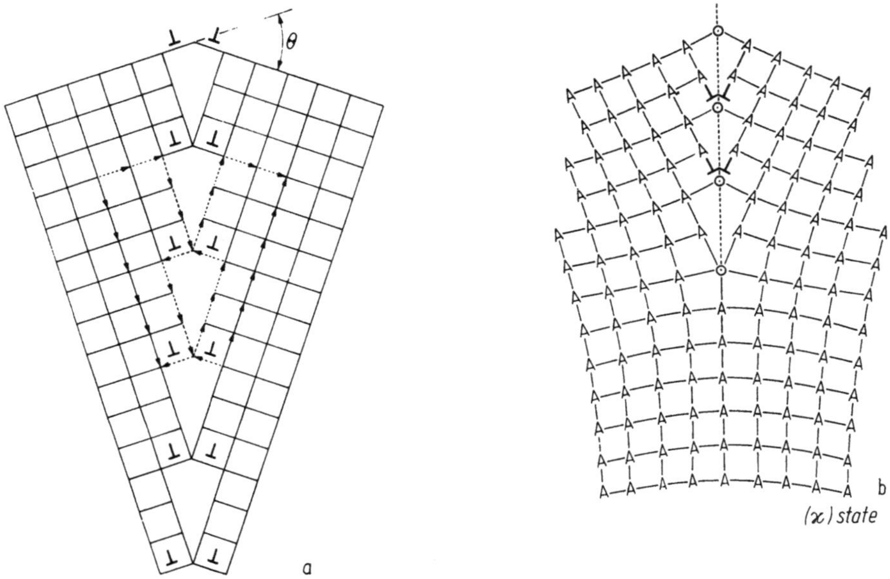

Fig. 1. a) Completely relaxed tilt-type symmetric grain boundary of misorientation angle $\theta = 36.9°$; b) completely relaxed wedge disclination formed by a grain boundary of misorientation angle $\theta = 53.1°$ ending inside a crystal

The tendency of the grain boundary surfaces to coalesce and thus reduce their surface energy was recognized in a recent grain boundary model constructed in terms of surface dislocations [3 to 5]. In particular, Fig. 2 shows the surface dislocation model of a grain boundary. A grain boundary element consists of the region ABCDEF and BGHEHI. The boundary consists of an array of such grain boundary elements. The grain boundary surfaces coalesce in the region EH and remain uncoalesced in the region BGHI. The coalescence of the grain boundary surfaces is described by intro-

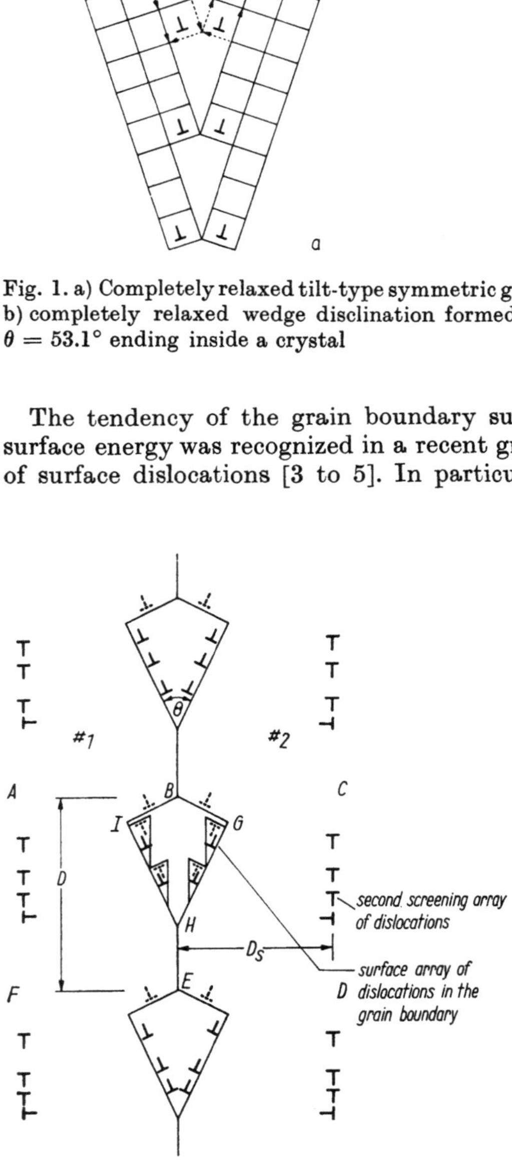

Fig. 2. Three units of a grain boundary region described in terms of the surface dislocation model. A set of surface dislocations which satisfy the free surface boundary conditions on the grain boundary surface shown together with a second set of surface dislocations which allow the screening of the stress field of the grain boundary array

ducing surface dislocations on the grain boundary surfaces. The surface array on the grain boundary surfaces shown by solid lines gives the stress field arising due to coalescense, while the arrays shown dotted indicate the ledge steps on the surface. The fully coalesced grain boundary, which is a hypothetical situation, is formed when the grain boundary surfaces coalesce completely and thus eliminate the surface. The Burgers vector of the surface dislocation array on the grain boundary surfaces has been chosen in the earlier models [3, 4] to be equal to the Burgers vector of the grain boundary lattice dislocation. However, further consideration of the coalescence of the grain boundary surfaces [5]. has shown that the total Burgers vector of the surface array on the grain boundary surfaces is less than the Burgers vector of the grain boundary lattice dislocation.

Coalescence of the grain boundary surfaces eliminates the surface area associated with ledge steps. The ledge steps eliminated during coalescence are in turn converted into elastic surface dislocations which have stress and strain field associated with them. This very important result shows the intimate relation between elimination of ledge steps and the formation of elastic surface dislocations [5]. The coalescence in the fully torn boundary is zero as shown in Fig. 1a and hence there are no elastic dislocations, and thus stress fields, in that boundary. A partially coalesced boundary has elastic dislocations and stress fields associated with it. The total Burgers vector of elastic surface array of dislocations on the grain boundary surfaces is equal to the ledge step eliminaed and thus it is directly proportional to the coalesced region of the boundary surfaces. The severe distortion resulting from coalescence does not permit full coalescence. On the other hand, the surface energy is reduced by the coalescence to the extent permitted by the distortion.

A more general principle followed is the conservation of the Burgers vector of dislocations in a solid. This requires the presence of a second dislocation array called the screening array as shown in Fig. 2 whose Burgers vector is opposite in sign and equal in magnitude to the total Burgers vector of the elastic grain boundary surface array. The position of the dislocations in the array and their Burgers vectors can be determined by minimization of the total energy of the configuration [5].

The total energy consists of the strain energy of the dislocation array and the surface energy of the uncoalesced grain boundary surfaces. Although the screening array is a two-dimensional distribution [5] surrounding the grain boundary, the present calculations are for simplicity carried out assuming that the array is present on a vertical plane. Earlier calculations of the grain boundary configuration assuming three grain boundary elements to calculate the nearest neighbour interaction have shown that the conservation of Burgers vectors is strictly satisfied [3 to 5]. The screening array has a well-defined equilibrium position away from the grain boundary [3 to 5]. The exact value of specific surface energy of the grain boundary surfaces could vary along the uncoalesced region due to the varying interaction of the atoms on the neighbouring surfaces introducing inaccuracy in the determination of the exact value of the total grain boundary surface energy. However, this inaccuracy does not lead to any conceptual errors. Also, the non-linear nature of distortion introduced into the solid has been represented by dislocation arrays described in terms of linear elasticity. The linear elasticity approach has given comparable results with experiments.

It is worthwhile at this point to state that the present surface dislocation model of the grain boundary is an extension of the more general idea that surface dislocations can be used to describe any type of distortion in a finite solid [7, 8]. The screening array satisfies the stress-free surface boundary condition outside the finite region. The position, D_S, of the screening array depends on the balance of the decrease in strain energy and the increase in surface energy of the uncoalesced grain boundary surfaces. It is the aim of the present calculations to extend the ideas developed earlier to grain boundaries and disclinations in finite solids. The presence of these configurations in a plastically bent crystal will also be illustrated.

The numerical analysis of the dislocation configurations throughout the present calculations is carried out assuming the elastic constants of iron, namely, shear modulus $\mu = 7.14 \times 10^{11}$ dyn/cm^2, and Poisson's ratio $\nu = 0.333$. The surface energy γ for Iron is ≈ 2000 erg/cm^2. The results are also illustrated at different values of γ.

2. Grain Boundaries of Infinite Length

The numerical calculations to determine the dislocation configuration of an infinite boundary of any orientation θ have been carried out using the surface dislocation model shown in Fig. 2. The total energy of the configuration which consists of the self energy and interaction energy of the dislocations and the surface energy of the uncoalesced grain boundary surfaces is minimized with respect to the positions of all the dislocations in each element namely ABCDEF and BGHI. The details of the calculations for infinite boundaries are presented elsewhere [5]. The numerical calculations showed that the total energy, E_T, of the configuration has a minimum at some value of D_S, the distance of the screening array from the grain boundary. The equilibrium configuration corresponds to the minimum energy configuration. The energy E_T per unit length at equilibrium has been found to increase with increasing θ. The equilibrium configuration is obtained at smaller values of D_S with increasing θ. The increase in energy E_T for a given θ and at values of D_S smaller than the equilibrium value arises due to the opening up of the grain boundary surfaces by the attractive force exerted by the screening array. On the other hand, the increase in E_T at values of D_S greater than the equilibrium value arises due to higher strain energy contribution. Numerical calculations carried out at different decreasing values of γ show that the total energy per unit length at equilibrium decreases with, although the minimum in the energy appears at the same value of D_S for all values of γ.

Subdividing the grain boundary element into smaller units composed of dislocations with smaller Burgers vectors without changing the misorientation angle θ of the boundary leads to an idealistic situation of a grain boundary with zero strain energy and zero surface energy. A grain boundary with zero strain energy and zero surface energy can be obtained by joining two crystals of proper orientation so that no dislocations are required to account for the misorientation. The results of an infinite boundary can be extended to that of a single dislocation from the limiting situation of θ approaching zero. This result, which cannot be computed, becomes obvious from increasing values of D_S at equilibrium with decreasing θ.

3. Grain Boundaries in a Finite Solid

The analysis carried out earlier for an infinite boundary [5] can now be extended to a grain boundary in a finite solid. The finite grain boundary configuration is defined by Fig. 3 and 4 where the uncoalesced regions of the boundary are approximately equally spaced from each other and from the horizontal surfaces of the finite body. Three elements of the grain boundary shown are considered in the configuration. The screening array is located on the horizontal as well as vertical planes, as shown in Fig. 3. The minimization of the total energy of the configuration shown in Fig. 3 is carried out with respect to the position and Burgers vector of the screening array and the grain boundary array and the distance, D_S, of the vertical array from the grain boundary surface [6]. The lattice dislocations forming the fully torn grain boundary

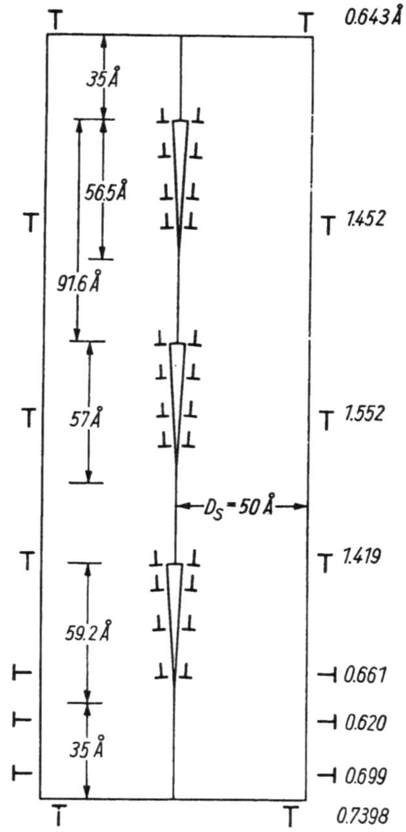

Fig. 3

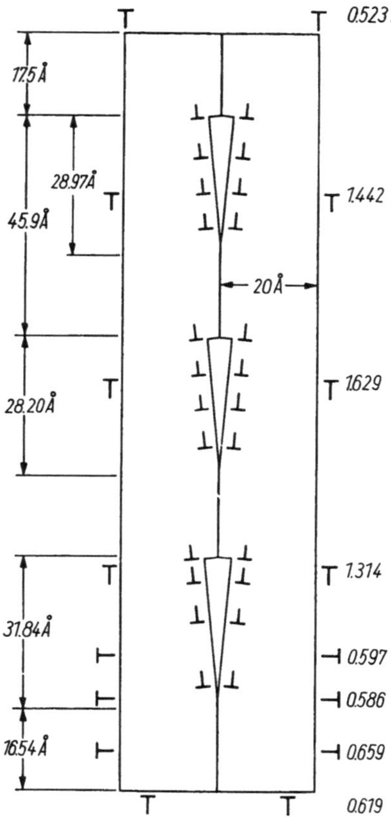

Fig. 4

Fig. 3. Surface dislocation configuration of an infinite boundary with misorientation angle $\theta = 5°$, $b_m = 4$ Å, and $\gamma = 2000$ erg/cm². The Burgers vectors of the surface array on the grain boundary surfaces in the three elements given by $b_{g_1} = 0.5113$, $b_{g_2} = 0.504$, and $b_{g_3} = 0.472$ Å are shown. The Burgers vectors of the screening array, b_I, are illustrated. D_S, the equilibrium distance of separation, is equal to 50 Å

Fig. 4. Same as Fig. 3 but with $\theta = 10°$. $b_{g_1} = 0.492$, $b_{g_2} = 0.514$, $b_{g_3} = 0.408$ Å

have a Burgers vector, $b_m = 4.0$ Å. The total energy of the configuration obtained as a result of minimization is shown in Fig. 5 as a function of D_S for two values of θ. Also, shown in the figure is the ratio of the sum of the Burgers vectors of the surface array of elastic grain boundary dislocations to the sum of the Burgers vector of grain boundary lattice dislocations, b_{mI}/b_m. The results show that the total energy, E_T, reaches a well-defined minimum as a function of D_S. Also, the minimum in the energy appears at a smaller value of D_S with increasing value of θ. As elucidated earlier, values of D_S smaller than the equilibrium value correspond to a grain boundary in a finite solid with surface dislocations satisfying the free surface boundary conditions. On the other hand, the equilibrium value of D_S corresponds to a grain boundary in a solid with finite dimension in the vertical direction only. Since the solid is not finite in the horizontal direction, the surface screening array of dislocations move towards the boundary. Physically, the screening array represents the relaxation of the distortion

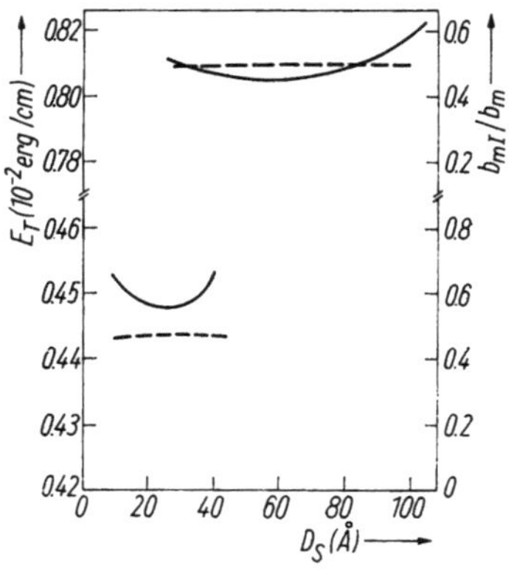

Fig. 5. The total energy E_T and the ratio b_{mI}/b_m shown as a function of D_S for two values of θ; $\gamma = 2000\,\text{erg/cm}^2$, $b_m = 4$ Å. The upper scale on either side corresponds to $\theta = 5°$ and the lower scale to 10°. The full line curve gives the value of E_T and the dashed curve, b_{mI}/b_m

around the grain boundary. The screening array exerts a force on the grain boundary surface array so as to separate the grain boundary surfaces. However, the surface energy has the opposing effect so as to coalesce the grain boundary surfaces. The Burgers vector of the elastic grain boundary surface array increases with increasing coalescence of the grain boundary surfaces. Since the screening array has a Burgers vector which is opposite in sign and equal in magnitude to the Burgers vector of the grain boundary array, the force opening up the grain boundary surfaces also increases with increasing coalescence. At the same time, the repulsive force between the dislocations in the grain boundary array increases with increasing coalescence since the Burgers vector of the surface dislocations is increased and their separation is decreased. Therefore, the ratio of the Burgers vector of the elastic grain boundary array to the Burgers vector of the grain boundary lattice dislocations reaches a maximum when the grain boundary surfaces coalesce to the largest extent possible.

The dislocation configurations shown in Fig. 3 and 4 correspond to grain boundaries in the finite solid with misorientation angle $\theta = 5°$ and 10°, respectively. The results show that the conservation of Burgers vectors is satisfied within the accuracy of discrete dislocation analysis. In particular, the sum of the horizontal component of the Burgers vectors of elastic grain boundary dislocations is equal to the sum of the horizontal component of the Burgers vector of the screening array. The vertical component of the grain boundary array on the two grain boundary surfaces forms a dipole whose stress field is short range. The second set of screening dislocations consists of those with Burgers vector in the vertical direction. This set of dislocations screen the σ_{xy}-component of stress. The opening up of the grain boundary surfaces is shown by the separation of surface array of dislocations from each other on the grain boundary surfaces. The total energy of the boundary is not shown per unit length. Therefore, E_T at equilibrium for $\theta = 10°$ is smaller than that for $\theta = 5°$. The ratio of b_{mI}/b_m for $\theta = 5°$ is larger than that for $\theta = 10°$. This result, which is also observed for infinite boundaries, shows that the elastic dislocation content of a grain boundary decreases with increasing misorientation angle [5].

It is useful to point out that the grain boundary surfaces can never be separated completely by the screening array unless an external stress is applied. Further separation of the grain boundary surfaces leads to a decrease in the Burgers vector of the elastic grain boundary array, thus decreasing the Burgers vector of the screening array. Therefore, the force opening up the grain boundary surfaces is reduced. The equilibrium distance of separation of the screening array from the grain boundary is smaller for increasing misorientation angle θ.

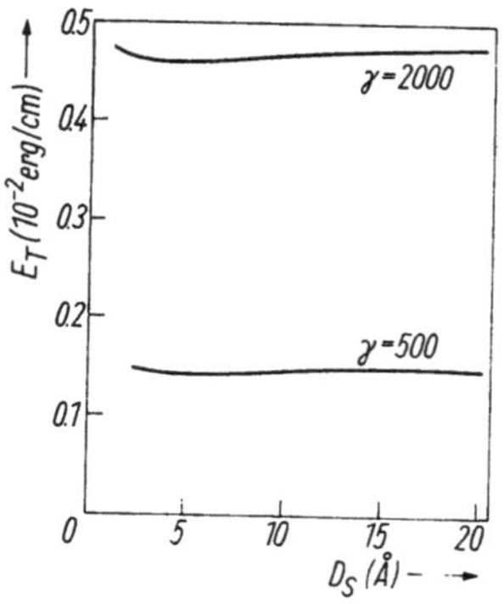

Fig. 6. The total energy E_T shown as a function of D_S for two values of γ; $\theta = 5°$ and $b_m = 2.5$ Å. Note that the minimum energy appears at the same value of D_S for the two values of γ

The results of the numerical calculations with $b_m = 2.5$ Å and two values of γ are shown in Fig. 6 for $\theta = 5°$. It is instructive at this point to note that the equilibrium value of D_S for $\theta = 5°$ is smaller with smaller values of b_m. Qualitatively, subdividing the boundary without changing the misorientation leads to a boundary with smaller energy per unit length and the equilibrium distance of separation of the screening array decreases, ultimately becoming zero for a boundary with zero dislocation content, i.e. a boundary which has been subdivided without limit [5]. It is also interesting to find that the equilibrium value of D_S is independent of the value of specific surface energy, γ, used in the calculations. Thus, the result of the attractive force exerted by the screening array on the grain boundary surface array of dislocations is independent of surface energy but depends only on the distance of separation, D_S.

4. Disclinations in a Finite Body

It is now well recognized that the distortion around a disclination is different from that around a dislocation or a grain boundary. However, a disclination can also be described in terms of these two defects. Only a uniform array of dislocations ending inside a crystal can constitute a disclination since the wedge angle introduced into the finite body is different from zero. On the other hand, the angle of the wedge for a dislocation is zero. This same difference exists between a dislocation and a grain boundary, as pointed out earlier. It is also important to know that a disclination in a finite body can be described as a grain boundary ending in the finite body. In other words, the uncoalesced regions are not separated by the same distance from the two horizontal surfaces of the finite body, as shown in Fig. 7. It is also important to recognize that the Burgers vector of the grain boundary surface array in the three elements representing a disclination, as shown in Fig. 7, is extremely non-uniform, namely b_{g_3} is much smaller than b_{g_1} or b_{g_2}. Comparison of Fig. 3 and 4 with Fig. 7 shows that b_{g_3} for a disclination is much smaller than b_{g_3} for a finite grain boundary. Also, the distribution of the surface array is uniform for a grain boundary. The above difference in the nature of the distortion is probably the most important distinction between a disclination and a grain boundary. The distribution of the screening array, which is an indication of the relaxation of the distortion, shows the difference between Fig. 3, 4 and Fig. 7, i.e. between a grain boundary and a disclination in a finite solid.

The dislocation configuration shown in Fig. 7 is obtained by minimizing the total energy with respect to the positions and Burgers vectors of all the dislocations in the configuration, as well as the distance, D_S, of the vertical array from the grain boundary and the distance D_W of the surface array at the bottom horizontal surface from the lower most element. The total energy E_T is found to reach a minimum almost at the

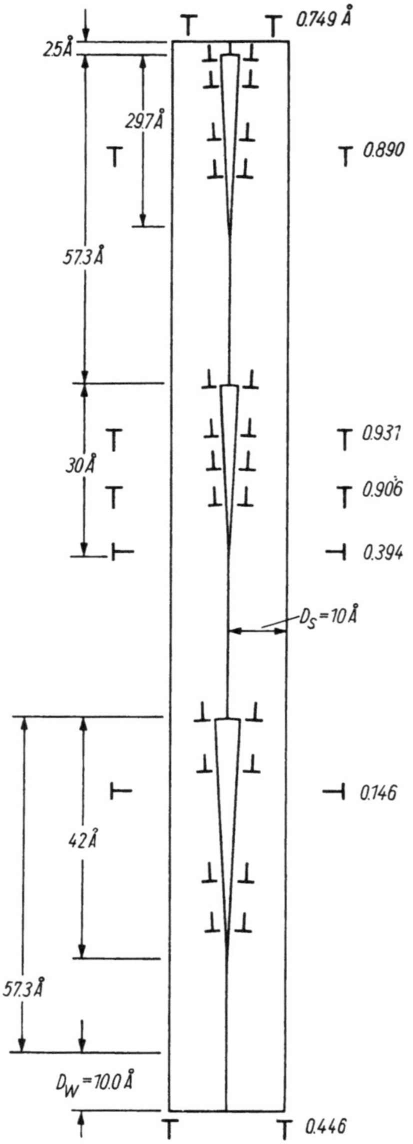

Fig. 7. The grain boundary array of dislocations and the screening array forming a grain boundary of misorientation angle $\theta = 5°$, $b_m = 2.5$ Å, and ending inside a finite body to create a disclination. The Burgers vector of the surface array on the grain boundary surfaces in the three elements $b_{g_1}=0.394$, $b_{g_2}=0.401$, and $b_{g_3} = 0.223$ Å are illustrated. The Burgers vectors of the surface array, b_I, are shown

same value of D_S for all values of D_W. E_T and b_{mI}/b_m, which have the same meaning as earlier, are shown in Fig. 8 as a function of D_W at $D_S = 10$ Å where the minimum energy configuration is obtained. The screening array which is in reality a two-dimensional distribution of surface dislocations surrounding the disclination is now chosen on well-defined vertical and horizontal planes. The force exerted by the screening array on the grain boundary surfaces constituting a disclination is the same as illustrated earlier for a grain boundary in a finite solid. Furthermore, the screening array on the lower horizontal surface exerts an opening up effect on the grain boundary surfaces of the lower most element, thus reducing the elastic dislocation content in that element given by b_{g_3}. Similarly, the screening array on the upper horizontal surface exerts an attractive force on the surface array on the grain boundary surfaces in the top most element and hence closing up the surfaces and increasing the elastic dislocation content, given by b_{g_1}. The equilibrium configuration corresponds to $D_W = 10$ Å as shown in Fig. 7. The ratio of the sum of the Burgers vectors of the elastic grain boundary array to the sum of the Burgers vectors of the grain boundary lattice dislocation is found to reach a maximum at the minimum energy configuration. The configuration of a disclination shown in Fig. 7 illustrates that the conservation of Burgers vectors is satisfied within the accuracy of discrete dislocation analysis.

In order to determine the distribution of the screening array of dislocations for a disclination of smaller length, the number of elements constituting the disclination is reduced to two from three. Numerical calculations with two elements have shown that the screening array has no equilibrium position, i.e. the total energy decreased with decreasing D_S. Increasing the number of elements in the disclination is expected to increase the equilibrium value of D_S and finally when the disclination is extending from the centre to infinity in an infinite body, the screening array remains at infinity.

The results of the surface dislocation analysis applied to a disclination are very useful in understanding the distortion around an indentation. The indentation is

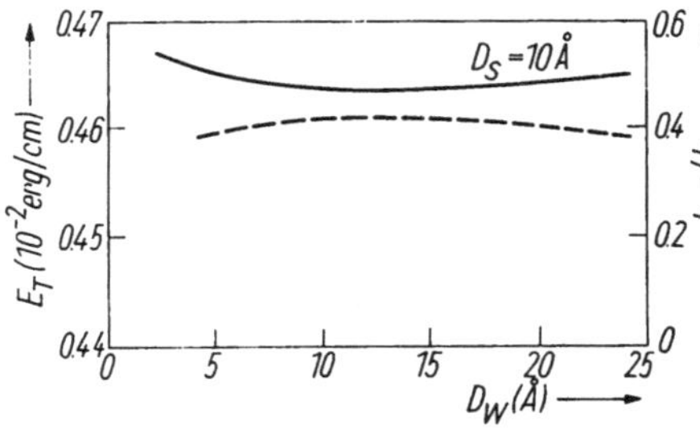

Fig. 8. E_T and the ratio b_{mI}/b_m shown as a function of D_W for $D_S = 10$ Å, $\theta = 5°$, $\gamma = 2000$ erg per cm², and $b_m = 2.5$ Å for a disclinition in a finite body. The significance of the terms D_W and D_S is shown in Fig. 7. The full line gives the value of E_T while the dashed line corresponds to b_{mI}/b_m. The minimum energy configuration is obtained at $D_S = 10$ Å for all values of D_W

produced by an indentor of constant wedge angle. The distortion around the indentor does not spread to very large distances in the solid. Instead, it is relaxed by the surface array of dislocations in the solid within a very short distance. Spreading screening array into a two-dimensional distribution will give a more accurate description of the distortion around the indentation.

5. Pair of Dislocations in a Finite Body

In order to further understand the nature of surface dislocations inside a body compared to the surface dislocations situated on the outer surface of the body, a finite solid containing a pair of unlike screw dislocations is considered as shown in Fig. 9. The free surface boundary conditions are satisfied by placing screw dislocations of opposite sign on the surface. The dislocation configuration is determined by minimizing the total energy of the configuration with respect to the positions and Burgers vector of the screw dislocations on the surface and the results are shown in Fig. 9. For every screw dislocation of positive sign, there is a dislocation of opposite sign and equal Burgers vector. In addition, the plane along the y-axis is a stress-free plane. It is now possible to cut the body along the y-axis. However, the cutting operation disturbs the free surface conditions unless an additional set of surface dislocations are placed on the cut surfaces.

An alternate method of describing the pair of unlike screw dislocations is shown in Fig. 10 where the body is divided into two regions along the y-axis. Surface dis-

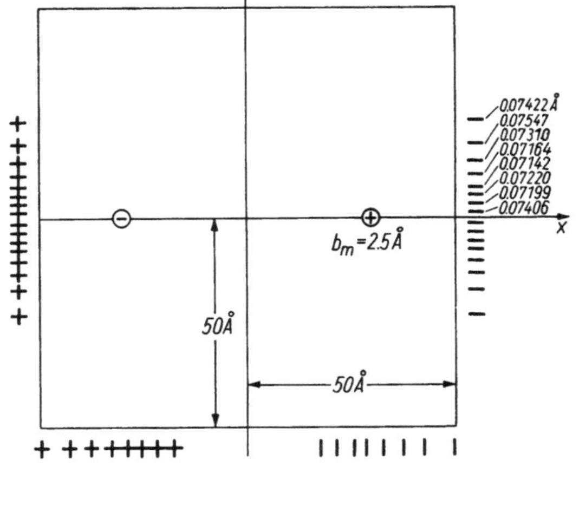

Fig. 9. Surface dislocation configuration satisfying the free surface boundary conditions for a finite solid containing a pair of unlike screw dislocations. The figure is symmetric about both the x- and y-axes. Therefore, the magnitudes of the Burgers vectors of the dislocations in one quadrant only are shown. A plus sign indicates a right-handed screw and a minus sign the left-handed screw

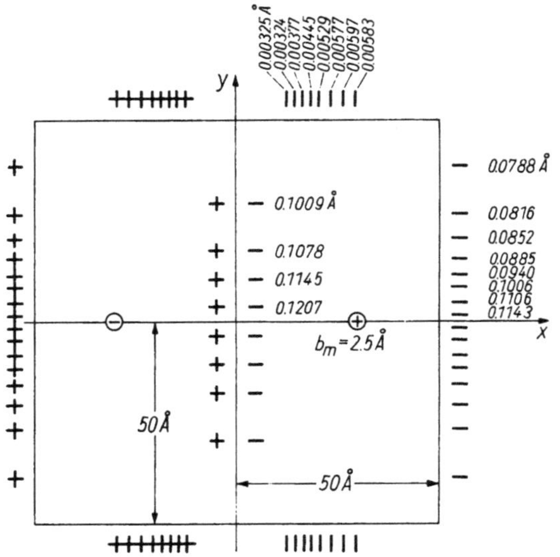

Fig. 10. Same as Fig. 9 but the body is now divided into two regions. Each region is considered separately. The surface dislocations on either side of the y-axis satisfy the free surface conditions. The surface arrays on the outer surface also satisfy the free surface boundary condition. The figure is symmetric about the x- and y-axes. The representation of the dislocations is the same as in Fig. 9

locations are located on either side of the y-axis in the two regions. The two regions are treated separately. In other words, the dislocations in one region do not contribute to any interaction with dislocations in the other region. The stress-free condition on all the surfaces is satisfied by the presence of surface dislocations. The dislocation configuration is once again determined by minimizing the total energy in each region with respect to the position and Burgers vector of surface dislocations and is shown in Fig. 10. The surface dislocation array located along the y-axis in any one region replaces the dislocation array in the second region so that the stress-free conditions are satisfied. The sums of the Burgers vectors of the dislocation array on the outer surface in Fig. 9 and 10 are equal, showing that these two descriptions are not different. The total energy of the configuration is also found to be equal in both descriptions. The array situated along the y-axis is similar to the surface arrays encountered earlier in grain boundaries and disclinations in finite solids. These surface arrays in homogeneous media represent planes, surfaces, or regions inside a body where the stress relaxation conditions are satisfied. Surface arrays could also be present across interfaces where stress continuity conditions should be satisfied [9]. The stress-free surface along the y-axis in Fig. 9 and 10 joins the free outer surface showing that stress-free surfaces do not end inside a crystal. It is necessary for the stress-free surface to join another free surface or else end outside the finite body on a free surface.

6. Surface Dislocation Analysis of an Elastically Bent Crystal

The surface dislocation analysis has hitherto been applied to disclinations and grain boundaries. It is well known that a crystal, when bent plastically, contains arrays of dislocations. In order to further understand the dislocation configurations in a bent crystal, it is important to determine the surface dislocation configuration representing the distortion arising in an elastically bent crystal. Fig. 11 shows the dislocation configuration on the surface of a finite body subjected to tension on the upper surface and equal compression on the lower surface and a linearly varying stress in between, becoming zero along the x-axis. The applied stress is assumed to be $\sigma y/D$ where $\sigma = 400 \times 10^8$ dyn/cm^2 and D is half the thickness of the finite body. Therefore, the applied stress reaches $+\sigma$ on the upper and $-\sigma$ on the lower surface. The stress

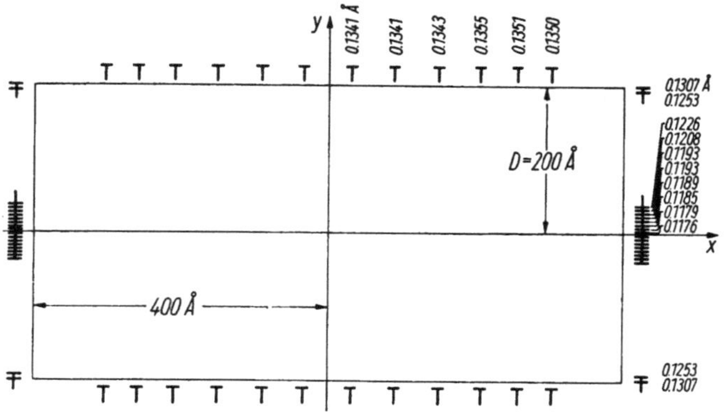

Fig. 11. Surface dislocation configuration representing the distortion in a finite solid subjected to uniform elastic bending. The stress is applied at the end faces. Note the uniform distribution of the surface dislocations on the horizontal surfaces. The Burgers vectors of the surface dislocations are shown on the figure

varies linearly across the cross-section. The dislocation configuration is obtained by minimizing the total energy of the configuration with respect to the positions and Burgers vectors of all the dislocations. The work done by the applied stress is calculated using

$$E_W = \int_{D_0}^{y} \frac{\sigma b y}{D} \, dy = \frac{\sigma b (y^2 - D_0^2)}{2D},$$

where D_0 is the arithmetic mean of the positions of the innermost dislocations of opposite sign on the surfaces of the finite body, b the Burgers vector of each dislocation, and y the coordinate position of the dislocations. The secondary array of dislocations with Burgers vector perpendicular to that of the primary array is also present on the surface but the Burgers vector of the secondary array is found to be negligibly small in magnitude and therefore the array is not shown in the figure. The conservation of Burgers vector of the surface dislocations is once again found to be valid within the discrete dislocation approximation. The surface dislocations on the two horizontal surfaces are distributed uniformly. This result is also found in the earlier calculations where a finite body is subjected to either constant tension or compression [8]. The distortion and stress field at the centre of the finite body corresponds to that of an elastically bent infinite crystal.

7. Surface Dislocation Analysis of a Plastically Bent Crystal

Previous surface dislocation analyses showed the nature of the distortion around an elastically bent crystal. A plastically bent crystal introduces lattice dislocations into the crystal in order to produce the deformation generated by the applied stress. When the crystal is bent plastically, the initial positions of the lattice dislocations in the crystal are random as shown schematically in Fig. 12a. The dislocations in the finite body are randomly distributed but the conservation of Burgers vectors requires surface dislocations of opposite sign on the surface of the finite body. The surface dislocations satisfy the free surface boundary conditions on the surface of the finite body. These surface dislocations cannot enter into the crystal and form a stress-free surface, as illustrated earlier for configurations of disclinations and grain boundaries. The necessary condition for the surface dislocations to enter into the crystal and form a stable configuration is that there is a well-defined minimum energy configuration at a distance from the array of lattice dislocations. A random array of lattice dislocations cannot satisfy this condition since their arrangement cannot have a well-defined equilibrium configuration for the screening array.

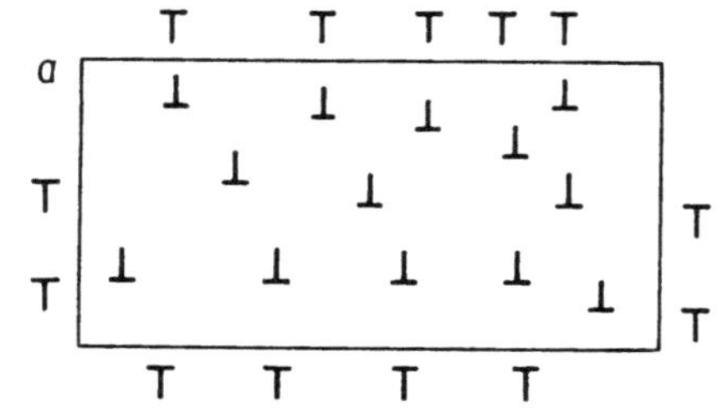

Fig. 12. a) Random array of lattice dislocations in a plastically bent crystal formed due to an applied stress. The surface array satisfies the free surface boundary conditions on the surface of the finite solid. b) A plastically bent crystal containing disclinations and grain boundaries surrounded by free surface boundaries. The free surface boundaries are formed by surface dislocations entering into the crystal and forming the screening array

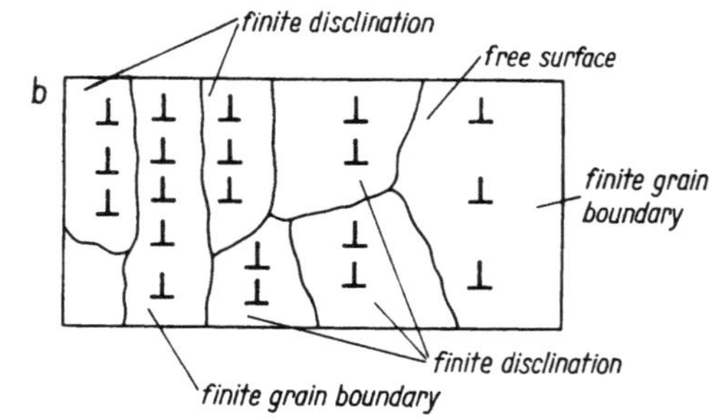

The surface dislocations can enter into the crystal when the random array changes into well-defined configurations of disclinations or grain boundaries. The surface dislocations will become the screening array surrounding the finite regions of disclinations and grain boundaries. A schematic illustration of a plastically bent crystal containing disclinations and grain boundaries surrounded by stress-free surfaces is shown in Fig. 12b. The stress-free surfaces are made of surface dislocations which screen the stress field around the lattice dislocation array. It is also important to note that the stress-free surfaces form a network and thus do not end inside a crystal. The rearrangement of lattice dislocations to form disclinations and grain boundaries may require additional thermal energy. Thus, a polygonized crystal contains these dislocation configurations. In reality, the screening array is a two-dimensional distribution of surface dislocations spread in a region around the distortion. Therefore, the stress-free surface is in general a region where the stress is relaxed by the two-dimensional distribution of the screening array.

It has been pointed out earlier [5] that the diffraction contrast due to lattice dislocations forming grain boundaries and disclinations in finite bodies can be seen because the distortion is strong and localized. However, the screening array is spread throughout a region and the distortion due to the screening array is not localized. Therefore, the contrast due to the screening array is not strong but presents only a nearly uniform background.

8. Summary and Conclusions

Earlier ideas of surface dislocation analysis of infinite grain boundaries are extended in detail to grain boundaries and disclinations in finite bodies. Detailed numerical calculations have been performed to determine the dislocation configurations comprising a grain boundary and a disclination in a finite body. The screening array of dislocations satisfies the conservation of Burgers vectors of dislocations. The coalescence of the grain boundary surfaces is related to the elastic dislocation content of the boundary. The equilibrium position of the screening array of dislocations, D_S, is found to be closer to the grain boundary at larger values of misorientation angle θ. The elastic grain boundary surface dislocation content of the boundary is found to decrease with increasing misorientation angle θ. These results are in agreement with those obtained earlier for infinite boundaries [5]. The equilibrium position of the surface array from the grain boundary does not depend on the specific value of surface energy used. Subdividing the boundary into finer elements containing dislocations with smaller Burgers vector leads to decreasing values of equilibrium distance, D_S, ultimately leading to $D_S = 0$ for a boundary with no dislocations.

A disclination is considered as a grain boundary ending in a finite body. The dislocation configuration of a disclination with three elements is determined along with their screening arrays. The most important difference between a grain boundary and a disclination in a finite body is the nature of distortion around the defects. The distribution of surface arrays of dislocations on the surfaces of the grain boundary elements constituting a disclination is extremely non-uniform, while this is not so for a grain boundary in a finite solid. Similarly, the screening array of dislocations responsible for the relaxation of the distortion is non-uniform for a disclination in a finite solid. The opening up of the surfaces of the grain boundary elements in a disclination is also non-uniform. The distortion around a grain boundary in a finite solid is however uniform; thus the opening up of the surfaces of the grain boundary elements is also uniform.

The screening array forms a stress-free surface. In general, however, the screening array consists of a two-dimensional distribution of surface dislocations. The nature of a stress-free surface in a finite solid containing a pair of unlike screw dislocations is also described by surface dislocations. The stress-free surfaces or regions formed in a medium cannot end inside a crystal and should therefore end on the outer free surface of the medium.

The distortion around an elastically bent crystal is obtained by surface dislocation analysis. The array on the bent surfaces remains almost uniform. Similar result was also obtained in earlier calculations where the surface dislocation analysis was applied to finite solids subjected to constant tension or compression. A plastically bent crystal containing a random array of lattice dislocations requires the surface dislocations to remain on the surface. However, when the random array rearranges and forms disclinations and grain boundaries, the surface screening array enters into the crystal and occupies equilibrium positions. The stress-free surfaces formed by the screening array form a network in the finite body and thus cannot end inside a crystal.

Acknowledgement

The computer time for the present study was supported in full through the facilities of the Computer Science Center of the University of Maryland.

References

[1] M. J. Marcinkowski, Unified Theory of the Mechanical Behaviour of Matter, John Wiley & Sons, 1979.
[2] M. J. Marcinkowski, K. Sadananda, and W. F. Tseng, phys. stat. sol. (a) 18, 361 (1973).
[3] M. J. Marcinkowski and K. Jagannadham, phys. stat. sol. (a) 50, 601 (1978).
[4] K. Jagannadham and M. J. Marcinkowski, Crystal Lattice Defects 8, 81 (1979).
[5] K. Jagannadham and M. J. Marcinkowski, J. Mater. Sci., in publication.
[6] M. J. Marcinkowski, in: Advances in Materials Research, Ed. H. Hermann, Wiley-Interscience Publishers, New York 1971 (p. 443).
[7] K. Jagannadham and M. J. Marcinkowski, phys. stat. sol. (a) 50, 293 (1978).
[8] K. Jagannadham and M. J. Marcinkowski, Mater. Sci. Engng. 38, 259 (1979).
[9] K. Jagannadham and M. J. Marcinkowski, J. Mater. Sci., in publication.

(Received May 17, 1979)

phys. stat. sol. (a) 54, 729 (1979)

Subject classification: 1.1

Department of Physics, Erevan State University

Fine Structure of X-Ray Interferograms

I. A Two-Crystal System with a Non-Diffracting Zone (a Narrow Gap)

By

P. KH. BEZIRGANYAN, KH. R. DRMEYAN, and G. M. ALADZHADZHYAN

X-ray interferometrical patterns obtained by diffraction of an X-ray beam in a crystal system, consisting of a two-crystal system having a narrow gap and a thick absorbing crystal, is studied. The spacing of interbranch scattering fringes is theoretically calculated. It is shown that the presence of the thick crystal brings to a five- to tenfold enlargement of fringe spacing. For the first time fringes of interbranch scattering and moiré fringes are obtained on the same topograph.

Экспериментально исследована рентгеноинтерференционная картина, полученная при дифракции узкого рентгеновского волнового пакета в системе, состоящей из двухкристальной системы с узким зазором и из толстого поглощающего кристалла. Теоретически рассчитан период полос межветвевого рассеяния, полученной от этой системы. Показано, что наличие толстой пластины дает возможность увеличить период полос примерно в 5 до 10 раз. Впервые одновременно наблюдались два динамических эффекта: на одной топограмме наблюдались полосы межветвевого рассеяния и муаровые узоры, образующиеся в двухкристальной системе.

1. Introduction

The method of moiré fringes [1 to 3] is a very sensitive method for studying lattice distortions in nearly perfect crystals. X-ray diffraction moiré fringes are formed by a superposition of two coherent waves, and they are the result of a phase difference between these beams. The difference arises as a result of lattice distortions in different blocks, their relative misorientation, and a certain phase distribution in the beams themselves. An interferogram, therefore, usually presents a complex set of fringes arising as a result of the strain dislocation field and different strain gradients on one hand, and as a result of a certain phase distribution in the superimposed beams, on the other hand.

When narrow X-ray wave packets are diffracted from a system containing a planar fault (a two-crystal system with a non-diffracting zone) [4, 5], this gives rise to interbranch scattering diffraction fringes. The spacing of the fringes depends strongly on the total thickness of the crystalline slabs, and also on the gap width (width of the non-diffracting zone). When one is working with relatively thin crystals, the fringe spacing decreases and they become difficult to observe experimentally. That is why in [6] only moiré fringes have been observed. From relatively thick and perfect crystals moiré fringes have not been observed [4, 5], since in the irradiated regions of these systems there have not been any lattice distortions.

In the present study by the aid of a third perfect crystal (crystalline enlarger), arranged in a reflecting position, we have succeeded in increasing sharply the spacing of the interbranch scattering fringes, and thus in observing on the same topograph the interbranch scattering fringes and moiré fringes formed in a two-crystal system. On the basis of the Takagi equations, in the framework of the geometrical optics approximation, the fringe spacing and the enlargement factor of the spacing of interbranch scattering fringes have been calculated as a function of thickness of the crystal enlarger.

2. Theory

Suppose a spatially inhomogeneous, narrow X-ray wave packet having a Ψ_0^i amplitude is incident on a crystal system, composed of a thick crystal and of a two-crystal system having a narrow gap (Fig. 1). After diffraction in the first crystal, the incident beam is split into two components: the transmitted beam and the diffracted beam having amplitudes Ψ_0 and Ψ_h, respectively (Fig. 2). Consider the successive diffraction of the beam, Ψ_h, in a two-crystal system with a narrow gap. At the exit of the system four beams emerge. The beam reflected in the first crystal and transmitted by the second one and having the amplitude Ψ_{hh0}, the double-diffracted beam having the amplitude Ψ_{hhh}, the beam transmitted by the first crystal and reflected in the second one having the amplitude Ψ_{h0h}, and the beam transmitted by the two crystals and having the amplitude Ψ_{h00}. When the gap is sufficiently narrow, the beams Ψ_{hhh} and Ψ_{h00} as well as the beams Ψ_{hh0} and Ψ_{h0h} interfere along the directions of reflection and transmission, respectively. In order to determine the distribution of the interference fringe maxima, one must know the value Φ_q of the phase inside the interfering beams:

$$\Psi_q = |\Psi_q| \exp (i\Phi_q) .$$

It is easy to calculate the phase using the Takagi equations [7] and the stationary phase approximation [8]. Omitting the terms relating to the modes that are strongly absorbed [9], in the symmetrical Laue case for the Ψ_{hhh} and Ψ_{h00} beams we have

$$\left.\begin{aligned} \Phi_{hhh} &= -\frac{\pi}{\Delta_0}(Z + Z_1 + Z_2)\sqrt{1 - p_1^2} + \frac{3\pi}{4}, \\ \Phi_{h00} &= -\frac{\pi}{\Delta_0}(Z + Z_1 + Z_2)\sqrt{1 - p_2^2} + \frac{3\pi}{4}, \end{aligned}\right\} \tag{1}$$

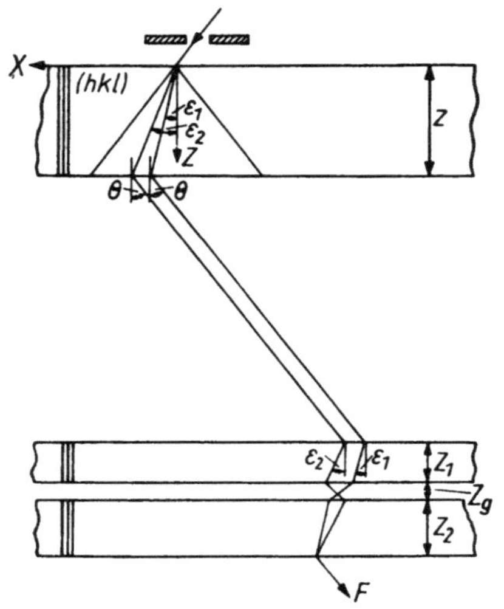

Fig. 1

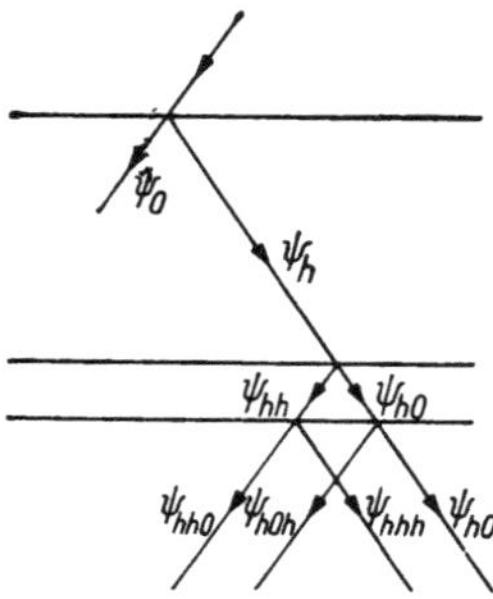

Fig. 2

Fig. 1. Scheme illustrating the ray paths in a three-crystal system. F is the interference point, Z_g is the width of the gap

Fig. 2. The beams arising in a three-crystal system

where Δ_0 is the extinction length, Z the thickness of the thick crystal, Z_1 and Z_2 are the thicknesses of the first and the second crystal, correspondingly, and the parameters $p_1 = \text{tg}\,\varepsilon_1/\text{tg}\,\theta_B$ and $p_2 = \text{tg}\,\varepsilon_2/\text{tg}\,\theta_B$ characterize the directions of the energy flow, θ_B being the Bragg angle.

The interference of the Ψ_{hh0} and Ψ_{h0h} beams will not be discussed, since it is similar to the one discussed.

Using (1) for the phase difference $\Delta\Phi = \Phi_{hhh} - \Phi_{h00}$, it is easy to obtain

$$\Delta\Phi = \Phi_{hhh} - \Phi_{h00} \approx \frac{\pi}{\Delta_0}(Z_1 + Z_2 + Z)\,\frac{p\,\text{d}p}{\sqrt{1-p^2}}\,,$$

where $p = (p_1 + p_2)/2$ and $\text{d}p = p_2 - p_1$ $\Delta\Phi$ is determined for the case of strong absorption [4], i.e., the calculations are made for the central part of a topograph.

The condition for the intersection of the energy flow lines in a certain point F is given by

$$Z + Z_1 + Z_2 = \frac{2Z_g}{\text{d}p}, \tag{2}$$

where Z_g is the width of the air gap (non-diffracting zone). Combining this with the condition for obtaining interference fringe maxima, $\Delta\Phi = 2\pi n$, one may obtain the surfaces of maximum intensity:

$$X_n = \pm\frac{\Delta_0 n\,\text{tg}\,\theta_B(Z + Z_1 + Z_2)}{Z_g\sqrt{1 + (n\Delta_0/Z_g)^2}}; \qquad n = 0, 1, 2, \ldots$$

The direction of the X-axis is shown in Fig. 1.

Without committing a great error one may assume that at the central part of the topograph $(n\Delta_0/Z_g)^4 \ll 1$; then the fringe spacing is given by

$$\Lambda = \frac{(Z + Z_1 + Z_2)\,\Delta_0\,\text{tg}\,\theta}{Z_g}\left[1 - \frac{3n^2}{2}\left(\frac{\Delta_0}{Z_g}\right)^3 - \frac{3n}{2}\left(\frac{\Delta_0}{Z_g}\right)^3 - \frac{1}{2}\left(\frac{\Delta_0}{Z_g}\right)^3\right]. \tag{3}$$

Thus, similar to the two-crystal system with a narrow gap, in the present case the diffraction image again represents a set of parallel fringes that are perpendicular to the scattering plane. Note the same character of the decrease of the spacing with the increase of fringe order. Comparing the fringe spacing for the two-crystal system (in that case in (3) Z is equal to 0) with (3), one may see that the presence of the crystal enlarger in a reflecting position causes an increase in the spacing of the interference fringes by the following linear coefficient in the direction of the diffraction vector:

$$D = 1 + \frac{Z}{Z_1 + Z_2}.$$

This enlargement is an interference phenomenon, and it sharply depends on the thickness of the thick crystal. By increasing its thickness one may obtain enlargements reaching a factor of ten and more and thus observe in the experiment fringes having large spacing. Extensive increasing of the thickness, however, increases the absorption and results in a sharp decrease in intensity of the reflected beams.

Note also that interchanging the places of the incidence point 0 and the interference point F (Fig. 1) (in the case when the beam is incident on the two-crystal side) does not change the spacing of the interference fringes, since in both cases the interference condition (2) remains the same.

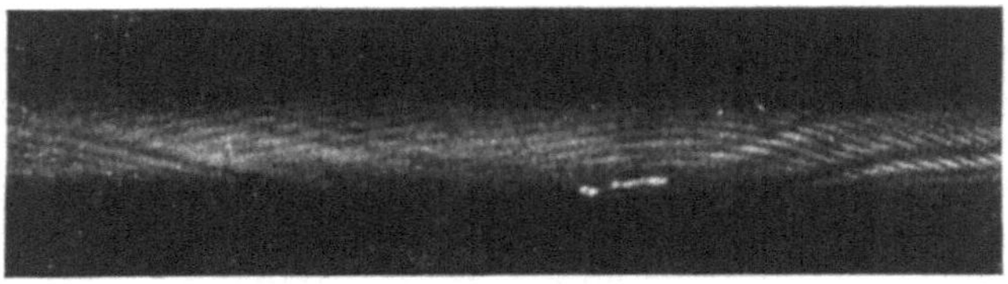

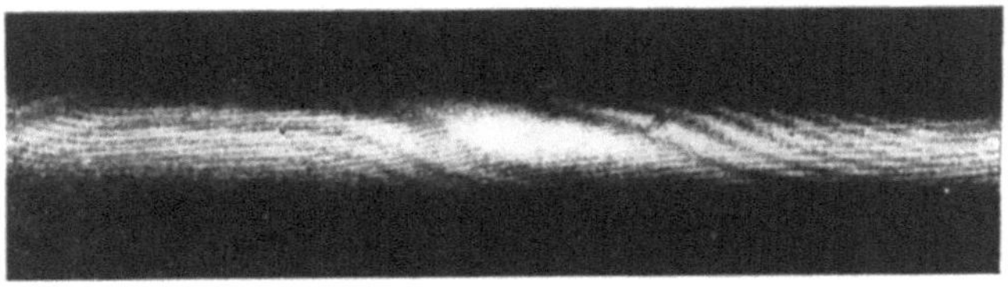

Fig. 3. X-ray topographic moiré patterns obtained by a three-crystal system shown in Fig. 1

3. Experiment

For an experimental study of the fine structure of X-ray interference patterns a three-crystal system having the same basis was prepared consisting of a thick parallel-faced crystal and of two thin crystals having a narrow gap between them. The Si crystal from which the system has been cut, was a nearly dislocation-free single crystal (the dislocation density was less than 10 cm^{-2}). After mechanical and chemical treatment the thicknesses of the crystals were $Z = 5550$ μm, $Z_1 = 420$ μm, $Z_2 = 1315$ μm, and the width of the gap $Z_g = 310$ μm. The large surfaces of the crystals were perpendicular to the growth axis [$\bar{1}\bar{1}1$], and the reflecting planes ($2\bar{2}0$) were perpendicular both to the large surfaces of the crystals and to the basis of the system. $MoK\alpha_1$ radiation was used. The spatial width of the incident beam after collimation was 40 μm.

Fig. 3a shows a topograph obtained from the crystal which is schematically illustrated in Fig. 1. As it is seen on the topograph, the moiré fringes have a complex structure. On each moiré fringe interbranch scattering fringes are superimposed, the superposition being a diffraction phenomenon. The fringes are inclined relative to the direction of the normal of the scattering plane, which is due to local distortions of the crystal lattice [10]. When the beam is incident on the two-crystal side of the system, the nature of the image is not changed (Fig. 3b). The moiré fringes illustrated in Fig. 3a and 3b are not identical, since by changing the side of the system, on which the beam incides, the irradiated regions of the two-crystal system are also changed.

On the topograph obtained only from the two-crystal system fringes of interbranch scattering have not been observed. They are observed only, when the beam passes through the enlarging (thick) crystal.

4. Discussion of the Results

1. By the aid of moiré patterns it is possible to determine the kind (extension, compression, rotation, etc.) and the value of the strain fields in a crystal lattice. Having the information contained in a single moiré fringe, however, one probably may determine the local distortion in a very limited region of the crystal lattice.

2. Since the crystals of the two-crystal system are rather thin ($\mu z_{1,2} < 1$), subsidiary dynamical lines (lines due to the margin effect, pendellösung fringes, etc.) also may arise. These lines, however, are absorbed, when passing through the thick crystal.

3. If the reflecting planes of the crystal enlarger are rotated relative to the reflecting planes of the two-crystal system, then on the topograph interbranch scattering fringes

of higher orders must appear. Actually, the centre of the reflected beam (corresponding to the fringes of low order) corresponds to the exact Bragg angle. When the reflecting planes of the crystal enlarger are rotated, the Bragg angle changes, and the beams corresponding to the exact reflection angle are strongly absorbed (Borrmann absorption), whereas other rays, which were not incident at the Bragg angle before, now do so. As a result of this in the experiment for the fringe spacing a value of $\Lambda = 130$ to 270 μm is obtained, while the calculated values are $\Lambda_{n=1} = 720$ μm and $\Lambda_{n=10} = 44.8$ μm. Thus, the presence of local rotations in the crystal enlarger causes an anomalous ab-sorption of the central fringes of order $n = 1, 2$ and an increase of spacing in fringes of order $n = 5, 6, 7$.

References

[1] G. Christiansen, L. Gerward, and A. L. Anderson, J. appl. Cryst. **4**, 370 (1971).
[2] M. Hart, Phil. Mag. **26**, 821 (1972).
[3] G. H. Schwuttke and K. Brack, Z. Naturf. **28a**, 654 (1973).
[4] A. Authier, A. D. Milue, and M. Sauvage, phys. stat. sol. **26**, 469 (1968).
[5] M. Hart, and A. D. Milue, Acta cryst. **A26**, 223 (1970).
[6] A. R. Lang, Nature (London) **220**, 652 (1968).
[7] S. Takagi, Acta cryst. **15**, 1311 (1962); J. Phys. Soc. Japan **26**, 1239 (1969).
[8] M. Born and E. Wolf, Principles of Optics, 2nd ed., Pergamon Press, 1964.
[9] G. M. Aladzhadzhyan, A. K. Kocharyan, and K. G. Truni, Kristallografiya, in the press.
[10] S. Tauemura and A. R. Lang, Z. Naturf. **28a**, 668 (1973).

(Received April 5, 1979)

phys. stat. sol. (a) **54**, 735 (1979)

Subject classification: 2 and 14.3.1

Institute of Physics, Academia Sinica, Nankang, Taipei,

A Theory of Threshold Switching in Amorphous Thin Films

By

CHUN CHIANG

Using the concept that two reversible forms exist in amorphous thin film and that the transition between them may be activated by the current, and these two forms have different conductivities, the threshold switching phonomenon may be explained. The threshold switching equation and time delay equation are also derived.

Mit der Annahme, daß zwei reversible Formen in dünnen amorphen Schichten existieren, der Übergang zwischen ihnen durch den Strom aktiviert werden kann, und diese Formen unterschiedliche Leitfähigkeiten besitzen, läßt sich das Phänomen des Schwellenschaltens erklären. Die Gleichung für das Schwellenschalten und die Zeitverzögerungsgleichung werden ebenfalls abgeleitet.

Threshold switching in amorphous thin films has recently attracted great attention [1 to 5]. Despite its intensive research no electronic switching theory has been developed to the stage that quantitative calculations and predictions can be made. Recently, a quantitative switching theory [6] has been given for organic thin films with dipoles, and a quantitative theory has also been given for the nerve excitation and action potential [7, 8].

This paper proposes some conditions in the amorphous thin film for showing threshold switching, and furthermore, threshold switching equation and time delay equation are derived and predictions are made.

The conditions required for the amorphous thin film are: (i) there are two stable and reversible configurations (configuration I and configuration II) co-existing in approximate equilibrium in the amorphous film. These two configurations may transform into each other through an energy barrier under electric excitation. The difference between configuration I and configuration II may either be due to the positional variation of the atoms, the variation of the number of the bonding atoms, the changing of the bonding state such as from p-orbital to d-orbital or from lone pair to bonding [9], or from neutral dangling bond to positive and negative dangling bond [10, 11] or a combination of the above situations; (ii) the conductivity of configuration II is much higher than that of configuration I; (iii) the energy levels of the configurations I and II may be elevated by current I by the amount $k_1 f(I)$ and $k_2 f(I)$, respectively, the elevation of the energy is due to the fact that the kinetic energy of the current may excite the system within the film (see Fig. 1). Thus k_1 and k_2 represent respectively the extent of electronic excitation of the energy level for configuration I and configuration II in the system by the current, and are temperature-dependent. This excitation is electrical in origin and may not necessarily be thermal. Also, the electronic energy may be elevated to such an extent that ionization may result and represent the occurrence of avalanche.

For a quantitative calculation, we write

$$\left.\begin{aligned} \frac{\mathrm{d}N_1}{\mathrm{d}t} &= -AN_1\,\mathrm{e}^{-[u-E_1-k_1 f(I)]/k_\mathrm{B}T} + AN_2\,\mathrm{e}^{-[u-E_2-k_2 f(I)]/k_\mathrm{B}T}\,,\\ \frac{\mathrm{d}N_2}{\mathrm{d}t} &= AN_1\,\mathrm{e}^{-[u-E_1-k_1 f(I)]/k_\mathrm{B}T} - AN_2\,\mathrm{e}^{-[u-E_2-k_2 f(I)/]/k_\mathrm{B}T}\,, \end{aligned}\right\} \qquad (1)$$

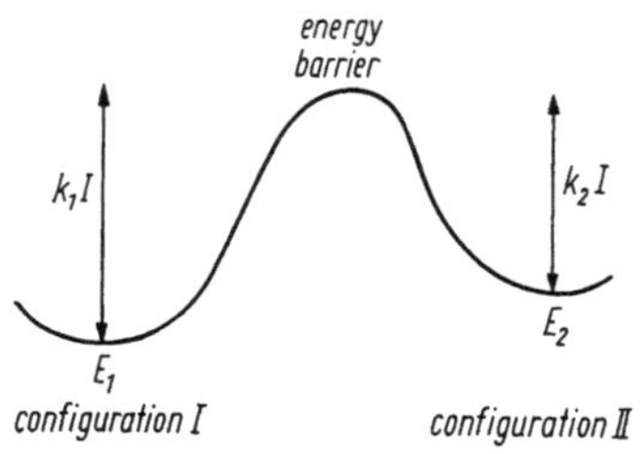

Fig. 1. Energy diagram of two reversible configurations in an amorphous thin film. The transition between the configurations is separated by the barrier and may be activated by the current

where N_1 or N_2 is respectively the number density of configuration I or configuration II per unit length within the film. A is the collisional frequency and k_1, k_2 are constant. E_1 or E_2 is the energy of configuration I and configuration II, respectively, u is the barrier energy. Equation (1) says that the changing of N_1 per unit time is proportional to the number of transformations from configuration II to configuration I minus the number of transformations from configuration I to configuration II. The number of transformations from configuration I to configuration II is equal to the number of N_1 multiplied by the collisional or vibrational frequency A and the probability of transformation per collision. The probability of transformation per collision from configuration I to configuration II is $\exp[-\{u-(E_1+k_1f(I))\}/k_BT]$, where $u-[E_1+k_1f(I)]$ is the energy difference between the barrier and the configuration I. The number of transformations from configuration II to configuration I is equal to $AN_2\exp[-\{u-[E_2+k_2f(I)]\}/k_BT]$. Similar reasoning applies to the changing of N_2 per unit of time. The effect of current in (1) may also consider to be similar to that of catalysis, passing the current across the film increases the rate of transformation.

Equation (1) may be solved to give

$$\left.\begin{aligned} N_1 &= \frac{N\,e^{[E_2+k_2f(I)]/k_BT}}{e^{[E_1+k_1f(I)]/k_BT}+e^{[E_2+k_2f(I)]/k_BT}} + \\ &+\left\{N_1^0-\frac{N\,e^{[E_2+k_2f(I)]/k_BT}}{e^{[E_1+k_1f(I)]/k_BT}+e^{[E_2+k_2f(I)]/k_BT}}\right\}\times \\ &\times\exp\left[-\left\{A\,e^{-[u-E_1-k_1f(I)]/k_BT}+A\,e^{-[u-E_2-k_2f(I)]/k_BT}\right\}t\right], \\ N_2 &= \frac{N\,e^{[E_1+k_1f(I)]/k_BT}}{e^{[E_1+k_1f(I)]/k_BT}+e^{[E_2+k_2f(I)]/k_BT}} - \\ &-\left\{N_1^0-\frac{N\,e^{[E_2+k_2f(I)]/k_BT}}{e^{[E_1+k_1f(I)]/k_BT}+e^{[E_2+k_2f(I)]/k_BT}}\right\}\times \\ &\times\exp\left[-\left\{A\,e^{-[u-E_1-k_1f(I)]/k_BT}+A\,e^{-[u-E_2-k_2f(I)]/k_BT}\right\}t\right], \end{aligned}\right\} \qquad (2)$$

where N_1^0 is the initial number of configuration I and N is the total number of configuration I and configuration II at any time, being a constant for a given film. Starting from the OFF state, configuration I predominates in the film and $N_1^0 = N$.

Thus, if G is the conductance of the film and g_1 or g_2 is the conductivity of the film in the pure state of configuration I or configuration II, respectively, then

$$G = \frac{N_1}{N}\frac{a}{d}g_1 + \frac{N_2}{N}\frac{a}{d}g_2 \qquad (3)$$

and

$$I = UG\,, \tag{4}$$

where a is the area and d the thickness of the film.

From (2) to (4) and $N_1^0 = N$, we have

$$\frac{I}{U} = \frac{g_1\, e^{[E_2+k_2 f(I)]/k_B T} + g_2 e^{[E_1+k_1 f(I)]/k_B T}}{e^{[E_2+k_2(I)]/k_B T} + e^{[E_1+k_1 f(I)]/k_B T}}\,\frac{a}{d} + \frac{a(g_1-g_2)}{d}\left\{1 - \frac{e^{[E_2+k_2 f(I)]/k_B T}}{e^{[E_1+k_1 f(I)]/k_B T} + e^{[E_2+k_2 f(I)]/k_B T}}\right\} \times \exp\left[-\left\{A\, e^{-[u-E_1-k_1 f(I)]/k_B T} + A\, e^{-[u-E_2-k_2 f(I)]/k_B T}\right\} t\right]. \tag{5}$$

The steady value is

$$U = \frac{e^{[E_1+k_1 f(I)]/k_B T} + e^{[E_2+k_2 f(I)]/k_B T}}{(a/d)\, g_1\, e^{[E_2+k_2 f(I)]/k_B T} + (a/d)\, g_2\, e^{[E_1+k_1 f(I)]/k_B T}}\, I\,. \tag{6}$$

This is the threshold switching equation for an amorphous thin film, and is valid for current controlled uniform switching. This equation shows that the threshold voltage is proportional to the thickness d in agreement with the experiment [13], whereas the "injection model" such as that proposed by Lucas [14] and Wright and Ibrahim [15] predicts it to be proportional to d^2. Equation (6) may also be transformed to be

$$V = \frac{1 + \exp\{[\Delta E + \Delta k\, f(I)]/k_B T\}}{a g_2/d + (a g_1/d)\exp\{[\Delta E + \Delta k\, f(I)]/k_B T\}}\, I\,. \tag{6a}$$

In the derivation of (1) and (6a) it is assumed that the excitation of the energy level by current is porportional to $f(I)$, this is because the process of the impact of molecules by electrons is very complicated, being structure-dependent, and may even vary from sample to sample. However, regardless of the exact form of the current dependence, the major switching property in (6a) is not appreciably changed, and the functional form of the current in the exponential term will only change of range of current for which the switching takes place. For simplicity, we assume that $f(I)$ equals $|I|$. If we let $a g_1/d = 0.5 \times 10^{-4}\ \Omega^{-1}$, $a g_2/d = 1.7 \times 10^{-3}\ \Omega^{-1}$, $\Delta E = 0.2$ eV, $\Delta k = -7.5 \times 10^{-2}$ eV/A, then (6a) appears as shown in Fig. 2. This is similar to the curve shown by Ovshinsky and Fritzsche [16]. A similar switching curve has also been observed in single crystals [17]. Independently, I suggest that measurements of absorption spectrum and emission spectrum of the film under electric excitation should be performed, and the intensity of the spectrum of the film in ON state and OFF state should be compared, to have an independent check of ΔE and Δk. This may be experimentally difficult, and also the identification of the spectrum with the structure of the film itself is a difficult task; nevertheless, the possibility exists that such works in optics and electronics may correlate and may be complementary to each other. From this theory, it would not be surprising to find that photo effect can also cause switching and structure change in thin films.

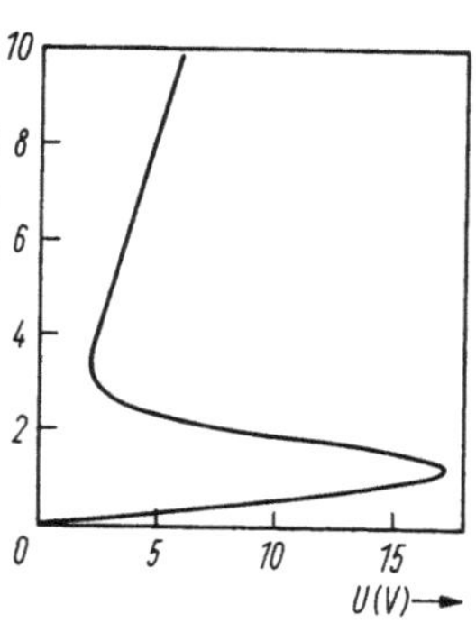

Fig. 2. I–U switching curve calculated according to equation (6a)

For a given voltage U, which is larger than the threshold voltage U_t by an amount ΔU, the delay time is the period required for the current to reach the threshold value I_t. Thus, from (5) we have

$$t = C_1(\ln C_2 - \ln \Delta U)\,, \tag{7}$$

where

$$C_1 = \left\{A \exp\left[\frac{E_1 + k_1\,|I_t| - u}{k_B T}\right] + \exp\left[\frac{E_2 + k_2\,|I_t| - u}{k_B T}\right]\right\}^{-1},$$

$$C_2 = \left\{1 - \exp\left[\frac{E_2 + k_2\,|I_t|}{k_B T}\right]\right\}\left\{\exp\left[\frac{E_1 + k_1\,|I_t|}{k_B T}\right] + \right.$$

$$\left. + \exp\left[\frac{E_2 + k_2\,|I_t|}{k_B T}\right]\right\}^{-1} \frac{a(g_2 - g_1)\,V_t^2}{d\,|I_t|}$$

also the approximation $\Delta U < U_t$ and the relation

$$U_t = \frac{d}{a}\,\frac{e^{(E_1+k_1|I_t|)/k_BT} + e^{(E_2+k_2|I_t|)/k_BT}}{g_1\,e^{(E_2+k_2|I_t|)/k_BT} + g_2\,e^{(E_1+k_1|I_t|)/k_BT}}\,I_t \tag{8}$$

have been used.

Equation (7) is the time delay equation and shows that the delay time decreases with increasing voltage, and if the voltage is at the threshold voltage, the delay time is infinity. However, the applied voltage is difficult to be fixed exactly at the threshold voltage, so a slight variation will cause a spread of the delay time. This fact is indeed observed in the experiment [12]. By comparison of (7) with the experimental data of time delay, E_1, E_2 and k_1, k_2 may also be estimated.

In conclusion, we have specified the conditions for the amorphous film to show the threshold switching behaviour, and a quantitative relationship has been derived for the threshold behaviour and the delay time, which no other electronic switching theory seems to be able to accomplish yet. I have also pointed out the way in which independent systematic experiments may be performed to check this theory.

References

[1] A. SZYMANSKI, D. C. LARSON, and M. M. LABES, Appl. Phys. Letters **14**, 88 (1969).
[2] I. BALBERG, Appl. Phys. Letters **16**, 491 (1970).
[3] M. SAJI, C. H. LEURG, and K. C. KAO, J. non-crystalll. Solids **23**, 147 (1977).
[4] W. VAN ROOSBROECK, Phys. Rev. Letters **28**, 1120 (1972).
[5] S. H. LEE and H. K. HENISCH, Appl. Phys. Letters **22**, 230 (1973).
[6] C. CHIANG, Appl. Phys. Letters **31**, 554 (1977).
[7] C. Y. LEE and C. CHIANG, Bull. Math. Biol. **38**, 59 (1976).
[8] C. CHIANG, Bull. Math. Biol. **40**, 247 (1978).
[9] M. KASTNER, D. ADLER, and H. FRITZSCHE, Phys. Rev. Letters **37**, 1504 (1976).
[10] P. W. ANDERSON, Phys. Rev. Letters **34**, 953 (1975).
[11] R. A. STREET and N. F. MOTT, Phys. Rev. Letters **35**, 1293 (1975).
[12] S. H. LEE and H. K. HENISCH, J. non-crystall. Solids **8/10**, 422 (1972).
[13] A. E. OWEN and J. M. ROLBERTSON, IEEE Trans. Electron Devices **20**, 105 (1973).
[14] I. LUCAS, J. non-crystall. Solids **6**, 136 (1971).
[15] G. T. WRIGHT and A. F. IBRAHIM, Electronics Letters **4**, 597 (1968).
[16] S. R. OVSHINSKY and H. FRITZSCHE, IEEE Trans. Electron Devices **20**, 91 (1973).
[17] T. TAKAHASHI and O. YAMADA, J. appl. Phys. **48**, 1258 (1977).

(Received May 2, 1979)

phys. stat. sol. (a) 54, 739 (1979)

Subject classification: 1.1 and 1.2; 22.8

Department of Physics, Silesian University, Katowice (a) and Institute for Low Temperature and Structure Research, Polish Academy of Sciences, Wrocław (b)

The Temperature Dependence of Lattice Parameters of $KMeF_3$ and $KMn_{0.9}Me_{0.1}F_3$ Compounds ($Me = Mn^{2+}$, Co^{2+}, and Ni^{2+})[1]

By

A. RATUSZNA (a), A. PIETRASZKO (b), A. CHEŁKOWSKI (a), and K. ŁUKASZEWICZ (b)

The temperature dependence of the lattice parameters of $KMnF_3$, $KCoF_3$, $KNiF_3$, and $KMn_{0.9}Me_{0.1}F_3$ ($Me = Co^{2+}$ and Ni^{+2}) are accurately measured using a single-crystal Bond diffractometer. New crystallographic results about phase transitions in $KMeF_3$ and the influence of the doped atoms on the rotation angle φ are reported. The linear thermal expansion coefficient α for the cubic phase is calculated.

Die Temperaturabhängigkeit der Gitterparameter von $KMnF_3$, $KCoF_3$, $KNiF_3$ und $KMn_{0.9}Me_{0.1}F_3$ ($Me = Co^{2+}$ und Ni^{2+}) werden mit einem Ein-Kristall-Bond-Diffraktometer exakt gemessen. Neue kristallographische Ergebnisse über die Phasenübergänge in $KMeF_3$ und der Einfluß der Dotierungsatome auf den Drehwinkel φ werden erhalten. Der Koeffizient der linearen thermischen Ausdehnung, α für die kubische Phase wird berechnet.

1. Introduction

All compounds $KMeF_3$ with $Me = Mn^{2+}$, Co^{2+}, and Ni^{2+} as well as mixed crystals $KMn_xMe_{1-x}F_3$ crystallize in ABO_3 perovskite structure and are cubic at sufficiently high temperature [1 to 5]. The low-temperature phase is tetragonally deformed. Benard and Walker [6] described this phase transition to be of weak first order coupled to second order, in agreement with earlier data [7, 8]. Although the structural phase transitions in perovskites have been studied in recent years extensively both experimentally and theoretically [9 to 11], it is difficult to describe exactly their mechanism. It seems that $KMnF_3$ crystals undergo a phase transition from cubic to tetragonal phase due to the condensation of one of the vibrational modes associated with the R-point of the cubic Brillouin zone [7].

The structural mechanism of the phase transition is connected with the rotation of MnF_6 octahedra around the [001] pseudocubic axis [12]. The rotation angle φ can be easily determined from the tetragonal distortion of the lattice cell [13]. The angle φ has been shown [13] to be an order parameter of the phase transition in $KMnF_3$.

In the previous paper [4] the influence of the doping ions Co^{2+}, Ni^{2+}, Fe^{2+}, and Mg^{2+} on the transition temperature T_c in $KMnF_3$ has been shown. In this paper new crystallographic results about the phase transitions in $KMeF_3$ compounds as well as the influence of the doping atoms on the rotation angles φ are reported.

[1]) Supported in part by the Polish Academy of Sciences, Wrocław.

2. Experimental Results

The lattice parameters were measured by using a Bond diffractometer with an accuracy of 10^{-5} Å. The temperature of the sample was kept constant within ±0.03 K in the range 130 to 295 K. The specimens were cooled by blowing cold nitrogen gas.

Single crystals of high purity grown by the Bridgman method were used [14]. The samples were cut parallel to (100) faces from the bulk crystal. The 500 and 510 diffraction profiles were selected for measurements. Below T_c, in the tetragonal phase the split reflections 500 and 005 were measured. For temperatures lower than $T_c - 1$ K the twinning reflections were well separated, but very near T_c a graphical analysis was necessary.

Fig. 1 to 5 illustrate the temperature dependence of the lattice parameters of $KMnF_3$, $KCoF_3$, $KNiF_3$, $KMn_{0.9}Co_{0.1}F_3$, and $KMn_{0.9}Ni_{0.1}F_3$. While the present measurements are in substantial agreement with the earlier works and our data from the powder xeperiment [4], the resolving power of the applied method shows up new details.

The lattice parameters of $KMnF_3$ as function of temperature (Fig. 1) show a temperature hysteresis with T_c(heating) = 187.5 K and T_c(cooling) = 186.5 K. At 195.5 K a small anomaly of the lattice parameter has been found. The tetragonal distortion c/a is greater than one below T_c.

The present results concerning the parameters versus temperature of $KNiF_3$ in the temperature region from 220 to 300 K are in close agreement with those obtained by Okazaki and Kawaminami [15]. A deviation from the linear temperature dependence $a(T)$ is obvious below the Néel temperature $T_N = 275$ K (Fig. 2). The crystal structure is apparently cubic in the whole range of temperature, although the crystal undergoes a para–antiferromagnetic phase transition.

It has been reported by Okazaki and Suemune [2] that $KCoF_3$ is cubic above $T_N = 114$ K without any anomalies. We have found, however, a distinct anomaly of

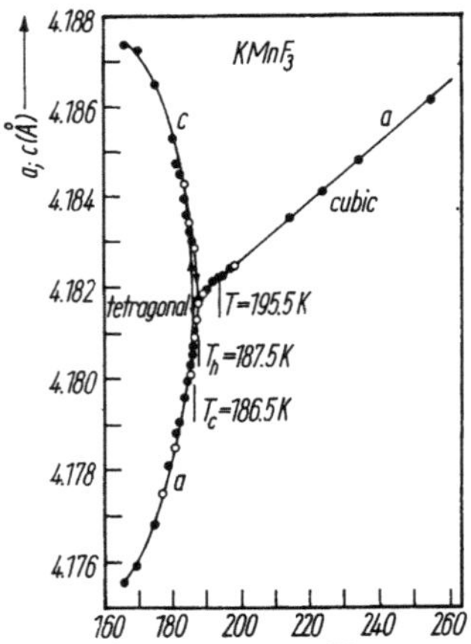

Fig. 1

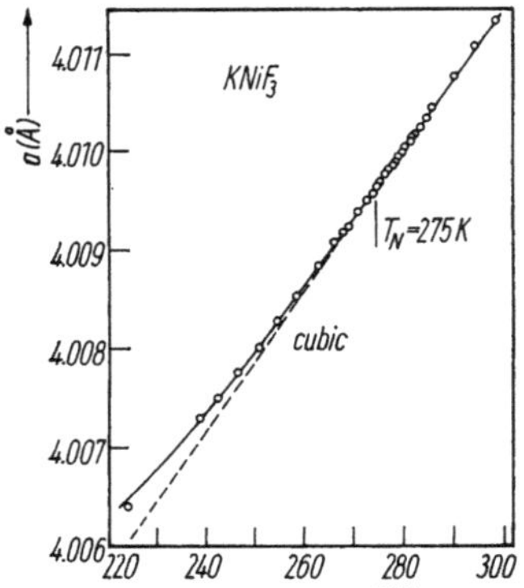

Fig. 2

Fig. 1. Lattice parameters of $KMnF_3$ measured by a single-crystal Bond technique in the temperature region from 160 to 270 K. The open circles represent experimental results obtained on heating, the full circles those on cooling

Fig. 2. Lattice parameter of $KNiF_3$ versus temperature. The mark indicates the Néel temperature. Below T_N a deviation from the linear temperature dependence of the lattice parameter has been observed

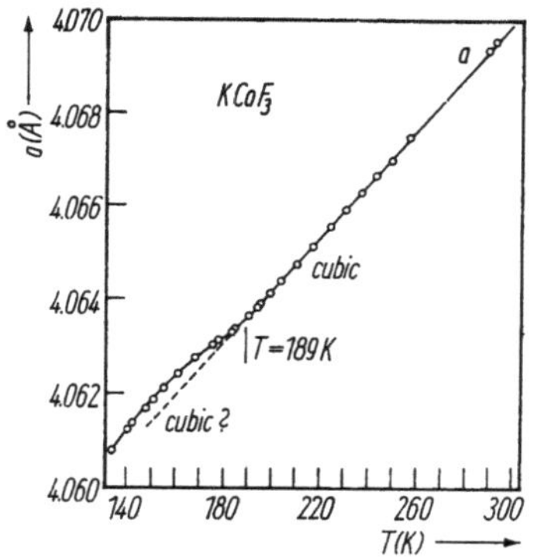

Fig. 3. Lattice parameter of $KCoF_3$ versus temperature. The mark indicates the temperature below which a deviation from the linear temperature dependence has been observed

thermal expansion at 189 K (Fig. 3). It is not quite clear whether this anomaly results from a deformation of the cubic lattice cell; in any case this deformation is very small because no splitting or even broadening of the diffraction line 510 could be detected.

The detailed study of lattice parameters of mixed single crystals $KMn_{0.9}Co_{0.1}F_3$ and $KMn_{0.9}Ni_{0.1}F_3$ in the vicinity of the transition point confirmed the decrease of T_c and distortion c/a for the crystals of $KMnF_3$ doped with Co^{2+} and Ni^{2+}. In case of $KMn_{0.9}Ni_{0.1}F_3$ the values of the c-parameter in the tetragonal phase have been taken from the powder measurements [4].

Similarly, as in pure $KMnF_3$ at some temperature above T_c anomalies have been observed resulting in the change of the slope of the curves $a(T)$ (see Fig. 4 and 5). Numerical values of lattice parameters of $KMeF_3$ crystals for a few temperatures are given in Table 1.

The thermal expansion coefficients $\alpha(T)$ defined by the formula

$$\alpha(T) = \frac{1}{a}\frac{\mathrm{d}a(T)}{\mathrm{d}T}$$

have been calculated in the form

$$a(T) = \sum_{n=0}^{N} A_n T^n .$$

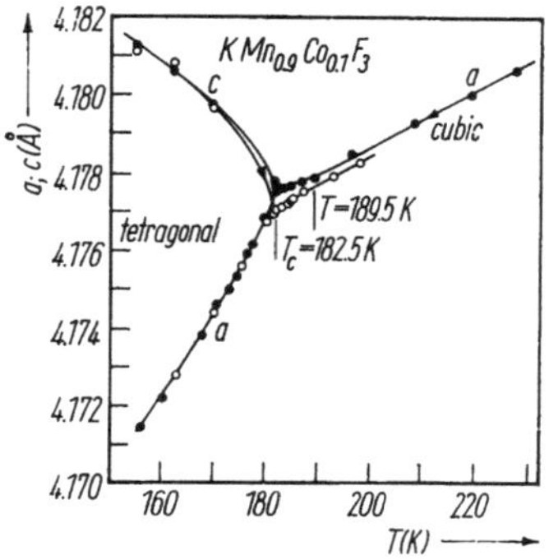

Fig. 4

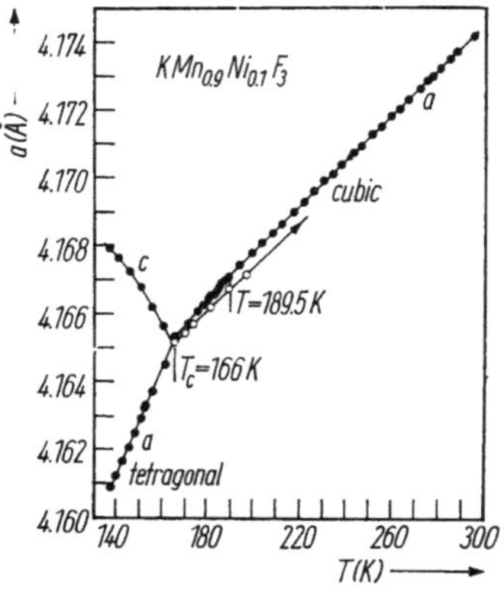

Fig. 5

Fig. 4. Lattice parameters of $KMn_{0.9}Co_{0.1}F_3$ from 155 to 270 K. ○ Heating, ● cooling

Fig. 5. Lattice parameters of $KMn_{0.9}Ni_{0.1}F_3$ versus temperature. ○ Heating, ● cooling

Table 1

Measured lattice parameters of $KMeF_3$ and $KMn_{1-x}Me_xF_3$ versus temperature

compounds	T (K)	cubic a (Å)	T (K)	cubic a (Å)	T (K)	tetragonal a (Å)	tetragonal c (Å)
$KMnF_3$	293.6	4.19267	192.4	4.18211	175.0	4.17684	4.18653
$KCoF_3$	294.6	4.06936	195.0	4.06383			
$KNiF_3$	295.6	4.01101	275.0	4.00959			
$KMn_{0.9}Co_{0.1}F_3$	289.0	4.18431	197.1	4.17837	162.9	4.17278	4.18128
$KMn_{0.9}Ni_{0.1}F_3$	296.5	4.17420	204.2	4.16812	150.0	4.16300	4.16695

This equation was fitted to the experimental data using least-squares methods to obtain a smooth representation of $a(T)$, which could be used for interpolations. The components A_n of the $a(T)$ functions and a few smoothed values of the linear thermal expansion coefficients for the cubic phase are given in Table 2.

Table 2

Values of the components A_n of the $a(T)$ function and of $\alpha(T)$ for the cubic phase of $KMeF_3$ and mixed crystals

	$KMnF_3$	$KCoF_3$	$KNiF_3$	$KMn_{0.9}Co_{0.1}F_3$	$KMn_{0.9}Ni_{0.1}F_3$
T (K)	293 to 192	295 to 195	299 to 275.3	249 to 197	296.5 to 204.2
A_0	4.16833	4.05017	3.99902	4.16535	4.16012
$10^{-4}\,A_1$	0.7573	0.7920	0.9327	0.6430	0.2151
$10^{-7}\,A_2$	−0.2249	−0.4748	1.0580	0.0896	0.8909
T (K)	293.1	294.6	295.6	249.4	296.5
α (10^{-6} K^{-1})	14.9	12.5	18.1	16.4	17.8
T (K)	192.4	195.0	275.3	197.1	204.2
α (10^{-6} K^{-1})	16.0	14.9	16.8	16.2	13.9

3. Discussion

It has been shown in [4] that the doping by ions Co^{2+}, Ni^{2+}, Fe^{2+}, and Mg^{2+} results in a decrease of T_c and the tetragonal distortion c/a in $KMnF_3$. These results have been fully confirmed in the present paper for Co^{2+} and Ni^{2+} doping.

Assuming Landau-type behaviour and applying a formula calculated by Slonczevsky and Thomas [16] in the mean field approximation, the rotation angle φ of the MeF_6 octahedron can been expressed as

$$\varphi^2 = A(T_c - T)^{2\beta} .$$

The angle φ can be estimated to a good approximation [4, 13] by assuming

$$\frac{a}{c} = \cos\varphi = 1 - \frac{\varphi^2}{2} + \dots ,$$

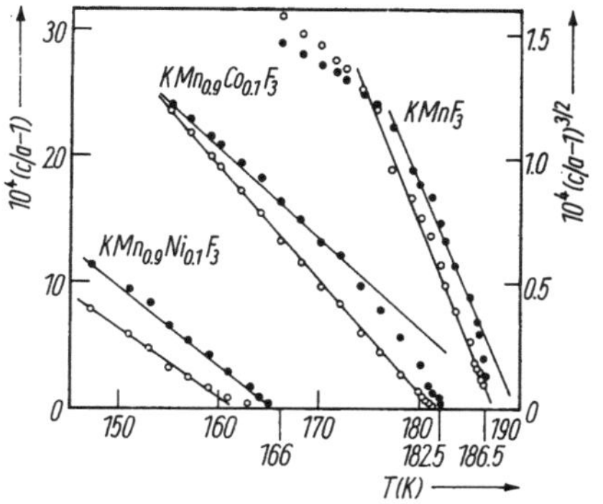

Fig. 6. Plots of $(c/a - 1)$ (●) and $(c/a - 1)^{3/2}$ (○) versus temperature for $KMnF_3$, $KMn_{.90}Co_{0.1}F_3$, and $KMn_{0.9}Ni_{0.1}F_3$

and consequently

$$\frac{c - a}{c} = A(T_c - T)^{2\beta'} .$$

In the Landau theory the critical exponent β is equal to $\frac{1}{2}$. Fig. 6 shows a plot of $(c/a - 1)$ — and $(c/a - 1)^{3/2}$ — versus temperature. For $KMnF_3$ and $KMn_{0.9}Co_{0.1}F_3$ to the accuracy available $\beta' = \frac{1}{3}$ clearly produces a better description than $\beta' = \frac{1}{2}$. For $KMn_{0.9}Ni_{0.1}F_3$, however, a better fit can be obtained for $\beta' = \frac{1}{2}$.

The precise measurement of lattice parameters of $KMnF_3$ in the region of the cubic-to-tetragonal phase transition performed on heating would lead one to regard this transition to be of second order, but a small hysteresis of about 1 K and the temperature dependence of the lattice parameters on cooling show that the phase transition in $KMnF_3$ is virtually of very weak first order.

Doping by Co^{2+} or Ni^{2+} results in smearing of the phase transition, which in mixed crystals becomes of second order. The small change of lattice parameter a in the cubic phase observed after the cooling cycle results probably from ordering processes.

References

[1] O. Beckman and K. Knox, Phys. Rev. **121**, 376 (1961).
[2] A. Okazaki and Y. Suemune, J. Phys. Soc. Japan **16**, 671 (1961).
[3] A. Okazaki and Y. Suemune, J. Phys. Soc. Japan **17**, 204 (1961).
[4] A. Ratuszna, Thesis, Katowice 1978.
[5] A. Ratuszna, A. Pietraszko, K. Łukaszewicz, and A. Chełkowski, XI. Internat. Congr. Crystallography, Warszawa 1978, printed in: Acta cryst. (Suppl.) **A34**, 281 (1978).
[6] D. J. Benard and W. C. Walker, Rev. sci. Instrum. **47**, 122 (1976).
[7] G. S. Shirane, V. Minkiewicz, and A. Linz, Solid State Commun. **8**, 1941 (1970).
[8] V. J. Minkiewicz and G. Shirane, J. Phys. Soc. Japan **26**, 674 (1969).
[9] W. Cochran, Adv. Phys. **9**, 387 (1960).
[10] R. A. Cowley, Phys. Rev. **134**, A 981 (1964).
[11] W. Cochran and A. Zia, phys. stat. sol. **25**, 273 (1968).
[12] M. Hidaka, J. Phys. Soc. Japan **39**, 180 (1975).
[13] F. Borsa, Local Properties at Phase Transitions, Compositori, Bologna (Italy) 1975.
[14] A. Chełkowski, P. Jakubowski, D. Kraska, A. Ratuszna, and W. Zapart, Acta phys. Polon. **A47**, 347 (1975).
[15] A. Okazaki and M. Kawaminami, Japan. J. appl. Phys. **12**, 783 (1973).
[16] J. C. Slonczevsky and H. Thomas, Phys. Rev. B **1**, 3599 (1970).

(Received May 24, 1979)

phys. stat. sol. (a) 54, 745 (1979)

Subject classification: 14.4; 15; 22.6

Department of Physics, Faculty of Technology, Split[1] *(a) and*
Institute of Electrochemistry, Faculty of Technology, Zagreb[2] *(b)*

The Temperature and Compositional Dependence of the Electrical Conductivity of Sintered Aluminium Oxide with Metal Excess

By

V. GOTOVAC (a) and B. LOVREČEK (b)

Samples of aluminium oxide with metal excess are made by sintering. Scanning electron microscopic examinations provide some useful compositional and structural information. The electrical phenomena in $Al_2O_3 + xAl$ are examined. Experimental data are obtained from measurements of thermoelectric power (TEP), dc and ac conductivity as functions of temperature and metal excess. A simple model is suggested on the basis of which the possible conductivity mechanisms are discussed and described in the high and low temperature ranges. The activation energies and pre-exponential factors are estimated.

Aluminiumoxidproben mit überschüssigem Metallgehalt werden durch Sintern hergestellt. Rasterelektronenmikroskopieuntersuchungen liefern einige nützliche Informationen über Zusammensetzung und Struktur. Die elektrischen Erscheinungen in $Al_2O_3 + xAl$ werden ebenfalls untersucht. Experimentelle Werte werden aus den Messungen der Thermospannung (TEP) und der Gleich- und Wechselstromleitfähigkeit in Abhängigkeit von der Temperatur und dem Metallüberschuß angegeben. Entsprechend diesen Untersuchungen wird ein einfaches Modell vorgeschlagen, auf dessen Grundlage der mögliche Leitungsmechanismus im Hoch- und Tieftemperaturgebiet beschrieben und diskutiert wird. Die Aktivierungsenergie un edxperimentelle Vorfaktoren werden abgeschätzt.

1. Introduction

In the last few years a great deal of attention has been paid to the studies of the properties of the sintered oxides with metal excess. The attractiveness of these studies lies in the possibility of investigation of a new class of disordered and very complex systems. In contrast to the structure and properties of crystals and amorphous solids little is known with high degree of credit about the structure and properties of such systems. A vast amount of experimental data on many other disordered systems has been accumulated by now and attempts have been made in explaining the observed properties both on the basis of various models and theoretical calculations.

We shall restrict our attention to electrical phenomena in aluminium oxide with metal excess [1]. In the further text the notation $Al_2O_3 + xAl$ will be used in which x means total metal excess over stoichiometric composition in Al_2O_3.

These systems are significantly different from simple oxides because they possess some specific properties typical of a given composite system itself, but completely different from pure Al_2O_3 and pure Al properties. At the same time some of its properties are common for many disordered systems. So many factors may affect their behaviour and this wide range of problems will require that we concentrate ourselves on only few aspects, leaving many details undiscussed. In this sense attempts will

[1]) 58000 Split, Yugoslavia.
[2]) 41000 Zagreb, Yugoslavia.

be made to find a correlation between electrical properties and chemical composition of the sintered specimens paying special attention to the conductivity mechanisms in the frame of the proposed energy diagram model.

2. Experimental

2.1 Preparation of specimens

The specimens were made by mixing aluminium and aluminium oxide powder in certain proportions according to [2]. The aluminium powder of 99.98% purity (Koch Light Labor.) and aluminium oxide powder of 99.999% purity (Fluka AG) were used in ratios 1:3, 1:5, 1:10, 1:20, 1:50, and 1:200. The mixtures were molded into the disks 12 mm in diameter and 2.5 to 10 mm long. The cylinders were fired in nonoxidizing atmosphere (argon or nitrogen) at about 1550 °C for 80 min. The notations A-3, A-5, A-10, A-20, A-50, and A-200 for the above-mentioned specimens will also be further used in the text.

2.2 Analytical determination of metal excess

The metal excess in samples was determined from the reaction between aluminium excess and MnO_4^- ions. The acidic sulphate 0.02 N permanganate solution was boiled and after cooling it was standardized with 0.02 N natrium oxalate solution. In the series of volumetric flasks with this solution the equal amounts of sintered aluminium oxide specimens were added. Starting from this moment MnO_4^- concentrations in volumetric flasks were titrated with standard natrium oxalate solution in subsequent time intervals. Simultaneously, MnO_4^- concentrations in the solutions without aluminium oxide were determined. After about 2 d the amount of natrium oxalate used became constant, and from this value the aluminium excess in the specimens was calculated.

2.3 Structural examinations

Topographic and compositional micrographs were otained by the scanning electron microscope JEOL JXA-50A with electron probe microanalyzer.

2.4 Thermoelectric power measurements

The experimental method of measuring TEP was described by Middleton and Scanlon [3] and Okamoto and Brener [4]. The sample holder was inserted vertically into a cylindrical electric furnace. A small electric heater, pressed against one end of the sample, provided the necessary temperature gradient. The copper–constantan thermocouples were attached in the holes, which were drilled into the sample near each end and filled with Ecobond solder 56C. The thermocouple voltages were measured by Pye Unicam potentiometer or by Keitley electrometer 610C.

2.5 Conductivity measurements

The electrical property measurements were carried out in an alumina tube furnace up to 1300 °C. Atmosphere control was maintained by flowing argon. The specimens were held between platinum electrodes with platinum lead wires. Electrical contacts to the specimens were made on their plane parallel surfaces using Engelhard platinum paste No. 05. Thick platinum coatings were obtained by repeated applications of platinum paste.

The dc conductivity measurements were made by the two-terminal method using Keitley electrometer 610C in its resistance mode. At the same time the measurements of the temperature dependence of the ac conductivity in the range 10^2 to 10^5 Hz were conveniently made using Hewlett-Packard impedance meter model 4800 A.

The dielectric properties of sintered specimens A-200 were measured only at room temperature. The specimen was held in a versatile two-terminal cell made according to Broadhurst and Bur [5], with which we were able to eliminate capacitive effects. It was assumed that at sufficiently high frequencies the high resistance contact between grains, represented by resistance R_c and capacitance C_c in parallel, was electrically shorted by its own capacitance. The remaining impedance was ascribed to the bulk impedance of the grains themselves represented by the parallel network of the resistance of grains R_s and capacitance of grains C_s. Both parallel networks were connected in series.

3. Results

3.1 Determination of metal excess

The quantities of natrium oxalate needed for titration of acidic permanganate solution with equal amounts of sintered aluminium oxide as a function of time are shown in Fig. 1. After about 20 h the quantity of natrium oxalate used becomes constant. During this experiment concentration of MnO_4^- ions in standard solution remains unchanged. From the final decrease of MnO_4^- concentration, the aluminium excess is calculated. The results of these calculations are given as values on the abscissa in Fig. 8 and 9.

3.2 Results of electron microscopic observation

By using scanning electron microscope and electron microprobe direct observation of surface morphology is done by backscattered electrons and COMPO and TOPO micrographs of the surface are obtained (Fig. 2). Due to the rather complex technique of fabrication the microstructure is complex: the surface exhibits grains of Al_2O_3 (position 1), second phase as precipitates (position 2), subgrain boundaries (position 3) with finely dispersed Al particles (Fig. 2c). When the sample is examined in the scanning microprobe a chemical map for Al and O can be obtained by rastering electron beam across the line on the surface and monitoring the intensity corresponding to Al and O as in Fig. 2b. An average composition over the line AA' is obtained. The peaks in O curve correspond to Al_2O_3 grains.

3.3 Temperature dependence of TEP

TEP measurements have only been done to get informations about the type of conductivity. Therefore, only measurement (Fig. 3) given for the A-200 sample has been taken as example. Q-value is of the order of 1 mV/K. Reproducible values are

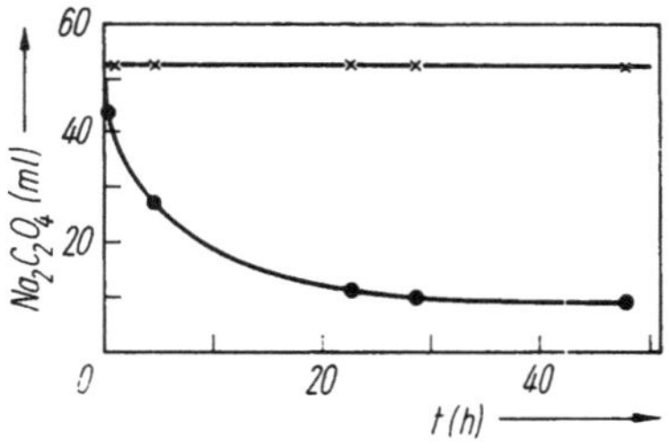

Fig. 1. Time dependence of 0.02 N $Na_2C_2O_4$ used during titration of 50 ml 0.02 N permanganate solution (dots) and of 50 ml pure 0.02 N permanganate solution (crosses)

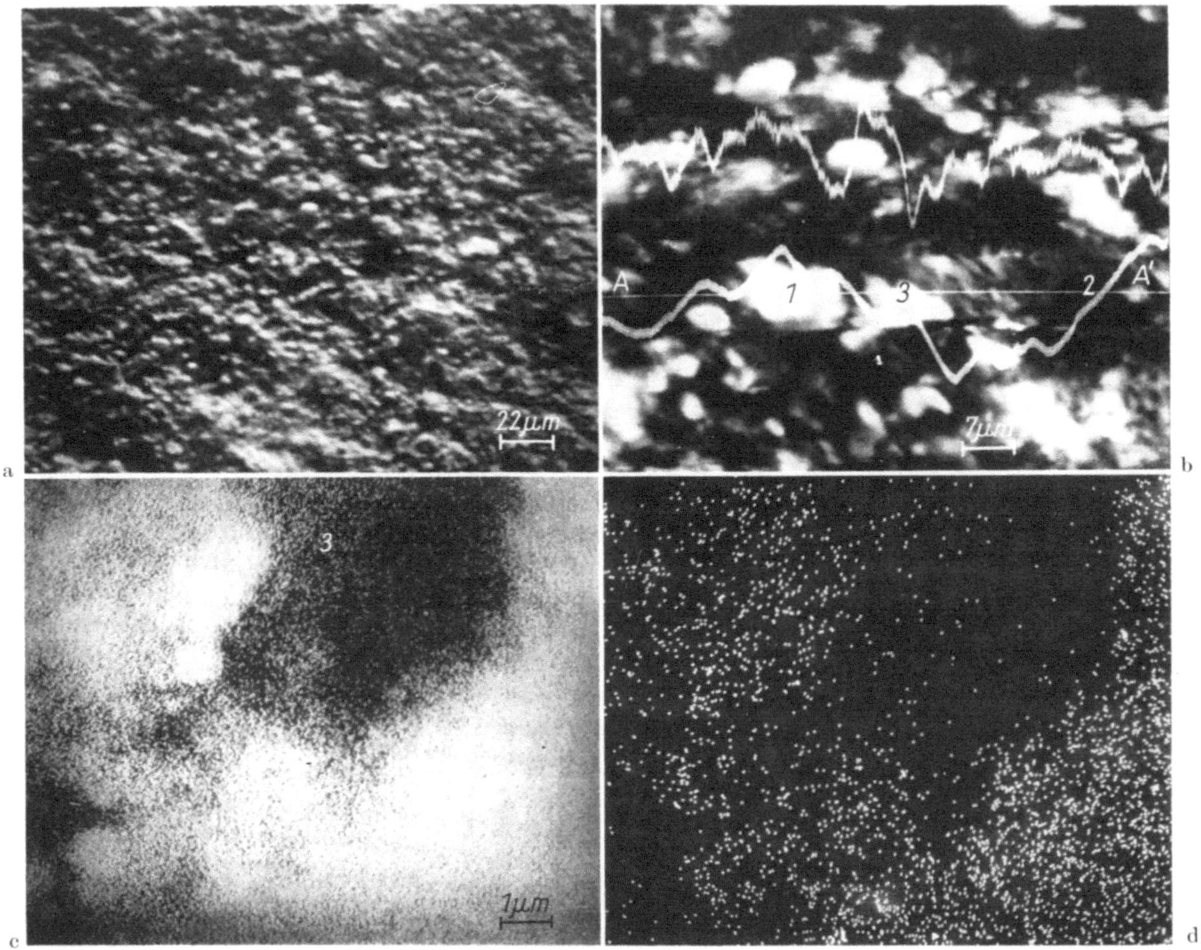

Fig. 2. Electron micrographs of sintered aluminium oxide surface (specimen A-5). a) TOPO, b) COMPO and the concentration profile along AA′ line for Al (upper curve) and O (lower curve). In c) and d) the surface is given in characteristic X-photographs for aluminium (AlK_α) and oxygen (OK_α), respectively

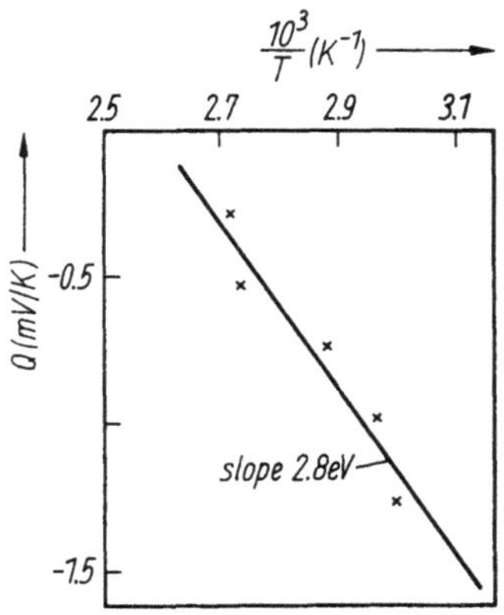

Fig. 3. Temperature dependence of thermoelectric power for specimen A-200

obtained in the temperature range from 300 to 600 K, regardless of the dimensions of the specimens. TEP value is negative indicating conductivity by negatively charged carriers. The slope of the experimental curve is indicated in the figure.

3.4 Temperature and frequency dependence of conductivity

In Fig. 4 and 5 the typical temperature dependences of dc and ac conductivity are chosen as examples among measurements of that kind performed on many aluminium oxide specimens with different metal excess. The frequency-independent

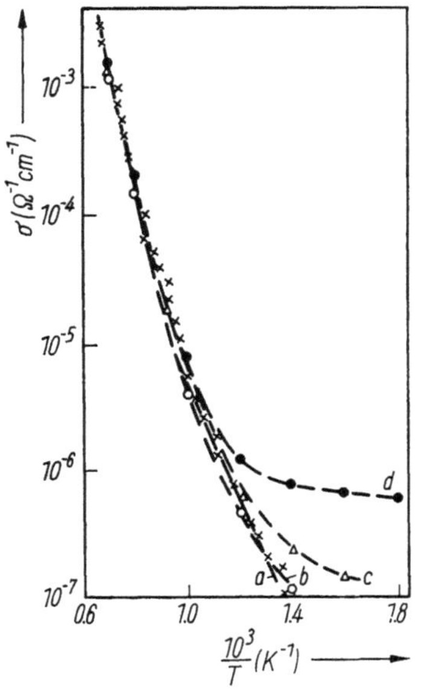

Fig. 4

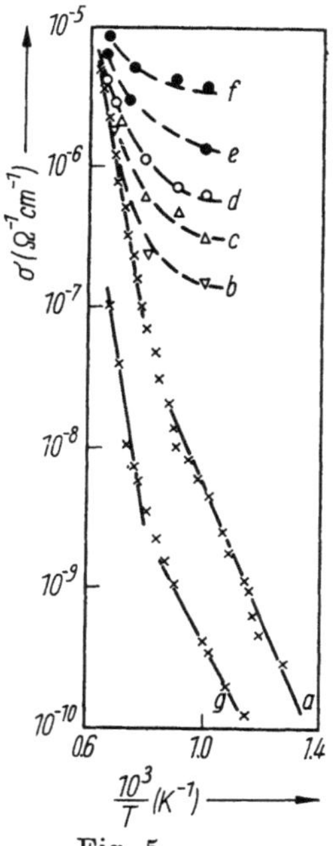

Fig. 5

Fig. 4. Temperature dependence of conductivity for specimen $Al_2O_3 + xAl$ with $x = 0.7$. (a) σ_{dc} and (b) σ_{ac} at 5, (c) 50, and (d) 500 kHz

Fig. 5. Temperature dependence of conductivity for specimen $Al_2O_3 + xAl$ with $x = 0.15$. (a) σ_{pc} and (b) σ_{ac} at 5, (c) 20, (d) 50, (e) 200, and (f) 500 kHz. Curve g shows the temperature dependence of conductivity for specimen $Al_2O_3 + xAl$ with $x = 0.04$

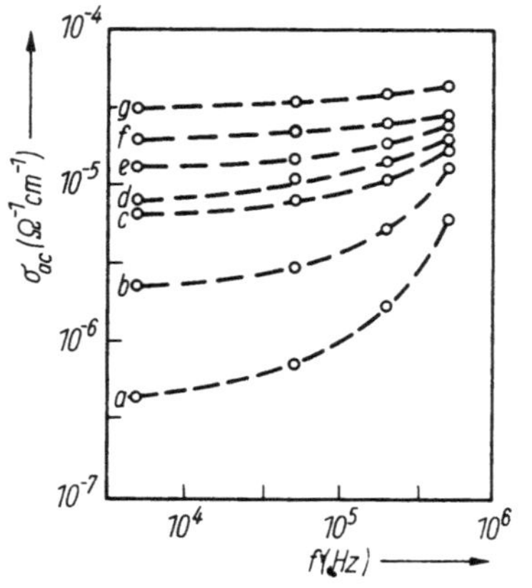

Fig. 6

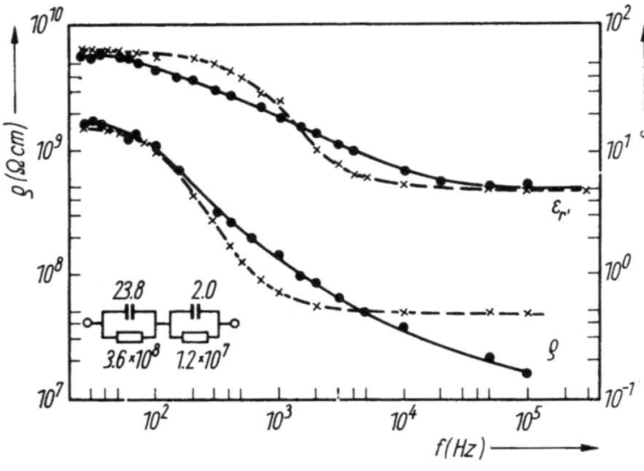

Fig. 7

Fig. 6. The ac conductivity of $Al_2O_3 + xAl$ with $x = 0.52$ over a wide frequency range at (a) $T = 540$, (b) 620, (c) 860, (d) 900, (e) 960, (f) 990, and (g) 1060 °C. For curves a and b σ_{ac} is proportional to $\omega^{0.9}$ and $\omega^{0.76}$, respectively

Fig. 7. Conductive properties of sintered specimens A-200 as functions of frequency. Units for illustrated network are Ω and pF. Dashed curve fitted as described in the text

conductivity

$$\sigma_1 = \sigma_{10} \exp\left(-\frac{E_1}{kT}\right) \tag{1}$$

prevails at higher temperatures above 1200 K. The values of experimental activation enerties E_1 are given in Fig. 9, from which the decrease in values E_1 with increasing metgl excess could be seen. It is interesting that from two different experimental maehods (TEP and conductivity measurements) very close values (2.8 and 2.3 eV) are estimated for two specimens with low metal excess (A-200 and A-50, respectively). The values σ_{10} range from 10 to 100 $\Omega^{-1}cm^{-1}$.

At lower temperatures the conductivity

$$\sigma_2 = \sigma_{20} \exp\left(-\frac{E_2}{kT}\right) \tag{2}$$

has also activated character. As seen from Fig. 9 E_2-values are approximately constant and lower than E_1. The changes of E_2 are within experimental error. The pre-exponential factor σ_{20} increases as the metal excess in specimens increases (Fig. 8).

The results of the two-terminal ac measurements are represented by dashed lines in Fig. 4 and 5. As seen from the figures dc conductivity is lower than ac conductivity. At higher temperatures conductivity σ_{ac} depends slightly on frequency but strongly on temperature. Fig. 6 shows changes of σ_{ac} with frequency at different temperatures. In the region above 10^5 Hz and at low temperatures σ_{ac} increases more slightly with T and exhibits a dependence on frequency as ω^s with $0.7 < s < 1.0$.

In Fig. 7 experimental values of resistivity $\varrho(f)$ and apparent dielectric constant $\varepsilon_{r'}(f)$ for one A-200 specimen are represented by dots. These values are calculated from the measured $R_p(f)$ and $C_p(f)$ ac values. Such frequency dependences of R_p and C_p are obtained if our samples are represented by the equivalent and the simplest possible network of C_c, R_c and C_s, R_s as shown in Fig. 7. The mathematical transformation of such network into its equivalent parallel circuit of R_p and C_p shows that R_p and C_p depend on frequency in the following way:

$$\frac{1}{R_p} = \frac{(R_s + R_c) + \omega^2 R_s R_c (R_s C_s^2 + R_c C_c^2)}{(R_s + R_c)^2 + \omega^2 R_s^2 R_c^2 (C_c + C_s)^2}, \tag{3a}$$

$$C_p = \frac{R_s^2 C_s + R_c^2 C_c + \omega^2 R_s^2 R_c^2 C_s C_c (C_s + C_c)}{(R_s + R_c)^2 + \omega^2 R_s^2 R_c^2 (C_c + C_s)^2}. \tag{3b}$$

It is possible to fit predicted curves to the data obtained experimentally by considering their limiting values at low and high frequencies. Fig. 7 also shows as dashed lines the predicted curves. The data are considered to be in satisfactory agreement with the model. The estimated values of the real part of the dielectric constant are $\varepsilon_{r'} = 57$ at 25 Hz and 5 at 5×10^5 Hz.

4. Discussion of Results

As composition and morphology play an important role in determining sample properties we shall first discuss the oxide structure. During the sintering of Al_2O_3 [6] the phenomenon known as discontinuous grain growth can occur and the resulting ceramic has residual porosity and complex structure of grain boundaries. Report [7] of Mg segregation in alumina is an excellent example of the ambiguity which can occur due to that. We have also been expecting an Al enrichment at grain boundaries second-phase particles located at grain corners, and precipitates as regions of high Al concentration. As seen from Fig. 2 the specimens actually have such complex structure with nonuniform distribution of the two components. A few reports [8 to 11] also discuss a modified structure as a consequence of the interaction of metal atoms in specimens with original Al_2O_3 structure. A part of aluminium excess is probably incorporated in Al_2O_3 as a nonstoichiometric excess. The observed frequency dependences of resistivity and apparent dielectric constant also point to that, because such dependences are characteristic of grain structures [12 to 14].

Experimental results show that the dc and ac conductivity, the activation energy E_1, and the pre-exponential factor σ_{20} are the functions of metal excess. Such compositional dependences are observed in many composite systems [15 to 21], but only the samples structurally and compositionally similar to our A-200 samples were investigated by others [6 to 10]. They obtained the same frequency dependence of resistivity and apparent dielectric constant as in Fig. 7 and their estimated values of ϱ and $\varepsilon_{r'}$ at low and high frequencies are not markedly different from ours. The observed conductivity of samples acquired an activation pattern similar to the behaviour of amorphous Ge [22 to 24] and Si [23, 25], amorphous and polycrystalline InSb, GaSb, and GaAs [26, 27]. This also indicates that specimens belong to the class of disordered systems. According to Ryvkin and Shlimak [28 to 30] a large scale component of random field plays an essential role in amorphous semiconductors, in heavily doped crystalline semiconductors, in systems with crystalline regions turned randomly one relative to another, in semiconductors of n-type with p-clusters, and in disordered systems. As a consequence the tails in the density-of-states function appear in the energy gap just as in the energy spectrum of amorphous semiconductors (Mott and Davis [31], Cohen et. al. therein). As we have seen in our sintered specimens microcrystals are oriented randomly with respect to one another and aluminium excess is also distributed randomly. Therefore, some predictions concerning the interpretation of experimental results and the conduction mechanisms in sintered aluminium oxide with metal excess are given in the frame of the "bended band model".

The high-temperature conduction process appears to be by the negatively charged carriers (TEP is negative) with an energy exceeding the corresponding "percolation"

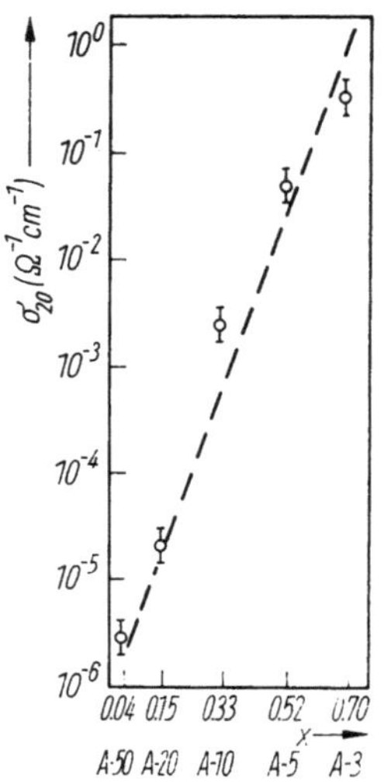

Fig. 8. Pre-exponential factor σ_{20} in equation (2) as a function of metal excess (circles experimental values). Equation (5) is represented by the dashed line

level, with which the charge carrier can just pass from one electrode to another in the "classical" way. The magnitude of pre-exponential factors σ_{10} suggests that. The conductivity is controlled by metal excess through the dependence of the activation energy E_1 on x. E_1 measures the energy difference between the percolation energy and the Fermi level and thus reflects the position of the Fermi level in the energy gap. Evidently, E_1 cannot exceed the value $E_g/2$ (as is the case for our specimens). In our opinion, the experimentally observed changes in the E_1-values, caused by increasing metal exceess, are the result of the corresponding variations in the amplitude and spatial extent of potential fluctuations, the amplitude of potential changes being in straightforward relationship with concentration of impurities as stated for highly doped semiconductors [29].

At lower temperatures (900 to 1200 K) the conduction is by the carriers with the energies between E_1 and E_2, their motion exhibiting elements of hopping conduction. The evidence for this is the magnitude of σ_{20}, which contrary to σ_{10} strongly depends on metal excess (Fig. 8). The values of the same order are obtained for a-Si [32] and a-As_2Te_3 [33, 34]. We suppose according to conclusions of Kocka et al. [34], that the hopping is between the "islands of allowed states". For lower metal excess their depth $E_1 - E_2$ is higher and the "islands" are smaller. While E_2 is independent of metal excess, the depth is dependent on metal excess. It may seem unusual that the position of the first allowed level E_2 is independent of metal excess but we have to mention that the similar behaviour is found in As_2Te_3 + Tl specimens [34], where the dependence of depth on metal excess is also observed. We shall dwell on this peculiarities in our future study of such systems, whose electrical properties point to a very complex electronic structure.

According to the above predictions the proposed model of $Al_2O_3 + x$Al specimens is drawn in the top right-hand corner of Fig. 9, where L means the length of the

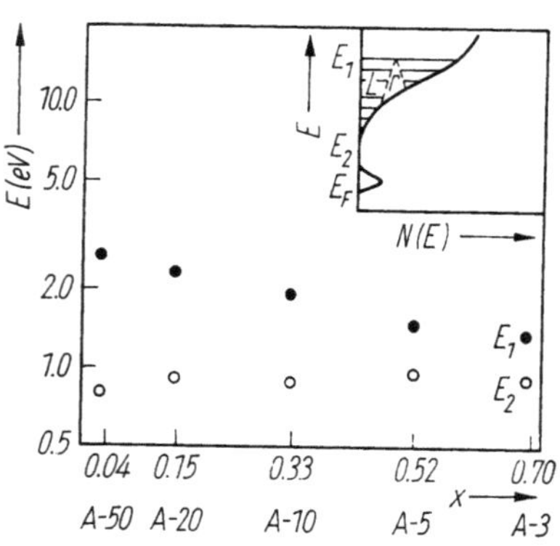

Fig. 9. Activation energies E_1 and E_2 in equations (1) and (2) as functions of metal excess. The proposed model of $Al_2O_3 + x$Al is also shown

"islands" and r their mean distance. The factor σ_{20} is proportional to $\exp[-\alpha r(x)]$, where $\alpha^{-1} = 8$ Å is the radius of localization. It is possible to explain the increase of σ_{20} with the increase of Al content by the growth of the "islands". If we suppose that $r(x)$ is the simple function of metal excess in the form

$$r(x) = r_0(1 - ax) , \tag{4}$$

then the experimental $\sigma_{20}(x)$ curve can be approximated by the equation

$$\sigma_{20} = K \exp[-\alpha r_0(1 - ax)] \tag{5}$$

with $K = 2.5 \times 10^{-2}\ \Omega^{-1}\text{cm}^{-1}$ and $a = 2$. It is known that the condition for localization is $\alpha r > 8$, therefore in our calculation the value $\alpha r_0 = 10$ is taken.

The temperature dependence of the thermoelectric power Q measures, in the frame of the proposed model [35], the energy difference between the Fermi level and the energy level where charge transport occurs. The slope of the experimental curve in Fig. 3 is calculated according to equation

$$Q = \frac{k}{e}\left(\frac{E - E_\text{F}}{kT} + \text{const}\right). \tag{6}$$

The conductivity activation energy reported by Brook et al. [10] and Pappis and Kingery [8], defined in their model as $E_g/2$ for Al_2O_3 with low metal excess (the samples compositionally very similar to our A-200 samples) is 2.8 and 2.6 eV, respectively. The Q-values are also of the same order.

From the results of the two-terminal ac measurements at low temperatures it is seen that dc conductivity is lower than ac conductivity. In the essence there is no marked difference between ac behaviour of our samples and some other amorphous compounds and evaporated semiconductors [24, 36 to 38]. In the case of amorphous semiconductors the before-mentioned experimental facts (especially weak temperature dependence) are taken as evidence that σ_{ac} is caused by a kind of hopping conduction between localized states near E_F [39] for which Austin-Mott relation [40] is frequently applied. The density of these states $N(E_\text{F})$ for a variety of amorpous specimens [41] ranges from 10^{18} to 10^{20} eV^{-1}cm^{-3}. In the Ryvkin's model of disordered system the carrier concentration at some points of the sample is large providing high conductivity at high frequencies while the dc conductivity is low because of the carriers being localized. It is interesting that the Austin-Mott relation for $T = 600$ K and $f = 500$ kHz gives the $N(E_\text{F})$ value of the order of 10^{19} eV^{-1}cm^{-3} for all investigated specimens, though the question about the application of Austin-Mott relation in this case remains.

In this article our purpose was to present very interesting properties of the composite system $Al_2O_3 + x$Al and to give explanation of some of them in the scope of the existing phenomenological models, but many questions considering type and origin of localized and extended states, changes of energy and pre-exponential factors with metal excess, etc. are still present here. In spite of the fact that the physical pattern of many processes is not completely clear, we believe that the study of the influence of metal excess in sintered disordered oxides would help to provide further and deeper explanation of some phenomena.

Acknowledgements

The authors would like to acknowledge the contribution made by Dr. E. Tkalčec for helpful discussion and R. Ruštinjac for performing electron microscopic investigations.

References

[1] V. Gotovac, Ph.D. Thesis, University of Zagreb, 1973.
[2] W. A. Miller, US Patent Office 2729880, patented January 10, 1956.
[3] A. E. Middleton and W. W. Scanlon, Phys. Rev. **92**, 219 (1953).
[4] Y. Okamoto and W. Brener, Organic Semiconductors, Reinhold Publ. Co., New York 1964.
[5] M. G. Broadhurst and A. J. J. Bur, J. Res. Nat. Bur. Standards **69C**, 165 (1965).
[6] W. D. Kingery, J. Amer. Ceram. Soc. **57**, 1 (1974).
[7] W. C. Johnson, Metallurg. Trans. **8A**, 1413 (1977).
[8] J. Pappis and W. D. Kingery, J. Amer. Ceram. Soc. **44**, 459 (1961).
[9] M. J. Pryor, Oxid. Metals **3**, 523 (1971).
[10] R. J. Brook, J. Lee, and F. A. Kröger, J. Amer. Ceram. Soc. **54**, 444 (1971).
[11] K. Eigenmann and H. H. Günthard, Chem. Phys. Letters **12**, 12 (1971).
[12] C. M. Huggins and A. H. Sharbaugh, J. chem. Phys. **38**, 393 (1963).
[13] J. Volger, in: Semiconductor Materials, Ed. H. K. Henish, Butterworths, London 1951 (p. 162).
E. J. Verwey, ibid. (p. 151).
[14] C. G. Koops, Phys. Rev. **83**, 121 (1951).
[15] B. T. Kolomiets, in: Electronics Phenomena in Non-Crystalline Semiconductors, Ed. B. T. Kolomiets, Nauka, Leningrad 1976 (p. 23).
[16] V. Trnovcovà, M. Pavlîkovà, D. Ležal, and E. Mariani, ibid. (p. 279).
[17] N. Van Dong, T. Hun Danh, B. Auguin, and A. Defresne, ibid. (p. 367).
[18] X. Le Cleach and J. F. Palmier, ibid. (p. 393).
[19] S. Sato, N. Yamaguchi, and H. Ozaki, J. Phys. Soc. Japan **33**, 1497 (1972).
[20] R. N. Blumenthal and R. L. Hofmaier, J. Electrochem. Soc. **121**, 126 (1974).
[21] R. N. Blumenthal and R. K. Sharma, J. Phys. Chem. Solids **13**, 360 (1975).
[22] A. H. Clark, Phys. Rev. **154**, 750 (1967).
[23] R. C. Chittik, J. non-crystall. Solids **3**, 255 (1970).
[24] A. Yoshida, T. Arizumi, K. Shimakawa, and S. Nitta, see [15] (p. 231).
[25] P. G. LeComber and W. E. Spear, Phys. Rev. Letters **25**, 509 (1970).
[26] J. Stuke, Proc. X. Internat. Conf. Phys. Semicond. Cambridge (Mass.) 1970 (p. 14).
[27] J. Stuke, Proc. Internat. Conf. Low Mobility Materials, Eilat (Israel) 1971 (p. 193).
[28] S. M. Ryvkin and I. S. Shlimak, see [15] (p. 203).
[29] S. M. Ryvkin and I. S. Shlimak, phys. stat. sol. (a) **16**, 515 (1973).
[30] S. M. Ryvkin, phys. stat. sol. (a) **19**, K173 (1973).
[31] N. F. Mott and E. A. Davis, Electronic Processes in Non-Crystalline Materials, Clarcudon Press, Oxford, 1971.
[32] P. G. Le Comber, A. Madan, and W. E. Spear, J. non-crystall. Solids **11**, 219 (1972).
[33] L. Štourač, see [15] (p. 44).
[34] J. Kočka, A. Triška, and L. Štourač see [15] (p. 249).
[35] A. Van der Plas and R. Bube, J. non-crystall. Solids **24**, 377 (1977).
[36] A. E. Oven and J. M. Robertson, J. non-crystall. Solids **2**, 40 (1970).
[37] E. V. Ivkin and B. T. Kolomiets, J. non-crystall. Solids **3**, 41 (1971).
[38] H. K. Rockstad, J. non-crystall. Solids **8–10**, 621 (1971).
[39] M. Pollak and T. H. Geballe, Phys. Rev. **122**, 1742 (1961).
[40] I. S. Austin and N. F. Mott, Adv. Phys. **18**, 41 (1969).
[41] H. Fritzsche, in: Amorphous and Liquid Semiconductors, Ed. J. Tauc, Plenum Press, London 1974 (p. 263).

(Received June 6, 1979)

phys. stat. sol. (a) 54, 755 (1979)

Subject classification: 19; 22

Institute of Physics, Technical University of Częstochowa[1]) *(a) and*
Institute of Physics, Pedagogical College, Opole[2]) *(b)*

Electron Paramagnetic Resonance of Mn^{2+} Ions in Aragonite Single Crystals[3])

By

A. ŚLĘZAK (a), J. LECH (a), and I. BÓJKO (b)

Electron paramagnetic resonance at X-band is studied in natural aragonite single crystals containing about 100 ppm Mn impurities. It is found that there are two inequivalent Mn^{2+} sites in the aragonite lattice. The principal axes Y of both the sites coincide. They are in ⟨100⟩ crystallographic directions. The principal axes X_I, Z_I, X_{II}, Z_{II} lie in the (b, c) crystal plane. The axes Z_I and Z_{II} are related at angles 65°. The crystallographic ⟨001⟩ direction bisects this angle. The EPR spectra of both these sites are well described with the same orthorhombic spin-Hamiltonian parameters.

Es wird die paramagnetische Elektronenresonanz in natürlichen Aragoniteinkristallen, die etwa 100 ppm Mn-Beimischung enthalten, untersucht. Es wird gefunden, daß im Aragonitgitter zwei ungleichwertige Lagen der Mn^{2+}-Ionen auftreten. Die Hauptachsen Y beider Lagen decken sich und sind der kristallographischen Richtung ⟨100⟩ parallel. Die Hauptachsen X_I, Z_I, X_{II}, Z_{II} liegen in der kristallinen (b, c) Ebene. Die Achsen Z_I und Z_{II} bilden einen Winkel von 65°. Die kristallographische Richtung ⟨001⟩ ist eine Winkelhalbierende dieses Winkels. Die mit diesen beiden Lagen verbundenen EPR-Spektren werden zufriedenstellend mit Hilfe von identischen Parametern des orthorhombischen Spin-Hamiltonoperators beschrieben.

1. Introduction

A very interesting phase transition takes place between two polymorphic modifications of $CaCO_3$; namely the transformation aragonite → calcite. The physical mechanism of this change is not quite clear. We decided to investigate the problem by the EPR of Mn^{2+} ions (substituting for Ca^{2+} ions). This communication includes some results of our studies on the EPR of Mn^{2+} in the aragonite phase. To the best of our knowledge only one paper on this subject has been reported so far [1]. Unfortunately, results in [1] have been obtained with microcrystalline and powdered clam shell samples.

2. Crystal Structures

Aragonite is a rhombic $CaCO_3$ modification. It is metastable at room temperature and low pressure, and spontaneously changes into a stable modification — calcite [2]. That is why well formed natural aragonite single crystals are found rather rarely, and if so they are in small quantities.

---

[1]) Częstochowa, Poland.

[2]) Opole, Poland.

[3]) This work was supported by the Institute for Low Temperatures and Structure Research, Polish Academy of Sciences, Wrocław.

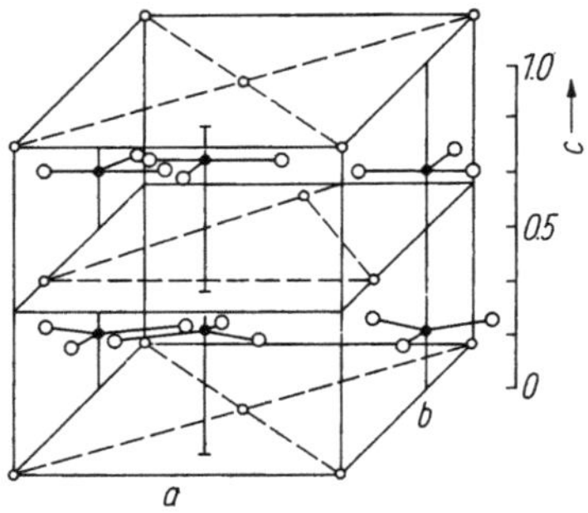

Fig. 1. Aragonite unit cell. ∘ Ca, ○ O, ● C (c in nm)

The unit cell of aragonite is drawn in Fig. 1. It is rhombic (pseudohexagonal) with $a = 0.4959$ nm, $b = 0.7968$ nm, and $c = 0.5741$ nm. The space group is $P_{mcn}(D_{2h}^{16})$ [3]. There are $Z = 4$ molecules per unit cell. Cations Ca^{2+} and anions CO_3^{2-} lie in separate layers normal to the axis c. CO_3^{2-} planes are at a distance $d = 0.17c$ from Ca^{2+} planes, which cut the axis c at the height $h = 0$ and $0.50c$. The coordination number is 6 if one treats the trigonous CO_3^{2-} ions as a whole. The assumption that oxygen ions are the nearest neighbours of Ca^{2+} yields the coordination number 9. (For comparison, there are six nearest oxygen ions coordinating Ca^{2+} in the rhombohedral calcite lattice.)

3. Experimental

Samples of natural aragonite single crystals having a form of hexagonal columns lengthened along ⟨001⟩ directions and set with (110), (010), and (001) faces were used in this work. They contained about 100 ppm Mn impurities and trace amounts of Fe.

Optically oriented specimens were mounted on a specially designed end of a quartz rod, which, in turn, was attached to a goniometer head. The measurements were performed by means of an X-band EPR spectrometer Model SE/X28 with 100 kHz magnetic field modulation and rectangular H_{112} cavity. Magnetic $\boldsymbol{B}$ field magnitudes were measured with a proton resonance magnetometer. A DPPH probe was used as standard g-value marker.

4. Results and Discussion

The full Mn^{2+} spectrum anisotropy has been investigated in this work, both experimentally and numerically. It has been pointed out that the experimental EPR signal is a sum of two signals raising from two inequivalent Mn^{2+} sites in the aragonite lattice. Detailed measurements and calculations carried out with a TI-59 calculator showed that the orthorhombic spin Hamiltonian in the form given by Posener [4]:

$$\mathcal{H} = \beta \boldsymbol{B} \cdot \mathbf{g} \cdot \boldsymbol{S} + \frac{D}{3} O_2^0 + E O_2^2 + \frac{a}{120} (O_4^0 + 5 O_4^4) + \frac{F}{180} O_4^0 + \\ + \boldsymbol{B} \cdot \mathbf{A} \cdot \boldsymbol{I} - \beta \boldsymbol{B} \cdot \mathbf{g}_I \cdot \boldsymbol{I} + \boldsymbol{I} \cdot \mathbf{P} \cdot \boldsymbol{I} \tag{1}$$

is applicable for a good description of these spectra. The coefficients and operators in (1) have their usual meanings [5] and are considered to be expressed in the principal axis system (x, y, z).

Basing on this general spin Hamiltonian and assuming the principal axes of tensors $\mathbf{g}$, $\mathbf{g}_I$, $\mathbf{A}$, and $\mathbf{P}$ to coincide, it has been proved by perturbation method in the usual high-field representation that up to third-order terms the energy levels of Mn^{2+} are the following [4, 6]:

$$E_{M,m} = g\beta BM + \left[M^2 - \frac{1}{3} S(S+1)\right]\sigma + AMm - g_I\beta Bm +$$
$$+ P\left[m^2 - \frac{1}{3} I(I+1)\right] + a'[35M^4 - 30S(S+1)M^2 + 25M^2 -$$
$$- 6S(S+1) + 3S^2(S+1)^2] + G^{-1}A^2\{m[M^2 - S(S+1)] +$$
$$+ M[I(I+1) - m^2]\} + G^{-2}|\lambda|^2 \frac{2Am}{M}\{[M^2 - S(S+1)]^2 - M^2\} +$$
$$+ 8G^{-2}|\xi|^2 AMm[2M^2 - 2S(S+1) + 1] +$$
$$+ G^{-2}A^2\sigma\{[S(S+1) - M(M+1)](2M+1)[I(I+1) - m(m-1)] -$$
$$- [S(S+1) - M(M-1)](2M-1)[I(I+1) - m(m+1)]\} +$$
$$+ G^{-2}A^3\{[S(S+1) - M(M+1)][I(I+1) - m(m-1)](m - M - 1)$$
$$+ [S(S+1) - M(M-1)][I(I+1) - m(m+1)](M - m - 1)\}, \quad (2)$$

where

$$G = 2g\beta B,$$
$$\sigma = \tfrac{1}{2}[D(3\cos^2\alpha - 1) + 3E\cos^2\alpha\cos 2\Phi],$$
$$\lambda = \sin\alpha\,[(D - E\cos 2\Phi)\cos\alpha + iE\sin 2\Phi]$$
$$= \tfrac{1}{4}[D\sin^2\alpha + E(\cos^2\alpha\cos 2\Phi - \sin^2\Phi) + 2iE\cos\alpha\sin 2\Phi],$$
$$a' = \tfrac{1}{192}[a\sin^4\alpha\cos 4\Phi + (a + \tfrac{2}{3}F)(7\cos^4\alpha - 6\cos^2\alpha + \tfrac{3}{5})],$$
$$P = \tfrac{3}{2}[P_z(3\cos^2\alpha - 1) + (P_x - P_y)\sin^2\alpha\cos^2\Phi],$$

$$\sin\alpha = \frac{g_\perp}{g}\sin\theta, \qquad \cos\alpha = \frac{g_z}{g}\cos\theta,$$

$$\sin\Phi = \frac{g_x}{g_\perp}\cos\varphi, \qquad \cos\Phi = \frac{g_y}{g_\perp}\sin\varphi,$$

$$g_\perp^2 = g_x^2\sin^2\Phi + g_y^2\cos^2\Phi, \qquad A_\perp^2 = A_x^2\sin^2\Phi + A_y^2\cos^2\Phi,$$
$$g^2 = g_\perp^2\sin^2\Theta + g_z^2\cos^2\theta, \qquad A^2 = A_\perp^2\sin^2\theta + A_z^2\cos^2\theta,$$

θ, φ are the polar angles which the static magnetic field $\boldsymbol{B}$ makes with the principal axis system (x, y, z), $S = 5/2$ and $I = 5/2$ for Mn^{2+}.

In the above formulae off-diagonal terms including a, F, and P have been omitted because they are negligible for Mn^{2+}.

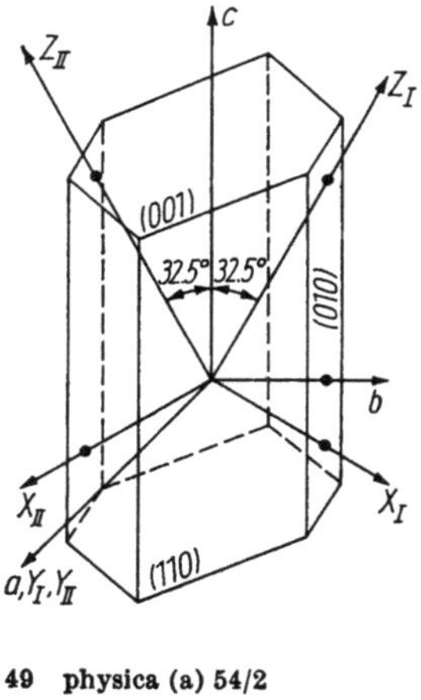

Fig. 2. The principal axes of two Mn^{2+} complexes in aragonite in relation to crystal axes and morphology

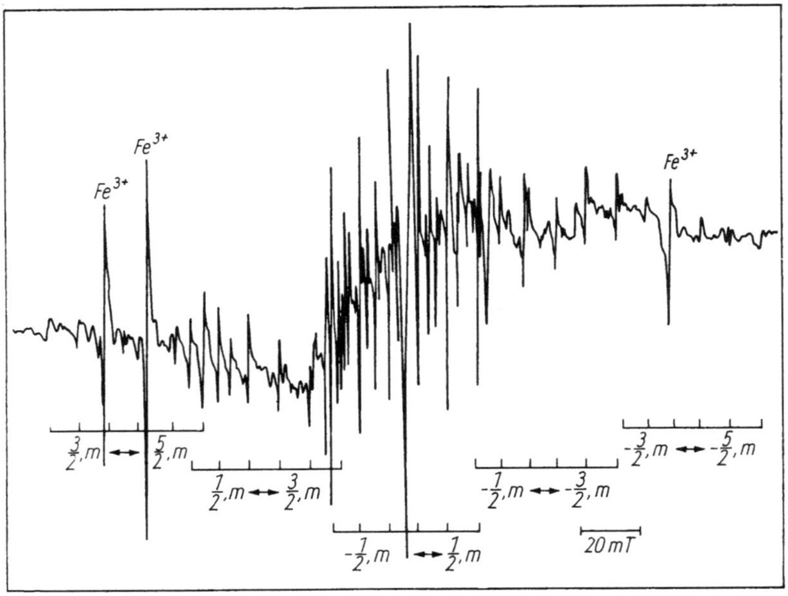

Fig. 3. Experimental EPR spectrum of Mn^{2+} in aragonite obtained at room temperature (295 K) and X-band for field $\boldsymbol{B}$ direction corresponding to $\theta_I = 0°$ and $\varphi_I = 90°$. The transitions due to "complex I" are assigned

As it was mentioned at the beginning of this section, two inequivalent Mn^{2+} sites in the aragonite lattice have been identified. Paramagnetic complexes relevant to these sites were simply assigned "complex I" and "complex II" in this work. Both complexes are characterized by the same spin-Hamiltonian parameters. Values of the parameters are listed in Table 1, which includes data for EPR of Mn^{2+} in the calcite lattice as well.

Table 1

Spin-Hamiltonian parameters of Mn^{2+} EPR spectra in $CaCO_3$ modifications. (Values of D, E, a_0, and A are in 10^{-4} cm^{-1})

	calcite		aragonite	
	this work	ref. [7]	this work	ref. [1]***)
g_x	2.0020	2.0018	2.0044	2.006
g_y	2.0020	2.0018	1.9910	± 0.001
g_z	2.0024	2.0022	2.0023	
D	−75.96	−76.45	218.8*)	234 ± 10
E	—	—	− 49.3*)	39 ± 5
A_x	−88.01	−88.04	− 88.7	
A_y	−88.01	−88.04	− 87.6	88 ± 1
A_z	−88.18	−87.43	− 90.2	
a_0**)	7.07	7.65	4.0	

*) The relative signs of D and E have been determined from line shifts brought about by the third-order terms.

**) $a_0 = (a - F)$ for calcite, $a_0 = (a + \frac{2}{3}F)$ for aragonite.

***) Results obtained with microcrystalline and powdered samples.

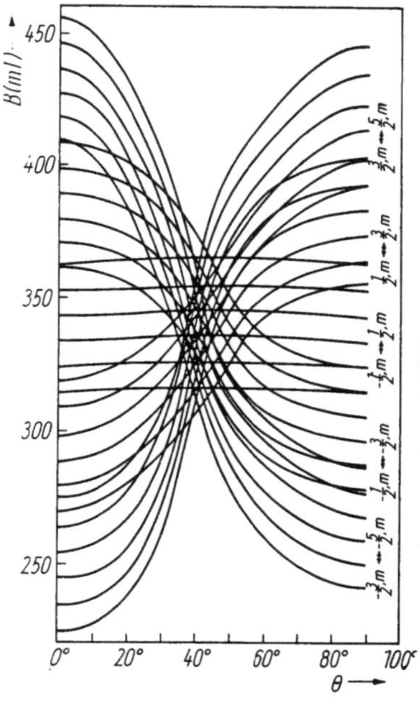

Fig. 4. Angular variation of "allowed" line positions of "complex I" in its Y, Z plane

The principal axis Y is common for both complexes. It is in the ⟨100⟩ crystallographic direction. The planes $(X_{\mathrm{I}}, Z_{\mathrm{I}})$ and $(X_{\mathrm{II}}, Z_{\mathrm{II}})$ coincide with the crystal plane (b, c). Details on the angles between the principal and crystal axes with respect to the crystal morpology are shown in Fig. 2.

In order to illustrate experimental pattern features, the spectrum obtained for direction of field $\boldsymbol{B}$ $\theta_{\mathrm{I}} = 0°$, $\varphi_{\mathrm{I}} = 90°$ is presented in Fig. 3. The spectrum due to the "complex I" includes 30 "allowed" lines corresponding to $\Delta M = 1$ and $\Delta m = 0$ transition rules at this configuration. These lines are assigned in Fig. 3. The spectrum of "complex II", however, contains "allowed" and "forbidden" lines. Five couples of "forbidden" lines under the transition rules $\Delta M = 1$ and $\Delta m = \pm 1$ are relatively strong at this field direction. All lines of the "complex II" are situated in the central part of the figure.

The angular variation of "allowed" line positions of "complex I" in its $(Y_{\mathrm{I}}, Z_{\mathrm{I}})$ plane is shown in Fig. 4. A good agreement between experimental and calculated resonance field values has been obtained.

Angular dependences of line intensity and width have been also observed. The intensity ratio of "allowed" fine structure components has been found to be near 5:8:9:8:5 at field directions coinciding with the principal axes only. Rough evaluations showed that the relative intensities of hyperfine structure components are described well enough by Bir's formula [8].

References

[1] L. K. White, A. Szabo, P. Carkner, and N. D. Chasteen, J. phys. Chem. 81, 1420 (1977).
[2] W. A. Deer, R. A. Howie, and J. Zussmann, Rock-Forming Minerals, Vol. V, Longmans, London 1963;
Mir, Moscow 1966 (p. 337) (in Russian).
[3] ASTM Publication PD1S, Philadelphia 1967.
[4] W. D. Posener, J. Magnetic Resonance 9, 128 (1973).
[5] A. Abragam and B. Bleaney, Electron Paramagnetic Resonance of Transition Ions, Clarendon Press, Oxford 1970.
[6] T. Takeda, J. Phys. Soc. Japan 23, 1314 (1967).
[7] W. C. Tennat, J. Magnetic Resonance 14, 152 (1974).
[8] G. L. Bir, Fiz. tverd. Tela 5, 2236 (1963).

(Received May 10, 1979)

phys. stat. sol. (a) 54, 761 (1979)

Subject classification: 14.3.3 and 14.3.4; 22.1.2; 22.6

Institute of Electron Technology, Warsaw

Potential and Carrier Concentration Distributions in MOS Transistors at Punch-Through

By

A. K. OWCZAREK

A two-dimensional (2D) model of an MOS transistor based on a complete set of steady-state transport equations is described. The model proposed enables a detailed analysis of punch-through in short-channel MOS transistors. By means of numerical solution of this model, the potential and carrier concentration distributions as well as current density distributions in the gate region are determined. It is shown that punch-through in MOS transistors does not take place at the silicon surface but in the bulk and the current densities along individual layers parallel to the $Si-SiO_2$ interface are not constant. This nature of punch-through is caused by the "surface accumulation wedge" which exist between the drain and source depletion region under the gate. The influence of the design parameters and supply voltages on the location and dimensions of the "accumulation wedge" is demonstrated.

Разработана двухмерная модель МОП-транзистора, которая дает возможность детального анализа явления сквозного пробоя. На основе решения этой модели определено распределения потенциала и концентрации свободных носителей заряда, а также распределения плотности тока в области под затвором. Доказано, что явление сквозного пробоя в МОП-транзисторах развивается не на поверхности полупроводника а в глубине прибора и что плотность тока в отдельных слоях параллельных граничной поверхности $Si-SiO_2$ непостоянна. Эта особенность сквозного пробоя является следствием поверхностного аккумуляционного клина образующегося под затвором между обедненными областями истока и стока. Показано как размеры и положение аккумуляционного клина зависят от параметров конструкции МОП-транзистора и от величины питающих напряжений.

1. Introduction

The most important problem in designing and manufacturing MOS IC's with smaller and smaller channel lengths is caused by the fact that the widening drain depletion region affects the field distribution under the gate causing changes in transistor characteristics. Among these short-channel effects special consideration has to be paid to the punch-through phenomenon.

In spite of the unquestionable progress in design and technology of MOS IC, the understanding of the physical mechanisms governing punch-through is not yet satisfactory. This is partly due to the fact that oversimplified models have been commonly used up to now. Conventional one-dimensional (1D) models take no account of a number of essential design features of the MOS transitor, and neglect the diffusion component of the drain current density, giving thereby no adequate agreement with experimental data [1 to 6]. Also, 2D electrostatic models [7, 8], neglecting the transport of carriers in a semiconductor, do not allow a proper simulation and interpretation of the punch-through phenomenon.

In this paper results are presented of an accurate 2D analysis of punch-through in MOS transistors operating below threshold. The model employed is based on a complete set of transport equations, and takes into account all related parts of current density as well as many important design features of MOS transistors. This approach gives a clear insight into the physics of punch-through, allowing for a proper interpretation of the dependence of punch-through voltage on design parameters and supply voltages.

List of symbols

U_{PT}	punch-through voltage
C_{ox}	gate-oxide capacitance
$C_N = N_D - N_A$	net impurity concentration
E	electric field
J, J_n, J_p	current density, electron, hole current density
k	Boltzmann's constant
$L = L_t - 2WLx_j$	channel length
$L_D = \sqrt{\frac{\varepsilon_0 \varepsilon_{Si} U_t}{q n_i}}$	intrinsic Debye length of silicon
L_t	channel length on the mask
N_A, N_D	acceptor, donor impurity concentration
n, n_B	electron concentration, electron concentration in the bulk
n_i	intrinsic carrier concentration
p, p_B	hole concentration, hole concentration in the bulk
R	generation-recombination rate
SY	source region dimension (AD according to Fig. 1)
T	absolute temperature
t_{ox}	gate-oxide thickness
$U_t = \frac{kT}{q}$	thermal potential
$U_{FB} = \varphi_{ms} - \frac{Q_{sr}}{C_{ox}}$	flat-band voltage
U_{GS}, U_{DS}, U_{BS}	gate-source, drain-source, substrate-source voltage
$U_{GS}^* = U_{GS} - U_{FB}$	effective gate-source voltage
$WL = \frac{x_l}{x_j}$	lateral diffusion factor
q	elementary charge
Q_{sr}	equivalent surface charge density
τ_{no}, τ_{po}	electron, hole lifetime (SRH recombination model parameters)
x_j	drain and source junction depth
x_l	lateral diffusion
ψ, φ_B	electrostatic, built-in potential
φ_n, φ_p	electron, hole quasi-Fermi potential
$\zeta_n = e^{(-\varphi_n)}, \zeta_p = e^{\varphi_p}$	auxiliary variables
φ_{ms}	metal-semiconductor work-function difference
$\varepsilon_0, \varepsilon_{ox}, \varepsilon_{Si}$	vacuum, oxide, silicon dielectric permittivity
μ_n, μ_p	electron, hole mobility
ϱ	substrate resistivity

2. MOS Transistor Model and Its Solution

Fig. 1 shows a cross-section of the MOS transistor under consideration.

In order to formulate the model, the following assumptions are made:

the temperature in the transistor is constant,

the impurities are fully ionized,

the semiconductor is not intrinsic and non-degenerate so that Boltzmann statistics applies,

the electric field is not too large so that mobility does not depend on the field, and Einstein's relation $D = \mu U_t$ holds,

avalanche multiplication of carriers does not occur,

the oxide is perfect; it has an infinitely large resistivity and does not contain uncompensated charges[1]),

Table 1

Normalizing constants for variables appearing in the transport equations

discription	symbol of normalized quantity	normalizing constant		
		symbol or expression	value ($T = 300$ K)	unit
coordinates	x, y	L_D	3.59454×10^{-3}	cm
electrostatic potential, quasi-Fermi potentials, voltage	$\psi, \varphi_n, \varphi_p, U$	U_t	0.025875	V
electric field	E	U_t/L_D	7.19842	V/cm
carrier concentrations, impurity concentrations	n, p, N_A, N_D, C_N	n_i	1.32817×10^{10}	cm^{-3}
carrier mobilities	μ_n, μ_p	D_0/U_t $D_0 = 1\ cm^2/s$	38.6473	cm^2/Vs
carrier lifetimes	τ_{no}, τ_{po}	L_D^2/D_0	1.29207×10^{-5}	s
current densities	J, J_n, J_p	qD_0n_i/L_D	5.91955×10^{-7}	A/cm^2

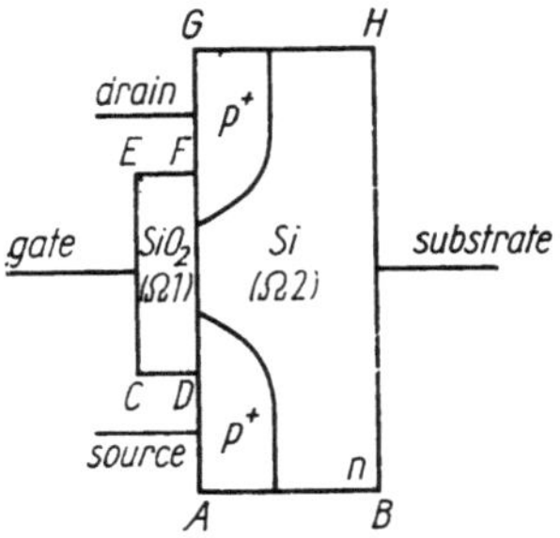

Fig. 1. Cross-section of the MOS transistor within indicated Ω1 ∨ Ω2 regions where the solution was searched for

1) It is assumed that Q_{sr} and φ_{ms} are independent of the surface potential. These parameters are included in the effective gate-source voltage U_{GS}^* (cf. list of symbols).

there are no fast surface states at the Si–SiO_2 interface,
the depths of source and drain regions are much smaller than their respective widths ($AD \gg x_j$ and $FG \gg x_j$ — see Fig. 1).

Under such assumptions a set of transport equations which describes potential and carrier concentration distributions as well as current density distributions in the structure under consideration at steady-state conditions may be written in the normalized form as (the normalizing constants are given in Table 1).

In the region Ω1 (according to the notations given in Fig. 1)

$$\nabla^2 \psi = 0 . \tag{1}$$

In the region Ω2

$$\nabla^2 \psi = \zeta_n e^{\psi} - \zeta_p e^{(-\psi)} - C_N , \tag{2}$$

$$\boldsymbol{J}_n = \mu_n(\nabla \zeta_n e^{\psi} - \zeta_n e^{\psi} \nabla \psi) , \tag{3}$$

$$\boldsymbol{J}_p = -\mu_p(\nabla \zeta_p e^{(-\psi)} + \zeta_p e^{(-\psi)} \nabla \psi) , \tag{4}$$

$$\nabla(\boldsymbol{J}_n) = R , \tag{5}$$

$$\nabla(\boldsymbol{J}_p) = -R , \tag{6}$$

$$\boldsymbol{J} = \boldsymbol{J}_n + \boldsymbol{J}_p . \tag{7}$$

The following boundary conditions were assumed for ψ, n, p, and $\boldsymbol{J}$:
on CE: Dirichlet condition for ψ;
on CD: linear interpolation of ψ between the gate and source or drain potential;
on AB and GH: Neumann conditions for ψ, n, and p:

$$\frac{\partial \psi}{\partial y} = \frac{\partial n}{\partial y} = \frac{\partial p}{\partial y} = 0 ; \tag{8}$$

on DF: normal component of current density vanishes:

$$\frac{\partial J_n}{\partial x} + \frac{\partial J_p}{\partial x} = 0 , \tag{9}$$

which is equivalent to the Newton condition for n and p

$$\mu_n \left(\frac{\partial n}{\partial x} - n \frac{\partial \psi}{\partial x}\right) - \mu_p \left(\frac{\partial p}{\partial x} + p \frac{\partial \psi}{\partial x}\right) = 0 \tag{10}$$

as well as

$$\nabla D = \varrho \tag{11}$$

for ψ, which for the applied normalization may be written as

$$\frac{\varepsilon_{ox}}{\varepsilon_{Si}} \frac{\partial \psi}{\partial x}\bigg|_{SiO_2} - \frac{\partial \psi}{\partial x}\bigg|_{Si} = p - n + C_N ; \tag{12}$$

on AD, FG, and BH: assuming an infinitely large carrier recombination rate (thermal equilibrium)

$$np = 1 , \tag{13}$$

and zero charge density

$$\varrho = 0 \tag{14}$$

the following form of the boundary conditions are obtained:

$$\psi = -\operatorname{arcsinh}\left(\frac{C_N}{2}\right) + U, \tag{15}$$

$$n = \frac{C_N}{2} + \sqrt{\left(\frac{C_N}{2}\right)^2 + 1}, \tag{16}$$

$$p = -\frac{C_N}{2} + \sqrt{\left(-\frac{C_N}{2}\right)^2 + 1}. \tag{17}$$

The impurity concentration distributions C_N in the source and drain regions are approximated by Gaussian distributions shifted relative to the edge of the diffusion mask by an appropriate value of the difference between vertical and lateral diffusions. This type of approximation, used among others in [9], reasonably approximates the exact solution of the 2D Fick's equation given in [10].

Although no essential limitations were made with respect to the form of R, μ_n, and μ_p, the SRH recombination model [11] and conventional approximations concerning the dependence of mobilities on total impurity concentration [12] were assumed for the calculations.

The formulated model was solved by the finite-difference method. The set of differential equations (1) to (7) was approximated by a set of difference equations describing the values of ψ, ζ_n, and ζ_p at discrete points (mesh points). In equations (1), (2), the standard approximations of derivatives by second-order central differences were employed, whereas in equations (3) to (7) the Scharfetter-Gummel approximation was used [13, 14]. A non-standard discrete mesh was used in this work; most dense at the Si–SiO_2 interface in the gate region and spread out in the regions of less importance for the problem considered (Fig. 2).

The solution of this model turned out to be not easy. First of all the difficulties resulted from the form of the boundary conditions at the Si–SiO_2 interface and to a lesser degree from a wide range of supply voltage variation resulting in a wide range of current densities obtained, as well as from the large number of variables (typical 3×1500). The approximate problem was solved by the block overrelaxation iterative method (LSOR). The computational program, written on the base of this method, calculated the values of ψ, n, and p at mesh-points as well as the punch-through current density at given values of the design parameters, among others: L, N_D, t_{ox}, x_j, U_{FB}, and of supply voltages: U^*_{GS}, U_{BS}, and U_{DS}. The form of the difference equations and a detaile ddescription of the algorithm (with the computer listing of procedures) was given in [15].

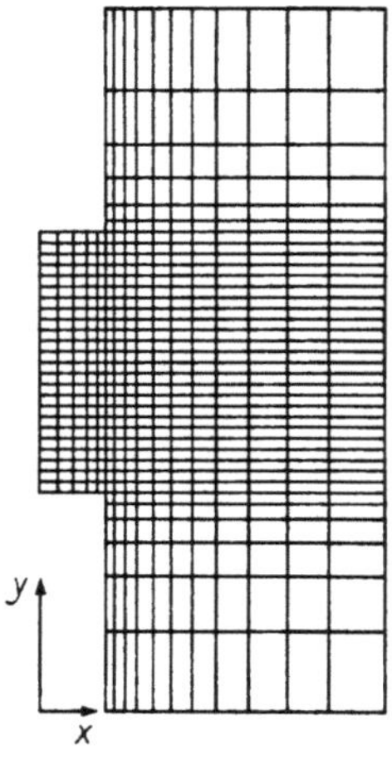

Fig. 2. Discrete mesh of the MOS transistor under consideration. Meshsizes in the left and right from the Si–SiO_2 interface as well as below and above of the gate edges increase in geometrical progression

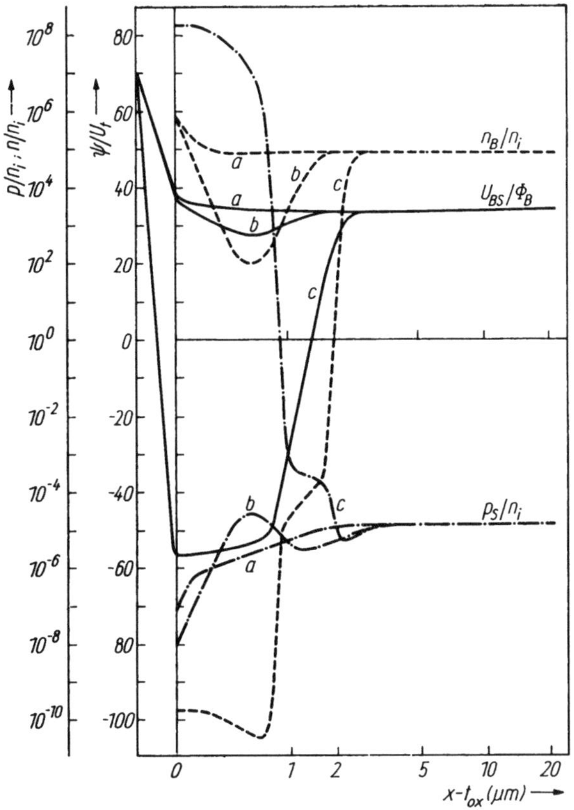

Fig. 3

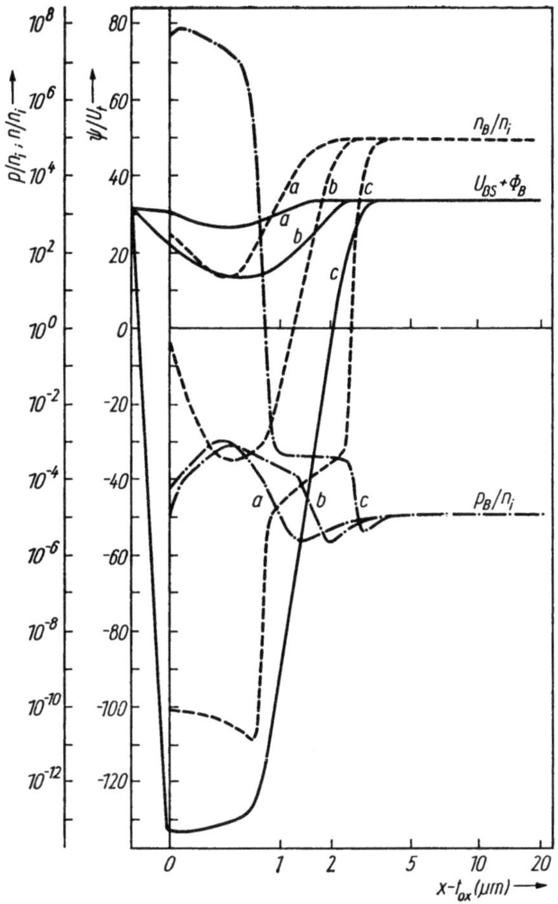

Fig. 4

Fig. 3. Potential and electron and hole concentration distributions in the gate region in the x-direction before punch-through; (a) along the axis crossing the centre of the a.w. ($y - SY = 1.5$ μm), (b) about 0.75 μm from the a.w. in the drain direction, (c) about 0.25 μm from the drain region contact. $N_D = 10^{15}$ cm^{-3} ($\varrho = 5\ \Omega$ cm), $t_{ox} = 1250$ Å, $L_t = 4$ μm, $x_j = 1.1$ μm, $L = 2.24$ μm ($WL = 0.8$), $U_{DS} = -1.5$ V, $U^*_{GS} = 1.8$ V, $U_{BS} = 0$ V. ——— ψ, ·—·—· p, — — — n. (Read $U_{BS} + \varphi_B$ and p_B/n_i instead of U_{BS}/Φ_B and p_S/n_i, respectively)

Fig. 4. Potential and electron and hole concentration distributions in the gate region in the x-direction at punch-through; (a) $y - SY = 1.5$ μm, (b) $y - SY = 2$ μm, (c) $y - SY = 3.75$ μm. $N_D = 10^{15}$ cm^{-3} ($\varrho = 5\ \Omega$ cm), $t_{ox} = 1250$ Å, $L_t = 4$ μm, $x_j = 1.1$ μm, $L = 2.24$ μm ($WL = 0.8$), $U_{DS} = -3.5$ V, $U^*_{GS} = 0.8$ V, $U_{BS} = 0$ V. ——— ψ, ·—·—· p, — — — n. (Read φ_B instead of Φ_B)

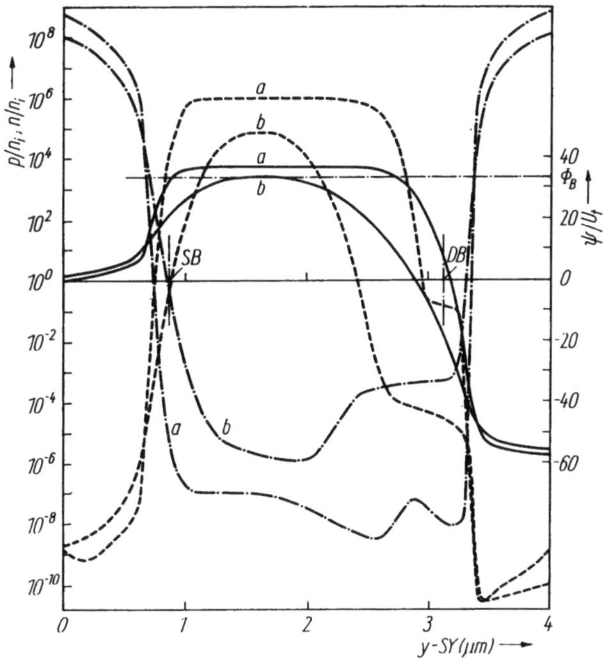

Fig. 5

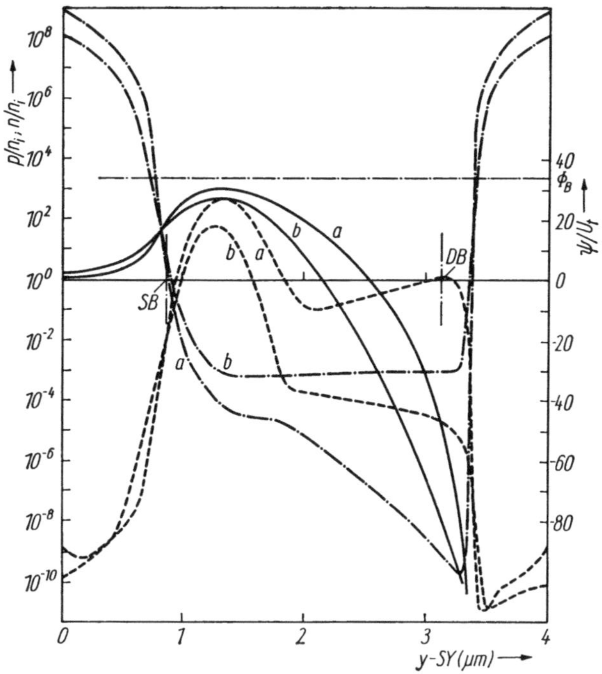

Fig. 6

Fig. 5. Potential and electron and hole concentration distributions in the gate region in the y-direction before punch-through; (a) at the surface, (b) about 0.4 μm below the surface. Symbols SB and DB are used to indicate the locations of the source and drain edge regions at the surface. $N_D = 10^{15}$ cm^{-3} ($\varrho = 5\ \Omega$ cm), $t_{ox} = 1250$ Å, $L_t = 4$ μm, $x_j = 1.1$ μm, $L = 2.24$ μm ($WL = 0.8$), $U_{DS} = -1.5$ V, $U_{GS}^* = 1.8$ V, $U_{BS} = 0$ V. ——— ψ, ·—·—· p, — — — n. (Read φ_B instead of Φ_B)

Fig. 6. Potential and electron and hole concentration distributions in the gate region in the y-direction at punch-through; (a) at the surface, (b) about 0.4 μm below the surface. Symbols SB and DB are used to indicate the locations of the source and drain edge regions at the surface. $N_D = 10^{15}$ cm^{-3} ($\varrho = 5\ \Omega$ cm), $t_{ox} = 1250$ Å, $L_t = 4$ μm, $x_j = 1.1$ μm, $L = 2.24$ μm ($WL = 0.8$), $U_{DS} = -3.5$ V, $U_{GS}^* = 0.8$ V, $U_{BS} = 0$ V. ——— ψ, ·—·—· p, — — — n. (Read φ_B instead Φ_B)

The program was run on the Odra 1305 computer (compatible with the ICL 1900 computer). The results, presented below, refer to PMOS transistors, but correlations and conclusions obtained here are valid in case of NMOS transistors as well.

3. Analysis of Results

The calculated distributions of ψ, n, and p in MOS transistors before and at punch-through indicate the 2D nature of this phenomenon which results mainly from the gate action. The gate field penetrates into the lightly doped substrate as well as into the heavily doped source and drain regions causing drastic changes of ψ, n, and p at the Si–SiO_2 interface (Fig. 3 to 6).

For not too high drain biases, an accumulation layer of electrons develops in the central part of the gate region and the potential rises above the $(U_{BS} + \varphi_B)$ value[2]) (Fig. 3, 5). Because of the specific shape of the surface accumulation region it will be called here "accumulation wedge" (a.w.). The a.w. repels the source and drain depletion regions from the surface into the bulk making the occurrence of punch-through difficult so that it develops at a certain distance from the Si–SiO_2 interface (Fig. 7, 8). The distributions of ψ, n, and p inside the a.w. in the x-direction are fairly irregular and are subject to appreciable changes even with small changes of U_{DS}, U^*_{GS}, and U_{BS}. Before the onset of punch-through, in the central section of a.w. "pure" accumulation of electrons occurs (Fig. 3, curve a) whereas in the lateral sections field-induced junctions result (Fig. 3, curve b).

The potential and electron concentration distributions throughout the field-induced junctions demonstrate the existence of characteristic wells which correspond to the maximum of hole concentration (Fig. 3, curve b). With increasing drain bias, the drain depletion region penetrates more and more in the gate region causing further lowering of the potential and electron concentration wells (Fig. 4, curves b).

At the onset of punch-through ($U_{DS} = U_{PT}$), the entire semiconductor surface in the gate region (including the central section of the a.w.) is covered by a field-

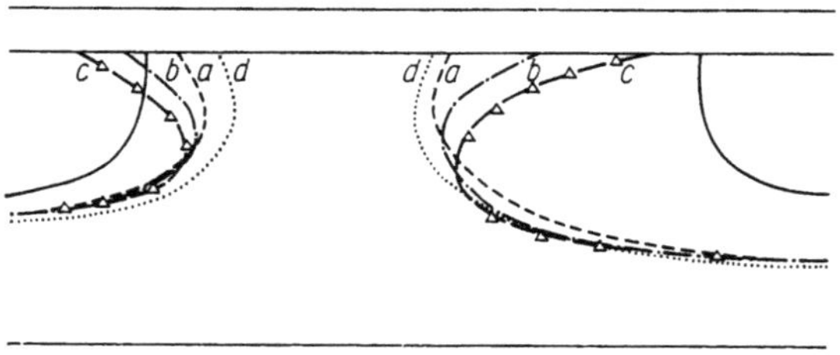

Fig. 7

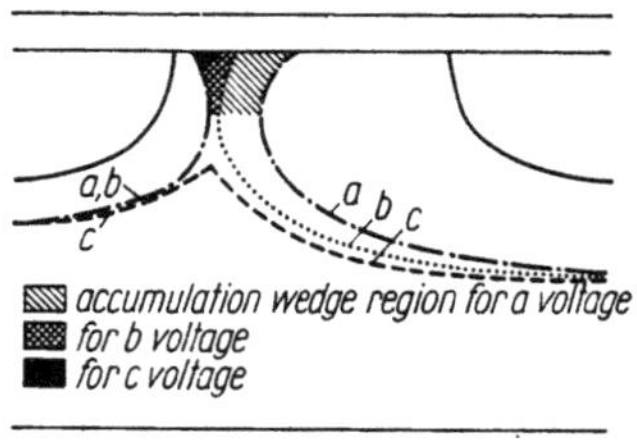

Fig. 8

Fig. 7. Shapes of the source and drain depletion regions at the different gate and substrate biases; (a) $U^*_{GS} = 0.5$ V, $U_{BS} = 0$, (b) $U^*_{GS} = 1.6$ V, $U_{BS} = 0$, (c) $U^*_{GS} = 2.8$ V, $U_{BS} = 0$, (d) $U^*_{GS} = 0.5$ V, $U_{BS} = 0.5$ V; $N_D = 1.3 \times 10^{15}$ cm^{-3}, $t_{ox} = 1450$ Å, $L_t = 6$ µm, $x_j = 125$ Å, $L = 4$ µm, $U_{DS} = -4$ V

Fig. 8. Changes of the location and dimensions of the a.w. with increasing drain bias; (a) $U_{DS} = -2$ V, (b) $U_{DS} = -2.5$ V, (c) $U_{DS} = -3$ V (a.w. disappears). $N_D = 1.3 \times 10^{15}$ cm^{-3}, $t_{ox} = 1250$ Å, $L_t = 4$ µm, $x_j = 1.1$ µm, $L = 2.24$ µm, $U^*_G = 0.8$ V, $U_{BS} = 0$ V.

[2]) It is assumed that the source contact is the reference electrode ($U_{SS} = 0$), thus the potential at the substrate contact is $(U_{BS} + \varphi_B)$.

induced junction. For sufficiently high drain bias, however, the a.w. disappears as shown in Fig. 4.

It is interesting to note that potential and carrier concentration distributions in the y-direction at the surface are different from the ones observed in the bulk (compare curves a and b in Fig. 5 and 6). At the surface, the hole concentrations in source and drain depletion regions are small and comparable to each other before the onset of punch-through (Fig. 5, curve a). With increasing drain bias, the hole concentration in the vicinity of the source increases while it decreases in the vicinity of the drain. Such a trend is maintained at the early stage of punch-through (Fig. 6, curve a). In punch-through conditions there is a potential maximum in the vicinity of the source region which corresponds to the maximum of electron concentration (Fig. 6, curves a and b). This potential barrier controls the rate of charge carrier injection and the magnitude of the current density. The rate of the potential barrier lowering with increasing drain bias after the onset of punch-through is not uniform.

At first the potential barrier decreases rapidly with increasing drain bias while for higher $|U_{DS} - U_{PT}|$ (corresponding to a higher punch-through current density), this effect is less pronounced. Under high punch-through current densities, corresponding to the decrease of the potential barrier in excess of $10U_t$, the hole concentration in the whole depletion region increases. This effect is most pronounced at the edge of the drain region.

Below the Si–SiO_2 interface, the hole concentration distributions have more regular shapes than at the surface, and the concentration of holes in the vicinity of the source is smaller than in the vicinity of the drain (Fig. 5, curve b). Because of the "weaker" action of the gate field and an "earlier" occurrence of punch-through, the hole concentrations in these regions are grater than at the surface.

The location and dimensions of the a.w. depend on the values of U_{DS}, U^*_{GS}, and U_{BS} as well as on the values of MOS transistor design parameters, in particular t_{ox}, N_D, and L. It should be noticed that the width of the a.w. at punch-through specifies the width of the punch-through current-path, and, as a consequence, the current density distributions in the x-direction (Fig. 9). With increasing gate bias the a.w. widens and the current-path moves away from the surface. Under sufficiently high gate bias, the surface charge of accumulated electrons is so large that it shields deeper lying regions from the gate action. In consequence the gate no longer controls the punch-through current and practically does not affect the value of U_{PT}.

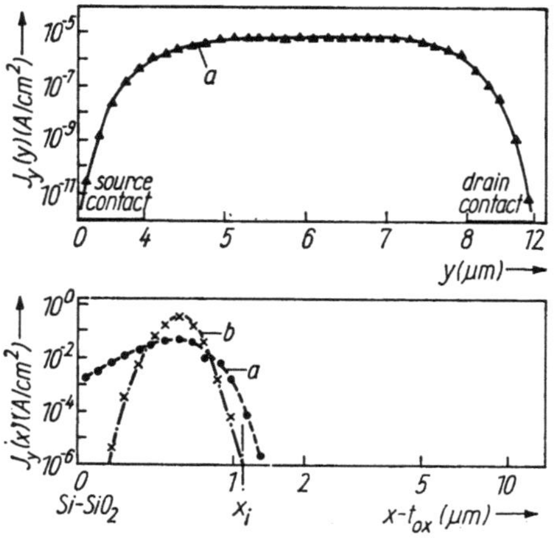

Fig. 9. Typical distributions of the y-component punch-through current density; (a) total current density, (b) current density distribution along the axis crossing the a.w., (c) current density distribution along the axis situated beyond the a.w. in the drain direction. $N_D = 10^{15}$ cm^{-3}, $L_t = 4$ μm, $L = 2.24$ μm, $U_{DS} = -2$ V, $U^*_{GS} = 0.8$ V, $U_{BS} = 0$ V. (Read curve c instead of curve a in the lower figure)

Reverse biasing of the substrate reduces accumulation at the surface, weakening the a.w. shielding action and restoring the gate control of the potential distributions in the gate region. Simultaneously the source and drain depletion regions widen and they join each other at lower drain bias (cf. Fig. 7). Nevertheless, the high potential barrier resulting from the substrate bias makes injection of holes in the drain depletion region impossible and punch-through does not occur. The drain bias necessary to reduce this potential barrier below the built-in potential to put the source junction in the conduction condition is higher than $|U_{PT}|$ at zero substrate bias.

Under increasing drain bias, the dimensions of the a.w. diminish and it moves closer to the source. Simultaneously the current-path widens. Under sufficiently high drain bias, the a.w. disappears and the current-path widens to such extent that it envelops the whole surface regions (cf. Fig. 8).

At lower substrate impurity concentration N_D and thinner oxide t_{ox} the gate field penetrates the gate region more easily and consequently the a.w. width is greater and the current-path lies deeper in the bulk.

The 2D nature of ψ, n, and p is responsible for the 2D nature of the current density distibutions at punch-through. Though the total current flowing in the central part of the gate region in y-direction is constant, the current densities along individual mesh-lines are not constant, because depth and width of the current-path along the gate are not constant (Fig. 9).

Under the centre of the a.w., the current-path width is the smallest and the local current density is higher than in the vicinity of source or drain regions (compare curve b and c in Fig. 9). This feature of the current density distribution practically excludes the possibility of assuming 1D nature of punch-through current density. We should notice, however, that this assumption was successfully used by some authors investigating the operation of MOS transistor above threshold [14, 16]. The non-constant values of the current densities along mesh-lines, obtained in this work, result probably from the fact that fairly realistic Newton boundary conditions have been used to approximate of the gate action in MOS transistors. In the literature there are no papers which might confirm this supposition. It is a well-known fact, however [17], that for bipolar transistors in which the boundary conditions at the Si–SiO_2 interface are approximated by the less realistic Neumann conditions, the relevant components of current densities are constant.

4. Conclusions

The punch-through phenomenon in MOS transistors consists in widening the drain depletion region to reach the source depletion region, and injection of charge carriers over the potential barrier of the source junction takes place directly into the drain depletion region. The potential barrier controls the rate of charge carrier injection and the magnitude of the current density. The necessary and sufficient condition for the occurrence of punch-through is the lowering of the potential barrier below the built-in potential φ_B resulting from the drain field action.

Joining of both depletion regions is not sufficient for the occurrence of punch-through.

Punch-through in a MOS transistor (at least in its initial stage) is not a surface phenomenon but develops in the bulk. The distance of the current-path from the surface is given by the a.w. width and increases for higher substrate resistivity, thinner oxide, and higher gate bias. The increase of gate bias enlarges the dimensions of a.w. which repels the source and drain depletion regions into the bulk and makes the development of punch-through more difficult. This action intensifies for higher substrate resistivity and thinner oxide and indicates a saturation effect especially pronounced for zero substrate bias.

Reverse biasing of the substrate increases the potential barrier and makes the development of punch-through more difficult. The higher the substrate resistivity, the more intensive is this action.

Acknowledgement

The author wishes to thank Dr. H. Przewłocki for stimulating discussions and his valuable comments in the course of this work.

References

[1] W. Shockley and R. C. Prim, Phys. Rev. **90**, 753 (1953).
[2] G. T. Wright, Solid State Electronics **15**, 381 (1972).
[3] S. Denda and M. A. Nicolet, J. appl. Phys. **37**, 2412 (1966).
[4] J. L. Chu, G. Persky, and S. M. Sze, J. appl. Phys. **43**, 3510 (1972).
[5] J. Shappir, Philips Report No. RNR-21-E/1970 (p. 23).
[6] R. R. Troutman, IBM Techn. Report TR19.0224, 1972 (p. 22).
[7] J. M. Bateman, G. A. Armstrong, and J. A. Magowan, Solid State Electronics **17**, 539 (1974).
[8] S. Muramoto, N. Miyahara, and Y. Sunohara, Electronics Commun. Japan **58**, 64 (1975).
[9] T. Yanawa, IEEE Trans. Electron Devices **19**, 1166 (1972).
[10] D. P. Kennedy and R. R. O'Brien, IBM J. Res. Developm. **9**, 179 (1975).
[11] W. Shockley and W. T. Read, Phys. Rev. **87**, 835 (1952).
[12] D. M. Caughey and R. T. Thomas, Proc. IEEE Letters **55**, 2192 (1967).
[13] D. L. Scharfetter and H. K. Gummel, IEEE Trans. Electron Devices **16**, 64 (1969).
[14] M. S. Mock, Solid State Electronics **16**, 601 (1973).
[15] A. K. Owczarek, Ph. D. Thesis, Institute of Electron Technology, Warsaw 1978 (p. 128).
[16] D. Vandorpe, J. Borel, G. Merckel, and P. Saintot, Solid State Electronics **15**, 547 (1972).
[17] J. W. Slotboom, IEEE Trans. Electron Devices **20**, 699 (1973).

(Received May 30, 1979)

phys. stat. sol. (a) **54**, 773 (1979)

Subject classification: 13.4 and 14.3.4; 22.2.3

Tblissi State University

Study of Electrical Breakdown Peculiarities in High-Resistivity Compensated p-InSb

By

Z. N. Chigogidze and N. P. Khuchua

The phenomenon of impact ionization at $T = 77$ K and 4.2 K in high-resistance compensated p-InSb obtained by heat treatment of n-InSb followed by a slow cooling is investigated. A deep impurity level characterized by an activation energy of $\Delta\varepsilon_0 \approxeq 0.12$ eV is inherent to the material. It is shown that at $T = 77$ K in the samples with the carrier concentration $p_0 \lessapprox 5 \times 10^{12}$ cm^{-3} the interband breakdown occurs, while at $p_0 > 5 \times 10^{12}$ cm^{-3} the interband breakdown is preceded by the impurity impact ionization. An interesting abnormal effect is found when applying a magnetic field H_α to the sample at an angle of 10° to 15° with respect to the direction of the electrical field. Some considerations are presented concerning the nature of the peculiarities observed in the electric breakdown.

Исследовано явление ударной ионизации при $T = 77$ и 4,2 K в высокоомном компенсированном p-InSb, полученном путем термообработки n-InSb с последующим медленным охлаждением. Образцы содержали глубокий примесный уровень с энергией активации $\Delta\varepsilon_0 \approxeq 0{,}12$ eV. Показано, что при $T = 77$ K в образцах с концентрацией носителей $p_0 \lessapprox 5 \times 10^{12}$ cm^{-3} возникает межзонный пробой, тогда как в случае $p > 5 \times 10^{12}$ cm^{-3} межзонному пробою предшествует ударная ионизация примесей. Интересный аномальный эффект обнаружен при приложении к образцу магнитного поля H_α, ориентированного под углом 10° до 15° относительно направления электрического тока. В работе приводятся некоторые соображения о природе наблюдаемых особенностей электрического пробоя.

1. Introduction

Electrical breakdown in InSb of both n- and p-type at $T = 77$ K has been extensively studied experimentally and theoretically. This phenomenon is associated with electron–hole pairs appearing as a result of impact ionization from the valence band to the conductivity band.

Impact ionization of impurities in InSb has been studied much less [1 to 3].

High-resistivity compensated material is of special interest for investigation: at high-impurity compensation the influence of deep impurity levels on the electrophysical properties of InSb [4, 5] and on the impact ionization therein [2, 3] can be revealed; the breakdown phenomenon can find practical application in producing high-speed switching devices based on a high-resistivity source material with high carrier mobility [6, 7].

In the present work some peculiarities of the breakdown in high-resistivity p-InSb are investigated at $T = 77$ and 4.2 K, in which compensation of a residual donor impurity by acceptor centres was achieved through heat treatment of n-InSb followed by slow cooling [6]. This method makes it possible to retain high values of carrier mobility and yields reproducible results. The concentration and carrier mobility of the compensated p-InSb thus obtained were $p_0 \approx 0.2$ to 2×10^{13} cm^{-3} and $\mu \approx 3$ to 7×10^3 cm^2/Vs, respectively.

The activation energy of impurities, $\Delta\varepsilon_0$, calculated from the temperature dependence of the Hall coefficient in the impurity region is ≈ 0.12 eV, which points to the presence of an impurity level in this material lying close to the mid-gap.

Note that this deep level in InSb depends neither on the way of producing the crystal, nor on the dopant nature, nor on the compensation method (see, for example, [2, 6, 8, 9] and other papers) and is considered to be due to structural lattice defects.

2. Experimental Procedure

The breakdown phenomenon was studied on high-resistivity samples of p-InSb in the pulse mode to avoid initiation of secondary thermal effects at high levels of power supply.

The measuring apparatus allowed us to take both *j–E* characteristics and time parameters of the breakdown.

To make measurements at $T = 77$ and 4.2 K the samples were placed into a special allmetal cryostat.

A rectangular pulse generator of maximum voltage up to 200 V at load resistance of 10 Ω served as a power supply. When measuring the breakdown parameters pulses of duration $\tau \gtrapprox 10$ μs and repetition frequency $\gtrapprox 10^3$ Hz were used.

The current through the sample was found from the voltage drop across the reference resistance connected in series with the sample. The current and voltage pulses were registered by means of a two-channel cathode-ray oscillograph.

Pre- and post-breakdown crystal resistances (R_0 and R_b, respectively) were measured by a substitutional method with the aid of a noninductive resistance box with time constant 1.5×10^{-7} s.

The following time characteristics of the breakdown were defined:

(i) delay time of the breakdown, τ_d, the time required for the free carrier concentration to rise up to a critical value necessary for breakdown initiation at $E \geqq E_c$ (critical field);

(ii) time of avalanche development, τ_a, the time in which the sample conductivity increases rapidly;

(iii) deionization time, τ_l, the time in which the steady-state carrier concentration decreases approximately by 95% of its initial value when the biasing voltage is removed.

The values of τ_d and τ_a were found from the oscillogram of the breakdown pattern (Fig. 1a) with time marks applied. The values of τ_a and τ_l were found by other similar methods.

As the avalanche develops, the voltage across the sample goes down according to the exponential law (Fig. 1a)

$$U = U_n + (U_g - U_n) \exp\left(-\frac{t}{\tau_a}\right). \tag{1}$$

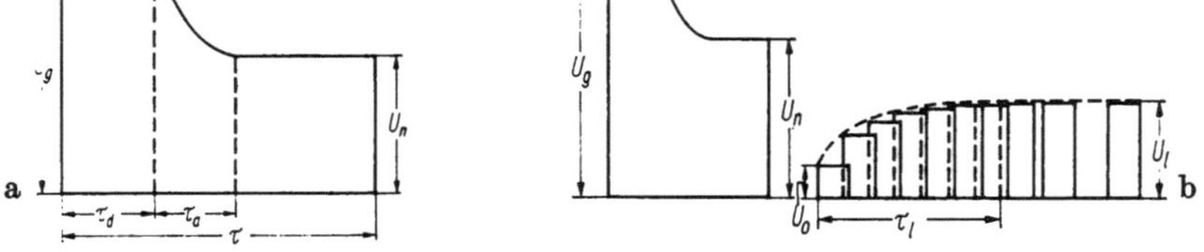

Fig. 1. Estimation of time characteristics of electrical breakdown. a) τ_d and τ_a, b) τ_l

From this

$$\ln\left(\frac{U_g - U_n}{U - U_n}\right) = \frac{1}{\tau_a} t \tag{2}$$

which is a straight line, whose reverse slope is

$$\frac{1}{\mathrm{d}\ln\left(\frac{U_g - U_n}{U - U_n}\right)\Big/\mathrm{d}t} = \tau_a\,. \tag{3}$$

To find τ_l, two pulses shifted in respect to each other were used, one of them being a bias pulse, while the other being a measuring or "probing" pulse (Fig. 1b). The latter following the first pulse finds the deionization process unfinished. The amplitude of this pulse at each moment of time will depend on the sample conductivity and will recover following the exponential law depending on the delay time of the "probing" pulse in respect to the bias pulse. Therefore, the dependence of the "probing" pulse amplitude on time can be written as

$$U = U_0 + (U_l - U_0)\left[1 - \exp\frac{t}{\tau_l}\right]. \tag{4}$$

Thus, τ_l will be defined from a graphical dependence similar to (2) from the formula

$$\frac{1}{\mathrm{d}\ln\left(\frac{U_l - U}{U_l - U_0}\right)\Big/\mathrm{d}t} = \tau_l\,. \tag{5}$$

When measuring the critical breakdown field, E_c, the dependence of this value on the bias pulse duration τ was taken into account, which in its turn is due to the field dependence of τ_d. Consequently, to find E_c for all the samples the dependences $E_c = f(\tau)$ were taken in the interval $\tau = 4$ to 200 μs and E_c was determined by extrapolation of the curves to the value $\tau = \infty$.

The experimental samples cut from a p-InSb single crystal ingot had the form of parallelepipeds 0.06 to 0.16 cm thick and 0.2 to 0.4 cm wide with In as contact material.

3. Results and Discussion

Typical results of the breakdown studies at $T = 77$ and 4.2 K in high-resistivity compensated p-InSb produced by a high-temperature annealing are presented in Fig. 2 to 8 and summarized in Table 1.

Fig. 2 shows the dependence of the current density j on the electric field strength E, as well as of the resistivity ϱ on the bias voltage U at $T = 77$ K for samples with various values of p_0.

At comparatively low values of electrical fields (portion I, curve 1 to 4) the dependence $j = f(E)$ is linear, while ϱ in the same field interval remains constant. Starting from a certain value of E, j increases monotonously with the increase of the field, a smooth decrease (portion II, curves 1 to 4) complying with this. Next, when a certain value $E = E_c$ is achieved, a portion of negative differential resistance occurs on the $j - E$ curve followed by an almost vertical rise of the current (portions III and IV, curve 2), a jump-like decrease of resistivity with following monotonous change of ϱ (portions III and IV, curve 4) corresponding to this.

In some cases two S-shaped portions appeared on the $j - E$ curves, i.e. following the first breakdown in the sample, after a bias voltage increase, the second breakdown

Table 1

sample No.	free carrier concentration p_0 (10^{12} cm^{-3})	carrier mobility μ_p (cm^2/Vs)	resistivity before breakdown ϱ_0 (Ω cm)	resistivity after breakdown ϱ_b (Ω cm)	critical field of breakdown E_c (V/cm)	product $E_c\mu_p$ (10^5 cm/s)	field strength E_W (V/cm)	delay time τ_b (s)	avalanche development time τ_a (s)	deionization time τ_l (s)
6*)	6.5	3500	275	1.72	18.0	0.63	60	1.80	0.3	1.0
7	1.7	5500	495	2.91	31.0	1.7	62	0.60	0.6	1.5
13	2.7	5100	450	4.8	37.2	1.9	80	0.40	0.6	1.9
15	3.0	4060	500	1.0	44.4	1.8	71	0.50	0.8	1.6
16	2.7	5520	420	3.7	38.0	2.1	80	0.40	0.8	1.8
17	2.5	5260	475	7.0	37.5	1.9	81	0.40	0.6	2.1
18	3.0	4000	404	6.0	50.2	2.0	84	0.35	0.8	2.1
19	3.8	3200	510	5.1	62.0	2.0	140	0.20	0.6	5.5
21	3.1	4500	440	8.5	44.5	2.0	90	0.40	0.8	2.5
24	5.6	3000	640	5.0	56.2	1.8	193	0.50	0.8	7.0
25*)	8.6	3100	233	1.5	20.0	0.62	70	1.75	0.4	1.2
27*)	7.8	3200	250	1.3	18.0	0.58	60	1.50	0.3	0.9

The values of the parameters ϱ_b, τ_b, τ_l, and τ_a are at E_W.

*) Samples revealing two types of the breakdown: the impurity and interband breakdown.

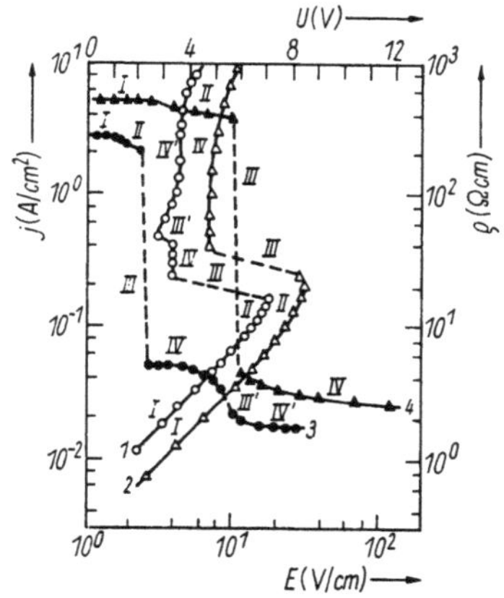

Fig. 2. Dependences of current density j on electrical field strength E (1, 2) and of resistivity ϱ on voltage U (3, 4) at $T = 77$ K for samples with different carrier concentration. (1), (3) for sample No. 6; (2), (4) for sample No. 7

took place which qualitatively repeated the pattern of the previous one (portions III′ and IV′, curve 1). The shape of the curves $\varrho = F(U)$ varied correspondingly (portions III′ and IV′, curve 3).

Typical breakdown oscillograms representing the change in the voltage across the sample when only a single breakdown is revealed during a single bias pulse, are shown in Fig. 3.

The obtained dependences $j = f(E)$ and $\varrho = F(U)$ are in good agreement with notions concerning current carrier heating and impact ionization in semiconductors in strong electric fields.

The linear portion on the curves indicates that at certain voltages Ohm's law is fulfilled. Subsequent deviation from this law which is observed up to critical fields is associated with a gradual increase of the concentration of carriers as they are heated by the electrical field, apparently at the expense of the decrease of the recombination

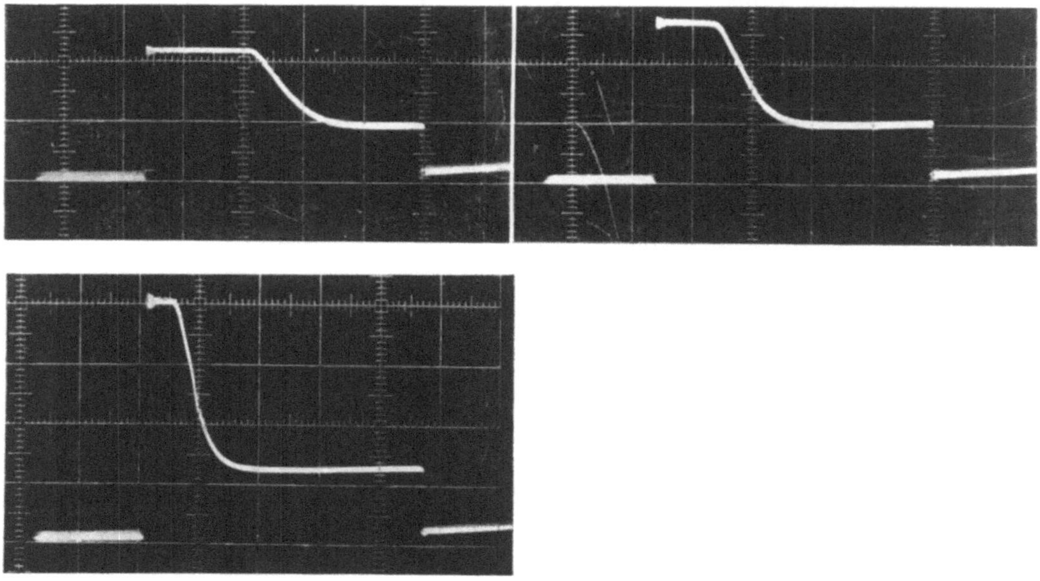

a b c

Fig. 3. Electrical breakdown oscillograms at $T = 77$ K for sample No. 7 at different biases. a) 3, b) 7.6, c) 11.3 V; pulse duration $\tau = 10$ µs

factor [10]. It is probable that there too a sublinear change of $j = f(E)$ takes place due to the carrier mobility decrease which is observed in some cases [11]. However, this change of the mobility disappears on the background of stronger effects caused by the field dependence of the free carrier concentration.

At $E = E_c$ a jump-like increase of the current density appears caused by avalanche multiplication of carriers as a result of the interband (curve 2) or impurity (curve 1) breakdown in the sample. As is known, the breakdown phenomenon is accompanied by a sharp conductivity switching, which in our case is well illustrated by curves 3 and 4 of Fig. 2 (see also Table 1).

As $E > E_c$ increases, the $j - E$ dependences continue to rise steeply. This is associated with a further increase of the free carrier concentration by the electric field and is expressed through the decrease of ϱ with increasing U.

In the case when the impurity breakdown is revealed earlier than the band-to-band breakdown occurs, portions III and IV of curves 1 to 4 (Fig. 2) correspond to the impurity level impact ionization, while portions III' and IV' correspond to the avalanche multiplication of electron–hole pairs at the expense of the band-to-band breakdown. Here one can consider that the impurity breakdown most probably takes place from the level $\Delta\varepsilon_0 \approx 0.12$ eV.

Fig. 4 shows the dependences $j = f(E)$ and $\varrho = F(U)$ for the p-InSb samples at $T = 4.2$ K.

The nature of the curves illustrated does not differ from the dependences obtained for the same samples at $T = 77$ K, however, at $T = 4.2$ K two breakdown regions following each other are revealed more distinctly. Similarly to the case of $T = 77$ K, the first region can be ascribed to the impact ionization of the impurity level, while the second one — to the interband breakdown.

The resistivity values of the samples before and after breakdown at $T = 4.2$ K are several times higher than the values of ϱ_0 and ϱ_b at $T = 77$ K, while the limits of the conductivity change are extended (see Fig. 2 and 4).

As is seen from Table 1, the values of E_c at $T = 77$ K in the case of interband breakdown are in the range $E_c \approx 30$ to 60 V/cm [1]), however, the product $E_c\mu_p \approx \approx 2 \times 10^5$ cm/s = const.

For impurity breakdown $E_c \approx 15$ to 20 V/cm, while $E_c\mu_p$, as one could expect, appears to be smaller ($\approx 0.6 \times 10^5$ cm/s).

Though taking into account all scattering processes leads to an expression different from the empirical dependence $E_c\mu_p =$ const, the latter also can serve as an argument in favour of the impact ionization mechanism, since the rate of carrier energy accumulation in the electric field is proportional to their mobility (see [2]).

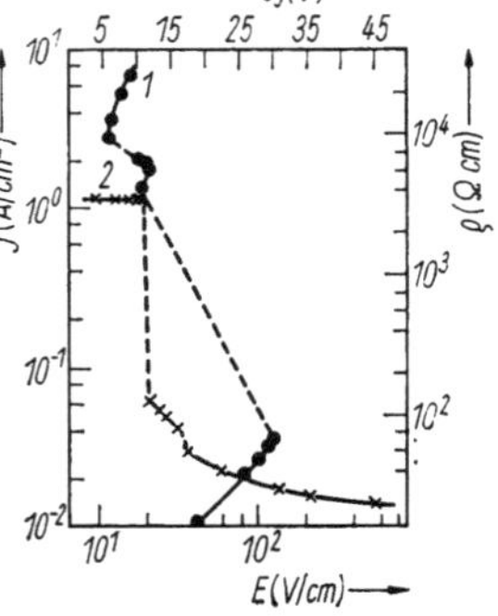

Fig. 4. Dependences of current density j on electrical field strength E (1) and of resistivity ϱ on voltage U (2) at $T = 4.2$ K for sample No. 24

1) The scatter in the values of E_c seems to be caused by various degrees of sample compensation [2].

At $T = 4.2$ K the critical fields are several times higher than E_c corresponding to $T = 77$ K. This must be caused by the temperature dependence of the carrier mobility.

The values of E_c for the interband breakdown in the case of compensated p-InSb at $T = 77$ K are much smaller than those obtained by other authors for the breakdown in p-InSb [12]. Since in our case the injection from the contacts can be neglected[2]) we are prone to explain the reduced value of E_c by a "facilitation" of the process of interband breakdown through the impact ionization of the impurity level even in the case when the impurity breakdown is not revealed on the $j - E$ characteristics [3].

Indeed, it is necessary to pay attention to the fact that the impurity breakdown preceding the interband breakdown is observed in the samples with $p_0 > 5 \times 10^{12}$ cm^{-3}, while for $p_0 \lesssim 5 \times 10^{13}$ cm^{-3} the band-to-band breakdown takes place. (At $p_0 > 2 \times 10^{13}$ cm^{-3} the breakdown is practically not revealed.)

For the impurity breakdown to appear before the interband breakdown sets in the impurity breakdown delay time τ_{di}, should be less than that of the band-to-band breakdown τ_{dl}.

If we use the estimates of [13], the values of τ_{db} for p-InSb at the breakdown point, i.e. at $E = E_c$, when $\tau \to \infty$ can be determined from

$$\frac{1}{\tau_{di}} = 2p_0\gamma_0\left(1 - \frac{N_A}{N_D}\right), \tag{6}$$

where γ_0 is the recombination factor, N_A/N_D the compensation degree. Assuming for holes $\gamma_0 \approx 10^{-6}$ cm/s [14], $N_A/N_D \approx 99\%$, we see that at $p_0 \approx 5 \times 10^{12}$ cm^{-3}, τ_{di} can become comparable with experimentally measured values of $\tau_{db} \approx 10$ μs for the case that only the interband breakdown occurs in the samples.

Thus, from these estimations one can see that with the increase of $p_0 > 5 \times 10^{12}$ cm^{-3} τ_{di} will decrease against τ_{db} and the impurity breakdown will be able to manifest itself earlier than the interband one appears. With the decrease of p_0, τ_{di} will increase against τ_{db} and only the band-to-band breakdown can be observed. However, this does not exclude the influence of the impurity breakdown on the interband breakdown at considerably close values of delay time.

Fig. 5 represents the dependence of delay time τ_{db} on the electric field strength at various load resistances. In this figure the solid lines show the calculated dependences $\tau \sim 1/E^2$ and $\tau \sim 1/E^4$.

As is seen from Fig. 5 at first as E increases over E_c, the data points lie very well on the curve $\tau \sim 1/E^4$. Then starting from a certain value of E, the law of delay time variation with field changes: τ_{db} becomes inversely proportional to E^2. In the given

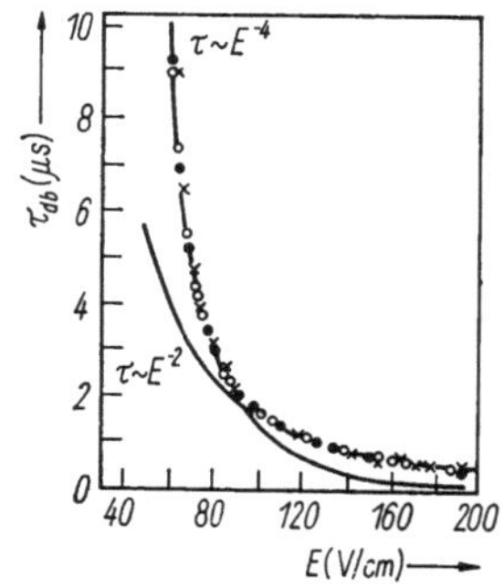

Fig. 5. Dependence of delay time τ_{db} on electrical field strength E for sample No. 24 at $T = 77$ K and various load resistances. ● 10, ○ 50, × 200 Ω

[2]) Measurements taken from the same sample with contacts of different areas and with different spacings between them show that the occurrence of the breakdown is due to E rather than to U.

range of fields the behaviour of $\tau_{db} = f(E)$ is in good agreement with the semi-empirical dependence established in [15] for the low-temperature impurity breakdown in Ge. The dependence plotted in Fig. 5, $\tau_{db} = f(E)$, is characteristic of all the samples with $p_0 \lesseqq 5 \times 10^{12}$ cm^{-3} and does not depend on the value of E_c.

The presence of two portions with different laws $\tau_{db} = f(E)$ can be, firstly, associated with the contribution of ionization impurities to the interband breakdown over a certain range of E-values, and secondly, caused by the field dependence of the recombination factor.

With electric field increasing from E_c up to E_w (see Table 1) the time of avalanche development decreases, while the deionization time increases monotonously by a half order of magnitude.

When a transverse magnetic field is applied to the sample with $p_0 \lessgtr 5 \times 10^{12}$ cm^{-3} at $T = 77$ K, an increase of magnetoresistance and of E_c was observed in accordance with published data [2]. In this case τ_{db} was practically independent of $H_\perp$ (Fig. 6). This result is one more evidence in favour of the fact that the value of τ_{db} is determined by the carrier concentration rather than by their mobility.

An interesting abnormal effect was observed for a number of samples in the case when a weak magnetic field, H^α, was applied at an angle $\alpha \approx 10°$ to $15°$ with respect to the direction of the electric current: on the j–E characteristics prior to the band-to-band breakdown the impurity breakdown appeared which did not exist at $H_\alpha = 0$, while E_c decreased monotonously with the increase of H_α (curve 1, Fig. 7). Here τ_{di} decreased approximately according to the same law as E_c did, while τ_{db} remained essentially constant (curves 2 and 3, Fig. 7, respectively).

The effect revealed is also well illustrated by the current oscillograms given in Fig. 8.

These data enable us to come to the conclusion that the occurrence of the impurity breakdown in the samples where only the interband breakdown was observed at $H_\alpha = 0$, is due to the decrease of τ_{db} with the increase of H_α.

The character of the impurity breakdown delay time is not quite understood and it requires further investigations. In accordance with (6) the reduction of τ_{db} can result from both the increase of concentration and recombination factor. It is worthwhile to note that in [5] an abnormal (with superlinear rise) dependence of the photomagnetoelectric effect signal on an arbitrarily oriented magnetic field is revealed in some strongly compensated p-InSb samples.

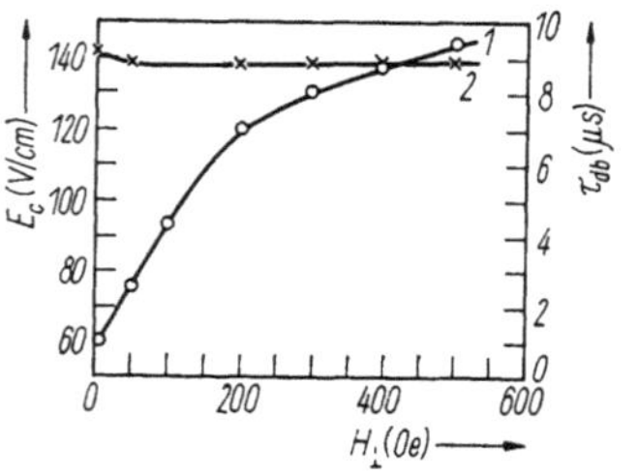

Fig. 6

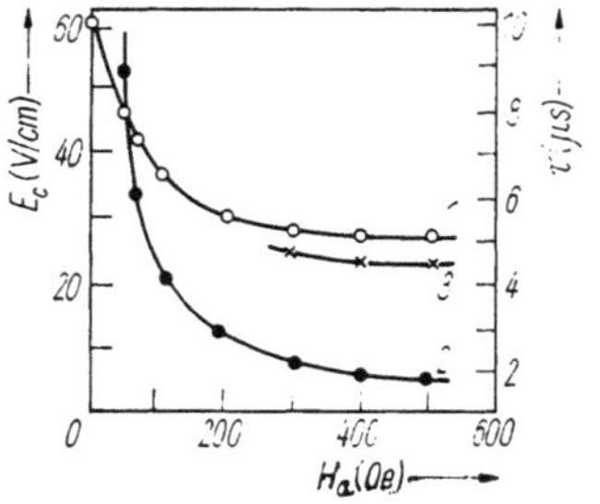

Fig. 7

Fig. 6. Dependences of critical breakdown field E_c (1) and of delay time τ_{db} (2) on transverse magnetic field $H_\perp$ for sample No. 19 at $T = 77$ K

Fig. 7. Dependences of critical brekdown field E_c (1), delay time τ_{di} (2), and delay time τ_{db} (3) on magnetic field H_α ($\alpha = 10°$) for sample No. 19 at $T = 77$ K

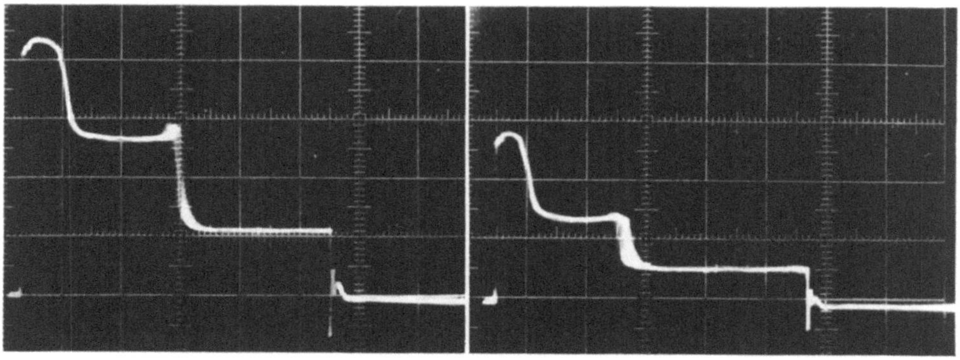

a b

Fig. 8. Oscillograms of electrical breakdown for sample No. 19 at $T = 77$ K and various values of H_α. a) 200, b) 300 Oe; bias pulse duration $\tau = 10$ μs

In conclusion note that when a magnetic field close to a longitudinal one at $E > E_c$ is applied to the samples, low-frequency current oscillations are observed (similar to those described in [16, 17].) The oscillations observed approached a sinusoidal shape, the generation frequency being on the average 30 kHz, and the maximum amplitude corresponding to angles $\alpha \approx 10°$ to $15°$.

Acknowledgements

The authors would like to express their gratitude to Prof. G. R. Khutsishvili and Prof. I. A. Mirtskhulava for the interest in the work and valuable advice. They also want to thank their colleagues A. Likhogrud, A. Tutundzhan, and N. Babitskaya for their help.

References

[1] R. J. Phelan and W. F. Love, Phys. Rev. **133**, A1134 (1965).
[2] Sh. I. Mirianashvili and D. I. Nanobashvili, Fiz. Tekh. Poluprov. **4**, 1872 (1970).
[3] Z. N. Chigogidze and N. P. Khuchua, Bull. Acad. Sci. Georgian SSR **90**, 341 (1978).
[4] R. W. Cunningham, E. E. Harp, and M. Bulis, Proc. Internat. Conf. Phys. Semicond., Exeter 1962 (p. 732).
[5] F. F. Kharakhorin and V. M. Petrov, Izv. Akad. Nauk SSSR, Ser. neorg. Mater. **7**, 2250 (1971).
[6] I. A. Mirtshkhulava, Z. N. Chigogidze, N. I. Kurdiani, L. V. Khvedelidze, and R. B. Dzhanelidze, Bull. Acad. Sci. Georgian SSR **35**, 299 (1964).
[7] K. E. Mortenson, A. L. Armstrong, G. M. Borrego, and G. F. White, Proc. IEEE **59**, 1191 (1971).
[8] A. D. Berkeliev, V. V. Salavatov, and D. N. Nasledov, Fiz. tverd. Tela **7**, 3685 (1965).
[9] W. M. Bulis and V. Harrap, Proc. VII. Internat. Conf. Phys. Semicond., Paris 1964 (p. 1847).
[10] M. Lax, Phys. Rev. **119**, 1502 (1960).
[11] M. Glicksman and M. C. Steel, Phys. Rev. Letters **2**, 461 (1959).
[12] M. C. Steel and M. Glicksman, Phys. Rev. **118**, 474 (1960).
[13] I. A. Mirtskhulava, Bull. Acad. Sci. Georgian SSR **38**, 535 (1965).
[14] S. H. Koening, R. D. Brown, and W. Schillinger, Phys. Rev. **128**, 1968 (1962).
[15] J. Bok and R. Veilix, Proc. IV. Internat. Conf. Phys. Semicond., Prague 1961 (p. 173).
[16] B. Anker-Johnson, Phys. Rev. Letters **9**, 485 (1962).
[17] B. Anker-Johnson, R. W. Cohen, and M. Glicksman, Phys. Rev. **124**, 1745 (1961).

(Received January 19, 1979)

See Short Note by B. PAVIS et al.

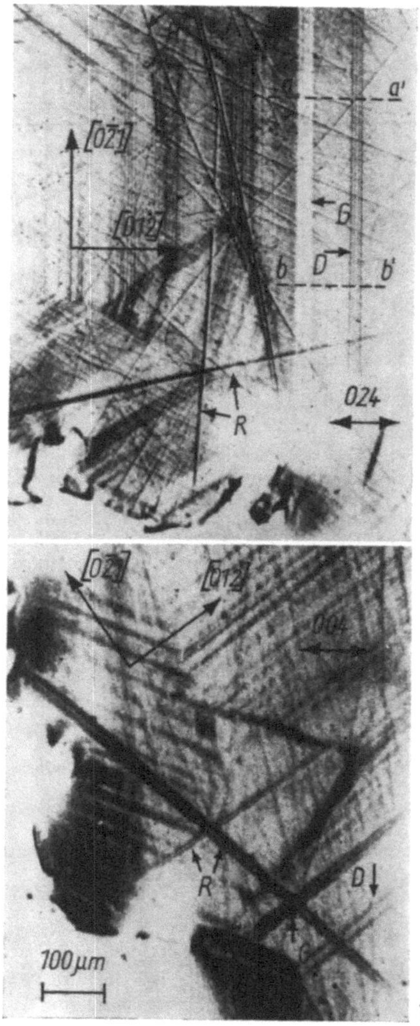

a

b

Fig. 2a) Topographie 024. Les rayures sont repérées par R. Les défauts *G* (6 interfranges) et *D* (3 interfranges) sont densitometrés entre *aa'* et *bb'*. b) Topographie 004. Les défauts *G* (3 franges) et *D* (2 franges) sont notés (les flèches indiquent la direction de la transition)

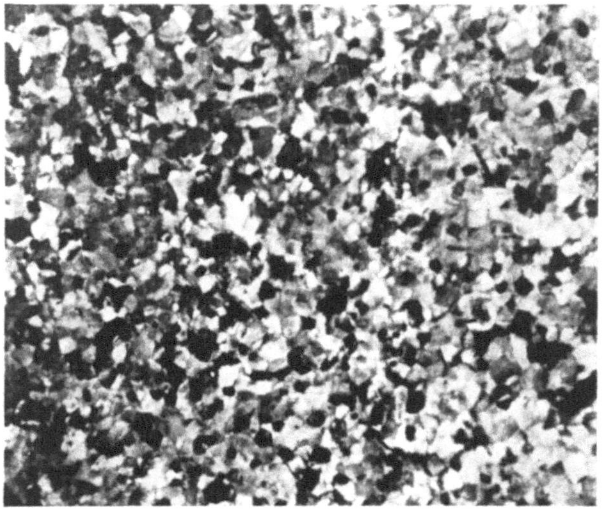

Fig. 1. Electron micrograph of a thin Bi_2O_3 film showing polycrystalline structure

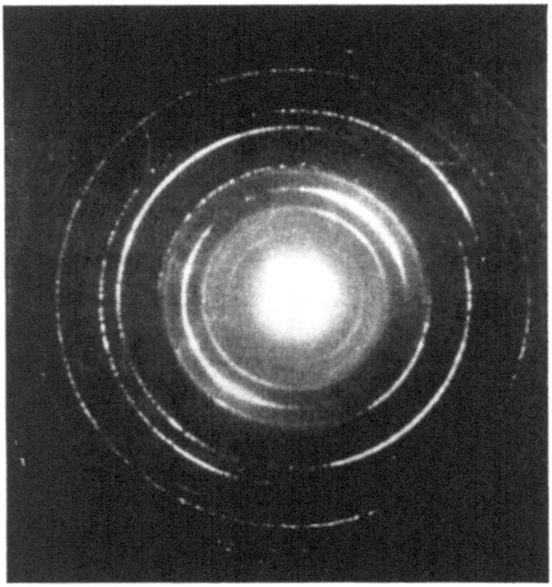

Fig. 2. Electron diffraction pattern of a thin Bi_2O_3 film after complete oxidation showing rhombic structure

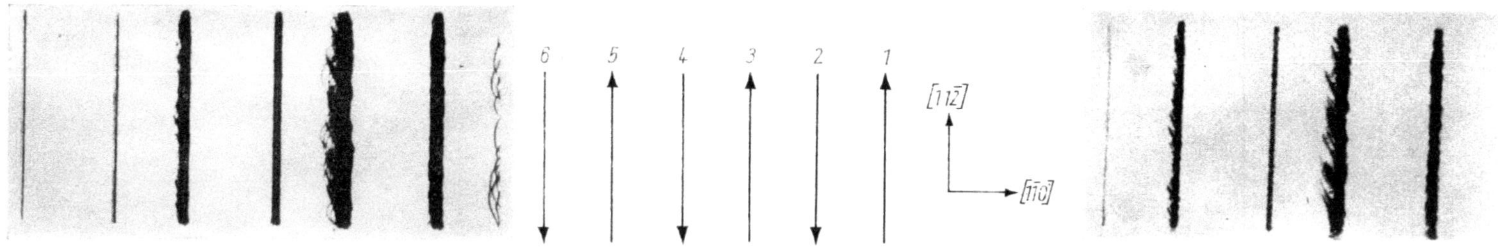

Fig. 1 a b c

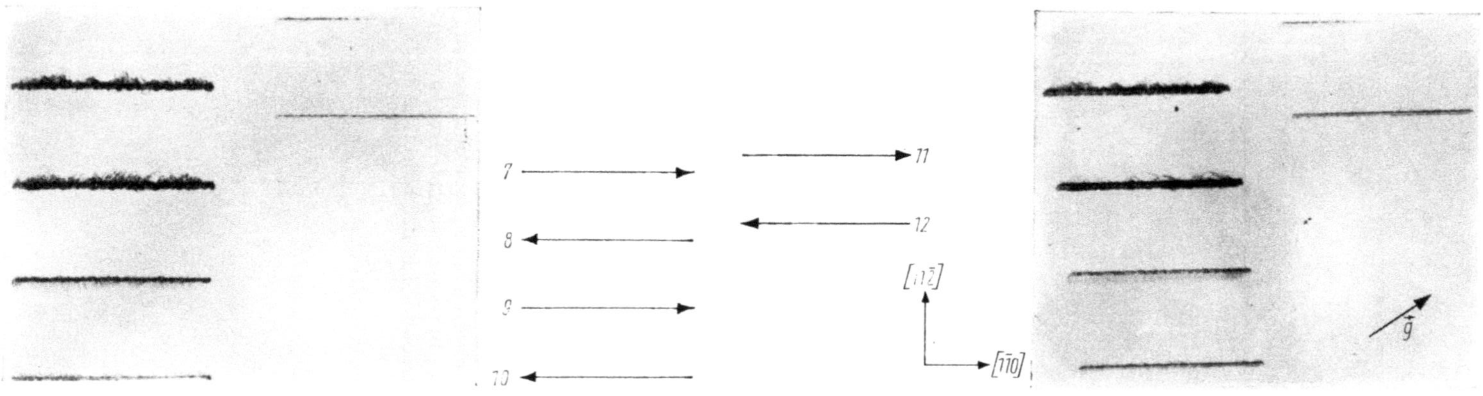

Fig. 3 a b c

Fig. 1. a) X-ray topograph of [11$\bar{2}$] and [$\bar{1}\bar{1}$2] scratches made under loads of (left to right) 5, 10, and 20 g. 2$\bar{2}$0 reflection in $MoK\alpha_1$. 14.4 ×; b) key to Fig. 1a, showing sense of scratching for scratches 1, 2 (20 g), 3,4 (10 g) and 5,6 (5 g); c) topograph of scratches shown in Fig. 1a. 11$\bar{1}$ reflection in $MoK\alpha_1$

Fig. 3. a) X-ray topograph of scratches from [$\bar{1}$10] and [1$\bar{1}$0]. 2$\bar{2}$0 reflection in $MoK\alpha_1$; b) key to Fig. 3a. Scratch loads 20 g (7 and 8), 10 g (9 and 10), and 5 g (11 and 12); c) topograph of scratches in Fig. 3a. 11$\bar{1}$ reflection in $MoK\alpha_1$. 14.4 ×

Fig. 2. X-ray topograph of scratches in Fig. 1 taken in $CuK\alpha_1$; a) $2\bar{2}0$, b) $\bar{2}20$ reflection. 14.4 ×

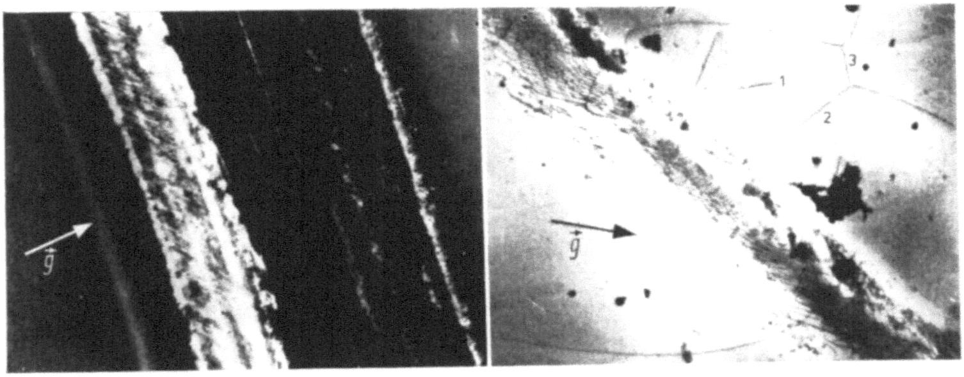

Fig. 4. a) Transmission electron micrograph of scratch in $[\bar{1}\bar{1}2]$ made under load of 5 g. Dark field, $2\bar{2}0$ reflection. 19800 ×; b) transmission electron micrograph of scratch made in $[\bar{1}\bar{1}2]$ after annealing at 1000 °C in argon for 1 h. 022 reflection. 69000 ×

See Short Note by O. Calzadilla et al.

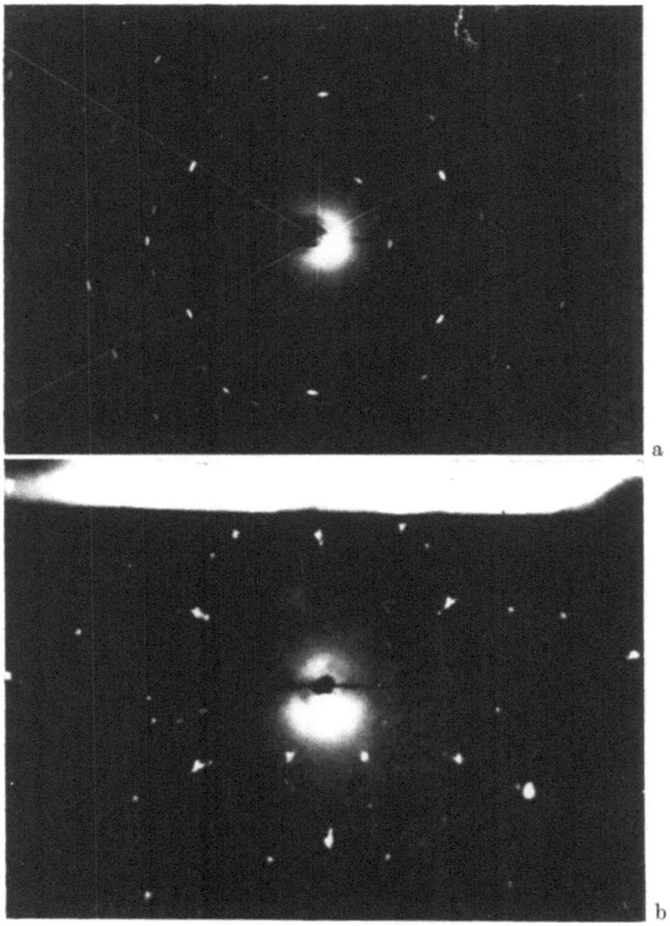

Fig. 1. Laue reflection patterns of $ZnInS_4$ III(a) polytyp, showing a) strong asterism, b) weak asterism

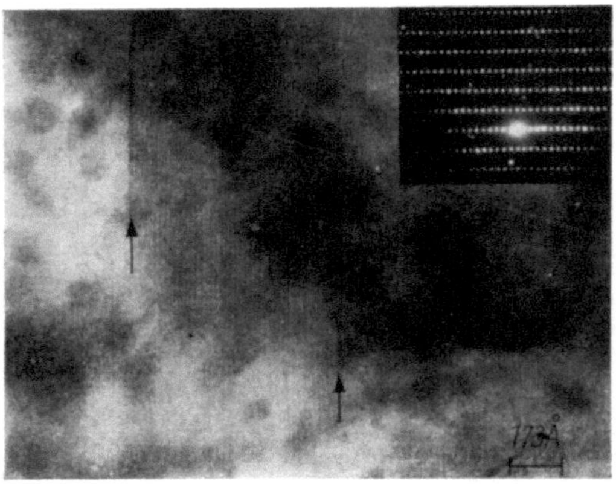

a

Fig. 1

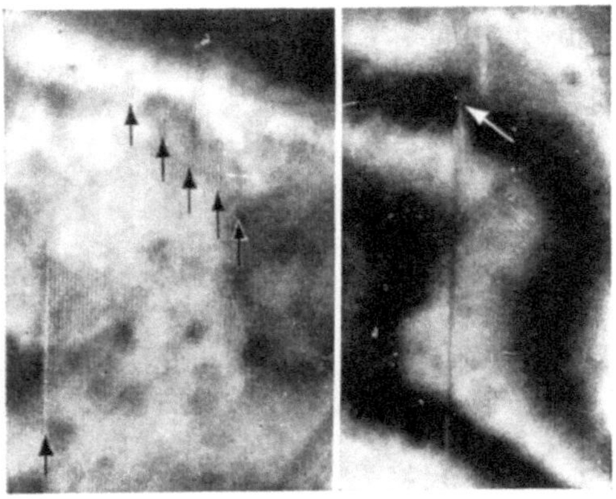

b

Fig. 2

Fig. 1. Dark field micrograph showing lattice fringes (fringe spacing 17.3 Å) along with the corresponding diffraction pattern. Notice the occurrence of stacking faults indicated by arrows

Fig. 2. a) Dark field micrograph showing random distribution of faults; b) dark field micrograph exhibiting two faults interconnected with a transition region marked by arrow

Fig. 3. Dark field electron micrograph corresponding to the disordered structure. Large number of stacking faults can be easily noticed. Inset shows the corresponding diffraction pattern

phys. stat. sol. (a) 54, K85 (1979)

Subject classification: 1.4 and 10.2; 22.8.2

Laboratoire de Métallurgie Physique, associé au C.N.R.S.,
Faculté des Sciences, Poitiers[1]

Observations de défauts dans GGG par topographie X

Par

B. PAVIS[2], G. RIVAUD, J. RABIER et P. VEYSSIERE

Cette note décrit une série d'observations directes de défauts plans dans un cristal de grenat de gallium et de gadolinium (GGG) par la technique de Lang /1/.

Le cristal, fabriqué par la méthode de Czochralski, nous a été fourni par le laboratoire L.E.T.I. de Grenoble (France). Son épaisseur est de 50 μm environ et il contient encore de nombreuses rayures R dues au polissage. Les résultats représentés ici ne concernent pas les dislocations, éventuellement masquées par ces rayures, mais des défauts plans dont l'origine est probablement liée à la croissance du cristal.

La normale au cristal est une direction [100] et la projection stéréographique (Fig. 1) indique les plans {112} mis en évidence par les topographies, les quatre directions faciles d'aimantation <111> et les réflexions étudiées dans cette note. Les topographies sont réalisées avec la raie Mo Kα_1, ce qui donne $\mu_0 t = 0.8$ (μ_0 coefficient linéaire d'absorption et t épaisseur).

Sur la topographie 024 présentée sur la Fig. 2a (voyez aux pages photographiques devant la part "Short Notes"), de défaut noté G, est dans le plan $(1\bar{1}\bar{2})$ parce que sa largeur est plus grande que le défaut noté D, qui se trouve dans

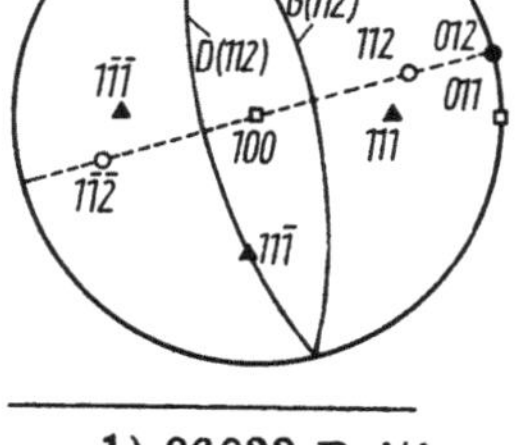

Fig. 1. Projection stéréographique indiquant les principales directions utilisées; ● directions correspondant aux topographies $0\bar{4}2$, $\bar{0}04$, 024

1) 86022 Poitiers, France.

2) Ce travail fait partie de la these de doctorat d'état de B. Pavis.

le plan (112). Les mêmes contrastes sont observés sur la topographie 0$\bar{4}$2 et les systèmes de franges peuvent être caractérisés de la manière suivante: pour le défaut G, la frange centrale est noir et les franges situées de part et d'autre sont distantes de 6 µm; pour le défaut D au contraire, la frange centrale est blanche et quatre franges noires, bien symétriques sont distantes de 12 µm; Ces deux défauts, sur la topographie de la Fig. 2b (voyez aux pages photographiques devant la part "Short Notes") ont toujours une frange centrale de même nature que précédemment, mais le nombre de franges est maintenant de trois franges noires pour le défaut G et de deux noires pour le défaut D. Une analyse densitomètrique (Fig. 3) révèle que la répartition d'intensité le long des franges du défaut G n'est pas symetrique pour la topographie 024 , alors qu'elle est symetriques (par rapport à la frange centrale) pour le défaut D.

Parallèlement, nous avons étudié ce cristal au microscope optique en lumiere polarisée et des régions de couleurs différentes apparaissent lorsque l'on tourne l'échantillon autour de l'axe [100] entre analyseur et polariseur croisés /2/. Le défaut G coïncide avec une limite de ces régions alors que au contraire, le défaut D n'est jamais mis en évidence dans un domaine de même coloration.

Dans les oxydes ferrimagnétiques de structure grenat, il est généralement admis que les directions d'aimantation facile sont proches des directions <111> /3/ et qu'elles peuvent avoir des composantes non nulles suivant les directions <110> /6, 7/. A la température ordinaire à laquelle ont été réalisées les topographies, nous sommes très nettement au-dessous du point de Curie /3/, et nous avons pu vérifiér expérimentalement que l'inverse de la susceptibilité magnétique suit une variation linéaire avec la température entre 70 et 300 K /5/. Nous sommes donc amenés à proposer que le défaut D est une paroi de domain magnétique. De part et d'autre de cette paroi, les directions d'aimantation ont une même composante suivant la direction [100] , de sort qu'optiquement (par effet Cotton-Mouton) il n'est pas possible de les observer. Les critères habituels d'extinction des images de parois de domaine, pour la topographie aux rayons X /4/, permettent d'éliminer la possibilité d'une paroi de domaine à 180° (qui devrait être visible par microscopie optique). Comme le plan de la paroi est bien défini, nous avons réuni dans le Tableau 1 les différentes possibilités de parois à 70^o53 et de parois à 109^o47 compatibles avec ce plan, ainsi que les directions d'aimantation possibles $\vec{M}_1$ et $\vec{M}_2$, et les directions $\Delta\vec{M} = \vec{M}_1 - \vec{M}_2$ utiles pour déterminer les extinctions en topographie.

Tableau 1

paroi à 70°			paroi à 109°		
$\vec{M}_1$	$\vec{M}_2$	$\Delta\vec{M}$	$\vec{M}_1$	$\vec{M}_2$	$\Delta\vec{M}$
$[111]$ $[1\bar{1}1]$	$[\bar{1}11]$ $[\bar{1}\bar{1}1]$	$[200]$ $[200]$	$[111]$ $[111]$	$[1\bar{1}\bar{1}]$ $[11\bar{1}]$	$[022]$ $[0\bar{2}2]$

Nous éliminons d'abord la possibilité de parois à 70° car elles devraient être visibles optiquement, puis celle d'une paroi à 109° avec $\Delta\vec{M} = [0\bar{2}2]$ parce qu' alors la topographie 024 devrait donner un contraste proche de l'extinction.

Nous pouvons donc conclure que dans le GGG il est possible d'observer des parois de domaines magnétiques à 109°, dans un plan (112) avec des directions d'aimantation de part et d'autre de cette paroi orientées suivant les directions $[111]$ et $[1\bar{1}\bar{1}]$. Pour confirmer la nature de cet défaut, l'analyse densitomètrique (Fig. 3) met bien en évidence que l'intensité de la frange centrale est la même que celle du cristal parfait de chaque côté de la paroi et, que plus on s' eloigne de cette frange centrale, plus les franges sont intenses. La distance d'extinction expérimentale est donc de 12 μm pour une topographie symétrique 024 et ce profile d'intensité est tout à fait celui d'un défaut $\delta(\alpha = 0)$ avec les notations habituellement utilisées en microscopie électronique /11/, et celui dans le cas ou l'absorption est faible.

En comparant les deux enregistrements suivant aa' et bb' (notés sur la Fig. 2a), on constate la disparition d'une frange sombre sur le côté gauche du défaut

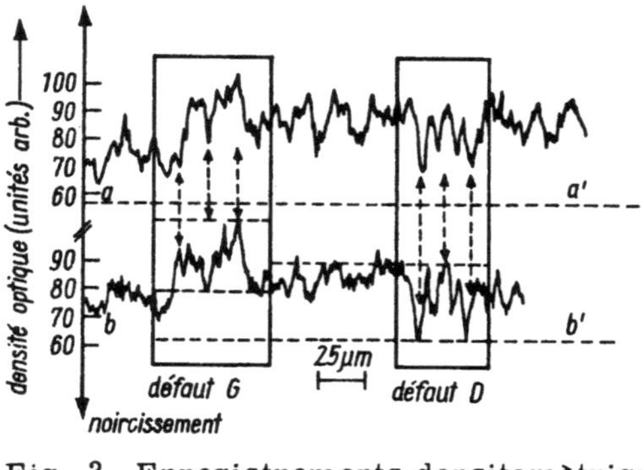

Fig. 3. Enregistrements densitomètriques suivant les directions aa' et bb' définies sur la Fig. 2a

G de la topographie 024 , bien que la frange centrale reste à la même position. On a donc une légère variation de l'épaisseur du cristal le long des défauts G. Une caractéristique d'un défaut $|\alpha| = \pi$ pur est que la frange centrale correspond toujours au milieu du cristal /11/, et dans ce cas on devrait aussi observer la disparition d'une frange du côté droit.

Le défaut G semble avoir les caractéristiques d'un défaut où $\alpha = 2\pi\vec{g}\cdot\vec{R}$ vaut $\pm\pi$ sur la topographie 004. On est donc amené à proposer un vecteur de faute $\vec{R}$, dans le plan (112) qui serait $\frac{1}{8}$ $[11\bar{1}]$ et non $\frac{1}{4}$ $[11\bar{1}]$. Ce dernier type de vecteur a déjà été proposé par Rabier et al. /8/ pour des défauts plans (dans le grenat de fer et d'yttrium) localisés dans des plans {100} et {110}. Des défauts avec $\vec{R} = \frac{1}{8}\langle 110\rangle$ ou $\frac{1}{8}\langle 100\rangle$ sont écartés devant la stabilité des défauts plans du type $\frac{1}{2}\langle 110\rangle$ ou $\frac{1}{2}\langle 100\rangle$.

Pour ce même défaut l'enregistrement densitométrique réalisé à partir de la topographie 024 confirme que le déphasage est $\pm\pi/2$. L'intensité de la frange centrale est inférieur à celle du cristal parfait déjà définie pour le défaut D. D'autre part, il existe une frange blanche très brillante du côté droit dont l'intensité est très inférieur à l'intensité du cristal parfait. Il semble donc que ce soit un défaut d'empilement "pur" dans la mesure ou la topographie aux rayons X n'est pas sensible à une paroi de domaine à 180°. De plus, comme le défaut est visible optiquement, on est amené à penser que les deux domaines qu'il sépare ont des directions d'aimantation antiparallèles.

Comme il a été montré par Belt et Moss /9/, des fluctuations locales de concentration peuvent justifier les stries de croissance observées dans des cristaux de GGG. Il est donc permis de penser que localement un écart à la bonne stoechiométrie autorise la création d'un défaut dans un plan (112) , avec un vecteur de faute $\frac{1}{8}[11\bar{1}]$, cette faute ne modifiant pas les directions d'aimantation mais leur sens. Dans le cas du YIG, Rabier et al. /8/ considèrent que le réseau des anions (O^{2-}) n'est pas globalement modifié avec des translations du type $\frac{1}{4}\langle 111\rangle$. Une translation du type $\frac{1}{8}\langle 111\rangle$ nécessite un réarrangement local des polyèdres sur lesquels sont situés les anions, ce qui peut expliquer la modification de la direction d'aimantation. Ou bien, les polyèdres sur lesquels se trouvent les anions et définis par les paramètres x, y et z /10/, sont favorables à ce type de faute dans les plans {112} pour le grenat de gallium et de gadolinium.

Les auteurs remercient Mr. Guisnet (L.E.T.L, Grenoble) qui leur a gracieusement fourni un monocristal de GGG, et J. Nogues (Service de mocrodensitomètrie du C.N.R.S. d'Orsay) pour sa coopération pour les enregistrements densitomètriques.

Bibliographié

/1/ A.R. LANG, Acta cryst. 12, 249 (1959).

/2/ E.J. TOROK, D.L. FLEMING, T.R. JOHANSEN, J.A. KRANWCCZAK et G.L. NELSON, IEEE Trans. Magnetics MAG-14, 31 (1978).

/3/ A. HERPIN, Théorie de magnétisme, P.U.F., Paris 1968 (p. 692).

/4/ M. POLCAROVA et J. KACZER, phys. stat. sol. 21, 653 (1967).

/5/ J. MIMAULT, communication personnelle.

/6/ J.R. PATEL, K.A. JACKSON et J.F. DILLON, J. appl. Phys. 39, 3767(1968). (1968).

/7/ A. MATTHIOT, J.-F. PETROFF et Y. BERNARD, phys. stat. sol. (a) 20, K1 (1973).

/8/ J. RABIER, P. VESSYIERE et J. GRILHE, phys. stat. sol. (a) 35, 259 (1976).

/9/ R.F. BELT et J.P. MOSS, Mater. Res. Bull. 8, 1197 (1973).

/10/ F. EULER et J.A. BRUCE, Acta cryst. 19, 971 (1965)

/11/ S. AMELINCKX, Proc. Internat. Summer School, Antwerp 1969 (p. 257).

(Received May 3, 1979)

phys. stat. sol. (a) 54, K91 (1979)
Subject classification: 14.4 ; 22.6

Thin Films Laboratory, Physics Department, Bareilly College[1)]

Dielectric Measurements of Thin Films of Bi_2O_3

By

S. SINGH, S. L. PANDEY, and R. N. SAXENA

Introduction Oxide films are widely used in thin film active as well as passive devices. A variety of metallic oxides has been tried and used as the dielectric material and their preparation conditions have been studied in great detail. In the present investigation the authors have taken relatively thin films of bismuth which could be completely oxidized by heating them in air at 175 °C for a longer period and then the dielectric parameters viz. breakdown field and dielectric constant have been determined for the films having thicknesses from 230 to 900 Å.

Experimental details Deposition of thin films has been done onto clean glass substrates in vacuum of the order of 10^{-5} Torr. For the measurements of dielectric properties, the coplanar sandwich structure, as reported earlier /1/, was used. Aluminium films were used as base electrodes and over them the metallic bismuth films were deposited. The films were heated in a furnace at 175 °C under atmospheric conditions for 5 to 15 days depending upon the thickness of the sample to ensure complete oxidation of bismuth into yellowish transparent Bi_2O_3. On the top of the dielectric film of Bi_2O_3, an aluminium counter electrode was deposited. The active area of the dielectric layer was of the order of 0.3 cm^2.

The process adopted for annealing the samples was as follows. The samples were kept in the oven at 175 °C for 2 h under atmospheric conditions. After it, the furnace was switched off leaving the samples resting inside it to cool down to room temperature. The next cycle of annealing was repeated after nearly 24 h. Before the next cycle of annealing, capacity and losses of the samples were measured. Several such annealing cycles were repeated for stabilization of capacitors.

For taking observations, these capacitors were mounted on a suitable base and leads were connected to the electrodes. The lead capacitance and (lead + electrode) resistance were less than 2 pF and 0.5 Ω, respectively. The capac-

---

1) Bareilly-243001, India.

itance and losses were measured at 1 kHz under atmospheric conditions at room temperature by a Marconi Universal Bridge TF2700 and the breakdown voltage of the films was obtained with the help of a Philips dc microvoltmeter model PP 9004 using the same circuit as reported in the previous paper /1/. A Philips EM 200 Electron Microscope was used for electron diffraction and electron microscopic studies of Bi_2O_3 films which were formed, under the same conditions, simultaneously on KCl crystals.

Results and discussion In the present investigation all the measurements have been made on dielectric films after complete oxidation and to ensure it, the deposition of Bi films was restricted to small thicknesses (230 to 900 Å), and they were annealed for a relatively longer period. Electron diffraction investigations of the films also confirmed the complete oxidation without revealing any trace of Bi in the oxide layers. Fig. 1 and 2 (see on the photo pages before the Short Notes part) show the electron micrograph and electron diffraction pattern of the Bi_2O_3 film. The electron diffraction pattern shows that Bi_2O_3 has rhombic structure. Also the complete oxidation was confirmed by resistivity measurements at various stages of oxidation.

It has been observed that on annealing the capacitance and loss decreased substantially in the beginning, and after 5 to 8 annealing cycles they approached stable values, Fig. 3. Hence in each case the capacitor was annealed repeatedly at 175 $^{\circ}C$ till it gave reproducible results before making further measurements. A remarkable point about these capacitors was that they did not show any appreciable aging effect even after 5 to 6 months.

The breakdown voltage U_b for the thinner films was smaller than that for the thicker ones, Fig. 4, while the breakdown field F_b decreased with increase in dielectric thickness, from 16×10^5 to 6×10^5 V/cm obeying the relation $F_b \sim t^{-\beta}$, where β is a constant and was found to be equal to 0.74 in the present case, Fig. 5. This variation of F_b was in accordance with the Forlani-Minnaja relation /2/. Similar results have been reported by earlier workers /3 to 5/ for other dielectrics.

The computed values of the dielectric constant for different film thicknesses in the present investigation, using the usual relation, have been found to vary from 16 to 30 for the range 230 to 900 Å, as shown in Fig. 4.

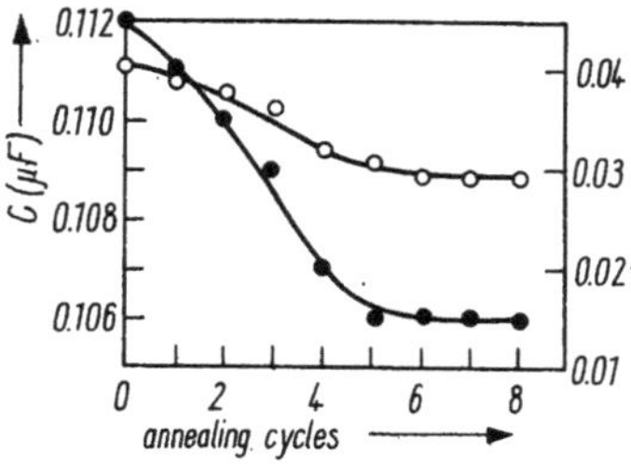

Fig.3

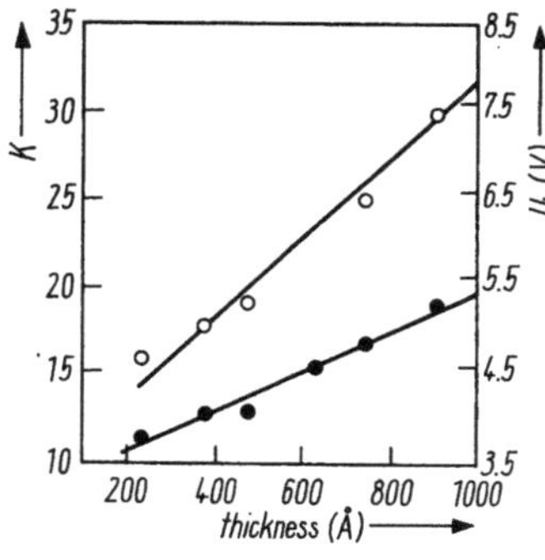

Fig.4

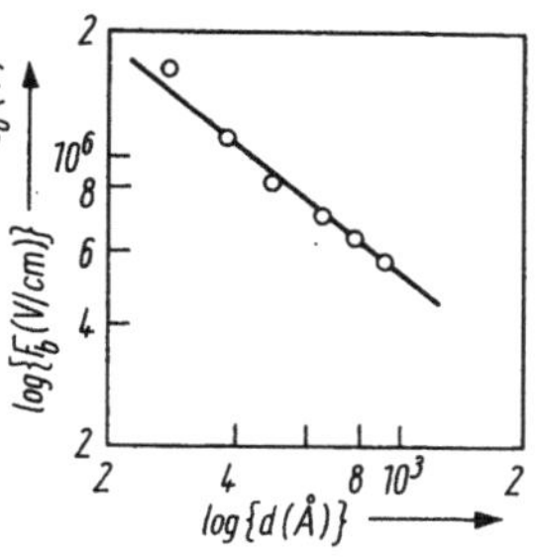

Fig.5

Fig. 3. Stabilization of capacitance (●) and loss factor (tanδ o) after several annealing cycles

Fig. 4. Variations of breakdown voltage U_b (●) and dielectric constant K(o) with thickness d of the dielectric layers

Fig. 5. Variation of breakdown field strength with thickness d. The slope of the line is negative and equal to 0.74

Conclusion The present study reveals that the capacitance and losses of completely oxidized Bi films are quite stable for fairly long time. The losses are moderate. The variation in the signal voltage applied for measurements has virtually no effect on capacitance and loss values of the capacitors. The breakdown field of these capacitors is of the order of 10^6 V/cm. So Bi oxide may be used as a good dielectric in research and industry. The study of the dielectric behaviour of the Bi oxide films under different controlled pressures and temperatures is in progress.

The authors thank Dr. S.K. Sharma, Electron Microscope Division, NPL, New Delhi, Prof. T.N. Agrawal, and Mr. Vinod Kumar Saxena, Physics Department, Bareilly College, for their help in the work.

References

/1/ R.N. SAXENA, S.L. PANDEY, SOORAJ SINGH, and S.K. SHARMA, Internat. Symp. Solid State Phys., IACS, Calcutta, January 10 to 14, 1977.

/2/ F. FORLANI and N. MINNAJA, phys. stat. sol. 4, 311 (1964).

/3/ F. FORLANI and N. MINNAJA, J. Vacuum Sci. Technol. 6, 519 (1969).

/4/ P.P. BUDENSTEIN and P.J. HAYES, J. appl. Phys. 38, 2837 (1967).

/5/ A.P. GOSWAMI and A. GOSWAMI, Indian J. pure appl. Phys. 12, 26 (1974).

(Received June 21, 1979)

phys. stat. sol. (a) 54, K95 (1979)
Subject classification: 10.2; 22.1.2

Physics Department, University of Surrey, Guildford[1)]

Dislocation Movement at Annealed Scratches on Near (111) Silicon

By
K.E. PUTTICK and M.A. SHAHID

Gerward /1/ investigated by X-ray topography the nature of stress relaxation during annealing of the strain field of a scratch on a silicon surface, and concluded that the Burgers vectors of the dislocations so produced were all in the (111) plane parallel to the surface, and in the <110> direction nearest to the scratch normal. Since then the nature of the strain field itself has been studied in some detail (Badrick et al. /2/) and it has been shown that it is strongly dependent on the crystallographic direction of scratching as well, probably, as (by analogy with the abrasion of diamond) the exact orientation of the specimen surface. In view of the technological importance of surface damage in silicon and its interaction with heat treatment it is worth reporting some observations of ours on the same relaxation process.

The material used was dislocation free silicon, phosphorus doped (6 to 10 Ωcm) in the form of industrial substrates 300 μm thick with the polished surface misoriented 3^{0} from (111) towards (110). These were divided into rectangular sections by scribing, and the specimens were scratched by a Vickers diamond pyramid mounted on a gramophone pick-up arm, using loads from 1 to 20 g ($\approx 10^{-2}$ to 0.2 N), in $[\bar{1}10]$, $[1\bar{1}0]$, $[11\bar{2}]$, and $[\bar{1}\bar{1}2]$ directions. The slices were subsequently annealed at a temperature of 1000 ^{0}C in a stream of high purity argon. X-ray topographs were taken before and after this process, using a Jarrell-Ash Lang camera, mainly in normal transmission with MoKα_1 radiation. The results on the unannealed scratches confirmed the conclusions of Badrick et al. /2/ on scratches made under the same conditions at a slightly higher load (50 g).

The topographic contrast field was found to be 10 to 30 times wider than the physical width of the scratch itself, but its size depended markedly on the orientation of the reflecting vector, being a minimum for vectors parallel to

---

1) Guildford, Surrey GU2 5XH, U.K.

the scratch length. Its width was also found to depend on the sense of scratching in <112> directions, being greater for $[\bar{1}\bar{1}2]$ than for $[11\bar{2}]$.

The annealing behaviour reflects the different strain fields produced by scratching in different directions. Fig. 1a (Fig. 1 see on the photo pages before the Short Notes part) is a topograph of scratches made in $[\bar{1}\bar{1}2]$ and $[11\bar{2}]$ taken in the $2\bar{2}0$ reflection (the key to the photograph is shown in Fig. 1 b). The maximum distance of dislocation movement has occurred round the $[\bar{1}\bar{1}2]$ scratches, though below 10 g load no detectable dislocation movement can be resolved. The dislocation array at the $[\bar{1}\bar{1}2]$ scratch appears to consist mainly of curved loops lying in the slip plane nearly parallel to the free surface, but there are also signs of a number of lines lying in $[01\bar{1}]$. These are more clearly seen in Fig. 1c, taken in the $11\bar{1}$ reflection in which the curved loops are invisible.

It was conjectured but not shown by Badrick et al. /2/ that the residual surface strain round a scratch is a state of plane strain compression. That this is indeed so is shown by Fig. 2a and 2b (see on the photo pages before the Short Notes part) of the annealed scratches of Fig. 1 taken in $2\bar{2}0$ and $\bar{2}20$ using $CuK\alpha_1$ radiation. Here the product μt of the absorption coefficient and the thickness is about 4 ($\mu t \approx 1$ for $MoK\alpha_1$) so that anomalous absorption has occurred producing strong contrast effects, as in the similar work of Frank et al. /3/ on abrasion of diamond. Comparison of these topographs shows that a dark contrast changes to white (and vice versa) when $\vec{g}$ is changed to $-\vec{g}$, i.e. in the dark wings of contrast the displacement vector $\vec{u}$ is parallel to $\vec{g}$ and anti-parallel in the white wing, so that the wings are in a state of compression with their displacement vectors directed along the normals to the scratch length.

Dislocation relaxation from <110> scratches is less extensive than from $[\bar{1}\bar{1}2]$, though greater than from $[11\bar{2}]$. In Fig. 3a (Fig. 3 see on the photo pages before the Short Notes part) taken in $2\bar{2}0$, the dislocations have mainly the form of loops intersecting the surface, and therefore presumably lying in slip planes inclined to it. The geometry of these loops suggests that the dislocations from $[\bar{1}\bar{1}0]$ have slipped in $(\bar{1}11)$ and those from $[\bar{1}10]$ in $(1\bar{1}1)$. Fig. 3c, in the $1\bar{1}\bar{1}$ reflection, reveals a few loops apparently nearly parallel to the free surface emanating from $[\bar{1}10]$.

The dislocations in Fig. 1 to 3 were examined systematically in a number of reflections, and the following Burgers vectors established. The vectors of the large loops from $[\bar{1}\bar{1}2]$ scratches lie parallel to $[\bar{1}10]$, but the straight dislocation lines in bands along $[01\bar{1}]$ from the same scratches possess vectors in $[\bar{1}01]$, and are thus 60° dislocations. The loops inclined to the free surface made by $[1\bar{1}0]$ and $[\bar{1}10]$ scratches had vectors lying in $[\bar{1}01]$ and $[01\bar{1}]$, respectively, while the loops nearly parallel to the free surface from $[\bar{1}10]$ scratches appeared to possess $[\bar{1}01]$ vectors.

Dislocation relaxation from the scratches under loads of less than 10 g was studied by transmission electron microscopy of foils prepared from scratched samples by thinning from one side. In general, the unannealed scratches consisted of grooves, free of cracks, within which plastic deformation had evidently produced a density of dislocations too high to resolve, as in Fig. 4a (Fig. 4 see on the photo pages before the Short Notes part), though there are suggestive bands of contrast along $[0\bar{1}1]$. Annealing at 1000 $^{\circ}$C for 1 h resulted in the formation of a well defined dislocation array within the groove, together with movement of dislocations outside it, as shown in Fig. 4b. Observation by stereomicroscopy revealed that while the long segments of line outside the groove appeared to have moved in the (111) plane nearly parallel to the free surface, many short segments of the groove network lay on slip planes inclined to (111). This impression received support from analysis of the Burgers vectors: the extra-groove dislocations possessed vectors lying in (111), while many of the groove dislocations had [110] and [101] vectors inclined to it. (These attributions are, however, provisional, as the specimen was broken before a complete series of reflections could be taken.)

Discussion It was shown by Badrick et al. /2/ that the main residual deformation at a scratch on near (111) silicon is in plane strain, with the displacement normal to the scratch length. Such a field would be expected to relax by the movement of like-signed dislocations in (111) under their mutual elastic repulsion, with Burgers vectors lying in $\langle\bar{1}10\rangle$ for $[\bar{1}\bar{1}2]$ scratches, and equal numbers of $[0\bar{1}1]$ and $[\bar{1}01]$ for $[\bar{1}10]$ scratches. Such a model of the relaxation is indeed found to give good agreement with observation (Badrick et al. /4/). The sense effect established by Badrick et al. /2/ would lead us to expect that the shorter ranged strain field associated with $[11\bar{2}]$

scratches should result in a correspondingly shorter distance of glide compared with $[\bar{1}\bar{1}2]$. Our results indicate the following:

1. The residual surface plane strain is indeed compressive.

2. The main relaxation at $[\bar{1}\bar{1}2]$ scratches is indeed greater than at $[11\bar{2}]$, and is in $[\bar{1}10]$ (111).

3. There is, however, a subsidiary relaxation in $[\bar{1}01]$ of which the slip plane is not clear.

4. The resolved direction of slip at ⟨110⟩ scratches is not perpendicular to the scratch direction, but is predominantly along a single close-packed direction in (111), in a plane inclined to the free surface.

5. Within the scratch groove some relaxation takes place in directions inclined to (111).

Thus in general the net displacement direction revealed by the dislocation geometry outside the scratch groove is not normal to the scratch but makes an angle with the normal between 0 and 30° for ⟨112⟩ scratches and between 0 and 60° for ⟨110⟩ scratches. This may reasonably be associated with the existence, also shown by Badrick et al. /2/, of a component of the residual strain field with a displacement parallel to the scratch length. The electron microscope studies indicate that within the groove plastic deformation occurs at room temperature under the hydrostatic compressive stress beneath the indenter, and that relaxation of this field involves displacements with components normal to (111).

During the period of this research M.A.S. was in receipt of a scholarship from the Government of Pakistan.

References

/1/ L. GERWARD, phys. stat. sol. (a) 2, 797 (1970).

/2/ A.S.T. BADRICK, F. El-DEGHAIDY, K.E. PUTTICK, and M.A. SHAHID, J. Phys. D 10, 195 (1977).

/3/ F.C. FRANK, B.R. LAWN, A.R. LANG, and E.M. WILKS, Proc. Roy. Soc. A301, 239 (1967).

/4/ A.S.T. BADRICK, F. EL-DEGHAIDY, and K.E. PUTTICK, J. Phys. D 12 (1979), in the press.

(Received June 21, 1979)

phys. stat. sol. (a) 54, K99 (1979)
Subject classification: 20.1;22.8

Department of Semiconductor Physics, Havana University

$ZnIn_2S_4$ III (a) Absorption Spectrum Tail Changes Due to Microstresses

By
O. CALZADILLA, J. FUENTES, and L. HERNÁNDEZ

Today, wide-spread research is being carried out on Zn-In-S compounds grown through chemical transport /1/. Promising photoconductive properties are observed, in particular, on $ZnIn_2S_4$ crystals /2, 3/. These crystals are orange plate-shaped crystals commonly from 20 to 120 µm thick.

As is known /4/, tails appear in the $ZnIn_2S_4$ absorption spectrum. In the tail regions, the absorption coefficient has an exponential dependence on photon energy. It has been proposed that the existence of microfields (which change the shape of the potential band) gives rise to optical transitions which are responsible for the absorption tail. These microfields are among others related to inclusions, vacancies, and lattice microdeformations.

Crystal imperfections may appear as the result of fast cooling - after single crystal growth. This practice may build up microstresses within the samples.

In the present paper, the single crystals grown were found to be of the $ZnIn_2S_4$ III(a) polytype /5/. X-ray analysis of the crystals showed that asterism was common in Laue reflection patterns, which means that microstresses are present in the samples.

The observed asterism (Fig. 1a, see on the photo pages before the Short Notes part) is undoubtedly related to stress and not to Wilson layer displacement /6/. A strong asterism is found in Fig. 1a, while weak asterism is present in Fig. 1b (see on the photo pages before the Short Notes part). These Laue patterns are obviously associated to heavily and lightly stressed samples, respectively. It is the authors' opinion that heavily stressed crystals are likely to appear closer to the ampoule wall during cooling.

It has been found that the absorption coefficient in the tail region of the absorption spectrum depends on the degree of stress within the crystal. It is seen in Fig. 2a and b that the reciprocal slope for heavily stressed samples diminishes through annealing. The slope reduction is found to be greater at longer and higher temperature annealing. Nevertheless, slope reduction is ir-

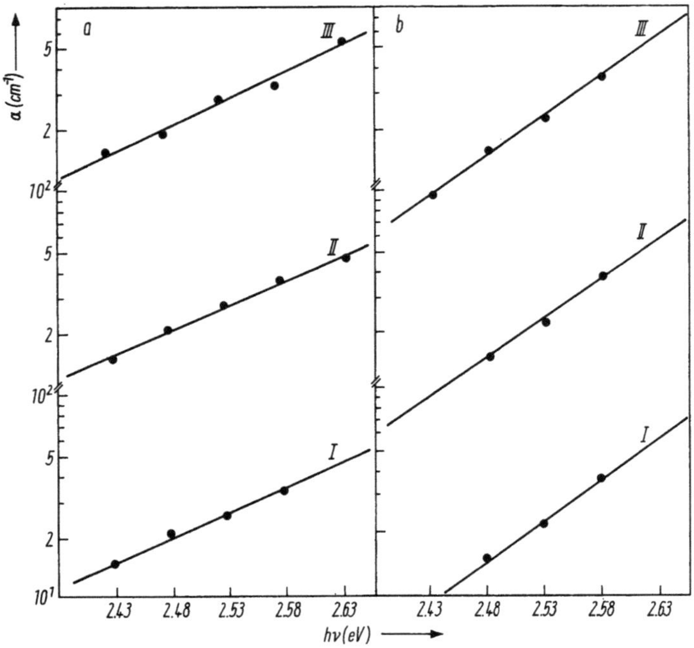

Fig. 2. Least-mean square plots of the dependence of α on photon energy for below threshold frequencies at 300 K for a) heavily tensioned, b) lightly tensioned crystal. I no annealing, II 8 h annealing at 100 °C, III 30 h annealing at 300 °C

relevant when annealing is performed on low-stress samples.

On the other hand, a gap enlargement has been found on heavily stressed samples, whereas no gap change occurs on low-stress crystals.

Through annealing lattice microstresses inherent to the growth of some samples have been eliminated. This means that potentials are less deformed and hence, the amount of virtual states is reduced.

In other words, X-ray viewed microstresses in $ZnIn_2S_4$ III(a) are associated to lattice deformations which alter the band potential and in consequence allow tunneling.

References

/1/ R. NITSCHE and W.J. MERZ, C.R. Soc. Swisse Phys. 35, 274 (1962).

/2/ J.A. BEUN, R. NITSCHE, and M. LICHTENSTEIGER, Physica 26, 647 (1960).

/3/ W. LUDWIG and G. VOIGT, phys. stat. sol. 24, K161 (1969).

/4/ S. SHIONOYA and A. EBINA, J. Phys. Soc. Japan 19, 1150 (1964).
S.I. RADAUTSAN, F.G. DONIKA, G.A. KYOSSE, I.G. MUŞRYA, and C.F. ZHITAR, phys. stat. sol. 34, K129 (1969).
L. HERNÁNDEZ, O. VIGIL, and F. GONZÁLEZ, phys. stat. sol. (a) 36, 33 (1976).

/5/ J. FUENTES and O. CALZADILLA, to be published.

/6/ A. WILSON, X-Ray Optics; The Diffraction of X-Rays by Finite and Imperfect Crystals, Metheun, London 1962.

(Received June 25, 1979)

phys. stat. sol. (a) 54, K103 (1979)
Subject classification: 1.2 and 1.5; 21.1

Department of Physics, Faculty of Sciences,
Banaras Hindu University, Varanasi[1)]

Electron Microscopic Studies of Phase Transformations in NiSe Thin Films

By

A.K. SINGH, C. SURYANARAYANA[2)], and O.N. SRIVASTAVA

Transition metal chalcogenides are important solid state materials. Many of them exhibit interesting physical properties /1, 2/. However, most of the studies on these materials have been carried out on the bulk samples only. Thin film forms of these materials have usually different crystallographic characteristics and physical behaviour compared to the bulk phases. The aim of the present note is to report the phase formation and transformation sequence in NiSe thin films.

Thin films (thickness being less than 500 Å) of NiSe were prepared by thermal evaporation under a vacuum of 10^{-6} Torr. Nickel and selenium of 5N purity were taken in a 1:1 atomic weight ratio. Selenium was evaporated first followed by nickel /1, 3/. Crystallographic characterization of these thin films was carried out in Philips EM200 and EM300 transmission electron microscopes by employing the imaging and diffraction modes.

As reported earlier /4/, the as-grown phase was found to have a hexagonal structure (NiAs type) with a = 3.71 Å and c = 5.40 Å. It may be noted here that these lattice parameters are quite close to the reported values for the bulk phase /5/. When these films were pulse annealed inside the microscope, using the technique described earlier /1/, an orthorhombic phase with a = = 6.26 Å, b = 3.71 Å, and c = 10.80 Å formed at about 150 °C. Continued pulse annealing transformed this orthorhombic phase at 250 °C to a primitive cubic phase with a = 4.30 Å. Both these transformations have been found to be irreversible. Another point of interest is the fact that the orthorhombic phase is closely related to the hexagonal phase observed in the bulk as well as in the as-

---

1) Varanasi-221005, India.

2) Department of Metallurgical Engineering, Institute of Technology, Banaras Hindu University, Varanasi-221005, India.

grown conditions. Further annealing of the films having the primitive cubic structure did not produce any other transformation.

Long annealing treatments given to the samples with the primitive cubic phase showed predominat grain growth. At this stage, the films often exhibited long-period superstructures which were based on the primitive cubic structure. Careful electron diffraction investigations showed that two types of superstructure existed /4/ and both of them were found to possess tetragonal symmetry. The crystallographic details and their relation to the basic primitive cubic lattice are as follows:

$$a_{tetra} = 17.32\ \text{Å} \approx 4\ a_{cubic};\ c_{tetra} = 6.08\ \text{Å} \approx \sqrt{2}\ a_{cubic}$$

for the first superstructure and

$$a_{tetra} = 1732\ \text{Å} \approx 4\ a_{cubic};\ c_{tetra} = 3.28\ \text{to}\ 3.46\ \text{Å}$$

for the second superstructure. It can be noted here that the c parameter for this superstructure varied considerably from specimen to specimen.

In order to obtain detailed crystallographic information and to characterize the films containing the long-period superstructure, direct lattice resolution technique has been resorted to. With a view to have better contrast, all the high-resolution micrographs presented here are taken under dark-field conditions. Fig. 1 (see on the photo pages before the Short Notes part) shows a typical micrograph with a corresponding diffraction pattern shown as an inset. $10\bar{1}$ row of spots was chosen to obtain the micrograph. Fig. 1 exhibits the lattice fringes corresponding to the long-period superstructure, the inter-fringe spacing being 17.3 Å. This provides direct confirmation in regard to the diffraction characterization of the superstructure. As can be seen in the micrograph, in addition to the lattice fringes, stacking faults (marked by arrows) also appear which correspond to the regions of perturbation embedded in the lattice. The stacking faults are exhibited more clearly in yet another high-resolution micrograph shown in Fig. 2 (see on the photo pages before the Short Notes part). Fig. 2a exhibits individual as well as a random distribution of faults. Fig. 2b on the other hand reveals two faults interconnected with a transition region which is perpendicular to the fault interface.

The existence of these faults suggests the possibility of occurrence of a disordered variant of the superstructure. In order to search for this, a large number of regions exhibiting the superstructure was analysed. The in-

inset of Fig. 3 (see on the photo pages before the Short Notes part) shows the diffraction pattern corresponding to the disordered phase. A dark-field micrograph obtained by enclosing the streak in the objective aperture is shown in Fig. 3. A large number of faults is visible in the micrograph. These faults by virtue of their random distribution result in the creation of a disordered variant.

In conclusion it may be said that the direct lattice resolution technique has been found to be helpful in the characterization of the superstructure phase which exists both in the ordered and in the disordered forms. In the ordered condition, in addition to well-resolved lattice fringes of 17.3 Å spacing, stacking faults were also occasionally seen. On the other hand, the disordered phase exhibits a high density of stacking faults at random interval.

References

/1/ M.M. SRIVASTAVA and O.N. SRIVASTAVA, Thin Solid Films 29, 275 (1975).

/2/ T.F. CONALLY, Solid State Physics Literature Guide, Vol. 3, T. 7, Plenum Press, New York 1972.

/3/ A. SINGH, O.N. SRIVASTAVA, and B. DAYAL, Z. Metallk. 61, 383 (1970).

/4/ A.K. SINGH, M.M. SRIVASTAVA, and O.N. SRIVASTAVA, J. less-common Metals 57, 225 (1978).

/5/ W.B. PEARSON, Handbook of Lattice Spacing and Structure of Metals and Alloys, Vol. 2, Pergamon Press, Oxford 1967).

(Received May 2, 1979)

phys. stat. sol. (a) 54, K107 (1979)
Subject classification: 1 and 2; 22.6

Sektion Physik der Wilhelm-Pieck-Universität Rostock

Diffraction Investigation on the Structure of SiO_x

By
H.-G. NEUMANN and TH. ROY

Structural investigations by means of diffraction methods can give important hints on the validity of existing structure models of amorphous SiO_x /1 to 6/ by the interpretation of radial distribution functions (RDF). By means of diffraction measurements (X-ray) in the range of large angles /7/ it was possible to exclude earlier assumptions of a macroscopic mixture model consisting of large-scale phase separated silicon- and silica-like regions in the amorphous SiO_x. A quantitative analysis of the peak areas in the RDF showed that the experimental results are consistent with a mixture model at atomic level /6/. In this microscopic mixture model phase separated regions have diameters between 5 and 10 Å. By means of optical measurements a statistical random bonding model was developed /5/ which includes all possible tetrahedra of the type $Si(Si_yO_{4-y})$ with $0 \leqq y \leqq 4$. The probabilities for finding these tetrahedra in the structure are given by the values of $c_0, \ldots, c_4$. A relation was found between the c_i and the ionicity of the Si-O bond /8/. In the microscopic mixture model /6/ partially oxidized tetrahedra ($Si(Si_3O)$, $Si(Si_2O_2)$, $Si(SiO_3)$) can be assumed in a thin transition zone between phase separated a-Si- and a-SiO_2-like regions. In contrast to the random bonding model /5/ the probabilities c_1, c_2, c_3 have much lower values here.

The present paper reports on X-ray diffraction measurements in the range of large angles on two different types of SiO_x samples. We investigated commercially available silicon oxide powder manufactored by Chemapol Prague and the same material after electron beam evaporation and deposition under high vacuum conditions (VD).

By using an X-ray generator with a rotating anode (Rigaku "Rotaflex") we were able to collect 2×10^5 counts per angle step in the measured intensity curves. Details of the experimental procedure and the data treatment will be published elsewhere /9/. Radial distribution functions for SiP powder sample and VD-prepared sample in the form

$$U(r) = 4\pi r^2 \left[\varrho(r) - \varrho_0\right]$$

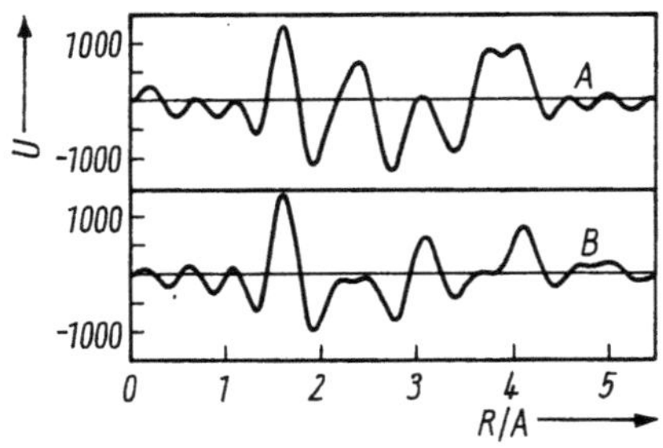

Fig. 1. RDF curves for SiO powder (A) and VD-prepared SiO (B). $s_{max}(A) = 15.3$ Å^{-1}, $s_{max}(B) = 14.9$ Å^{-1}

are shown in Fig. 1. Here $\varrho(r)$ represents the structure sensitive part of the electron density; ϱ_o is the mean electron density in the sample. Significant differences due to the preparation method can be seen in the curves A and B. The stoichiometry x of both samples can be obtained from the area of the first maximum in the RDF. From the number of nearest oxygen neighbours around a silicon atom ($n_o = 1.97 \approx 2$) the stoichiometry $x = 0.98 \approx 1$ is obtained in silicon oxide powder. In the vacuum deposited material (VD silicon oxide) $n_o = 2.21 \approx 2.2$ leads to $x = 1.1$ demonstrating the oxygen implantation during the preparation procedure under high vacuum

Table 1

Locations of RDF peaks and contributing atoms

silicon oxide (powder)

i	1	2	3	4	5	6
r_i (Å)	1.62	2.44	3.12	3.80	4.12	-
pair	$Si-O^I$ SiO_2	$Si-Si^I$ $O-O^I$ Si SiO_2	$Si-Si^I$ SiO_2	$Si-Si^I$ Si	$Si-O^{II}$ SiO_2	-

silicon oxide (VD-prepared)

i	1	2	3	4	5	6
r_i (Å)	1.62	2.27	2.51	3.12	3.75	4.11
pair	$Si-O^I$ SiO_2	$Si-Si^I$ Si	$O-O^I$ SiO_2	$Si-Si^I$ SiO_2	$Si-O^{II}$ $Si-Si^{II}$ SiO_2 Si	$Si-O^{II}$ SiO_2

conditions using additional LN_2-cooled cryosurfaces. An overview of the different peaks in both RDF curves and possible contributing atom pairs is given in Table 1. In contrast to /7/ we found in all cases the length of a Si-O bond at 1.62 Å - a value well-known from investigations on amorphous silica (see, e.g., /10/). If the possible angle range of the scattering function is used completely ($s_{max} = 4\pi \sin \Theta_{B\,max}/\lambda = 15$ Å^{-1}; Θ_B Bragg angle, λ wavelength) the fourth peak of the RDF splits into a double maximum at 3.8 and 4.12 Å. The 3.8 Å peak is in good agreement with the edge length of a $Si(Si_4)$ tetrahedron (i.e. 3.84 Å assuming a Si-Si bond length of 2.35 Å). RDF curves known from amorphous silica show a contribution at about 3.8 Å also (Si-2nd O), but this peak is very different in height from the state mentioned above. So the interpretation as a Si-2nd Si contribution from a-Si seems to be possible. In the RDF of VD-prepared SiO this peak at 3.8 Å occurs drastically decreased in height without any significant change in its radius. This can be understood as a result of a decreasing number of pure $Si(Si_4)$ tetrahedra due to the evaporation procedure. Furthermore we found that the second RDF peak splits into two peaks at 2.27 and 2.51 Å after vacuum deposition. The second maximum for SiO powder is located at 2.44 Å and can be explained as a superposition of contributions from the Si-Si distance (a-Si, 2.35 Å) and the O-O distance (a-Si, 2.65 Å). The visible separation of these two peaks after evaporation seems to be possible because of the decreased Si-Si contribution (atomic scattering factors). The differences in comparison to the exact locations at 2.35 and 2.65 Å are due to ripples from the Si-O and the Si-Si peak. Peak areas in atoms - evaluated from experimental data - are compared with those obtained from different structure models (see Table 2). It can be seen without any difficulty that the assumption of a macroscopic mixture model for both SiO powder and VD-prepared SiO is not consistent with the experimental results and has to be excluded. The description of SiO powder by means of the microscopic mixture model /6/ leads to better results than the use of the statistical random bonding model /5/. The evaluation of the probabilities $c_0, \ldots, c_4$ for finding the different types of tetrahedra shows that in SiO powder about 2/3 of the whole structure is built up by $Si(Si_4)$ and $Si(O_4)$ tetrahedra (about 1/3 each); partially oxidized tetrahedra altogether give a contribution of about 1/3. These results are in best agreement with earlier investigations on SiO powder /6, 7/. The microscopic mixture model is also useful for the VD-prepared SiO sample (i.e. $SiO_{1.1}$),

T a b l e 2

Peaks areas in atoms obtained experimentally in comparison with the results of different structure models (macroscopic and microscopic mixture model (MM), random bonding model)

SiO (powder)	$A(r_1)$	$A(r_2)$	$A(r_3)$
macroscopic MM	1.70	2.48	1.79
microscopic MM /6/	1.70	2.46	3.03
random bonding model /5/	1.70	2.29	4.80
experiment	1.70	2.63	3.03

$SiO_{1.1}$ (VD)	$A(r_1)$	$A(r_2, r_3)$	$A(r_4)$
macroscopic MM	1.95	2.49	2.08
microscopic MM /6/	1.95	2.27	4.46
random bonding model /5/	1.95	2.30	5.28
experiment	1.95	2.25	4.46

but a distinct tendency toward the random bonding model can be observed. We found the vapour deposited sample to consist of $Si(Si_4)$ and $Si(O_4)$ tetrahedra at about 1/10 each; the remaining part of the structure is built up by partially oxidized tetrahedra ($Si(SiO_3)$, $Si(Si_2O_2)$, $Si(Si_3O)$). This result is a possible explanation of optical measurements on SiO powder and VD-prepared SiO /11/. Properties which can be predicted by a random bonding model were found in the vapour deposited sample only. From the optical point of view commercially available SiO powder shows a behaviour like a simple mixture of a-Si and a-SiO_2. The strong influence of the preparation method on the atomic structure of SiO_x is much better recognizable by comparing the experimentally obtained structural data with a modified random bonding model /12/ which employs changed tetrahedra statistics (see also /9/).

We are very thankful to H. Steil and C. Müller for helping with the X-ray diffraction measurements and to G. Zuther for helpful discussions.

References

/1/ G.W. BRADY, J. phys. Chem. 63, 1119 (1959).
/2/ S.C.H. LIN and L. YOSHI, J. Electrochem. Soc. 116, 1740 (1969).
/3/ M.V. COLEMAN and D.J.D. THOMAS, phys. stat. sol. 22, 593 (1967).
/4/ B.J. MOLNAR and D.B. DOVE, Bull. Amer. Ceram. Soc. 51, 374 (1972).
/5/ H.R. PHILIPP, J. non-crystall. Solids 8-10, 627 (1972).
/6/ R.J. TEMKIN, J. non-crystall. Solids 17, 215 (1975).
/7/ J.A. YASAITIS and R. KAPLOW, J. appl. Phys. 43, 995 (1972).
/8/ K. HÜBNER, phys. stat. sol. (a) 40, 133 (1977).
/9/ H.-G. NEUMANN, TH. ROY, K. HÜBNER, and R. ENGELKE, to be published.
/10/ R.L. MOZZI and B.E. WARREN, J. appl. Cryst. 2, 172 (1969).
/11/ G. ZUTHER, Wiss. Z. W.-Pieck-Univ. Rostock, to be published.
/12/ K. HÜBNER and R. ENGELKE, to be published.

(Received May 9, 1979)

phys. stat. sol. (a) 54, K113 (1979)
Subject classification: 18.2; 21.1; 21.4

Institute of Metal Physics, Academy of Sciences of the USSR, Sverdlovsk

Magnetic Heterogeneities in $Y(Co_{1-x}Ni_x)_5$ Alloys and Their Relation to Coercive Force

By
A.S. ERMOLENKO, A.Z. MENSHIKOV, and YU.A. DOROFEEV

An interesting phenomenon has been shown to exist in quasi-binary compounds of the $R(Co_{1-x}Ni_x)_5$ type, where R is a rare earth, yttrium or thorium /1, 2/. The coercive force H_c at low temperatures is equal to 10^3 to 10^5 Oe for the bulk single crystals of compounds with intermediate x values. This coercive force is connected with domain wall pinning. Though many explanations were suggested for the mechanism of this pinning, their essence can be reduced to two models /3/. One of them /4/ is based on the intrinsic coercive force mechanism. The crystal structure is considered to be ideal in this model. An alternative model was proposed by Ermolenko et al. /5, 6/ who underlined the role of atomic structure heterogeneities as thin domain wall pinning centers in quasi-binary compounds.

The statistical distribution of Co and Ni atoms among the crystal lattice sites provides such heterogeneities to be an inevitable property of $R(Co_{1-x}Ni_x)_5$ compounds in the quasi-binary region. Therefore, one can expect spatial variations of magnetic moment, exchange parameter, and magnetocrystalline anisotropy constants in these alloys. The purpose of the present paper is to give experimental evidence for the existence of magnetic heterogeneities in $Y(Co_{1-x}Ni_x)_5$ compounds and to reveal their relation to the coercive force.

The method of low-angle neutron scattering was used. The intensities of scattered neutrons were measured at temperatures of 4.2, 80, and 300 K for angles 2Θ between 1° and 7°. The wavelength of neutrons was $\lambda = 1.165$ Å.

The ingots of $Y(Co_{1-x}Ni_x)_5$ alloys studied earlier /6/ were used in this paper. They were annealed at 1100 °C for 24 h, and consisted of single phase with $CaCu_5$ structure type according to metallography and X-ray analysis. The ingots were crushed into powders with particle sizes ≤ 100 µm. These powders were used for the neutron diffraction experiments.

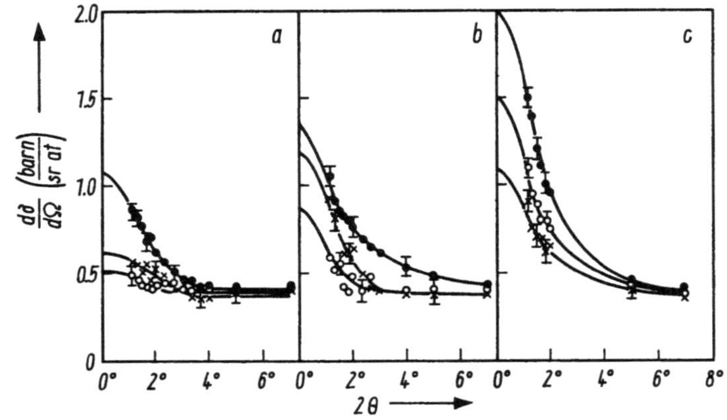

Fig. 1. The angular dependence of the neutron scattering intensity for $Y(Co_{1-x}Ni_x)_5$ alloys at temperatures a) 4.2, b) 80, and c) 300 K. x = ● 0.6, × 0.7, o 0.5, points experiments, — calculation

The angular dependence of the scattered neutron intensity is plotted in Fig. 1. One can see that the effect of the low-angle scattering is observed for the alloys $YCo_{2.5}Ni_{2.5}$, YCo_2Ni_3, and $YCo_{1.5}Ni_{3.5}$. There is no much effect for alloys with $x < 0.5$ and $x > 0.7$. The low-angle scattering effect is very sensitive to the temperature T and composition x. It is a strong evidence in favour of the magnetic nature of the low-angle scattering.

The low-angle scattering does not vanish at 4.2 K. This fact shows that the effect is not connected with critical scattering on thermal dynamical magnetization fluctuations because Curie points of alloys are high enough (130, 280, and 440 K for x = 0.7, 0.6, and 0.5, respectively).

It is more likely that the low-angle neutron scattering is caused by the occurrence of static magnetic heterogeneities which is a fundamental property of these compounds. The curves of Fig. 1 were used for the calculation of the linear size of heterogeneities. The expressions given in paper /7/ were applied for this purpose. The estimations are rather crude and do not allow to reveal the temperature and composition dependences of the heterogeneity sizes which are equal by an order of magnitude to ≈ 10 Å. This estimation is somewhat lower in comparison with ≈ 20 Å for the compound $SmCo_{2.5}Ni_{2.5}$ /8/. The latter figure was obtained from a magnetic reversal kinetics study at 50 K.

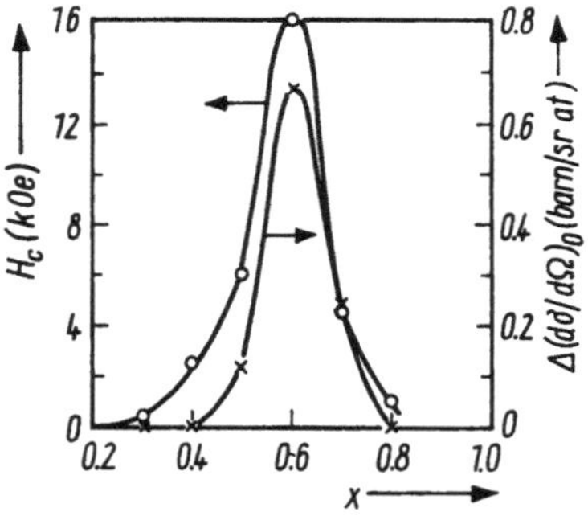

Fig. 2. The composition dependence of the coercive force (o) and the neutron scattering intensity extrapolated to $2\Theta = 0$ (×) at 4.2 K for $Y(Co_{1-x}Ni_x)_5$ alloys

The existence of magnetic heterogeneities in $Y(Co_{1-x}Ni_x)_5$ alloys can be understood on the basis of the Jaccarino-Walker model /9/. According to this model one can suppose that Co and Ni atoms possess magnetic moments only in those crystal lattice sites which have among their nearest neighbours n_k or more Co atoms. The probability for the reference site to have among its nearest neighbours $n \geq n_k$ Co atoms is determined by the binomial distribution. The composition dependence of the magnetic moment calculated by Buschow et al. /10/ for the $Y(Co_{1-x}Ni_x)_5$ system in terms of this model has the best fit with the experiment if $n_k = 4$.

It is difficult to calculate the composition dependence of coercive force and intensity of low-angle neutron scattering because these parameters should be complex functions of the density of magnetic heterogeneities, their size, and size distribution. We only pay attention to a good correlation between experimental composition dependences of coercive force /6/ and the low-angle neutron scattering intensity extrapolated to $2\Theta = 0$ at $T = 4.2$ K (Fig. 2). It gives some reasons to consider the magnetic heterogeneities revealed by low-angle neutron scattering in $Y(Co_{1-x}Ni_x)_5$ alloys as having direct relation to the high fields of domain wall pinning in $R(Co_{1-x}Ni_x)_5$ compounds. The mechanism of this pinning was discussed earlier /3, 5, 6, 8/.

References

/1/ A.S. ERMOLENKO and A.V. KOROLYOV, Zh. eksper. teor. Fiz., Pisma 21, 34 (1975).

/2/ K.H.J. BUSCHOW, M. BROUHA, and A.G. RIJNBEEK, Solid State Commun. 16, 31 (1975).

/3/ A.S. ERMOLENKO, A.V. KOROLYOV, and A.F. ROZHDA, Izv. Akad. Nauk USSR, Ser. fiz. 42, 1762 (1978).

/4/ K. H. J. BUSCHOW and M. BROUHA, J. appl. Phys. 47, 1653 (1976).

/5/ A. S. ERMOLENKO, A. V. KOROLYOV, and A. F. ROZHDA, Fiz. Metallov i Metallovedenie 42, 518 (1976).

/6/ A. S. ERMOLENKO and A. F. ROZHDA, Fiz. Metallov i Metallovedenie 43, 312 (1977).

/7/ A. Z. MENSHIKOV, S. K. SIDOROV, and A. E. TEPLYKH, Fiz. Metallov i Metallovedenie 45, 979 (1978).

/8/ A. S. ERMOLENKO and A. F. ROZHDA, Fiz. Metallov i Metallovedenie 44, 535 (1977).

/9/ V. JACCARINO and L. R. WALKER, Phys. Rev. Letters 15, 258 (1965).

/10/ K. H. J. BUSCHOW, M. BROUHA, J. W. M. BIESTERBOS, and A. G. DIRKS, Physica B/C91, 261 (1977).

(Received April 2, 1979)

phys. stat. sol. (a) 54, K117 (1979)
Subject classification: 6 and 20.1; 20.2; 21.4

Department of Physics, University of Sulaimaniyah

Optical Properties of Laser Materials: Solid State Vibrational Spectra of Rare Earth: p-Nitrobenzoic Acid in the Region 800 to 200 cm^{-1}

By

S.P. TANDON[1)], P.P. VAISHNAVA[2)], and K. TANDON[3)]

The optical properties of a number of rare earth laser materials have been reported by Tandon and coworkers /1 to 4/. The absorption spectra of rare earth p-nitrobenzoic acid complexes, which can be used as laser materials, were first reported by /5/. With a view to determine the position of bands accurately in the region 800 to 200 cm^{-1}, their spectra have been recorded again on a Perkin-Elmer model 120.

Samples in the form of CsI pellets were used, since CsI absorbs very little in this region. The bands have been recorded with an accuracy of ± 1 cm^{-1}. The results of measurements have been given in Fig. 1. About eighteen bands have been recorded.

Most of the bands occur at the same frequencies as in the ligand. The bands due to C-H out-of-plane bonding occur at 788, 509, and 297 cm^{-1}; those due to C-C out-of-plane bending at 717, 449, and 210 cm^{-1}; C-C-C in-plane bending at 626 cm^{-1}; C-C in-plane bending at 204 cm^{-1}; combination band (443 + 263 = 706 cm^{-1}) at about 702 cm^{-1}; NO_2 symmetric bending at 697 cm^{-1}; NO_2 out-of-plane rocking at 315 cm^{-1} (however, this band is shifted to 339 cm^{-1} in Pr^{3+} and to 336 cm^{-1} in Nd^{3+} complexes); COO symmetric bending at 669 cm^{-1}; COO out-of-plane rock at 562 cm^{-1} and COO in-plane bending at 260 cm^{-1}. The COO in-plane frequency (414 cm^{-1}) and C-H out-of-plane bending at 788 cm^{-1} in ligands do not appear on complexation. The bands observed in complexes at 290 and 287 cm^{-1} can be assigned to M-O stretching, since they do not appear in the ligand. The band at 732 cm^{-1} in complexes also

---

1) On leave from Physics Department, University of Jodhpur, India.

2) Same as 1). Present address: Department of Chemistry, Heriot-Watt University, Edinburgh, U.K.

3) Department of Chemistry, University of Jodhpur, India.

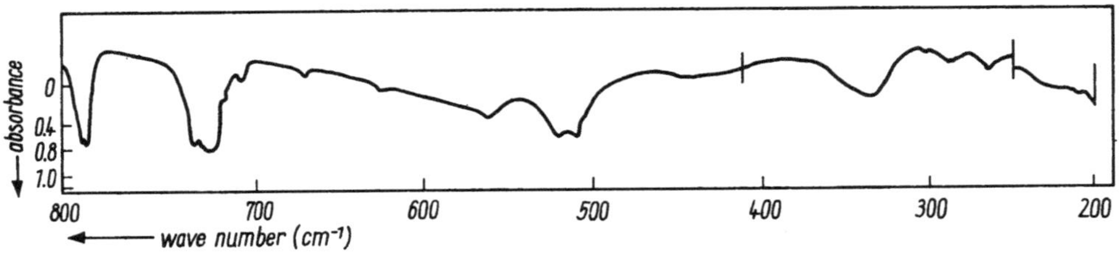

Fig. 1. Absorption spectrum of Nd^{3+}: p-NBA in the region 800 to 200 cm^{-1}. The absorption spectrum of Pr^{3+}: p-NBA was also similar

does not appear in ligand. This may be due to C-C out-of-plane bending. The above assignments are in agreement with others /6 to 9/ made for the ligand.

The authors are very thankful to the Universität Dortmund, for providing laboratory facilities and to the Alexander von Humboldt Stiftung for awarding Dozent fellowships to S. P. T. and K. T. Thanks are also due to Prof. Dr. A.N. Nigam and Prof. Dr. R.C. Kapoor, University of Jodhpur for discussion. One of us (S. P. T.) is also thankful to the Chairman, Physics Department, University of Sulaimaniyah, Iraq, for encouragement where a part of the data was analysed.

References

/1/ S. P. TANDON and P.C. MEHTA, J. chem. Phys. 52, 4313, 4896, 5417 (1970); 53, 414 (1970).

/2/ P.C. MEHTA, S.S.L. SURANA, and S. P. TANDON, Canad. J. Spectrosc. 18, 55 (1973).

/3/ S.S.L. SURANA, R.C. MATHUR, and S.P. TANDON, J. Phys. C 8, 2323 (1975).

/4/ S. P. TANDON and P. P. VAISHNAVA, Canad. J. Spectrosc. 22, 17 (1977).

/5/ I. P. TANDON, P.P. VAISHNAVA, and M. P. BHUTRA, Spectrosc. Letters 6, 239 (1973).

/6/ J.H.S. GREEN, W. KYNASTON, and A.S. LINDSAY, Spectrochim. Acta 17, 486 (1961).

/7/ M.D. TAYLOR, C.P. CARTER, and C.I. WYNTER, J. inorg. nuclear. Chem. 30, 1503 (1968).

/8/ S. PISCHES, D. SAMUEL, and M. WEISS, J. Chem. Soc. 2382 (1961).

/9/ K.C. MEDHI, Spectrochim. Acta 20, 675 (1964); Indian J. Phys. 39, 390 (1965).

(Received May 4, 1979)

phys. stat. sol. (a) 54, K119 (1979)
Subject classification: 11; 22.5.2

Institute of Physics, Academy of Sciences of the Latvian SSR, Riga[1)]

Annealing of NaCl Crystals Irradiated by High Electron Doses at High Temperatures

By
J. Z. KRISTAPSON, V.I. OZERSKII, and V.A. CHADAEV

In most of the studies on the thermal stability of color centers in irradiated alkali halides the crystals are irradiated at or below room temperature. This can partly be attributed to the fact that increasing the irradiation temperature causes a reduction of the total production efficiency of color centers in alkali halide crystals. In this case, color centers are not stable at low irradiation rates. Powerful radiation sources, however, allow to produce color centers which remain stable at higher irradiation temperatures /1, 2/.

In the present paper we report on results with synthetical NaCl crystals irradiated by 3.5 MeV electrons at a dose rate of the order of 1 Mrad/s to doses up to 2.6×10^4 Mrad at temperatures between 60 and 300 $^{\circ}$C. In these crystals annealing processes were studied by change of absorption spectra during annealing, electron micrographs of surface cleaning, and change of density (Fig. 1). The measurements provided information about the defects produced by radiation and their thermal stability. In particular for the first time, they allowed to establish two clear stages of high temperature annealing: 340 and 520 $^{\circ}$C. This allows us to make assumptions as to the peculiarities of the metal and halogen radiolysis products of NaCl crystals.

These two stages of annealing are observed only in crystals irradiated at temperatures above 100 $^{\circ}$C. If the irradiation temperature is below 100 $^{\circ}$C, complete annealing of defects is obtained at 200 to 250 $^{\circ}$C, as has been reported previously /2/, whereas the second stage has not been observed.

An increase in irradiation temperature from 100 to 300 $^{\circ}$C brings about a relative increase in the number of defects annealed in the second stage, and a corresponding reduction of the number of defects annealed in the first stage. Increasing irradiation temperature and dose causes a shift of both annealing stages to the high temperature range.

---

1) Riga-Salaspils, 229021, USSR.

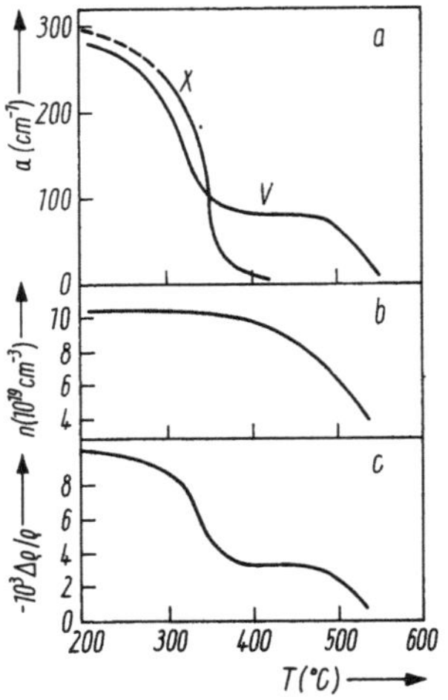

Fig. 1. Annealing curves for a NaCl crystal irradiated with 3.5 MeV electrons at 200 °C up to 2.5×10^4 Mrad. Dose rate 0.95 Mrad/s. Annealing is performed in temperature steps of 20 °C by heating in the dark during 10 min. a) Absorption coefficient in the maximum of the X band (560 to 580 nm) and in the maximum of the V band (230 to 210 nm); b) total concentration of atoms of colloidal metal (according to electron microscopy with the Tesla Bs-61 3 using gilded carbon replica); c) relative change in density (measured by a thermogradient tube)

The results (Fig. 1) show that in the first stage (340 °C) the annealing of the largest part of the V color centers is followed by a complete annealing of X color centers, i.e. recombination V + X takes place. In the second stage, recombination of the remaining part of V centers and large colloidal particles of Na occurs, which is confirmed by the annealing curves representing data obtained by electron microscopy. The density measurements particularly show that different V centers are annealed in the first and in the second stage. Here we shall consider only the nature of these V centers.

The combination of high temperature and high irradiation dose produces thermally stable radiolysis products, i.e. aggregates of halogen and metal centers. The coagulation of metal centers is assumed to proceed as follows: F → X → large colloidal particles of Na /3/. The coagulation of halogen centers probably occurs in a similar way: V → large aggregates of V centers. Formation of these large aggregates of V centers is enhanced by high local concentrations of these centers which is similar to the F center clusters by the reaction F → X /2/.

The activation energy of the thermal decomposition of X centers in these crystals colored either by radiation at 200 °C or additively is the same 1.6 eV /4/. This and the identical temperature range of annealing show that the recombination of metal and halogen radiolysis products is due to the decay of the metal components of radiolysis. Thus high temperature irradiation (200 °C) produces predominantly thermally stable aggregates of halogen centers. It should be noted that the annealing temperature of dislocation loops /5/ coincides with the second annealing stage described here.

References

/1/ S. YOSHIDA and T. IKEDA, J. Phys. Soc. Japan 14, 473 (1959).

/2/ J. Z. KRISTAPSON and S.S. KALIMULIN, Izv. Akad. Nauk Latv. SSR, Ser. fiz. tekh. Nauk No. 5, 14 (1971).

/3/ V.I. GOTLIB, J. Z. KRISTAPSON, K.K̇. SHVARTS, and I.A. EKMANIS, in: Radiatsionnaya fizika, Vol. 7, Izd. Zinatne, Riga 1973 (p. 143).

/4/ I.H. KALNINS and V.I. OZERSKII, Izv. Akad. Nauk Latv. SSR, Ser. fiz. tekh. Nauk No. 2, 39 (1979).

/5/ L.W. HOBBS, A.E. HUGHES, and D. POOLEY, Proc. Roy. Soc. (London) A 332, 167 (1973).

(Received May 8, 1979)

phys. stat. sol. (a) 54, K123 (1979)
Subject classification: 12.1; 21

O. Yu. Schmidt Institute of Physics of the Earth,
Academy of Sciences of the USSR, Moscow

Micro-Indentation of Aluminium under Hydrostatic Pressures up to 8500 kp/cm^2

By
L. D. LIVSHITS and A. G. FOMICHEV

Microhardness test is successfully used to study the effect of structure, composition, and temperature on the physical and mechanical properties of solids. The attractiveness of employing this method under high pressures lies in the possibility of testing miniature specimens and determining the properties of the components in multiphase systems and aggregates, which in other types of tests are masked by the integral properties, as well as the possibility of obtaining direct evidence about the effect of uniform compression on the micromechanism of plasticity. The study of the effect of pressure on the relationship between microhardness and indentation duration is also of considerable interest.

Despite the attractiveness mentioned above, among the innumerable publications devoted to microhardness, it seems that only one paper by Brauer et al. /1/ deals with the application of this method under high pressure. They found that a 2.3-fold increase of the microhardness of barite at 10^4 kp/cm^2 compared to that at 1.0 kp/cm^2. Unlike in standard tests, they made indentations, using the high kinetic energy of a falling indenter.

In order to examine the potentialities of the micro-indentation technique under high pressures to study the plastic deformation and fracture phenomena, in the present work we investigated the variation of hardness of aluminium by static indentation[1)] under hydrostatic pressures $p \leq 8.5 \times 10^3$ kp/cm^2.

The experiment consisted in creating several indentations on a specimen under four different pressures and then measuring the recovered impressions under atmospheric conditions.

---

1) The technique and the preliminary results of this work were reported orally at the IV All-Union Conference (Tbilisi, October 9 to 13, 1974) and at the Annual Meeting at GEOKhI of the Academy of Sciences of the USSR (Moscow, March 28 to 29, 1978).

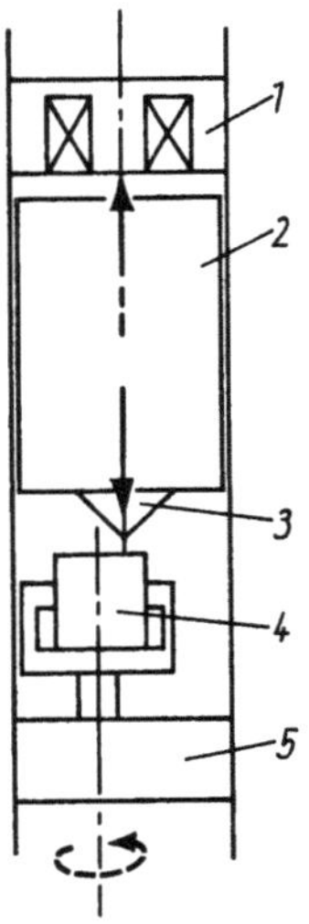

Fig. 1. High pressure microhardness device

Cylindrical specimens of size ⌀ =6x5.5 mm^2 were electro-spark cut from annealed (500 ^{o}C, 2 h) billets of 99.97% pure cast metal. The test surfaces were electropolished in $HNO_3 + CH_3OH$ (3:7) solution which also etches the grain boundaries. The sections of most specimens revealed large (0.5 mm or larger), mainly, isometric grains. The average microhardness of aluminium determined, using a conventional microhardness meter PMT-3, from n = 4 to 20 impressions in one series was in the range from 20.2 to 22.5 kp/mm^2; the coefficients of variation of these measurements were $W_n \leq 6.3\%$; the loads applied were from 35 to 50 p. The impression diagonals d were measured by a microscope of 0.45 µm resolution.

Experiments were carried out on a general-purpose setup comprising M12 and M20 multipliers and a pressure vessel /2/.[2)] The pressure was transmitted by glycol-glycerine mixture ($\rho_o = 1.2$ g/cm^3) and was measured by spring pressure gauges. Control tests were also carried out in organo-silicon liquid having a different density ($\rho_o = 0.97$ g/cm^3), compressibility, and viscosity under pressure than those of glycol-glycerine mixture. No difference was observed in H_μ of aluminium measured in air and in these fluids.

In pressure tests, indentations were made by means of the device shown in Fig. 1. It is designed for use with a vessel of inner size ⌀ =20x200 mm^2 and comprises an electromagnet 1 with a vertical armature 2 carrying an indenter 3, a holder 4 mounted on the shaft of a rotary step motor 5, magnetoelectric regulator to control the armature drop rate, and armature position transducer, both not shown in the figure. The indentation device is all-electric operated, it can lower the armature at a preset rate until the indenter contacts the specimen surface, gradually increase the indenter load up to the full armature weight, apply full load for a given time period, lift and arrest the armature, and shift the specimen. The indenter used was a standard diamond Vickers

2) A setup of this type of construction was used by several workers and unfortunately, they do not always give the proper references.

pyramid. The indenter speed (for all test pressures) was 30 to 50 or 140 to 230 μm/s; the full load application time was 10 to 15 s.

The effective indenter load p_{eff} (about 40 p) was measured under atmospheric pressure by hydrostatic weighting of the armature; it was additionally checked by H_μ measurement in air, water, and the working fluid.

Four indentations were made with each pressure (i.e. 16 in all) near the centre of the polished section and spaced 400 to 500 μm (7 to 10 d) apart.

The sequence of application of four predetermined pressures was chosen at random. The treatment of the measurement results for all specimens showed that the sequence of pressure application and repeated pressurization have no effect on H_μ. This is not contradictory to the data reported in /3 to 5/, wherefrom it follows that the hydrostatic pressure does not cause any residual changes in the mechanical properties of annealed polycrystalline aluminium but may reduce the yield stress of prestressed metal.

The H_μ measurements made at a given pressure have a w_n = 4.4 to 6.6%, therefore twenty impressions are sufficient to evaluate $H_{\mu p}$ (mean microhardness under a given pressure), to 3% accuracy. The correction in $H_{\mu p}$ for decreased p_{eff} under pressure due to the liquid compressibility and for increased d due to additional impression recovery resulting from the compressibility of aluminium do not exceed -2.5% (-3.5% in the control test) and +0.8%, respectively.

The $H_{\mu p}$ data measured on six specimens (including one in organo-silicon fluid) are presented in Table 1, which also lists $\Delta H_{\mu p}$ (halfwidth of 95% confidence interval for $H_{\mu p}$). From Table 1 it follows that for non-prestressed aluminium, the application of uniform pressure approximately four times greater than the mean contact stress (which is about 1.08 H_μ) increases the latter by 17±5%.

Table 1

p (10^3 kp/cm^2)	$H_{\mu p}$ (kp/mm^2)	$\pm\Delta H_{\mu p}$
0.1	22.2	0.6
2.4	23.4	0.5
6.2	24.0	0.5
8.5	26.0	0.75

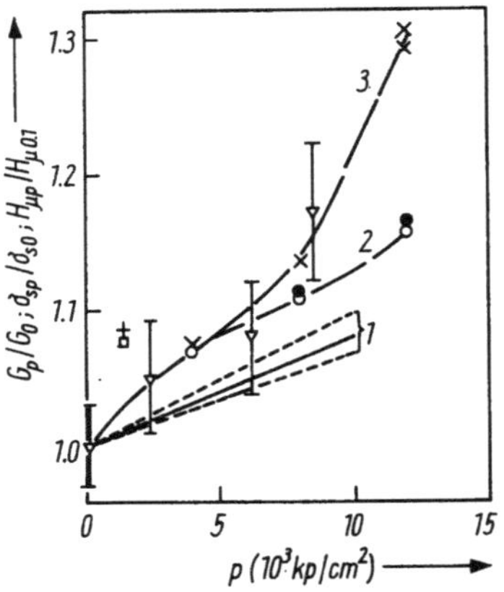

Fig. 2. Pressure dependence of shear modulus (G_p/G_o), yield stress (σ_{sp}/σ_{so}) at different values of deformation in extension (e) and microhardness ($H_{\mu p}/H_{\mu 0.1}$). The subscripts "p", "0.1", and "o" signify the values of corresponding quantities at pressure p, 100 kp/cm^2, or atmospheric pressure, respectively. The values of σ_{sp}/σ_{so} were calculated from the "stress-strain" curves for p = const given in /4, 6/: o , □ σ_{sp}/σ_{so} at e = 8% according to /4/ and /6/, respectively; ● σ_{sp}/σ_{so} at e = 10% according to /4/; ×, + σ_{sp}/σ_{so} at e = 20% according to /4/ and /6/, respectively; ∇ $H_{\mu p}/H_{\mu 0.1}$ present experiment, the vertical bars represents the 95% confidence limits. (1) G_p/G_o according to /7, 8/; (2) σ_{sp}/σ_{so} at e = 8 to 10%; (3) σ_{sp}/σ_{so} at e = 20%

In Fig. 2 the relative increase in aluminium microhardness with pressure is compared with the respective increase in the shear modulus G_p/G_o and tensile yield stress in polycrystalline specimens under three-axial stressed state (under hydrostatic pressure) /4, 6 to 8/. Such a comparison is possible because there exists a proportionality between the Vickers hardness and the yield stress σ_s under simple extension; for annealed metals σ_s corresponds to the axial strain e = 8 to 10% /9/. To evaluate the mean normal stress σ under indentation and under extension, representing an indentation by centre-symmetrical compression of the material /10/, and taking account of the uniformity of the aluminium strain under extension /4/, we can show that the difference in σ for these two tests varies with pressure like $H_{\mu p}$ and is equal to ca. (1.1 to 1.2)x10^3 kp/cm^2 at p = 0, and (1.3 to 1.4)x10^3 kp/cm^2 at p = = 10^4 kp/cm^2. So the pressure in Fig. 2 may be interpreted as the increment of σ.

Thus, from two independent measurements it follows that an increase in the mean normal stress up to 10^4 kp/cm^2 markedly (as compared to the variations in the shear modulus) increases the yield stress of aluminium. Moreover the corresponding variations in $H_\mu(p)$ and $\sigma_s(p)$ suggest that the effect of p (or σ) is controlled by the changes in the resistance to plastic deformation in crystallites.

References

/1/ K.H. BRAUER, R. HARTUNG, and S. WINKLER, Kristall und Technik 7, 1239 (1972).

/2/ L.D. LIVSHITS and E.D. MARTYNOV, Prib. i Tekh. Eksper. 3, 161 (1963).

/3/ A.I. PETROV and V.I. BETEKHTIN, Fiz. tverd. Tela 12, 2587 (1970).

/4/ S. YOSHIDA and A. OGUCHI, Trans. Japan. Inst. Metals 11, 424 (1970).

/5/ M. OHNAMI et al., Proc. 16th Japan. Congr. Mater. Res., Kyoto 1973 (p. 115).

/6/ M. OHNAMI, K. SHIOZAWA, and A. KAMITANI, Proc. 4th Internat. Conf. High Pressure, Kyoto 1974 (p. 228).

/7/ F.F. VORONOV and L.F. VERESHCHAGIN, Fiz. Metallov i Metallovedenie 11, 443 (1961).

/8/ M.W. GUINAN and D.J. STEINBERG, J. Phys. Chem. Solids 35, 1501 (1974).

/9/ D. TABOR, The Hardness of Metals, Oxford 1951.

/10/ T.O. MULHEARN, J. Mech. Phys. Solids 7, 85 (1959).

(Received April 9, 1979)

phys. stat. sol. (a) 54, K129 (1979)

Subject classification: 1.4; 21

Institut National des Sciences et Techniques Nucléaires, Gif-sur-Yvette[1] (a) and Institute of Nuclear Research and Nuclear Energy, Sofia[2] (b)

Dynamic Study of Cu Precipitation in Al + 4 wt% Cu by Positron Annihilation

By

B. G. HOGG (a), R. PAULIN (a), and T. D. TROEV (b)

Recently, two different groups have reported /1, 2/ that the presence of copper precipitates in Al + 4 wt% Cu alloys can be detected by positron annihilation. Both the Guinier-Preston zones 1 and 2 /3/ in precipitation hardening and the Θ and Θ' precipitates in the subsequent softening are found to induce changes, relative to the quenched sample, in the positron annihilation characteristics. In these experiments, the Doppler broadening technique was used to detect, as variations of the line shape parameter /4/, the modifications in the state of the alloy after various isothermal annealing. We study here how the development of G. P. zones can be followed kinetically through the positron annihilation characteristics measured in situ during the isothermal annealing.

We have measured the peak intensity of the angular correlation curve to characterize /5/ the positron behaviour in the alloy samples. After annealing for 2 h at 550 °C, our samples are quenched in ice water. They are then fixed to a holder equipped with an electric heater and the peak position of the angular correlation curve is experimentally determined. The peak counting rate corresponding to the quenched sample is first measured with an accuracy of ≈ 0.1%. That sample is then heated in situ and maintained at a constant temperature for the duration of the aging process while the peak counting rate is continuously measured. Aging was generally 1 or 2 days but in one instance a run was extended to 10 days. For isothermal aging at 130 °C, the data obtained with single and polycrystalline samples of Al + 4.28 wt% Cu are presented on Fig. 1. After a fast decline during the first hour, the main effect observed in the single crystal, as the hardness corresponding to G. P. 1 zone develops,

1) BP No. 6, 91190 Gif-sur-Yvette, France.

2) Sofia 13, Bulgaria.

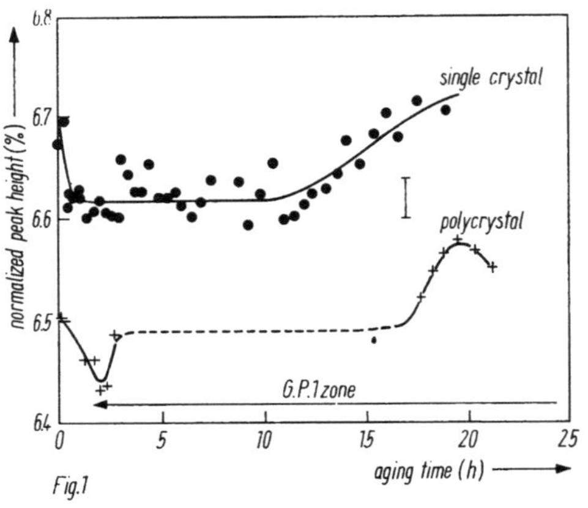

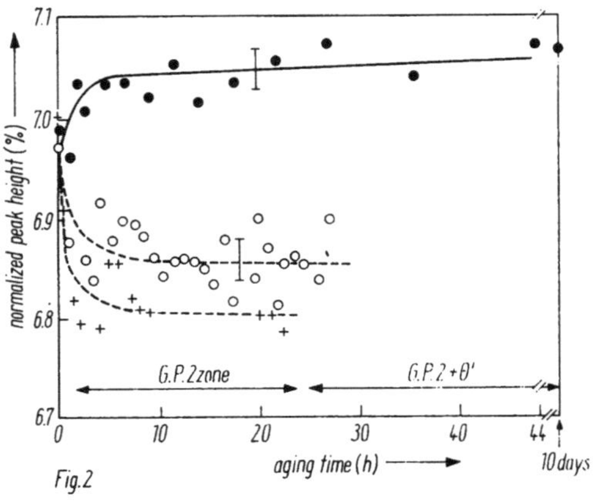

Fig. 1. Peak height of the angular correlation curve measured versus time for positron annihilating in an Al+4.28% Cu alloy during its isothermal aging at (130 ± 2)°C. The curves are only drawn to guide the eye

Fig. 2. Peak height of the angular correlation curve measured versus time for positron annihilating in Al-Cu alloys during isothermal aging at (190 ± 2)°C; ● 4.28 wt% Cu, single crystal; o 4.00 wt% Cu, polycrystal; + 4.28 wt% Cu, polycrystal

is a small but continuous increase of the peak height after an aging of 10 h. The same trends appear for the polycrystal despite the incomplete set of data. It must be noticed that the peak parameter for the polycrystalline sample always remains significantly lower than for the single crystal. To study the Cu precipitations producing the G. P. 2 zone hardness and the subsequent Θ' softening, these two samples and a new polycrystalline sample containing only 4.00 wt % Cu were isothermally heated at 190 °C. The data are presented on Fig. 2. Again the single crystal results lie above the polycrystals and a fast decline is detected for the three samples in the first hour. Then, during the entire time of the G. P. 2 zone precipitation, the data exhibit no significant variation, the very small fluctuations underlined by the drawn curves on Fig. 2 have magnitudes which remain within the experimental uncertainties of ±0.3%. Thus, the limits of the positron method to detect the copper precipitations clearly appear from these in situ experiments. Instead of a positron sensitivity to all processes of Cu precipitation as concluded from previous experiments /1, 2/ only the possibility to follow the G. P. 1 zone formation as a narrowing of the angular (or energetic) distribution of annihilation γ quanta is confirmed. The other variations observed of the peak height parameter cannot be correlated to copper precipitation.

The fast broadening which occurs as soon as the quenched samples are heated results certainly from the annealing of points defects frozen during the quenching treatment as already suggested /1/. The difference in peak heights between polycrystals and single crystals, noticed by the Utah group /2/, can only be understood by assuming a larger concentration of stable trapping defects in single than in polycrystals. Instead of point defects as assumed by this group, the presence of dislocations seems to us a more realistic hypothesis because of their stability.

To explain the limitation of positron sensitivity to the G. P. 1 zone precipitation, various arguments can be advanced. Since the maximum magnitude of the hardness is smaller in the G. P. 2 than in the G. P. 1 zone /6/, one expects a smaller positron sensitivity to the G. P. 2 precipitation because the positron trapping by defects generally increases with the hardness. This qualitative argument supposes that the positron is simply trapped by defects resulting from the stresses produced in the crystal during the copper precipitation. But the reality of that hypothesis where dislocations and vacancies are suspected to be the positron traps induced by the precipitation process, is difficult to establish. We examine here to what extent a process where the positrons are directly trapped at the precipitate-crystal interfaces could eventually explain the present data.

In that model, the trapping effect has to be proportional to the probability that a positron during its diffusion path encounters a precipitate boundary which is assumed to act as a perfect absorbing barrier. Then the fraction F of positrons trapped at precipitate-crystal interfaces can be written as /7/

$$F = \frac{K\tau_c}{1 + K\tau_c} , \qquad (1)$$

where $\tau_c = 1.5 \times 10^{-10}$ s (is the positron lifetime in the alloy and K the positron trapping rate at the interfaces which can be expressed in terms of the β^+ diffusion constant D_+, the volume V, and the concentration C of the precipitates as

$$K \sim D_+ V^{1/3} C \qquad (2)$$

From (1) and (2), for the same quantity of precipitated copper, a larger value of F is expected in the G. P. 1 states where the concentration of pure copper clusters ($V_1 \approx 2x10^4$ Å^3) is about 10^4 times larger that the concentration of $CuAl_2$ precipitates ($V_2 \approx 5x10^8$ Å^3) in the G. P. 2 states /8/. In metals $D_+ \approx 0.1$ cm^2/s, so that for complete copper precipitation F is found $\approx$ 90% for G. P. 1 clusters and only $\approx$ 4% for G. P. 2 precipitates. That calculation which predicts that only the G. P. 1 precipitates will be efficiently probed by positrons is in good agreement with the present data.

We conclude that a model of positron trapping at the crystal-precipitate interfaces is not unrealistic and cannot be a priori ruled out in a quantitative analysis of the positron sensitivity to precipitation phenomena.

References

/1/ R. NAGAI, S. TANIGAWA, and M. DOYAMA, Proc. 4th Internat. Conf. Positron Annihilation, Helsingor (Denmark) 1976 (paper E27).

/2/ S. PANCHANADRESWARAN, R.W. URE, and J.S. BYRNE, phys. stat. sol. (a) 48, K83 (1978).

/3/ A. GUINIER, Ann. Physique 12, 161 (1939).

/4/ J.L. CAMPBELL, Appl. Phys. 13, 165 (1977).

/5/ W. BRANDT and R. PAULIN, Phys. Rev. B 8, 4125 (1973).

/6/ T. SATO, Y. KOJIMA, and T. TOKAHASHI, Japan.J. Light Metals 27, 488 (1977).

/7/ W. BRANDT and R. PAULIN, Phys. Rev. B 5, 2430 (1972).

/8/ J.M. SIBROCK, T.J. HEAL, and H.K. HARDY, J. Inst. Metals 82, 239 (1953).

(Received February 13, 1979)

phys. stat. sol. (a) 54, K133 (1979)
Subject classification: 1.1; 22.8.1

Physikalisch-Technische Bundesanstalt, Braunschweig

Lattice Parameters of $(Ba_xSr_{100-x})SO_4$ Doped with Europium

By
M. KRYSTEK

In chemical combinations with a common anion and two different cations, the existence of a mixed crystal series depends strongly on differences in the chemical character of these cations, as indicated by ionic radii, electronegativities, etc. If these parameters are similar one may expect complete solid solution in a wide temperature range. Furthermore the influence of high doping concentrations (1 mol% or more) on the parameters of the mixed crystal system should not be remarkable, if even a hundred-per-cent-substitution of the cations by foreign ions does not change the space group of the host material.

Since the ions Ba^{2+} and Sr^{2+} are very similar and barium sulphate ($BaSO_4$), strontium sulphate ($SrSO_4$), and europium sulphate ($EuSO_4$) crystallize in the same space group D_{2h}^{16}-Pnma, one expects no miscibility gap in the mixed crystal series of undoped and Eu-doped $(Ba_xSr_{100-x})SO_4$ in the whole range of x.

To prove this expectation, some samples were prepared by the following method. The material (Merck, p.a.) was weighted to the nearest mg and then dissolved in hot concentrated sulfuric acid. For the doped samples 0.5 mol% Eu_2O_3 was added (yielding an Eu concentration of 1 mol%). After complete dissolution of the components, the solution was diluted with pure water. Thus the mixed crystal powder was precipitated. Then the sample material was washed and dried at 400 °C for 2 h, to take away the remaining acid.

X-ray diffraction analysis was made according to Guinier's method. Ni-filtered Cu-Kα radiation (40 kV, 25 mA) was used and reduced to Cu-Kα_1 radiation (λ = 0.154056 nm) by a quartz monochromator crystal. As reference material As_2O_3 was taken. The experiments were done at room temperature.

In order to eliminate systematic errors, the measured X-ray reflection angles ϑ were corrected with help of the known ϑ values of As_2O_3, yielding ϑ_{korr} values with an uncertainty of $\pm 2.5 \times 10^{-3}$ degrees. There were found no

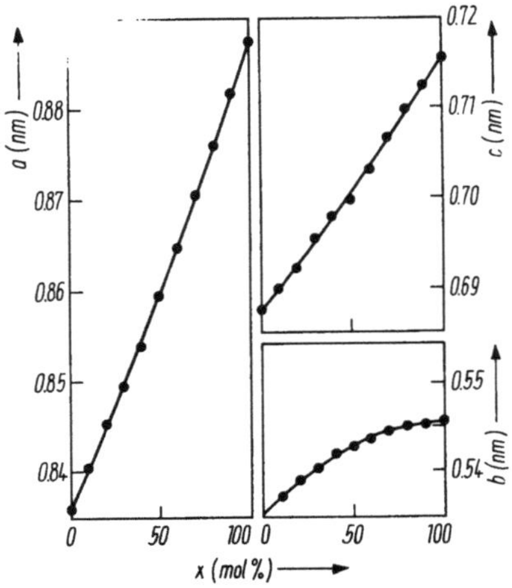

Fig. 1. Composition dependence of lattice parameters in $(Ba_xSr_{100-x})SO_4$:Eu

significant differences within the uncertainty limits between ..oped and undoped samples. From the ϑ_{korr} values the lattice parameters were calculated.

To the results quadratic functions are fitted by least squares approximations. These functions are given in Table 1 and also presented graphically together with measurement points in Fig. 1 and 2. The correlation is sufficient.

As one can see from Fig. 2, there is only a very small deviation of the unit cell volume from a linear x-dependence, i.e. Vegard's rule is nearly fulfilled. Therefore no miscibility gap exists in the mixed crystal system $(Ba_xSr_{100-x})SO_4$ doped with Eu up to 1 mol%. The influence of Eu doping up to these concentrations is not remarkable, because there was found no significant difference between lattice parameters of doped and undoped samples within an uncertainty limit of $\pm 10^{-4}$ nm. This result points out that Eu must be incorporated on regular lattice places, i.e. it is a real substitution defect of the crystal.

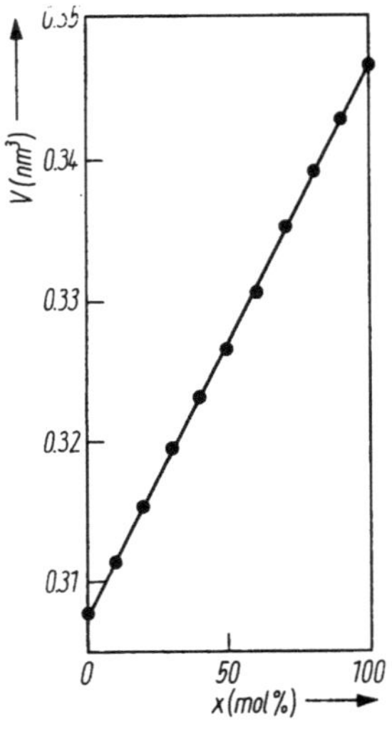

The author has to thank Prof. Tennyson from the Technische Universität Berlin (West) for the permission to do the measurements in her institute.

Fig. 2. Composition dependence of unit cell volume in $(Ba_xSr_{100-x})SO_4$:Eu

Table 1

Relationships for the composition dependence of lattice parameters and unit cell volume. x in mol%; a, b, c in nm; V in nm^3

$a = 0.8358 + 4.34 \times 10^{-4}\,x + 8.9 \times 10^{-7}\,x^2$	
$b = 0.5353 + 1.99 \times 10^{-4}\,x - 9.5 \times 10^{-7}\,x^2$	
$c = 0.6873 + 2.49 \times 10^{-4}\,x + 3.5 \times 10^{-7}\,x^2$	
$V = 0.3075 + 3.85 \times 10^{-4}\,x + 0.6 \times 10^{-7}\,x^2$	

(Received April 19, 1979)

phys. stat. sol. (a) 54, K137 (1979)
Subject classification: 13.4; 22

Forschungsbereich Physik, Kern- und Werkstoffwissenschaften der Akademie der Wissenschaften der DDR, Dresden[1]

Temperature Dependence of the Pre-Exponential Factor in the Glow Curve Theory

By
M. BALARIN

In the theory of glow curves sometimes it is supposed that besides a usual Boltzmann factor also the pre-exponential factor can exhibit some temperature dependence attributed to carrier concentration, cross section, or mobility. This dependence, for instance, is expressed in the form T^a, where a is expected to be an integer or a half-integer mainly between a = - 2 and + 2 /1 to 3/. In this note a generalized integral (equation (1)) will be considered, which is necessary for the evaluation of the activation energy E from experimentally observable glow curves.

So far such an integral has been considered only for the special case a = 0 /4 to 10/, that means a neglect of the above mentioned temperature dependence. Therefore it is to be shown that some features of this integral retain their general meaning. After separation of the variables concentration C and temperature T in the kinetic equation for an experiment with constant heating rate q ≡ ≡ dT/dt the corresponding integrals are

$$\exp\left(Z_l(C)\right) \equiv -\int_{C_0}^{C(T)} \frac{dC}{C^l} = \frac{K_0}{q}\int_0^T \left(\frac{T'}{T''}\right)^a e^{-E/kT'}\,dT' \equiv \left(\frac{T}{T''}\right)^a S\,\eta_a(y) \qquad (1)$$

with the previous notation K_0 frequency factor, l order of the reaction kinetic, T'' any fixed temperature, only for conservation of the correct dimension; $y \equiv$ $\equiv kT/E$, usually $y < 1/15$.

The main temperature dependence is carried by the tempering function /4 to 8/

$$S(T;\, E,\, K_0/q) \equiv \frac{K_0\, kT^2}{qE}\, e^{-E/kT}. \qquad (2)$$

1) DDR-8027 Dresden, DDR.

It can be seen that with respect to S the correction factor /6 to 8/

$$\eta_a(y) \equiv y^{-(2+a)} e^{1/y} \int_0^y y'^a e^{-1/y'} dy' = 1 + \sum_{n=1}^{n'} (-y)^n \frac{\Gamma(a+n+2)}{\Gamma(a+2)} \quad (3)$$

which is always beneath but close to unity (see Fig. 1), is without any serious importance; therefore sometimes crude approximations or different numerical representations had been admitted /3, 8, 10/.

Although a priori η_a is known to be close to unity, it is necessary to determine its exact value to an order of accuracy which is intended for the evaluation of the activation energy (see Table 1). One of the methods for this evaluation of E uses the fact that $Z_1(C)$ versus reciprocal temperature is nearly a straight line, whose slope is governed by the activation energy /9/, but directly also by η_0 /6, 8/. Also for $a \neq 0$ one deduces with (1) to (3), in quite the same way as previously for the special case $a = 0$,

$$\frac{d Z_1}{d(1/kT)} = - \frac{E}{\eta_a} \ . \quad (4)$$

Because η_a shows only a slight variation with temperature ($d \ln \eta_a / d \ln T \approx \approx - (a + 2)y \ll 1$), mean values may be taken from experimental curves for a consistent result

$$E = - \langle \eta_a \rangle \left\langle \frac{d Z_1}{d(1/kT)} \right\rangle . \quad (5)$$

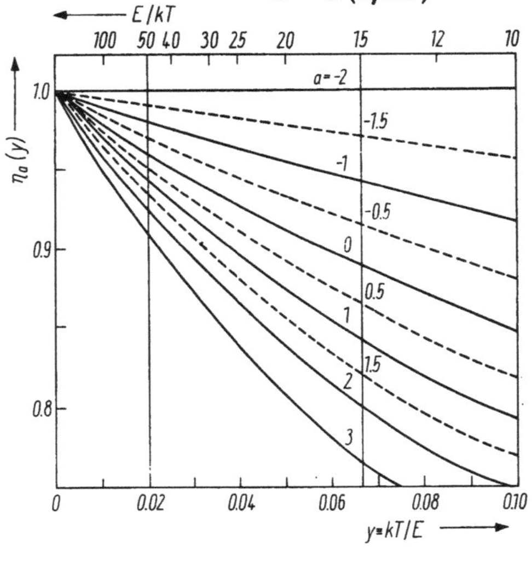

Fig. 1. Behaviour of the correction factor and slope contribution $\eta_a(y)$ for small $\bar{y} < 0.1$

Some comments are suitable for practical purposes: The semiconvergent and alternating polynomial (3) approaches the finite existing integral (1) optimally when the summation is restricted to $n' \leqq 1/y - 2$; obviously for $y < 1/15$ it converges quickly enough and if the

Table 1

$\eta_a(y)$ for selected E/kT

$\frac{E}{kT}$	$y \equiv \frac{kT}{E}$	$\eta_{-2}(y)$	η_{-1}	$\eta_0(y)$	η_{+1}	η_{+2}	η_{+3}
10	0.100	1.0	0.915	0.844	0.783	0.746	0.707
15	$0.06\bar{6}$	1.0	0.940	0.088	0.841	0.797	0.762
20	0.050	1.0	0.954	0.912	0.874	0.839	0.807
25	0.040	1.0	0.963	0.928	0.896	0.866	0.838
30	$0.03\bar{3}$	1.0	0.969	0.939	0.911	0.885	0.860
40	0.025	1.0	0.976	0.953	0.932	0.911	0.891
50	0.020	1.0	0.980	0.962	0.945	0.928	0.913

demanded accuracy is within 1% to 1‰ only few of the first terms of the series (3) must be taken into account. For convenience further

$$\eta_a(y) = 1 - (a + 2)y + (a + 2)(a + 3)y^2 - (a + 2)(a + 3)(a + 4)y^3 +-\ldots \quad (6)$$

$$= \eta_0 - ay + (a^2 + 5a)y^2 - (a^3 + 9a^2 + 26a)\,y^3 +-\ldots . \quad (6')$$

Numerical values η_a for various small y can also be approximated simply by

$$\eta_a(y) \approx \frac{1}{\sqrt{1 + 2(a + 2)y}} \quad (7)$$

whose error is of the order $\Delta\eta_a = ((a + 2)/2)y^2 \left\{a + (3a^2 + 6a - 4)y\right\}$, especially $\Delta\eta_2 \approx 4y^2$, $\Delta\eta_0 \approx - 4y^3$, $\Delta\eta_{-2} = 0$. It is evident that another special case is naturally a = -2 with $\eta_{-2} = 1$ identically, when integral (1) can be taken directly.

Another remark to the choice of a: this is not an object of an evaluation procedure; a comes out of the considered models. Now if any value a has been chosen the linearity of the curves Z(1/T) can be finally improved by one order in y extending a suggestion by Coats and Redfern /11/ for their case a = 0. Therefore instead of Z versus T^{-1} (notice $Z_1(C)$ is defined differently for every kinetic order l) one has to take $\ln \exp(Z_1)(\hat{T}/T)^{a+2} = Z_1 - (a + 2)\ln(T/\hat{T})$ versus T^{-1}; $\hat{T}$ arbitrarily chosen temperature, desirably close to the temperature of the glow peak. Then in the slope of this new curve the term with y^1 is eliminated automatically and that means

$$E = -\left\langle \frac{\eta_a}{1-(a+2)y\eta_a} \right\rangle \frac{d(Z_1 - (a+2)\ln(T/\hat{T}))}{d(1/kT)} \approx - \frac{d(Z_1 - (a+2)\ln(T/\hat{T}))}{d(1/kT)} \left\{ 1 + (a+2)y^2 - + \ldots \right\}.$$

This slope is influenced only in the relative order of few y^2, so that the activation energy can be taken directly from this slope.

References

/1/ R.H. BUBE, Photoconductivity of Solids, New York/London 1960 (Chap. 9).

/2/ P.M. KEATING, Proc. Phys, Soc. 78, 1408 (1961); J. Phys. Chem. Solids 22, 25 (1961).

/3/ R. CHEN, J. appl. Phys. 40, 570 (1969).

/4/ P. MURRAY and J. WHITE, Trans. Brit. Ceram. Soc. 54, 204 (1955).

/5/ G. BONFIGLIOLI, P. BROVETTE, and C. CORTESE, Phys. Rev. 114, 951, 958 (1959).

/6/ M. BALARIN and A. ZETZSCHE, phys. stat. sol. 2, 1670 (1962); 3, K387 (1963).

/7/ M. BALARIN, phys. stat. sol. (a) 31, K111 (1975).

/8/ M. BALARIN, J. Therm. Anal. 12, 169 (1977).

/9/ M. KRYSTEK, phys. stat. sol. (a) 40, K65 (1977).

/10/ E. MLITZKE, phys. stat. sol. (a) 47, K181 (1978).

/11/ A.W. COATS and J.P. REDFERN, Nature 201, 68 (1964).

(Received May 11, 1979)

phys. stat. sol. (a) 54, K141 (1979)

Subject classification: 10.2; 22.1.2; 22.2.1

Sektion Chemie und Werkstofftechnik
der Technischen Hochschule Karl-Marx-Stadt

Quantitative Electron Probe Microanalysis of Noble Gases in Sputtered Layers of Semiconductors

By

L. KÜCHLER and H.-J. HUNGER

Principles of the method The electron probe microanalysis (EPMA) is an important method for the analysis of sputtered layers of semiconductors. Whereas for the analysis of the main elements of semiconducting layers sufficient standards are available, up to now EPMA investigations of noble gases contained in the layers are not known, because of the absence of suitable standards. One possibility of a quantitative analysis of noble gases consists in producing hypothetical noble gas standards.

Green and Cosslett /1/ have shown that the intensity of the characteristic X-rays of each series is a continuous function of the atomic number Z. The intensities measured under identical conditions on standard specimens with different atomic numbers permit to interpolate/extrapolate the intensity of noble gases under investigation. Because not all of the necessary standards were available as pure elements, compounds of known composition were also used. Non-conducting compounds were evaporated with carbon films of about 30 nm thickness. The intensities measured on these compound standards were converted into the intensities of the corresponding pure elements after correction for effects of atomic number, absorption, and characteristic fluorescence /2/. To see an apparatus drift during the measurements, control measurements on germanium standard were made after measuring the line and background intensity of each specimen. A significant discernible drift was corrected linearly.

The intensities of several sets of measured data taken at different days were normalized and the mean values were plotted as a function of the atomic number. From these diagrams the intensities of hypothetical argon and krypton standards could be evaluated. These intensities were related to a suitable pure element standard (in our case Ge or Ti) and in this way we had produced a hypothetical noble gas standard.

The method demonstrated in this paper can be applied in principle to each electron probe microanalyzer. The given graphs, however, are specific for our instrument and are valid only for a special set of conditions (spectrometer, crystal and adjustment as well as acceleration voltage). The following measurements were made with the electron probe microanalyzer JEOL JXA-3A.

Results - argon The intensities were taken as function of Z with quartz as analyzing crystal and an acceleration voltage of 10 kV for elements from Z = 19 to Z = 26. As standards we had pure elements for Ti, V, Cr, Mn, and Fe and compounds for K (natural crystal of KCl) and Ca (natural crystal of $CaCO_3$). For Sc no suitable standards were available.

As result from five sets of data measured at several days and normalized to Ti (Fig. 1) we got for the hypothetical argon standard using Ti as reference

$$I(100\%\ Ar) = (0.79 \pm 0.08) I(100\%\ Ti).$$

The semiconducting layers which should be analyzed quantitatively with respect to argon were sputtered silicon layers on silicon substrate. The film thickness amounted to 1.3 μm. The exact spectrometer angle for the ArK_α line was evaluated by registering the X-ray spectrum near the ArK_α line. After this the line and background intensities were determined with an electron probe of 20 μm diameter on specimen and standard. It follows from Castaing's formula /3/ that the layers under investigation can be taken as massive at an incident beam energy of 10 keV. The evaluated intensity ratios from specimen and standard were corrected after /2/ and were listed in Table 1.

Table 1

Results of analysis of noble gases

specimen	element	concentration c (wt%)		
		LIMS (after /4/)	PIXA (after /5/)	EPMA
62	Ar	0.078	0.196	0.137
69	Ar	0.131	0.208	0.150
209	Kr	-	-	0.071
255	Kr	0.271	-	0.166

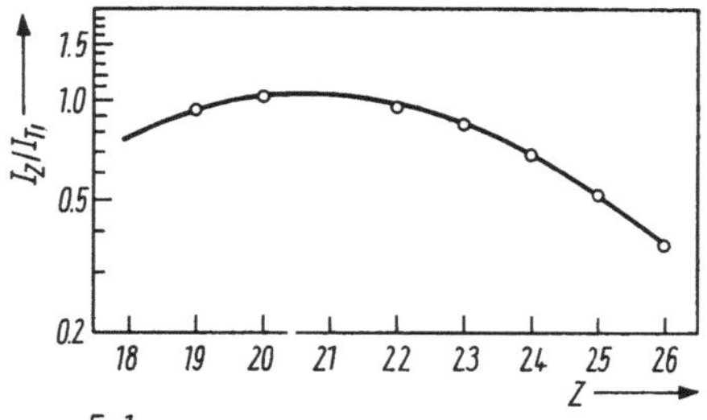

Fig.1

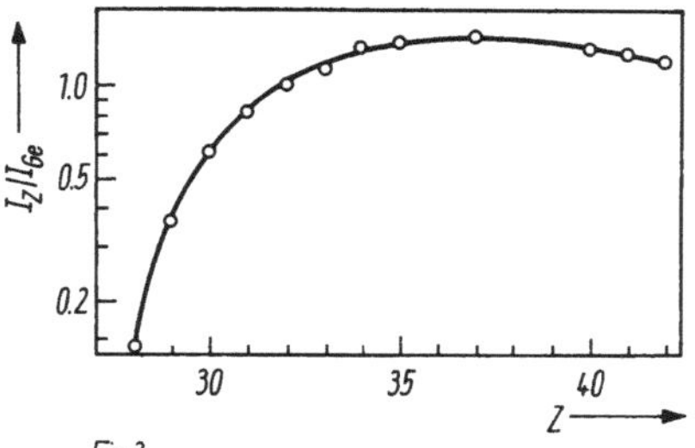

Fig.2

Fig. 1. Relative X-ray intensities as a function of the atomic number Z for JEOL JXA-3A with quartz as analyzing crystal at E_0 = 10 keV

Fig. 2. Relative X-ray intensities as a function of the atomic number Z for JEOL JXA-3A with KAP as analyzing crystal at E_0 = 10 keV

Krypton For the evaluation of the KrK_α intensity the X-ray intensities were measured for the elements with atomic number from Z = 28 to Z = 42 with KAP as analyzing crystal and an incident beam energy of 10 keV. As standards we had pure elements for Ni, Cu, Zn, Ge, As, Se, Zr, Nb, and Mo. Compounds were used for Ga (GaAs) and Rb and Br (RbBr). For Sr and Y no suitable standards were available.

Three sets of data measured at several days were normalized to Ge and the mean values are shown in Fig. 2. For the hypothetical krypton standard with Ge as reference we got

$$I(100\%\ Kr) = (1.40 + 0.04)I(100\%\ Ge)$$

The analyzed layers consisted of GaAs on silicon substrate and were produced by sputtering. By varying the incident beam energy it was found that the specimen can be taken as massive at a beam energy of 10 keV. The line and background intensities were measured with an electron probe of 20 μm diameter for Kr, Ga, and As on specimen and standard. The evaluated intensity ratios from specimen and standard were corrected after /2/ and are listed also in Table 1.

Comparison of the results with other methods Other methods have also been used for specimens which were produced simultaneously: laser induced mass spectrometry (LIMS) /4/ and proton induced X-ray analysis (PIXA) /5/. In the case of PIXA and EPMA the specimens could be taken as massive.

In the case of LIMS craters of several micrometer depth were generated (e.g. 7.1 μm for specimen 69). To convert the analyzed volume into the layer thickness it was assumed that the crater volume was cylindrical. The results of the various methods show good agreement in the order. If one takes into account that the ratio of the argon content of specimens 62 and 69 must be constant for all three methods, then one can see that PIXA (c_{62}/c_{69} = 0.94) and EPMA (c_{62}/c_{69} = 0.91) give nearly the same ratios and therefore the same relative contents. The results of LIMS (c_{62}/c_{69} = 0.60) deviate from the other two methods, because the layer cannot be taken as massive under conditions of LIMS and because the crater volume is not known exactly.

References

/1/ M. GREEN and V.E. COSSLETT, Proc. Phys. Soc. 78, 1206 (1971).

/2/ H.-J. HUNGER, L. KÜCHLER, G. MARX, and J. MÜLLER, 4th Conf. Mikrosonde, Dresden 1978 (p. 22).

/3/ R. CASTAING, Adv. Electronics, Electron Phys. 13, 353 (1960).

/4/ L.I. SAPRIKIN and G. YUDELEVICH, Inst. Anorg. Chem., Siberian Branch Acad. Sci. USSR, Novosibirsk , private communication.

/5/ W. RUDOLPH, (Central Inst. Nuclear Res., Acad. Sci. GDR, Rossendorf), private communication.

(Received April 16, 1979)

phys. stat. sol. (a) 54, K145 (1979)
Subject classification: 14.4.1; 22.8.1

Higher Institute of Civil Engineering and Architecture (a) and Institute of Electronics (b), Sofia

Electric Conductivity and Dielectric Properties of Ferroelectric PZT Ceramics Doped with Cr_2O_3

By

M.M. NADOLIISKY (a), S.D. TOSHEV (a), and T.K. VASILEVA (b)

The ferroelectric lead-zirconate-titanate ceramics exhibit useful piezoelectric properties in the range of the morphotropic phase boundary. In this range the solid $Pb(Zr-Ti)O_3$ solution represents a two-phase system with a tetragonal and a rhombohedral structure. The properties of the ceramics depend strongly on their technology and on the impurities. Substituting some other ions for Pb, Zr, or Ti ions improves the piezoelectric properties of the ceramics. Addition of some oxides makes the ceramics softer or harder. Addition of Cr_2O_3 improves their temperature and aging characteristics.

The electric dc resistivity of PZT ceramics doped with different impurities was studied by Takahashi /1/. He has found that some oxides change slightly its resistivity (e.g. Fe_2O_3, Al_2O_3); others (e.g. Nb_2O_5, La_2O_3) increase the resistivity enormously because of the electron-hole compensation; and a third group of oxides increases the resistivity at low concentration of the impurities but decreases it at higher concentration (e.g. Cr_2O_3, MnO). The increase of resistivity in the last case is connected with the electron-hole compensation while its decrease is connected with the generation of segregated impurity layers which enclose the grains of the solid solution.

The solubility limit of Cr_2O_3 in $Pb(Zr-Ti)O_3$ was determined /2/ to about 2.2 mol% of $(1/2)(Cr_2O_3)$ and later to 0.325 mol% /3/. In /2/ it was found by X-ray diffractometry that on increasing the amount of Cr_2O_3 three phases were present in the doped solid solution; a perovskite type (the grains themselves), a baddelite type ZrO_2, and a third phase of $2PbO \cdot CrO_3$ which later on was found to consist mainly of Pb_2CrO_5 /4/. The last formula shows that the chromium ions are six-valent. They can replace the Zr or Ti ions and in such a way create vacancies in the place of Pb ions. The six-valent Cr ions are donors. They decrease the conductivity of the p-type solution $Pb(Zr-Ti)O_3$

due to the electron-hole compensation at concentrations less than 0.325 mol%. At higher concentrations the segregated impurity layers increase the conductivity. The experiments show /5/ that at concentrations less than 0.325 mol% chromium ions are six-valent and act as donors decreasing the conductivity of the p-type ceramics. At higher concentrations these ions are trivalent and act as acceptors, increasing the conductivity of the p-type material. This change in valency of Cr ions, with the change of concentration, creates the possibility of chromium dioxide to play an important role in the production of ceramics with good temperature and aging characteristics.

In the present work the dc and ac conductivity as well as the dielectric properties of PZT ceramics doped with Cr_2O_3 were studied. The ceramics were produced by one of us (T.K. Vasileva) applying some technological improvements.

Experiment, results, and discussion The studies were carried out on samples of $Pb_{0.9}Sr_{0.1}(Zr_{0.51}Ti_{0.49})O_3$ ceramics doped with different concentrations of Cr_2O_3. The samples were discs of about 5 mm in diameter and about 0.8 mm thick. The dielectric measurements were made by means of a digital LCR meter, Hewlett-Packard type 4271 A, at a frequency of 1 MHz. The ac conductivity was measured at the same frequency. The strength of the ac field was 2 V/cm. The dc conductivity was measured by a Teraohmmeter, Orion TR 1300 type. The strength of the dc field was about 625 V/cm. The temperature runs were from RT to 700 K. The Cr_2O_3 concentration was 0.1, 0.25, 0.50, 0.80, and 1.00 wt%.

The temperature dependence of the dielectric constant of samples with the above mentioned Cr_2O_3 concentrations is shown in Fig. 1. On increasing the concentration the dielectric constant decreases continuously. The maxima of the curves are at a temperature of about 620 K above which the ceramics go gradually into a paraelectric phase. The Curie temperature T_c for the samples with 0.25 wt% is the lowest. This result correlates well with our X-ray data /5/ according to which, at a concentration of about 0.3 wt%, the ceramics consist almost only of the rhombohedral phase, which has a lower T_c than the tetragonal phase. At higher concentrations there is a slight increase of T_c.

The behaviour of the loss tangent is shown in Fig. 2. In the range of T_c there are well expressed maxima connected with the ferroelectric losses in

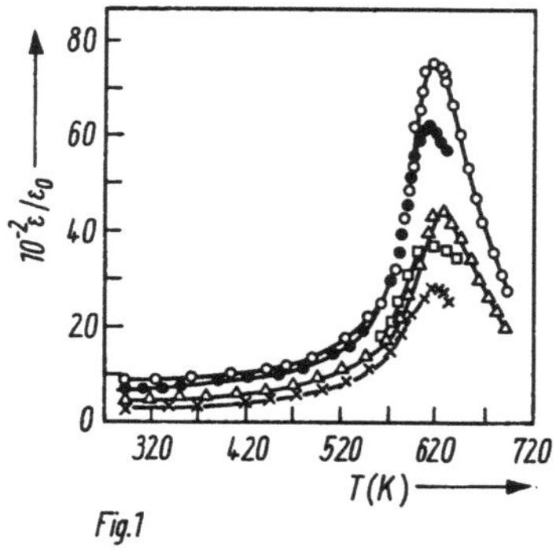

Fig.1

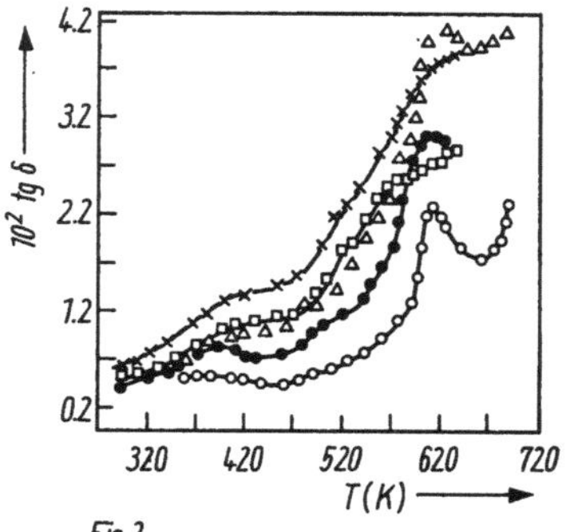

Fig.2

Fig. 1. Temperature dependence of the dielectric constant of PZT ceramics doped with Cr_2O_3. The measuring field has a frequency of 1 MHz and a strength of 2 V/cm. ○ 0.1; ● 0.25, △ 0.5, □ 0.8, × 1.0 wt% Cr_2O_3

Fig. 2. Temperature dependence of the dielectric losses of the ceramics in a field with strength of 2 V/cm and frequency of 1 MHz

the ceramics. The maxima are best expressed for samples of lower Cr_2O_3 concentration. The losses are greatest for samples with the highest Cr_2O_3 concentration in the whole temperature range with the exception of the range of the diffuse phase transition from the ferroelectric to the paraelectric state.

Fig. 3 shows the temperature dependence of the ac conductivity for samples of the same Cr_2O_3 concentrations as in Fig. 1. The pronounced differences in the values of the ac conductivity of the samples were observed mainly in the temperature range of the ferroelectric phase transition. On increasing the concentration we have a slight increase of the ac conductivity and then a more pronounced (about 40%) decrease. In the range from RT to about 470 K the ac conductivity almost does not depend on the Cr_2O_3 concentration.

The dc conductivity as a function of reciprocal temperature is shown in Fig. 4. On all curves there are three ranges with a linear dependence of dc conductivity on 1/T, corresponding to three different activation energies for the conductivity. The dependence of the activation energies on the Cr_2O_3 concentration for two of them (at lower temperatures) are shown in Fig. 5. The activation energies for the conductivity in the middle temperature range are

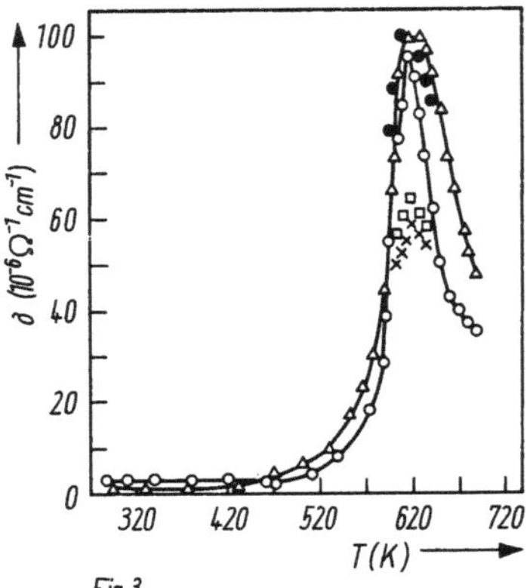

Fig.3

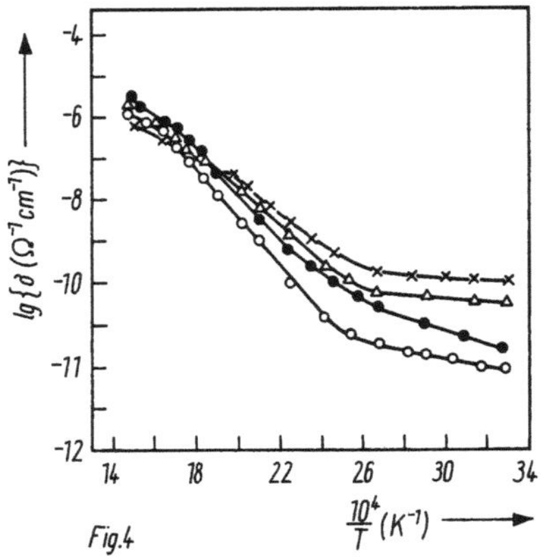

Fig.4

Fig. 3. Temperature dependence of the ac conductivity of PZT ceramics doped with Cr_2O_3. The frequency and the strength of the field are as in Fig. 1 and 2

Fig. 4. Temperature dependence of the dc conductivity. The field strength is 625 V/cm. o 0.1, • 0.25, △ 0.5, × 1.0 wt% Cr_2O_3

between 1.6 and 2.4 eV. Those for the RT range run between 0.15 and 0.75 eV. The activation energies for the highest temperature range are less than those for the middle temperature range and are about 1.1 eV.

The dc resistivity of PZT ceramics doped by thermal diffusion with Cr_2O_3 was studied by Takahashi /1/. He did not found the third linear range on the $\ln \varrho = f(1/T)$ curve which we found in the paraelectric temperature range. This difference is probably due to the lower temperature of his experiment and to the different way of doping (thermal diffusion in his case and additon of Cr_2O_3 before the sintering in our case).

To account for the dc conductivity Takahashi has proposed a model /1/ which was made more realistic by Wersing /6/. According to this model in the range of RT the electric conductivity of the ceramics is due mainly to the layers between the crystalline grains and at higher temperature mainly to the grains themselves. Therefore the activation energies between 1.6 and 2.4 eV correspond to the $Pb_{0.9}Sr_{0.1}(Zr_{0.51}Ti_{0.49})O_3$ grains doped with Cr_2O_3. The activation energies between 0.15 and 0.75 eV are related to the conductivity of the layers. The energies at the highest temperature of our experiment correspond to the conductivity of the paraelectric phase of the ceramics. The ferro-

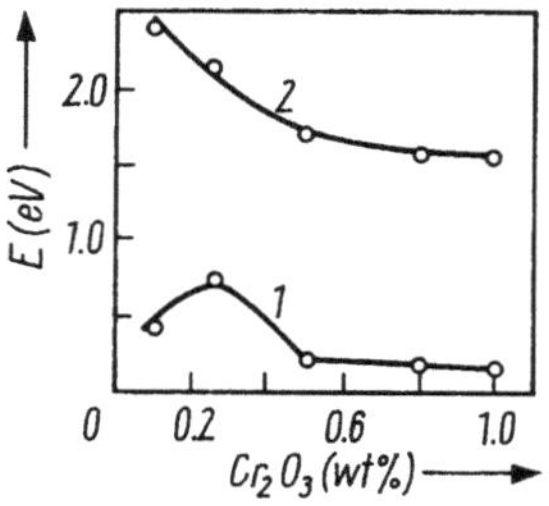

Fig. 5. Dependence of the activation energy of the conductivity on the Cr_2O_3 concentration

electric-paraelectric phase transition in the ceramics at about 620 K is connected with a change in the band structure of the material. As a result there is a change in the concentration and/or the mobility of the free current carriers which leads to a new value for the activation energy of conductivity in this temperature range.

Our results (see Fig. 1 and 3) show that at a concentration higher than 0.3 wt% the ceramics become harder. These results agree well with that of Uchida and Takeda /2/ who have found that on increasing the concentration the dielectric and elastic anisotropy of the ceramics decreases but the coercive field increases. As a result the displacement component $\partial D/\partial t = \varepsilon_0 \partial E/\partial t + \partial P/\partial t$ of the total current density $j = \sigma E + \partial D/\partial t$ decreases due to the decrease in the mobility of the ferroelectric domains in the heavily doped samples. Taking into account the low value of the field in our measurement (2 V/cm) we can in such a way understand the behaviour of the loss tangent at higher Cr_2O_3 concentration (Fig. 2).

The pronounced difference between dc and ac conductivity of the ceramics (see Fig. 3 and 4) shows that the dielectric relaxation is very important. It is related to the ferroelectric domains as well as to the creation and rearrangement of space charges in the layers between the grains. The doping with Cr_2O_3 creates defect dipoles as well which also give a contribution to the dielectric relaxation. Taking into account that the losses have to be proportional to the concentration n_D of the defect dipoles /7/ we see that the effective concentration of the defect dipoles is lower than that found from the values of the loss tangent (Fig. 2). Therefore we can conclude that part of the Cr_2O_3 segregates in the layers between the grains at any concentration of the oxide.

The peculiar behaviour of the curve corresponding to 0.25 wt% Cr_2O_3 (Fig. 4 and 5) is connected with the structure of the ceramics which at this concentration consist mainly of a rhombohedral phase.

References

/1/ M. TAKAHASHI, Japan. J. appl. Phys. 10, 643 (1971).

/2/ N. UCHIDA and T. TAKEDA, Japan. J. appl. Phys. 6, 1292 (1967).

/3/ D.J. PERDUIJN and H. C. VERJANS, Ber. Dt. Keram. Ges. 52, 227 (1975).

/4/ K. NEJEZCHLEB, L. PLOCEK, and S. KOSINA, Silikaty 3, 215 (1976).

/5/ T.K. VASILEVA, Thesis, Sofia 1978.

/6/ W. WERSING, IMF 4, Leningrad 1977, to be published in Ferroelectrics.

/7/ G. BUSCH and H. SCHADE, Lectures in Solid State Physics, Pergamon Press, London 1977.

(Received March 8, 1979)

phys. stat. sol. (a) 54, K151 (1979)
Subject classification: 17 and 20; 10.2; 13.1; 22.1.1

Institute of Physics, Polish Academy of Sciences, Warsaw[1)]

X-Ray Bremsstrahlung Isochromat and X-Ray Photoelectron Investigations of Ion Implanted Germanium

By
E. SOBCZAK

The unoccupied electron distribution in the conduction band of solids can be investigated by the X-ray bremsstrahlung isochromat method described e.g. in /1/ and the valence electron states in solids by using an X-ray photoelectron spectrometer. These two methods have been used in this work to study the ion implantation effects on both valence and conduction bands of ion implanted germanium monocrystals.

The (111) monocrystalline platelets of 1 mm thickness were cut from n-type pure germanium. They were mechanically polished and chemically etched. B^+ ions of 40 keV energy and 10^{16} ions/cm^2 dose and P^+ ions of 65 keV energy and 7.5×10^{15} ions/cm^2 dose were implanted into Ge platelets at room temperature. All the samples have been fixed in the photoelectron spectrometer as well as in the X-ray tube of the isochromat spectrometer at the same crystallographic orientation against electron and X-ray beams.

The X-ray photoelectron valence band spectra of germanium were investigated by means of a GCA McPherson ESCA 36 spectrometer for $MgK\alpha$ radiation. The spectral resolution of the instrument was estimated to be 1.5 eV. The intensity of the Ge 3d photoelectron peak from the GeO_2 surface layer was smaller by 20 % than that of pure germanium. Oxygen and carbon contaminations were removed by argon ion sputtering inside the sample chamber. The time and conditions of ion sputtering were the same for all germanium samples.

The X-ray bremsstrahlung isochromats of germanium have been measured point-by-point in 1 eV steps for $CrK\alpha_1$ wavelength ($h\nu$ = 5416 eV). There was used a Johann-type vacuum spectrometer with a bent quartz ($10\bar{1}0$) crystal of 1000 mm curvature radius set up in the second order of reflection. The sample was fixed to the target of an Ohlin-type X-ray tube /2/. The spectral resolution of the isochromat spectrometer was estimated to about 3.5 eV. Owing to the

---

1) Al. Lotników 32, 02-668 Warszawa, Poland.

oil-free pumping system the vacuum of 10^{-4} Pa pressure in the X-ray tube was relatively clean with regard to hydrocarbon contaminations.

In Fig. 1 and 2 the measured photoelectron valence band spectra and isochromats for pure and implanted germanium are presented. In the case of photoelectron spectra one cannot separate argon ion sputtering effects from B^+ and P^+ ion implantation, however, some conclusions are available. The intensity of the lower valence subband decreases from pure Ge to P^+ ion implanted Ge due to partial amorphization /3/ as a consequence of ion implantation. This effect is stronger for P^+ ions than for B^+ ions. The photoelectron valence spectrum of B^+ ion implanted Ge shows a 0.5 eV energy shift of the upper subband toward the Fermi level.

The X-ray isochromats are not so sensitive to surface contaminations as photoelectron spectra, because of the high energy of electrons and electron impact cleaning of the surface. The bremsstrahlung isochromats of the Ge monocrystal and B^+ ion implanted Ge have two weak maxima spaced about 5 eV in energy scale, which reflect maxima in the density of electron states in the conduction band. A plasmon energy loss for primary high-energy electrons is visible at energy of 16 eV above the threshold of the isochromat. The P^+ ion implanted Ge isochromat does not show the second maximum. Auleytner and

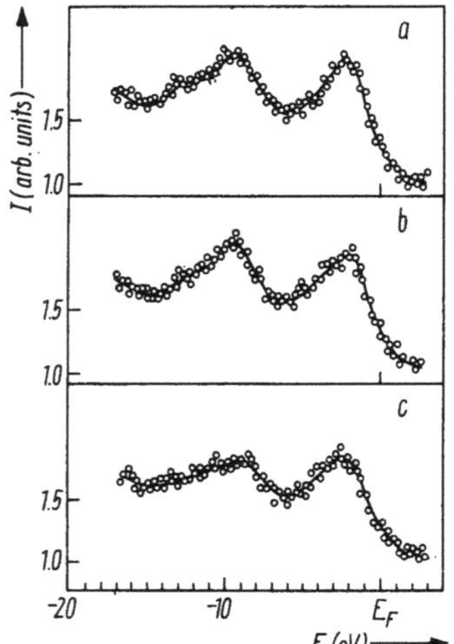

Fig.1

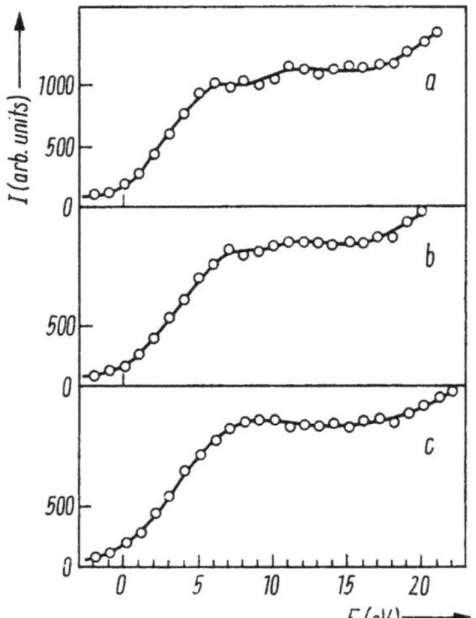

Fig.2

Fig. 1. a) XPS valence band spectra of pure germanium, b) B^+ ion implanted germanium, c) P^+ ion implanted germanium

Fig. 2. a) X-ray bremsstrahlung isochromats of pure germanium, b) B^+ ion implanted germanium, c) P^+ ion implanted germanium

Lidén /4/ measured X-ray isochromats for crystalline and evaporated Ge and observe this second maximum for the evaporated isochromat. It supports the above conclusion that P^+ ion implantation of 65 keV ion energy giving comparable ion distribution to the B^+ ion implantation of 40 keV energy more strongly forms an amorphous Ge layer. This ion implantation effect has been observed for valence and conduction bands of germanium.

The author would like to thank Prof. H. Ebel (Vienna) for making available the photoelectron spectrometer used for the present work, Dr. M. Drwiega (IFK Kraków) for help in ion implantation, M.Sci. K. Godwod for germanium samples, and Prof. J. Auleytner for his continuous interest in this work.

References

/1/ H. CLAUS and K. ULMER, Z. Phys. 185, 139 (1965).

/2/ Å. NILSSON, Ark. Fys. 6, 513 (1953).

/3/ L. LEY, S. KOWALCZYK, R. POLLAK, and D.A. SHIRLEY, Phys. Rev. Letters 29, 1088 (1972).

/4/ J. AULEYTNER and B. LIDÉN, Ark. Fys. 23, 41 (1962).

(Received June 12, 1979)

phys. stat. sol. (a) 54, K155 (1979)
Subject classification: 13.3; 22.2.1

Indian Institute of Technology, Kharagpur

Inversion Layer Formation and Surface States at the GaAs-Electrolyte Interface

By

D.N. BOSE, U. SEN GUPTA, and H.N. ACHARYA

Despite the expected advantages of GaAs MIS devices, progress in their development has been slow due to i) the high surface state density observed on all GaAs interfaces and ii) the lack of a suitable passivating layer analogous to the oxide in the case of silicon. Recently the fabrication of GaAs MOSFETs /1/ and microwave MISFETs /2/ have been reported using anodically grown oxide as insulator. However, these devices operate in the deep depletion mode since due to the high surface state densities and large band-gap an inversion layer formation at the GaAs surface has not been reported. This note describes the observation of inversion layer conditions from the study of C-U characteristics on etched and aged GaAs GaAs-NaOH interfaces. The presence of an interface level at the surface is also demonstrated.

The study of the interface properties between (100) n-GaAs ($N_D = 2x10^{17} cm^{-3}$) and NaOH using a Pt counter-electrode included the study of I-U, C-U solar cell and spectral characteristics /3/. The forward direction of current flow in dark and with illumination occurred with GaAs biased positive indicating hole transfer to the electrolyte. The $1/C^2$-U characteristics on specimens freshly etched in $1H_2O_2:3H_2SO_4:1H_2O$ solution for a few minutes showed depletion layer conditions existing at the surface with the flat-band potential U_{FB} = 0.50 V and $N_D = 2x10^{17}$ cm^{-3} using 0.1N NaOH. This is similar to the results in 1N KOH reported by Myamlin and Pleskov /4/. U_{FB} and the cell open circuit voltage U_{OC} were found to be increased with pH in going from 0.01N to 1.0N NaOH.

Electrochemical dissolution of the GaAs surface occurred /5/ as evidenced by a small zero bias dark current due to the possible reaction

$$GaAs + 5\ H_2O + 6\ H^+ \longrightarrow Ga(OH)_3 + HAsO_2 + 6\ H^+.$$

Gerischer /6/ has shown that the contribution of holes to the dissolution current is about 99.9 %. In an alkaline medium $HAsO_2$ is neutralised by the $(OH)^-$ ions and H_2 is liberated at the Pt electrode.

When specimens were biased negatively with U > 0.8 V, the current was found to increase rapidly due to decreased band bending allowing electrons to reach the surface and to take part in the reaction

$$GaAs + 3\ H^+ + 3\ e^- \rightarrow Ga + AsH_3.$$

This resulted in etching of the specimen. Thereafter the specimen was kept in the electrolyte and the surface was allowed to age for a period of few days. The C-U plot obtained after such aging showed a pronounced minimum both in 0.1N and 1.0N NaOH. In Fig. 1a the capacitance minimum was found to occur at U = -0.354 V. According to the theory of space-charge capacitance at a semiconductor-electrolyte interface /6/ a minimum capacitance should occur when the surface potential $U_S = -2\ U_B$, where U_B is the bulk Fermi level and is 0.61 eV for $N_D = 2 \times 10^{17}\ cm^{-3}$. Since in a semiconductor-electrolyte system the entire voltage drop occurs in the semiconductor space-charge layer, changes in barrier height are given directly by the applied voltage. Thus theoretically C_{min} should occur at U = - 1.22 V with respect to the flat-band condition. Thus there is evidence of considerable band-bending at the GaAs-electrolyte interface in the field-free condition since only - 0.3 V applied was sufficient to obtain the capacitance minimum.

The C-U characteristics measured at 10 kHz in 1.0N NaOH (Fig. 1b) shows, in addition to the capacitance minimum, the presence of a surface state, since the total effective capacitance is the sum of the space-charge and surface-state capacitances C_{sc} and C_{ss}, respectively. Such characteristics have been reported for the Ge-vacuum interface by Many et al. /7/ and Ge-electrolyte interfaces by Gerischer /6/. In III-V compounds the occurrence of a capacitance minimum at the KOH-electrolyte interface has been reported in low band-gap materials like InAs /8/.

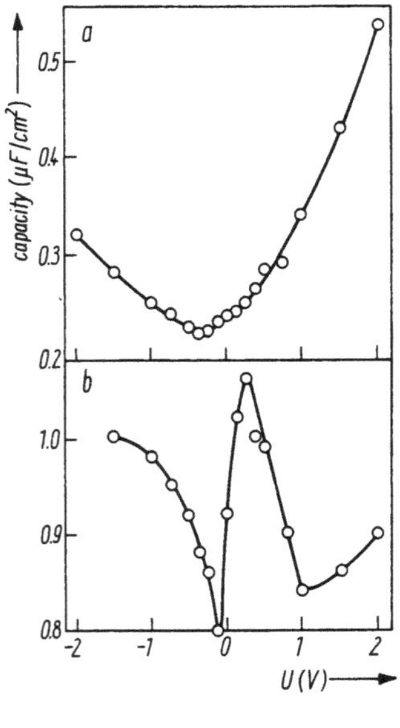

Fig. 1. C-U plots after aging in a) 0.1N and b) 1.0N NaOH for negative voltages applied to the electrode

The maximum of the effective differential capacitance occurs when the surface state level crosses the Fermi level. Using the capacitance minimum as a reference, the surface state level is found to lie 0.353 eV above this. Hence it is $(U_B + \frac{1}{2}E_g - 1.244 + 0.353) = 429$ eV above the valence band. Since it is now believed that there are no intrinsic surface states in the band gap of GaAs both from experimental and theoretical considerations /9/, the observed state must be due to the adsorbed layer, although its exact origin is not known. From the relation /5/ $C_{ss} = (e^2/4\,kT)\,N_{ss}$ the surface state density N_{ss} is found to be 7x $\times 10^{11}\,cm^{-2}$. The large shift in the flat-band potential observed in these experiments may be due to a high concentration of $(OH)^-$ ions adsorbed on the surface as found in the case of Ge by Brattain and Boddy /10/. As the experiments were conducted in the presence of ambient illumination, this may be responsible for the generation of minority carriers necessary to maintain equilibrium at the surface.

References

/1/ H. TAKAGI, G. KANO, and I. TERAMOTO, IEEE Trans. Electron Devices ED-25, 551 (1978).

/2/ D.L. LILE, D.A. COLLINS, L. MESSICK, and A.R. CLAWSON, Appl. Phys. Letters 32, 247 (1978).

/3/ U. SEN GUPTA, H.N. ACHARYA, and D.N. BOSE, Proc. Nat. Solar Energy Conv., Bhavnagar 1978 (p. 22).

/4/ V.A. MYAMLIN and Y.U. PLESKOV, Electrochemistry of Semiconductors, Plenum Press, New York 1967.

/5/ A. BARD and M. WRIGHTON, J. Electrochem. Soc. 124, 11 (1977).

/6/ H. GERISCHER, Physical Chemistry, Vol. 9A, Academic Press, New York 1970.

/7/ A. MANY, Y. GOLDSTEIN, and N.B. GROVER, Semiconductor Surfaces, North-Holland, Amsterdam 1965.

/8/ C.W. WILMSEN, L.G. MEINERS, and D.A. COLLINS, Thin Solid Films 46, 331 (1977).

/9/ A. HUIJSER and J. VAN LAAR, Surface Sci. 52, 202 (1975).

/10/ W.H. BRATTAIN and P.J. BODDY, Proc. 6th Internat. Conf. Phys. Semicond., Exeter 1962 (p. 796).

(Received May 2, 1979)

phys. stat. sol. (a) 54, K159(1979)
Subject classification: 2 and 14.3; 22

Central Research Institute for Physics, Hungarian Academy of Sciences, Budapest[1)]

Electron and Hole Drift Mobilities in $GeSe_2$ Glass

By
L. TÓTH

The optical and electrical properties of $GeSe_2$ glass have recently been subjected to extensive investigation /1/.

We report here our first experimental results on drift mobility measurements of electrons and holes in $GeSe_2$ glass.

The sample was prepared by melting the corresponding weight of components at 900 ^{o}C in a sealed silica ampoule. The ampoule was kept at this temperature for 50 h after which it was plunged into cold water. Disks of 2 and 7 mm thickness were cut from the bulk rod of 5 mm in diameter. Both sides of the disks were polished and supplied with evaporated electrodes of Au and Al.

The drift mobilities of electron and hole were measured by the time-of-flight method; illumination of the sample was by a strongly absorbed light flash of 100 μs duration through the semitransparent Au electrode. The transient photocurrent was amplified and displayed on an oscilloscope. The RC time constant of the electrical circuit was 10 μs.

The amplitude of the photocurrent signal was rather small but could be enhanced by exposing the sample to the light flash several times before the electrical field was applied ($7x10^2 \leqq E \leqq 10^3$ V/cm).

The shapes of the transient photocurrent traces for both polarities were similarly very dispersive. To get a current decay over wide time range a series of measurements was made under identical conditions except for gradually increasing the time base on the oscilloscope and varying the sensitivity of the amplifier.

To extract the transit time, the photocurrent decay was plotted in log I-log t scale in accordance with the theory of Scher and Montroll /2/. For each decay in this plot two straight lines were obtained with different slopes.

In Fig. 1 we show a typical transient photocurrent decay in log I-log t representation for electron mobility obtained in a 2 mm thick sample with a field of

1) P.O.B. 49, H-1525 Budapest, Hungary.

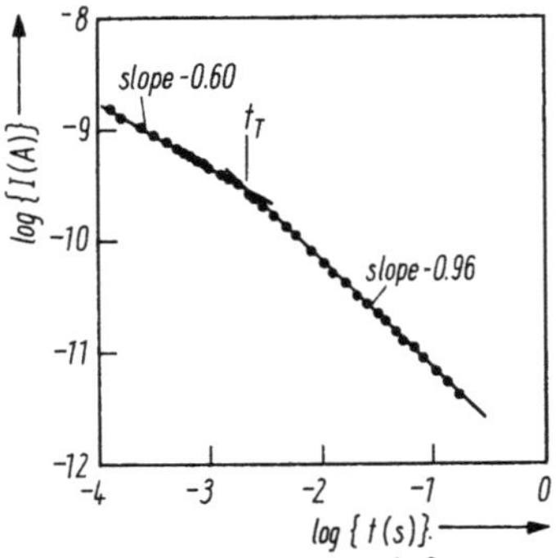

Fig. 1. Typical log I-log t plot of the transient current in electron mobility measurements on $GeSe_2$ glass (d = 2 mm, E = $7x10^2$ V/cm)

$7x10^2$ V/cm. The slope values of the initial and final straight lines fall into the intervals (- 0.60) to (- 0.75) and (- 0.95) to (- 1.05), respectively. The time belonging to the intersection of the straight lines was taken as the transit time (t_T). The drift mobility (μ) was calculated from the equation

$$\mu = \frac{d^2}{V t_T} , \quad (1)$$

where d is the thickness of the sample and V is the applied voltage across the sample.

Identical slopes both for electron and for hole were obtained in log I-log t plots of transient current decays at different bias fields referring to similar dispersive non-Gaussian transport phenomena in $GeSe_2$ glass for both charge carriers.

Our measurements - at room temperature and at an electric field of 10^3 V/cm - showed the electron drift mobility in $GeSe_2$ glass to be $1.4x10^{-1}$ $cm^2V^{-1}s^{-1}$; the hole drift mobility in $GeSe_2$ glass was $4x10^{-2}$ $cm^2V^{-1}s^{-1}$. We estimate our experimental error to be $\pm$ 20 %.

The results we obtained are in good agreement with those recently reported for Ge_xSe_{1-x} glasses /3/. In that work three compositions were investigated: x = 0.15, 0.20, and 0.40, in which higher mobilities were also reported for electrons than for holes.

References

/1/ See for example: P. TRONC, M. BENSOUSSAN, A. BRENAC, and C. SEBENNE, Phys. Rev. B 12, 5947 (1973).

E. DELORME, P. TRONC, M. BENSOUSSAN, A. BRENAC, and C. SEBENNE, Rev. Phys. appl. 12, 949 (1977).

M. BENSOUSSAN, Rev. Phys. appl. 12, 753 (1977).

T.T. NANG, M. OKUDA, T. MATSUSHITA, S. YOKOTA, and A. SUZUKI, Japan. J. appl. Phys. 15, 849 (1976).

J. SHIRAFUJI, G.I. KIM, and Y. INUISHI, Japan. J. appl. Phys. 16, 67 (1977).

M. LANNOO and M. BENSOUSSAN, Phys. Rev. B 16, 3546 (1977).

V.A. VASSILYEV, M. KOÓS, and I. KÓSA SOMOGYI, Solid State Commun. 22, 633 (1977).

/2/ H. SCHER and E.W. MONTROLL, Phys. Rev. B 12, 2455 (1975).

/3/ G.I. KIM and J. SHIRAFUJI, Japan. J. appl. Phys. 17, 1789 (1978).

(Received May 23, 1979)

phys. stat. sol. (a) 54, K163 (1979)
Subject classification: 20.1; 22

University of Chemical Technology, Pardubice

Temperature Dependence of the Exponential Absorption Edge in $Ge_{20}Sb_{25}S_{55}$ Chalcogenide Glass

By
H. TICHÁ, L. TICHÝ, and M. FRUMAR

The Ge-Sb-S system has been studied previously in a number of papers /1 to 6/ where the glass forming region of rapidly /1, 2/ (cold water) and slowly /3/ (air) quenched samples was determined besides other physical properties /4 to 6/.

In the present communication the results of /5/ are verified from view points of the position and shape of the absorption edge of $Ge_{20}Sb_{25}S_{55}$ glass. The temperature dependence of the edge is discussed semi-quantitatively.

Samples of $Ge_{20}Sb_{25}S_{55}$ glasses for measurements of optical properties were prepared using the method described in /2 to 5/. Optical transmittance (I) in the temperature range 92 to 424 K was measured on samples of three different thicknesses so as to keep the condition $Kd \approx 1$.

The absorption coefficient (K) was calculated using

$$K = d^{-1} \ln \frac{(1 - R)^2 \left[(1 - R)^4 + 4 R^2 I^2\right]^{1/2}}{2 I} \tag{1}$$

assuming the reflectivity (R) in the studied spectral region to be independent of both the radiation wavelength and the temperature. The spectral and temperature dependence of the absorption coefficient is given in Fig. 1. It can be described using the well-known Urbach relation

$$K = K_o \exp\left\{\frac{\sigma}{K_B T}\left[\hbar\omega - E_g^{opt}(T)\right]\right\}, \tag{2}$$

where σ is the slope of the Urbach edge, $E_g^{opt}(T)$ the temperature dependent optical gap, K_B the Boltzmann constant.

It follows from /6/ that the empirical relation $E_g^{opt} = \hbar\omega_{(K = 10^2)}$ is fulfilled in the Ge-Sb-S glass system. Using this relation the temperature dependence of E_g^{opt} and the slopes of the Urbach edge (σ) are shown in Fig. 2. Like in ionic crystals the spectral "width" of the exponential edge of the studied glasses

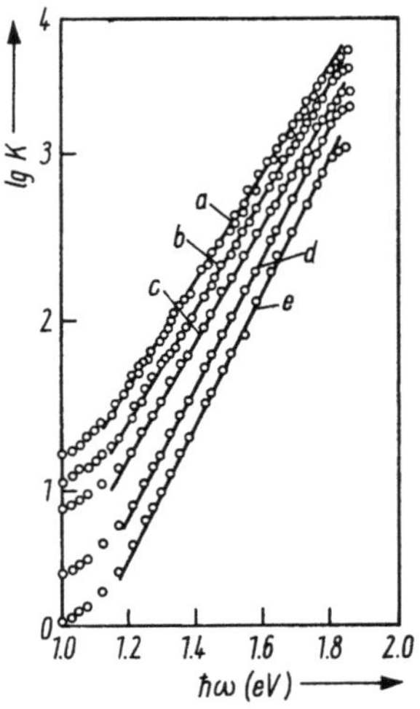

Fig. 1. Optical absorption edge of $Ge_{20}Sb_{25}S_{55}$ glasses. (a) 424, (b) 370, (c) 296, (d) 201, (e) 92 K

amounts to ≈0.6 eV. Therefore we assume that the slope σ can be described by means of the relation /7/

$$\sigma = \sigma_o (2 K_B T/\hbar \omega_p) \tanh (\hbar\omega_p/2K_B T); \tag{3}$$

σ_o and $\hbar\omega_p$ are constant and a mean phonon energy, respectively. By fitting the experimental dependence σ versus T we obtained $\sigma_o = 0.30$ and $\hbar\omega_p = 0.0642$ eV which are close to those obtained in other chalcogenide glasses /8/.

A unified explanation of the exponential absorption edge was suggested by Dow and Redfield /9/ based on the assumption of exciton ionization by inner electrical microfields, where the electron tunnels away from the hole through the Coulomb potential barrier. Optical phonons, impurities, fluctuations of density,and/or other imperfections might act as sources of the internal electric fields.

Electrical microfields due to phonons with some mean energy $\hbar\omega_p$ (at the temperature T) can be described /9/ by

$$\langle F^2 \rangle = B \, \hbar\omega_p \coth (\hbar\omega_p/ 2K_B T) \tag{4}$$

with constant B given in /9/. According to the Franz theory, the shift of the exponential absorption edge (ΔE_g) in an electric field (F), is directly proportional to the square of the electrical field /10, 11/:

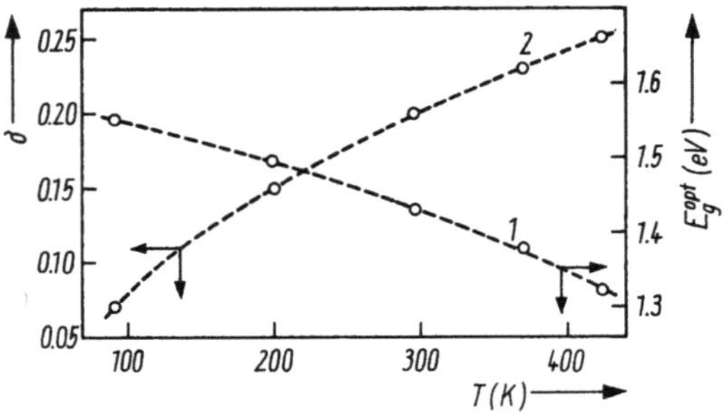

Fig. 2. Temperature dependence both of $E_g^{opt} = \hbar\omega_{(K=10^2)}$ and slopes of the Urbach edge (σ)

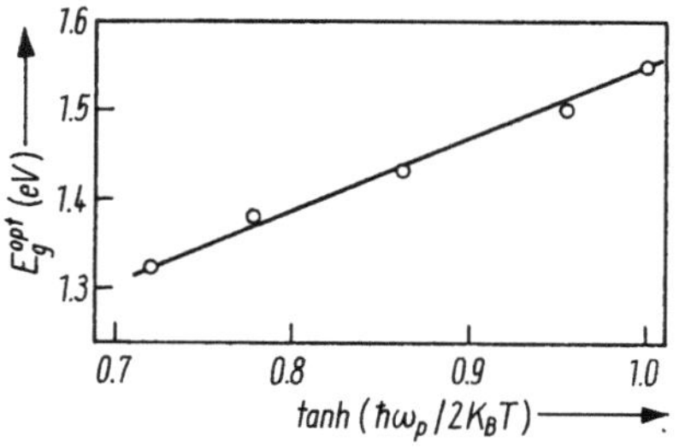

Fig. 3. Dependence of $E_g^{opt} = \hbar\omega_{(K = 10^2)}$ versus tanh ($\hbar\omega_p/2K_BT$) for $\hbar\omega_p = 0.0642$ eV, tg $\alpha = M \approx 0.78_7$

$$\Delta E_g = \frac{\hbar^2 e^2}{24 E_a^2 m^*_{exc}} F^2 , \tag{5}$$

where m^*_{exc} is the exciton effective mass, $E_a = K_BT/\sigma$ is the reciprocal of the slope of the Urbach edge /11/. Dow and Redfield /9/ showed that the exponential shape of the absorption edge is not sensitive to the details of the distribution of microfields. Therefore we assume that $\langle F^2 \rangle = F^2$. Thus it follows from (3) to (5)

$$E_g(T_2) = E_g(T_1) - M \left[\tanh(\hbar\omega_p/2K_BT_1) - \tanh(\hbar\omega_p/2K_BT_2)\right] \tag{6}$$

($T_2 > T_1$) and where $M = H\hbar^2 e^2 \sigma_o^2/6 m^*_{exc}\hbar\omega_p$.

In Fig. 3 are correlated our values of $E_g^{opt} = \hbar\omega_{(K = 10^2)}$ versus tanh ($\hbar\omega_p/2K_BT$) and good agreement is evident in this temperature region. Values of $E_g(T)$ calculated by means of (6) and using M = 0.787 (see Fig. 3) are compared with $E_g^{opt} = \hbar\omega_{(K = 10^2)}$ values in Table 1.

Table 1

T (K)	$E_g^{opt} = \hbar\omega_{(K=10^2)}$ (eV) /6/	$E_g(T)$ (eV)
92	1.55	-
201	1.50	1.517
296	1.43	1.441
370	1.38	1.376
424	1.32	1.336

The good agreement of the $E_g(T)$ values from relation (6) and the $E_g^{opt} = \hbar\omega_{(K = 10^2)}$ values indicates that in the case under study the temperature shift of the exponential absorption edge can be described by means of the Dow-Redfield model. As the main sources of the electrical microfields can be taken phonons with some mean energy $\hbar\omega_p = 0.0642$ eV.

References

/1/ M. FRUMAR, H. TICHÁ, M. BUREŠ, and L. KOUDELKA, Z. Chem. 15, 199 (1975).

/2/ M. FRUMAR, H. TICHÁ, L. KOUDELKA, and J. KLIKORKA, Proc. Conf. Amorphous Semicond. 1975, Ed. P. SÜPTITZ, Acad. Sci. GDR, Berlin 1974 (p. 236).

/3/ D. LINKE and I. BÖCKEL, Z. anorg. allgem. Chem. 419, 97 (1975).

/4/ H. TICHÁ, M. FRUMAR, L. TICHÝ, and J. KLIKORKA, Perspective and Special Anorganic Materials and Their Applications, Dům techniky, Pardubice 1975 (p. 140).

/5/ M. FRUMAR, H. TICHÁ, L. TICHÝ, J. KLÍMA, and L. KOUDELKA, in: Structure and Properties of Non-Crystalline Semiconductors, Ed. B.T. KOLOMIETS, Nauka, Leningrad 1976 (p. 335).

/6/ H. TICHÁ, PhD Thesis, University of Chemical Technology, Pardubice 1977.

/7/ H. MAHR, Phys. Rev. 125, 1510 (1962); 132, 1880 (1963).

/8/ J. TAUC, in: Optical Properties of Non-Crystalline Solids, Ed. F. ABELES, North-Holland, Amsterdam 1972 (p. 297).

/9/ J.D. DOW and D. REDFIELD, Phys. Rev. B 5, 594 (1972).

/10/ W. FRANZ, Z. Naturforsch. 13a, 484 (1958).

/11/ J. TAUC, in: Amorphous and Liquid Semiconductors, Ed. J. TAUC, Plenum Press, London/New York 1974 (p. 159).

(Reveived July 4, 1979)

phys. stat. sol. (a) 54, K167 (1979)

Subject classification: 8; 21; 21.1

Institute for Low Temperature and Structure Research, Polish Academy of Sciences, Wrocław (a), Laboratoire de Spectrométrie Physique, associé au C.N.R.S., U.S.M.G., Grenoble (b), and Service Basses Températures, C.E.N., Grenoble (c)

Low-Temperature Heat Capacity Studies of the Dihydride Phase of Titanium-Niobium-Hydrogen Alloys

By

B. NOWAK (a, b) and R. LAGNIER

The present communication gives preliminary results of an investigation of the low temperature specific heat of the dihydride phase of $Ti_{1-x}Nb_xH_{1.94}$ (where x = 0.25 and 0.50). The preparation of the samples, their structure, and lattice parameters are described elsewhere /1/.

The magnetic susceptibility data /2/ as well as the proton-spin lattice relaxation time measurements /3/ suggest that the density of states at the Fermi level for the nearly stoichiometric dihydride phase $Ti_{1-x}Nb_xH_{1.94}$ decreases with increasing niobium concentration x. To confirm these expectations it seemed very promising to measure the low temperature specific heat of this phase. The measurements were performed on powder samples (1.5 to 1.8 g) at C.E.N., Grenoble, by using a differential-type adiabatic calorimeter /4/ in the temperature range 2 to 15 K. For one of the samples, i.e. $Ti_{0.75}Nb_{0.25}H_{1.94}$, a small specific heat anomaly at 12 K whose origin is not clear as yet, has been observed (Fig. 1).

Fig. 2 shows the C/T versus T^2 plots for both samples. The deviation from a simple linear dependence of C/T with T^2 in the lowest temperature range is indicative of an unsufficient good thermal contact between the grains and/or the adsorption of helium exchange gas in the powder sample.

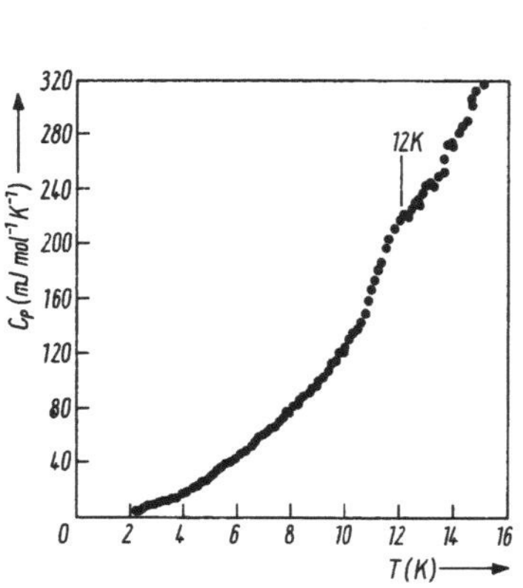

Fig. 1. Heat capacity as a function of temperature for $Ti_{0.75}Nb_{0.25}H_{1.94}$

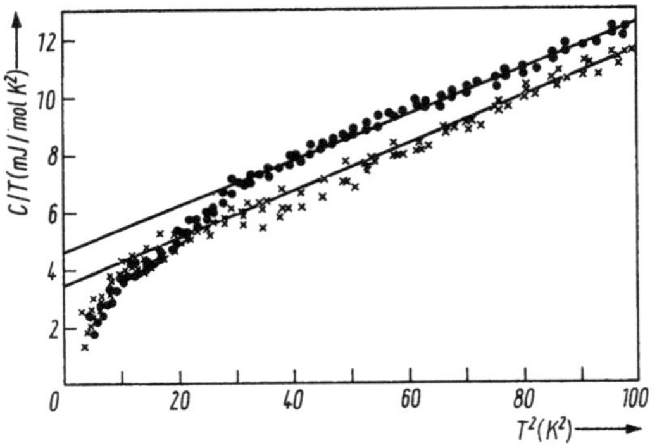

Fig. 2. C/T versus T^2 plot for $Ti_{1-x}Nb_xH_{1.94}$; ● x = 0.25, × x = 0.50

The values of γ and Θ_D estimated from the slopes of the straight lines presented in Fig. 2 are listed in Table 1. The densities of states derived from the coefficient of the electronic specific heat, γ, are equal to 1.0 and 0.74 states/eV atom for x = 0.25 and x = 0.50, respectively, and are in good agreement with expectations.

Table 1

sample composition	γ (mJ/mol K^2)	Θ_D (K)
$Ti_{0.75}Nb_{0.25}H_{1.94}$	4.7	291
$Ti_{0.50}Nb_{0.50}H_{1.94}$	3.5	288

It must be pointed out that the deduced values of $N(E_F)$ contain the enhancement due to the electron-phonon interaction but, unfortunately, the enhancement factor $(1 + \lambda)$ is not available at the present time.

The authors thank Dr. E. Bonjour for his interest in this work. One of us (B.N.) would like to express his gratitude to Laboratoire de Spectrométrie Physique, U.S.M.G., for financial support during this experiment.

References

/1/ B. STALIŃSKI and B. NOWAK, Bull. Acad. Polon. Sci., ser. sci. chim. 25, 451 (1977).

/2/ B. STALIŃSKI and B. NOWAK, Bull. Acad. Polon. Sci., ser. sci. chim. 25, 65 (1977).

/3/ B. NOWAK, N. PIŚLEWSKI, and W. LESZCZYŃSKI, phys. stat. sol. (a) 37, 669 (1976).

/4/ E. BONJOUR, Rapport CEA-r-3181 (1967).
R. LANGNIER, Rapport CEA-r-4419 (1973).

(Received April 17, 1979)

phys. stat. sol. (a) 54, K171 (1979)
Subject classification: 10.2 and 20.3; 22.8.1

Health Physics Division, Babha Atomic Research Centre, Bombay[1)]

Thermoluminescence Process in $CaSO_4$:RE

By
V.N. BAPAT

The process of thermoluminescence (TL) in $CaSO_4$:RE is said to be associated with the reduction and oxidation of the rare earth impurity. The rare earth impurity normally exists in the phosphor host in trivalent form. On irradiation, these ions are reduced to divalent rare earth ions by trapping electrons, holes are trapped at other sites. In the TL readout process, holes are said to be released and these released holes recombine with the electron at the divalent rare earth sites converting them back to the trivalent form. The reconverted trivalent ions are excited by the recombination energy of the electron-hole recombination. These excited ions relax to ground states giving characteristic trivalent rare earth emission /1/. Recently a direct proof of the oxidation and reduction process in TL of $CaSO_4$:Eu was given by Bapat in his studies on fluorescence /2/. Work on fluorescence of $CaSO_4$:Ce has also shown the existence of valence conversion of cerium on irradiation and TL readout /3/.

Detailed fluorescence work on these two phosphors is carried out. The fluorescence of irradiated phosphors is studied. The irradiated phosphor sample is bleached at different temperatures and the fluorescence of the bleached phosphor sample is taken again at room temeprature without disturbing the experimental geometry.

The fluorescence spectrum of $CaSO_4$:Eu shows the presence of a characteristic fluorescence of both Eu^2 and Eu^3 /2/. The variation of both Eu^2 and Eu^3 fluorescence with the bleaching temperature is shown in Fig. 1. A characteristic fluorescence of Ce^3 is observed in $CaSO_4$:Ce /3/. The variation of this fluorescence with bleaching temperature is shown in Fig. 2. In both these figures the TL glow curve of the corresponding phosphor is shown. It is seen from these figures that the distruction of divalent europium ions in $CaSO_4$:Eu and the regeneration of trivalent Eu and Ce in $CaSO_4$:Eu and $CaSO_4$:Ce, respectively, is not associated with the bleaching of the 100 ^oC TL peak. A clear conversion of Eu^2 to Eu^3

1) Bombay - 400085, India.

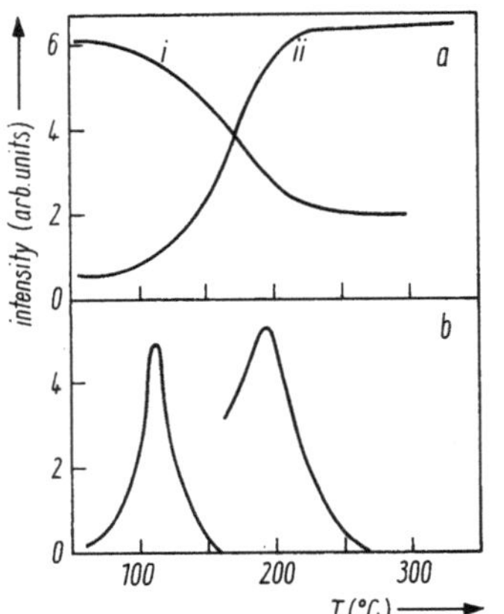

Fig.1

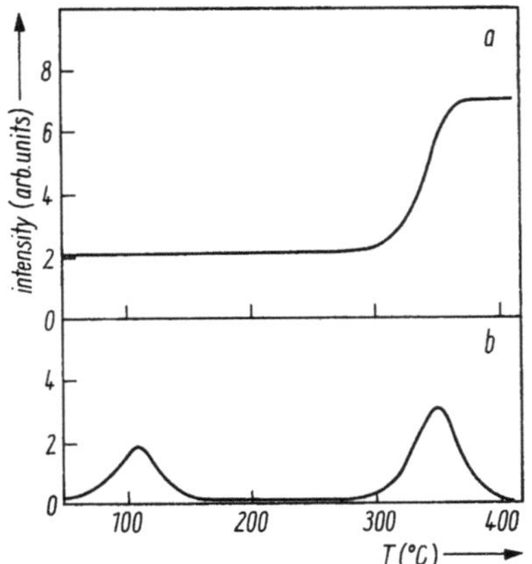

Fig.2

Fig. 1. a) Variation of fluorescence with bleaching temperature after irradiation; (i) characteristic Eu^2 fluorescence at 385 nm, (ii) characteristic Eu^3 fluorescence at 600 nm. b) TL glow curve of irradiated $CaSO_4$:Eu

Fig. 2. a) Variation of fluorescence with bleaching temperature after irradiation; characteristic Ce^3 fluorescence at 325 nm. b) TL glow curve of irradiated $CaSO_4$:Ce

is seen corresponding to the bleaching of the 200 °C TL peak in $CaSO_4$:Eu while reconversion of Ce^3 is observed corresponding to bleaching of the 360 °C peak.

It is interesting to note that TL of $CaSO_4$:Mn gives only a high 100 °C TL peak and other higher temperature peaks are absent /3/. In $CaSO_4$:Eu we have both 100 and 200 °C peaks equally high. The TL emission of the 100 °C peak in this phosphor consists of both Eu^2 and Eu^3 emission while the TL emission corresponding to the 200 °C TL peak is predominantly that due to Eu^3 /3/. The TL glow curve of nominally pure $CaSO_4$ has a strong 100 °C peak. On addition of a rare earth impurity like dysprosium, the 200 °C TL peak and other higher temperature peaks rise very rapidly as compared to the 100 °C TL peak /4/. Similar effects are observed with other rare earth dopants, too.

In view of the above observations it is proposed that reduction/oxidation of the rare earth impurity is not associated with the traps corresponding to the 100 °C TL peak. The TL emission in this peak is due to energy transfer of recombination to the luminescent impurities. The 200 °C TL peak is associated with the reduction and oxidation of the trivalent rare earth ion impurity and the

TL emission is due to the excitation of the reconverted trivalent ion in the process of recombination as predicted earlier.

The author is thankful to Dr. A.K. Ganguly for his interest in this work.

Refernces

/1/ K.S.V. NAMBI, V.N. BAPAT, and A.K. GANGULY, J. Phys. C 7, 4401 (1974).

/2/ V.N. BAPAT, J. Phys. C 10, L465 (1977).

/3/ V.N. BAPAT, Ph. D. Thesis, Bombay University, 1977.

/4/ K.S.V. NAMBI, Ph. D. Thesis, Gujarat University (India), 1977.

(Received July 6, 1979)

phys. stat. sol. (a) 54, K175 (1979)

Subject classification: 10.1 and 12.1; 21; 21.1

Department of General Physics, Purkyně University, Brno[1] (a), Department of Metal Physics, Charles University, Prague [2] (b), and Physico-Technical Institute of Low Temperatures, Academy of Sciences of the Ukrainian SSR, Kharkov (c)

The Critical Resolved Shear Stress of Cd-Zn Single Crystals at Low Temperatures

By

V. NAVRÁTIL (a), P. LUKÁČ (b), and V.P. SOLDATOV (c)

Introduction Considerable attention has been given recently to the effect of solute atom on the critical resolved shear streee (CRSS), τ_0, of single crystals. An increase in the CRSS with the solution concentration, c, has been observed for single crystals of hexagonal metals. Usually solid solution hardening in hexagonal alloys has been investigated at temperatures above 77 K /1 to 4/. However, little information is available on the CRSS of hexagonal alloy single crystals deformed at very low temperatures /5 to 7/. Stulíková et al. /7/ have reported on the investigation of the CRSS of cadmium single crystals alloyed with silver at temperatures between 1.55 and 45 K and found an anomalous peak in the temperature dependence of the CRSS. A positive temperature dependence of the CRSS ($d\tau_0/dT > 0$) in a certain temperature range has also been observed for Zn and Cd single crystals by Vladimirova and Lavrentev /8/.

In the present paper the influence of zinc as solute on the temperature dependence of the CRSS of cadmium single crystals is investigated.

Results and discussion Single crystals (25 mm in length and 4 mm in diameter) were grown from the melt in a glass tube using a modified Bridgman method. Specimens with an orientation (determined by the Laue back reflection technique) allowing only basal slip were used. The handling of the samples before deformation is described elsewhere /3/. The deformation tests were made using a creep technique /9/. One sample was deformed successively at several temperatures in the range from 1.52 to 80 K.

Fig. 1 shows the temperature dependence of the CRSS for Cd-Zn single crystals containing 0.0022, 0.012, 0.053, 0.215, 0.348, 0.567, and 0.645 at% Zn

---

1) Kotlářská 2, 61137 Brno, CSSR.

2) Ke Karlovu 5, 12112 Prague, CSSR.

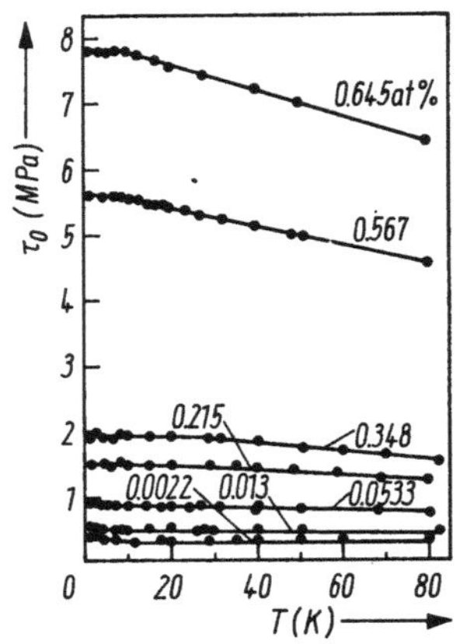

Fig. 1. Temperature dependence of the CRSS for Cd-Zn single crystals

in Cd. It is obvious that the CRSS does not increase with decreasing temperature in the whole temperature range studied. At very low temperatures below approximately 10 K in the CRSS for higher concentration of Zn is practically independent of temperature. In contrast to Cd-Ag single crystals /7/ no peak in the temperature dependence of the CRSS of Cd-Zn single crystals is observed. This can be explained if it is taken into account that the temperature dependence of the CRSS for Cd-Zn single crystals has been estimated using one specimen while the CRSS of Cd-Ag single crystals has been determined using one sample at any temperature.

It is possible to consider that the dislocation after the overcoming of obstacles can glide under an applied stress lower than the stress which is necessary to start the motion of the dislocation from rest /10/. A model describing these experiments in more detail will be published later.

The authors are very grateful to Prof. V.I. Startsev for the possibility to make these experiments in his laboratory.

References

/1/ H. SCHARF, P. LUKÁČ, M. BOCEK, and P. HAASEN, Z. Metallk. 59, 799 (1968).

/2/ A. AKHTAR and E. TEGHTSOONIAN, Acta metall. 17, 1339 (1969).

/3/ M. ROJKO and P. LUKÁČ, J. Mater. Sci. 8, 1065 (1973).

/4/ P. LUKÁČ and I. STULÍKOVÁ, Czech. J. Phys. B24, 648 (1974).

/5/ P. LUKÁČ and L.D. WILL, phys. stat. sol. (a) 5, K179 (1971).

/6/ A. AKHTAR and E. TEGHTSOONIAN, Phil. Mag. 25, 897 (1972).

/7/ I. STULÍKOVÁ, P. LUKÁČ, and G.I. SHKLAREVSKAYA, phys. stat. sol. (a) 41, K197 (1977).

/8/ V.L. VLADIMIROVA and F.F. LAVRENTEV, Fiz. Metallov i Metallovedenie 35, 801 (1973).

/9/ K. KAMADA and I. YOSHIZAWA, J. Phys. Soc. Japan 31, 1056 (1972).

/10/ R.B. SCHWARZ and R. LABUSCH, J. appl. Phys. 49, 5174 (1978).

(Received June 6, 1979)

Pre-Printed Titles

Papers to be published
in "physica status solidi (a) - applied research"
Vol. 55, No. 1

Review Article

J. FOUSEK and J. PETZELT, Prague
Changes of Refractive Indices of Crystals Induced by Structural Phase Transitions

Original Papers

D. BENSAHEL and M. DUPUY, Grenoble
SEM and TEM: Diffusion of Lithium in ZnTe

T.J. MAGEE, J. PENG, and J.D. HONG, Sunnyvale, California
Gettering of Cr in GaAs by Back Surface Mechanical Damage

YU.A. BAIKOV, YU.V. ZELENEV, and W. HAUBENREISSER, Moscow and Jena
A Fluctuation Theory of Normal Crystal Growth for Metallic One-Component Systems

O. BUCK, L.A. AHLBERG, L.J. GRAHAM, G.A. ALERS, C.A. WERT, and K.C. HSIEH, Thousand Oaks, California and Urbana, Illinois
Harmonic and Anharmonic Elastic Properties of the α-Nb-H System

G. VAN TENDELOO, M. VAN SANDE, and S. AMELINCKX, Antwerpen
Imaging Hexagonal Based Superstructures

H. NEUMANN, E. BUTTER, P. KRAMER, H.-G. ERNST, and W. KOHL, Leipzig and Frankfurt/Oder
Simultaneous Doping of GaP with Sn and Zn

E.K. HENNER and I.G. SHAPOSHNIKOV, Perm
Clustering of Magnetic Ions in Diluted Solid Paramagnets

G. MARTENS, P. RABE, G. TOLKIEHN, and A. WERNER, Kiel
EXAFS in Photoelectron Yield Spectra at K Edges of Cu, Ni, and Ge

J.M. YELLUP and B.A. PARKER, Ascot Vale, Victoria and Clayton, Victoria
The Determination of Compositions in Non-Homogeneous Ferromagnetic Materials by Curie Temperature Measurement

V.A. GOLTSOV, N.I. TIMOFEEV, and I.YU. MACHIKINA, Donetsk
The Phenomenon of Controllable Hydrogen Phase "Naklep" of Metals and Alloys

YU.D. CHISTYAKOV, S.V. POTAPOV, D.N. GULIDOV, and H.G. SCHNEIDER, Moscow and Karl-Marx-Stadt
The Effect of Residual Mechanical Stresses and Strains on the Superconductivity of Heteroepitaxial Layers of Nb on α-Al_2O_3

R.D. GOULD and C.A. HOGARTH, Uxbridge
The Time Dependence of Circulating and Emission Currents in Evaporated Thin Film Sandwich Structures of Au-SiO_x-Au

D. BENSAHEL, M. DUPUY, and J.C. PFISTER, Grenoble
Electrical, SEM, and TEM Studies in ZnTe (I)

L.M. SANDLER, V.P. POPOV, and YA.V. NAGLYUK, Kiev
Magnetic Anisotropy Induced by Rolling Single Crystals of Ferromagnetic Alloys (I)

J.F. GRACZYK, Yorktown Heights, N.Y.
Structure of Glow Discharge Amorphous Silicon

H.-G. UNRUH, W. 'ELLER, and G. KIRF, Saarbrücken
Spectroscopic and Dielectric Investigations of K_2SeO_4

A.A. GUZEV, V.A. GURTOV, A.V. RZHANOV, and A.A. FRANTSUZOV, Novosibirsk
Value and Gradient Distribution of the Quasi-Fermi Level in MOS Transistor Channels

K. HESS, J.F. DETRY, and C.T. SAH, Urbana, Illinois
Trapping Noise in Charge Coupled Devices

P. GAWORZEWSKI and K. SCHMALZ, Berlin
On the Electrical Activity of Oxygen in Silicon

M. DUBIEL, G. BERG, and F. FRÖHLICH, Halle (Saale)
The Formation of Dimer and Trimer Aggregates in Sr-Doped NaCl Crystals

S. RADHAKRISHNAN and P.C. MEHENDRU, New Delhi
Electronic Conduction in Dye-Sensitized Selenium Films

H.C. KIM and T. KISHI, Tokyo
The Grain Size and Flow Stress Dependence of Acoustic Emission Energy Release in Polycrystalline Aluminium

Z. LILIENTAL, Warsaw
The Effect of Ag Impurity on the Real Structure of ZnTe

P. LUKÁČ and V. JESENÁK, Bratislava
Distribution of Ion-Bombardment Implanted ^{85}Kr in Si, Cu, and SiO_2

M. TAKEUCHI, Hitachi, Ibaraki
Photoconductive Properties of Sputtered TiO_2 Films

L.S. KHAZAN, L.A. MATVEEVA, G.N. SEMENOVA, and YU.A. TKHORIK, Kiev
Account of Dislocation Structure at Stress Measurements in Heteroepitaxial Systems (II)

C. MANOLIKAS, J. VAN LANDUYT, R. DE RIDDER, and S. AMELINCKX, Antwerpen and Mol
Electron Microscopic Study of the Domain Structure and of the Transition State in $Cu_{0.5}In_{2.5}Se_4$

A.YU. LEIDERMAN, Tashkent
On the Theory of Double Injection in Semiconductors with Continuous Distribution of Trapping Levels within the Gap

D. BONCHEV, O. MEKENYAN, and H.-G. FRITSCHE, Burgas and Jena
A Topological Approach to Crystal Vacancies Studies (I)

P. BARTUŠKA, R. KROGGEL, H. LÖFFLER, M. SIMERSKÁ, V. SYNEČEK, and G. WENDROCK, Prague and Halle
On the Room Temperature Stability of Metastable Phases Formed at 200 °C in Al-Zn Alloys

R. HAUG, G. KAPPEL, and A. JAÉGLÉ, Mulhouse
Electrical Resistivity Studies of the System Mn_5Ge_3-Mn_5Si_3

C. DE BLASI, S. GALASSINI, C. MANFREDOTTI, G. MICOCCI, L. RUGGIERO, and A. TEPORE, Lecce
Thermally Stimulated Currents in n-GaS

M. GODLEWSKA, J. ŚCIESIŃSKI, and E. ŚCIESIŃSKA, Cracow
Spectroscopic Study of Phase Polymorphism and Vibrational Assignments for 2,2-Dinitropropane

Z. DOBROVOLSKIS, W. HOERSTEL, and A. KROTKUS, Vilnius and Berlin
Determination of the Impact Ionization Rate of Holes in Tellurium at 77 K

J.J. JACKSON, E.A. RYAN, and Y.N. LWIN, Argonne and Macomb, Illinois
A Study of Thermal Conversion of Interstitials in Platinum by Means of Electrical Resistivity Measurements

M. SATO and K. SUMINO, Sendai
Motion of Extended Dislocations in Silicon Crystals Observed by HVEM

J. ECHIGOYA and S. HAYASHI, Sendai
Anomaly of Dislocation Motion in Fe - 35% Ni Invar Alloys

L. NIKOLOVA, T. TODOROV, D. POPOV, and K. TERSIISKI, Sofia and Russe
Laser Induced Fast Reversible Processes in Additively Coloured KCl Containing F_A Centers

E.A. BAKULIN, V.L. HILAROV, and V.V. SHCHERBININA, Leningrad
Characteristic Energy Losses and Optical Properties of Intermetallic Compounds in the Fe-Ge System

N.D. STOJADINOVIĆ and R.S. POPOVIĆ, Niš
A New Method for Elimination of Emitter Edge Dislocations of Silicon Planar NPN Transistors

Papers to be published
in "physica status solidi - basic research"
Vol. 95, No. 1

Review Article

M. AVEROUS, Montpellier
Symmetry-Induced Zero-Gap Semiconductors

Original Papers

T. SOMA, Akita
The Electronic Theory of Si-Ge Solid Solutions

J. SCHUMANN and D. ELEFANT, Dresden
Tunneling Investigations on V_3Si

T. SUSKI and W. MÜLLER-LIERHEIM, Warsaw and Aschaffenburg
Anharmonic Effects in Ferroelectric $Pb_5Ge_3O_{11}$

Y.P. JOSHI, Varanasi
Scattering of Phonons by Density Fluctuations and Thermal Conductivity of Amorphous Solids

G. VÖLKEL, W. BRUNNER, W. WINDSCH, and D. STEUDEL, Leipzig
Electron Spin-Echo (ESE) Investigations of Critical Paramagnetic Relaxation Anomalies in X-Irradiated Triglycine Sulphate (TGS:X)

K.K.P. SRIVASTAVA and H.C. SINGH, Bhagapur and Nalanda
Effect of Second-Order Spin-Orbit Coupling $\lambda'(L.S.)^2$ on Elextronic Spin Relaxation of Paramagnetic Fe^{2+} in $CaCO_3$ and $CdCO_3$

F. LOPEZ-AGUILAR, Barcelona
A Simple Linear Method in Band Theory

H. KÖHLER and M. ROOS, Würzburg
Evidence for a Landé Factor g = 2 in n-Type Inverted Silicon MOSFET Surfaces

D.P. BHATTACHARYA and Z.S. KACHLISHVILI, Kalyani and Tbilisi
Nonradiative Multiphonon Carrier Capture by Deep Neutral Traps

G. AMBRAZEVIČIUS, G. BABONAS, and A. ŠILEIKA, Vilnius
Reflectance Spectra of $CdSiP_2$ and $ZnSiP_2$

W.J. GUNNING and A.J. HEEGER, Philadelphia, Pennsylvania
Effect of Controlled Defects on the Microwave Transport and Dielectric Constant of Tetrathiofulvalene-Tetracyanoquinodimethane

A. HANDSTEIN, B. PIETRASS, and G. BEHR, Dresden
Magnetic Susceptibility of the A-15 Compound System $(V_{1-x}Cr_x)_3Si$

O. GRESCHNER and R. WEDELL, Berlin
Comparison between the Spontaneous Radiation in Channeling, Coherent Bremsstrahlung, and Synchrotron Radiation

K. SZLENK, Warsaw
Temperature Dependence of Electron Concentration in Intrinsic-Like HgTe

G. BORSTEL, W. BRAUN, M. NEUMANN, and G. SEITZ, Osnabrück
Band Structure of Rhodium and Photoemission from Its Low Index Surfaces

A.V. TURIK and N.B. SHEVCHENKO, Rostov-on-Don
Dielectric Spectrum of $BaTiO_3$ Single Crystals

M. CIFTAN and V. RUCK, Durham
The Adsorptive Chemostress Effect

A.S. RAO and R.J. KEARNEY, Moscow, Idaho
Logarithmic Derivative Reflectance Spectra of BrO and SrO

M. NISHIDA, Kanazawa
Electronic States for the Al Overlayer on the Si (111) Surface

A. SINGH and K.N. SHRIVASTAVA, Simla and Hyderabad
Optical-Acoustic Two-Phonon Relaxation in Spin Systems

I. BROSER and M. ROSENZWEIG, Berlin (West)
A Quantitative Study of Excitonic Polariton Reflectance in CdS (II)

H. OPYRCHAŁ and K.D. NIERZEWSKI, Wrocław
Photostimulated Low-Temperature Recombination Luminescence in γ-Irradiated $KCl:Eu^{2+}$ Crystals

P. BYSZEWSKI, K. SZLENK, J. KOSSUT, and R.R. GAŁĄZKA, Warsaw
Energy Levels at Γ-Point in $Hg_{1-x}Mn_xTe$ in Intense Magnetic Fields

A.V. MITIN and V.A. TARASOV, Kazan
Excitation of Magneto-Acoustic Surface Waves by Bloch Walls

K.I. WYSOKINSKI and M. PIŁAT, Lublin
Effect of Temperature on the Off-Diagonal Alloy Properties

H. DRULIS and N.M. SULEIMANOV, Kazan
Electron Paramagnetic Resonance of Er Localized Electron Magnetic States in Thorium Hydrides

S.W. KOCH and H. HAUG, Frnakfurt (Main)
Treatment of the Electron-Hole Nucleation in the Fokker-Planck Approximation

P. KLEINERT, Berlin
Cluster Bethe Lattice Treatment for Phonons in Pseudobinary Alloy Systems

I.P. ZVYAGIN, Moscow
On the Hopping Mechanism of Dispersive Transport

E.A. VINOGRADOV, G.N. ZHIZHIN, N.N. MELNIK, S.I. SUBBOTIN, V.V. PANFILOV, K.R. ALLAKHVERDIEV, E.YU. SALAEV, and R.KH. NANI, Troitsk and Baku
Raman Scattering and Phase Transformations of $TlGaSe_2$ and $TlGaS_2$ under Pressure

F. BECHSTEDT and R. ENDERLEIN, Berlin
Electronic Polarization (Relaxation) Effects in the Core Level Spectra of Semiconductors (II)

S. UEDA and T. SHIMIZU, Kanazawa
NMR Studies of the Effect of Alloying Cu, Ag, Tl, I, and Ge to Amorphous As_2Se_3

F. HARBACH, Münster
Disorder-Induced Raman Scattering in Non-Crystalline Films of Alkaline-Earth Halides

L. BOSI and M. NIMIS, Milano
To what Extent are the Observed Lifetimes of F-Centres Intrinsic?

H. FINKENRATH, G. FRANZ, and N. UHLE, Darmstadt
Reststrahlen Spectra of $PbSe_{1-x}Te_x$

M. RAZEGHI, J.P. BUISSON, and B. HOULIER, Orsay
EPR Investigation of Gd^{3+} and Eu^{2+} in the α- and β-Phases of Lead Phosphate

K.K. PUKHOV and V.P. SAKUN, Moscow
Theory of Nonradiative Multiphonon Transitions in Imputity Centres with Extremely Weak Electron-Phonon Coupling

A. MÖBIUS, F. GOEDSCHE, D. ELEFANT, K.-H. BERTHEL, and G. VOJTA, Dresden
Deviations from Matthiessen's Rule in the Low-Temperature Resistivity of Iron

J. D'HERTOGHE and G. JACOBS, Gent
Self-Trapped Exciton and F Centre Formation by Picosecond Laser Pulses in Alkali Bromides and Iodides

G.D. PRIFTIS, J. BOVIATSIS, A. VRADIS, and K. KOUNTOURIS, Patras
On the X-Ray Plasmon Scattering from Small Particles of Graphite at k = 0

V.Z. POLINGER and G.I. BERSUKER, Kishinev
Multimode Jahn-Teller Effect for an E Term with Strong Vibronic Coupling (I)

L. KUBLER, A. JAÉGLÉ, and J.J. KOULMANN, Mulhouse
The Effect of Oxygen on the Properties of Evaporated Amorphous Silicon

KIN-ICHI MASUDA, Tokyo
Chemisorption of Ordered Atomic Layers on a Model Transition Metal

CH. MÜLLER, H. WONN, W. BLAU, P. ZIESCHE, and V.P. KRIVITSKII, Dresden and Kiev
Band Structure and X-Ray Investigation of Transition Metal Aluminides

19. Magnetic Resonance

20. Optical Properties
    - 20.1 Spectra. Optical Constants (X-Ray Spectra See 20)
    - 20.2 Lasers
    - 20.3 Luminescence (X-Ray Spectra See 20)

21. Metals. Alloys
    - 21.1 Transition Metals (Group Sc to Zn) and Their Alloys
        - 21.1.1 Fe and Fe Alloys
    - 21.2 Alkali Metals
    - 21.3 Alkali-Earth Metals
    - 21.4 Rare-Earth Metals
    - 21.5 Actinides
    - 21.6 Noble Metals
    - 21.7 Semi-Metals (Semiconducting Alloys 22.7)

22. Semiconductors and Ionic Crystals
    - 22.1 Elements
        - 22.1.1 Germanium
        - 22.1.2 Silicon
        - 22.1.3 Group VI Elements
    - 22.2 III–V Compounds
        - 22.2.1 Arsenides
        - 22.2.2 Phosphides
        - 22.2.3 Antimonides
        - 22.2.4 Mixed Crystals
    - 22.3 IV–IV, II–V, and III–VI Compounds
        - 22.3.1 Mixed Crystals
    - 22.4 II–VI Compounds
        - 22.4.1 Sulphides
        - 22.4.2 Selenides
        - 22.4.3 Tellurides
        - 22.4.4 Mixed Crystals
    - 22.5 Halides
        - 22.5.1 Silver Halides
        - 22.5.2 Alkali Halides
        - 22.5.3 Alkali-Earth Halides
        - 22.5.4 Mixed Crystals
    - 22.6 Simple Oxides (CdO, ZnO See 22.4)
        - 22.6.1 Mixed Crystals
    - 22.7 Semiconducting Intermetallic Compounds
        - 22.7.1 Mixed Crystals
    - 22.8 Tertiary and Higher Compounds
        - 22.8.1 Salts of Oxy-Acids (Carbonates, Phosphates, Silicates Including Mica etc.)
        - 22.8.2 Spinels, Garnets, Ferrites
    - 22.9 Organic Semiconductors

23. Solidified Gases

Attention

The preceding classification scheme is closely akin to the scheme used by Physics Abstracts to classify the various fields of solid state physics. A paper will either be classified under one or more of the subsections or, if none of these are relevant, it will just appear under the general section heading (i. e. sections and subsections are equivalent in this sense).

The *subject classification* given at the head of each paper begins with the main subject matter of the paper, and this is followed, if necessary, by figures referring to subsidiary subject matter and the type of solids investigated.

The papers published in this issue are listed in the order of their subject classification at the end of the table of contents.

32914

Manuscripts and letters for physica status solidi (b) — basic research and physica status solidi (a) — applied research

should be sent to

DDR-102 Berlin, Neue Schönhauser Str. 20

or

Internationaler Buchversand GmbH, 1 Berlin 19, Eschenallee 21.

The object of "physica status solidi (b)" and "physica status solidi (a)" is to publish rapidly important and topical results concerning fundamental research in the field of solid state physics and aspects of application and experimental techniques of fundamental solid state research, respectively. The issues of "physica status solidi (b)" and "physica status solidi (a)" will be published on the 1st and the 16th day of each month, respectively. Either series will contain the following categories:

1. Review articles written by arrangement with the editors. They will be remunerated.
2. Original papers selected from those which have not been submitted for publication elsewhere, after being reviewed. (The latest date for acceptance of contributions is the 8th and the 23rd day of each month for "physica status solidi (b)" and "physica status solidi (a)", respectively, 50 days before the delivery of the journal.)
3. Short notes published in the off-set part of the journal. They are, if necessary after being reviewed, selected by the editorial staff. (The latest date for acceptance of manuscripts for this part is the 12th and the 27th day of each month for "physica status solidi (b)" and "physica status solidi (a)", respectively, 16 days before the delivery of the journal.)
4. Pre-printed titles of papers to be published in the next issue of both series.

Editorial Information

Manuscripts should be submitted preferentially in English. Besides, papers in German, French, and Russian are admitted. The final decision on the incorporation of the papers in the series (b) or (a) is taken by the editorial staff after consultation of the referee. Authors are asked to observe the following rules:

1. Manuscripts have to be written as concisely as possible. As most, 20 and 4 pages (28 lines per page; abstracts, tables, references, figures, and captions of figures included, where 3 figures are counted as about one page) are admitted for original papers and short notes, respectively. All manuscripts should be type-written on one side only. Two copies are required. Badly arranged manuscripts are returmed.
2. The author's name, initials, and institution should be given.
3. One abstract of about ten lines in English and one in German, French, or Russian are required for original papers.
4. Because of lack of space conclusions (if any) must not have the character of a summary (i.e. substantially a repetition of the contents of the abstract).
5. Formulae should not be written too small and not with pencil. Separate lines for formulae are desirable. Indistinct writing will lead to a return of the manuscript.
6. Drawings should be enclosed on separate sheets of paper, not exceeding the maximum and minimum sizes of 20×30 and 10×15 cm^2, respectively. Photographs should be submitted, on white paper, only for review articles and original papers. Duplicates are required.
7. Captions for the figures should be submitted on separate sheets of paper, and special instructions with regard to the size of illustrations should be given on these.
8. References should be numbered in order of appearance and should be given, on a separate sheet, in the form used in physica status solidi.
9. SI-units should be used.
10. Each paper should contain a proposal for the subject classification according to the classification scheme appearing on the second and third cover of each issue.

The authors will be sent galley proofs of review articles and original papers. Proof reading of short notes and pre-printed titles will be done by the editorial staff.

For each original paper and short note a total of 75 reprints is available free of charge. The number of reprints free of charge per review article is 25. Requests for additional reprints (against payment) should be given together with the submitted manuscript. **The deadline for orders of additional reprints is the 15th and the 30th of each month** for "physica status solidi (b)" and "physica status solidi (a)", respectively, two weeks before the delivery of the journal.

Bei Fragen zur Produktsicherheit wenden Sie sich bitte an:
If you have any questions regarding product safety,
please contact:

Walter de Gruyter GmbH
Genthiner Straße 13
10785 Berlin
productsafety@degruyterbrill.com